LOGIC, THOUGHT AND ACTION

LOGIC, EPISTEMOLOGY, AND THE UNITY OF SCIENCE

VOLUME 2

Logic, Epistemology, and the Unity of Science aims to reconsider the question of the unity of science in light of recent developments in logic. At present, no single logical, semantical or methodological framework dominates the philosophy of science. However, the editors of this series believe that formal techniques like, for example, independence friendly logic, dialogical logics, multimodal logics, game theoretic semantics and linear logics, have the potential to cast new light on basic issues in the discussion of the unity of science.
This series provides a venue where philosophers and logicians can apply specific technical insights to fundamental philosophical problems. While the series is open to a wide variety of perspectives, including the study and analysis of argumentation and the critical discussion of the relationship between logic and the philosophy of science, the aim is to provide an integrated picture of the scientific enterprise in all its diversity.

Logic, Thought and Action

Edited by

Daniel Vanderveken

University of Quebec, Trois-Rivières, QC, Canada

Springer

A C.I.P. Catalogue record for this book is available from the Library of Congress.

ISBN 1-4020-2616-1 (HB)
ISBN 1-4020-3167-X (e-book)

Published by Springer,
P.O. Box 17, 3300 AA Dordrecht, The Netherlands.

Sold and distributed in North, Central and South America
by Springer,
101 Philip Drive, Norwell, MA 02061, U.S.A.

In all other countries, sold and distributed
by Springer,
P.O. Box 322, 3300 AH Dordrecht, The Netherlands.

Cover image:
Adaptation of a Persian astrolabe (brass, 1712-13), from the collection of the Museum of the History of Science, Oxford. Reproduced by permission.

Printed on acid-free paper

Printed in the Netherlands.

In Memoriam
J.-Nicolas Kaufmann
1941–2002

Contents

Contributing Authors xi

1
Introduction 1
Daniel Vanderveken

Part I Reason, Action and Communication

2
The Balance of Reason 27
Marcelo Dascal

3
Desire, Deliberation and Action 49
John R. Searle

4
Two Basic Kinds of Cooperation 79
Raimo Tuomela

5
Speech Acts and Illocutionary Logic 109
John R. Searle and Daniel Vanderveken

6
Communication, Linguistic Understanding and Minimal Rationality in Universal Grammar 133
André Leclerc

Part II Experience, Truth and Reality in Science

7
Truth and Reference 153
Henri Lauener†

8
Empirical Versus Theoretical Existence and Truth 163
Michel Ghins

9
Michel Ghins on the Empirical Versus the Theoretical 175
Bas C. van Fraassen

Part III Propositions, Thought and Meaning

10
Propositional Identity, Truth According to Predication and Strong Implication 185
Daniel Vanderveken

11
Reasoning and Aspectual-Temporal Calculus 217
Jean-Pierre Desclés

12
Presupposition, Projection and Transparency in Attitude Contexts 245
Rob van der Sandt

13
The Limits of a Logical Treatment of Assertion 267
Denis Vernant

Part IV Agency, Dialogue and Game-Theory

14
Agents and Agency in Branching Space-Times 291
Nuel Belnap

15
Attempt, Success and Action Generation: a Logical Study of Intentional Action 315
Daniel Vanderveken

16
Pragmatic and Semiotic Prerequisites for Predication 343
Kuno Lorenz

17
On how to be a Dialogician 359
Shahid Rahman and Laurent Keiff

18
Some Games Logic Plays 409
Ahti-Veikko Pietarinen

19
Backward Induction Without Tears? 433
Jordan Howard Sobel

Part V Reasoning and Cognition in Logic and Artificial Intelligence

20
On the Usefulness of Paraconsistent Logic 465
Newton C.A. da Costa, Jean-Yves Béziau, and Otávio Bueno

21
Algorithms for Relevant Logic 479
Paul Gochet, Pascal Gribomont and Didier Rossetto

22
Logic, Randomness and Cognition 497
Michel de Rougemont

23
From Computing with Numbers to Computing with Words — From Manipulation of Measurements to Manipulation of Perceptions 507
Lofti Zadeh

Contributing Authors

Nuel Belnap is Alan Ross Anderson Distinguished Professor of Philosophy and Professor of the History and Philosophy of Science at the University of Pittsburgh. He has written chiefly in philosophical logic, having co-authored *The Logic of Questions and Answers* (Yale University Press, 1976) with Thomas Steel, *Entailment: the Logic of Relevance and Necessity* (Princeton University Press) vol. I (1975) with Alan Ross Anderson and vol. II (1992) with Anderson and J. Michael Dunn, *The Revision Theory of Truth* (MIT Press, 1993) with Anil Gupta, and *Facing the Future: Agents and Choices in our Indeterminist World* (Oxford University Press, 2001) with Michael Perloff and Ming Xu.

Jean-Yves Béziau is now Professor of the Swiss National Science Foundation at the Institute of Logic of the University of Neuchâtel. He got a PhD in mathematical logic in Paris and a PhD in philosophy in São Paulo. He worked as a research fellow in Brazil, Poland and California (UCLA, Stanford). His main interests are paraconsistent logic, universal logic, philosophy of logic and philosophy of mathematics. He has written with Newton da Costa and Otavio Bueno the book *Elementos de teoria paraconsistente de conjuntos*, Cle-Unicamp, Campinas, 1998, 188pp. and more than 50 papers in Journals and collective books.

Otávio Bueno is Associate Professor of Philosophy at the University of South Carolina. His main research area is in philosophy of science, philosophy of mathematics, philosophy of logic, and paraconsistent logic. He has published papers in many journals and collections including *Philosophy of Science, Synthese, Journal of Philosophical Logic, Studies in History and Philosophy of Science, British Journal for the Philosophy of Science, Analysis, Erkenntnis, History and Philosophy of Logic,* and *Logique et Analyse.* He is the author of two books: *Constructive Empiricism: A Restatement and Defense* (CLE, 1999), and *Elements* *of Paraconsistent Set Theory* (CLE, 1998) with Newton da Costa and Jean-Yves Béziau.

Newton da Costa is retired Professor of Philosophy at the University of São Paulo, Brazil, and currently Visiting Professor of Philosophy at the Federal University of Santa Catarina, Brazil. His main research interests are non classical logic, model theory, foundations of inductive inference and the philosophy of science. His publications include *Logiques Classiques et non Classiques* (Paris, Masson, 1997), *El Conocimiento Cientifico* (Mexico, UNAM, 2000) and *Science and Partial Truth* (Oxford University Press, 2003), as well as some other books and numerous papers in specialized journals.

Marcelo Dascal is a Professor of Philosophy at Tel Aviv University, Israel. He has taught in major universities in Europe, the Americas, and Australia. He has been a fellow of the Netherlands Institute of Advanced Studies (Wassenaar), of the Institute of Advanced Studies (Jerusalem), Leibniz Professor at the Center for Advanced Studies (Leipzig), and is currently Gulbenkian Professor at the University of Lisbon (PT). He was awarded the Alexander von Humboldt Prize for 2002-2003. His main research areas are the philosophy of language and communication, the philosophy of mind, pragmatics, the history of modern philosophy, and the study of controversies. As a Leibniz specialist, he has published *La sémiologie de Leibniz* (Paris, 1978), *Leibniz: Language, Signs and Thought* (Amsterdam, 1987), and has co-edited *Leibniz and Adam* (Tel Aviv, 1991) and *Leibniz the Polemicist* (Amsterdam, forthcoming). In the area of pragmatics and the philosophy of language he has published *Pragmatics and the Philosophy of Mind, vol. 1: Language and Thought* (Amsterdam, 1983) and *Interpretation and Understanding* (Amsterdam, 2003); he has edited *Dialogue — An Interdisciplinary Approach* (Amsterdam, 1985) and co-edited *Philosophy of Language — A Handbook of Contemporary Research* (Berlin, 1991, 1995). He is the founder and editor of the journal *Pragmatics & Cognition* and of several book series.

Jean-Pierre Desclés is professor of Computer Sciences and Linguistics at Sorbonne University. He is vice president of the Académie Internationale de Philosophie des Sciences. His research interests are in the domain of the logic and natural languages; cognition and language; time, tense and space; logic of object determination. His publications include *Langages applicatifs, langues naturelles et cognition* (Paris, Hermès 1990) and articles in journals dealing with combinatory Logic, theoretical linguistics, artificial intelligence and epistemology of human sciences in relation with computers.

Bas van Fraassen is McCosh Professor of Philosophy at Princeton University. His research interests straddle philosophical logic and philosophy of science, with special interests in empiricism, (anti-)realism, probability, foundations of relativity and quantum physics, and philosophy of literature. His publications include *The Scientific Image* (Oxford 1980), *Laws and Symmetry* (Oxford 1989), *Quantum Mechanics: An Empiricist View* (Oxford 1991), and *The Empirical Stance* (Yale 2002).

Michel Ghins is professor at the Université Catholique de Louvain. He has taught at the Universities of Pittsburgh, Campinas (Brazil), Turin and the Catholic University of America (Washington, D.C.) He has published *L'inertie et l'espace-temps absolu de Newton à Einstein. Une analyse philosophique.* (Académie Royale de Belgique, 1990) and a number of papers on history and philosophy of physics, scientific realism and the semantic view of theories. He is the editor of the *Revue philosophique de Louvain.*

Paul Gochet, Pascal Gribomont and Didier Rossetto are respectively Professor Emeritus at Université de Liège, Professor at Université de Liège and at Faculté universitaire de Namur (Belgium). Among their publications are *Outline of a Nominalist Theory of Propositions* (Dordrecht, Reidel 1980); *Ascent to Truth* (Munich, Philosophia Verlag 1986); "Concurrency without toil: a systematic method for parallel program design" (in *Science of Computer Programming, 21*, pp. 1–56, 1993); *Logique I : Méthodes pour l'informatique fondamentale* (Paris, Hermès 1990, 1991, 1998) ; *Logique II : Méthodes pour l'étude des programmes* (Paris, Hermès 1994) ; *Logique III : Méthodes pour l'intelligence artificielle* (Hermès 2000) ; *Eléments de programmation en Scheme* (Paris, Dunod 2000); "Simplification of Boolean Verification Conditions" (in *Theoretical Computer Science, 239*, pp. 165–189, 2000); "Quantification, Being and Canonical Notation" (in *a companion to philosophical logic* pp. 265–280, Oxford, Blackwell 2002); "Epistemic Logic" (forthcoming in the *Handbook of History and Philosophy of Logic*, Dov Gabbay and John Woods (Eds), Amsterdam, Elsevier.)

Laurent Keiff is currently chargé de cours at the university of Lille 3, where he teaches logic and history of philosophy. He also teaches philosophy in the Lycée E. Thomas (Nord-Pas de Calais). He is a PhD student in philosophy of logic, under the direction of S. Rahman (Lille 3), working on a dialogical way to give a satisfying account of logical pluralism.

Henri Lauener was initially an expert in the philosophy of Hegel. He studied philosophy in his country (Switzerland) and later at the University of Paris. He was appointed to a chair of philosopy at the University of Bern. Meanwhile he had moved to analytic philosophy. He was visiting professor at the University of Helsinki and several times at the University of California at San Diego. Editor of *Dialectica* from 1977 to 2001, he established it as a prominent forum for all kinds of debates within analytic theoretical philosophy. He organized the now legendary colloquia in Bienne (Swtizerland). Henri Lauener's own philosophy is an original blend of both Kantian and Quinean influences. His more recent wiritings, apart from his monograph *W.v.O.Quine* (Beck, Munich, 1982), were assembled under the title *Offene Transzendentalphilosophie* (Hamburg, Verlag Dr.Kovac, 2001). See *Dialectica 56*:4, 2002, pp.293-298.

André Leclerc is Professor of Philosophy at the Federal University of Paraíba at João Pessoa, Brazil. He got his Ph.D. from the University of Québec at Trois-Rivières with a doctoral thesis on illocutionary aspects of meaning in the tradition of Universal Grammar. He is working and publishing regularly in the philosophy of language and of mind, and is teaching epistemology and philosophy of science. He has publications on the history of linguistics, philosophy of language and on mental causation and externalism in the philosophy of mind.

Ahti-Veikko Pietarinen is a Post-Doctoral Research Fellow of the Academy of Finland at the Department of Philosophy, University of Helsinki. He has published articles on logical, philosophical, game-theoretical and semantic/pragmatic topics in several journals, including *Acta Analytica, Cognitive Systems Research, Foundations of Science, Interaction Studies, Linguistic Analysis, Logic Journal of the IGPL, Logique et Analyse, Nordic Journal of Philosophical Logic, Notre Dame Journal of Formal Logic, Open Systems & Information Dynamics, Quality & Quantity, Semiotica, Theoretical Linguistcs*, and *Transactions of the Charles S. Peirce Society.*

Shahid Rahman is full professor of logic and epistemology at the University Lille 3 (Humanities), France. He works on philosophy and history of logic and epistemology. He published among others *Uber Dialogue, protologische Kategorien und andere Seltenheiten* (Peter Lang, 1993); *Wege zur Vernunft. Philosophieren zwischen Tätikgeit und Reflexion*, edited with Kai Buchholz and Ingrid Weber (Campus, 1999); "*New Perspectives in Dialogical Logic*" edited with Helge Rückert (*Synthese, 127*, nos. 1-2, 2001); *Logic, Epistemology and the Unity of Science*, edited with Dov Gabbay, John Symons and Jean-Paul van Bendegem (Kluwer, 2004). He is editor with John Symons of the new Kluwer series: Logic, Epistemology and the Unity of Science.

Michel de Rougemont is a Computer Science Professor at the University of Paris II (Panthéon-Assas) since 1995 and a researcher at LRI (Laboratoire de Recherche en Informatique) of Paris-South University at Orsay. He received his Ph.D. in Computer Science from UCLA in 1983, was a researcher at the European Computer Industry Research Centre (ECRC) in Munich from 1984 until 1987 and then a Professor at the Ecole Nationale Supérieure de Techniques Avancées (E.N.S.T.A.) in Paris until 1995. His main interests are logic, complexity and the analysis of problems with uncertainty where both Logic and randomized algorithms interact. He is co-author of the book *Logic and Complexity* at Springer in 2004.

Rob van der Sandt is professor in Philosophy of Language and Logic at the University of Nijmegen. His main field is natural language semantics and pragmatics. His primary reseach interests are dynamic semantics, discourse processing and dialogue. His publications include *Context and Presupposition* (Routledge 1988), *Focus. Linguistic, Cognitive and Computational Perspectives* (Cambridge University Press, 1999) (edited with Peter Bosch) and numerous articles on presupposition and discourse processing. He is associate editor of the *Journal of Semantics.*

John R. Searle is Mills Professor of Philosophy at the University of California at Berkeley. He is well-known for his theory of speech acts, his critique of strong AI, his work on intentionality and consciousness in the philosophy of mind and his theory of social reality and institutions. Among his publications are *Speech Acts* (Cambridge University Press, 1969), *Expression and Meaning: Studies in the Theory of Speech Acts* (Cambridge University Press, 1979), *Intentionality: An Essay in the Philosophy of Mind* (Cambridge University Press, 1983), with Daniel

Vanderveken *Foundations of Illocutionary Logic* (Cambridge University 1985), *The Rediscovery of the Mind* (MIT Press, 1992), *The Construction of Social Reality* (Free Press, 1995), *The Mystery of Consciousness* (New York Review Press, 1997), *Mind, Language and Society* (Basic Books 1998) and *Rationality in Action* (MIT Press, Bradford Books, 2001).

Jordan Howard Sobel is Professor Emeritus at the University of Toronto, and Visiting Professor at Uppsala University, works in logic, probability, rational choice, ethics, and philosophy of religion. His publications include *Taking Chances: Essays on Rational Choice* (Cambridge 1994), *Puzzles for the Will: Fatalism, Newcomb and Samarra, Determinism and Omniscience* (Toronto 1998), *Logic and Theism: Arguments for and against Beliefs in God* (Cambridge 2004), "Utilitarianisms: Simple and General" (*Inquiry* 1970), "Ramsey's Foundations Extended to Desirabilities" (*Theory and Decision* 1998), and "Blackburn's Problem: On its not Insignificant Residue" (*Philosophy and Phenomenological Research* 2001). Projects in advanced stages include papers on liar paradoxes, on probable *modus ponens* and *modus tollens* and updating on uncertain evidence, and on Fregean and Gödelian "collapsing arguments", a book on metaethics, Moore through Mackie, and books on the moral philosophies of Plato, Aristotle, Hume, and Kant.

Raimo Tuomela is Professor at the Department of Philosophy of Helsinki University in Finland. His main field of research is the philosophy of social action. He was recipient of several grants and awards, including the von Humboldt Foundation Research Award, awarded in recognition of achievements in research. He is member of the editorial board of several journals and book series. Among Tuomelas recent books, let us mention: *The Importance of us: A Philosophical Study of Basic Social Notions.* (Stanford UP, 1995), *Cooperation: A Philosophical Study* (Kluwer, 2000), *The Philosophy of Social Practices: A Collective Acceptance View* (Cambridge UP, 2002).

Daniel Vanderveken is Professor at the University of Québec at Trois-Rivières where he is the director of a Research Group on Communication. His research interests are illocutionary and intensional logics, the logic of action, formal semantics and pragmatics of discourse and semiotics. He is the co-author with John Searle of *Foundations of Illocutionary Logic* (Cambridge UP 1985). His other main books are: *Les actes de discours* (Mardaga, Liège Brussels 1988) and *Principles of Language Use* *and Formal Semantics of Success and Satisfaction* Volumes 1 and 2 of *Meaning and Speech Acts* (Cambridge U P 1990-91). He is the guest editor of the special issue *Searle with his Replies* of *Revue internationale de philosophie* (n°216, 2001) and co-editor with Susumu Kubo of *Essays in Speech Act Theory* (P&b ns 77, John Benjamins, 2001). His webpage is: `www.vanderveken.org`.

Denis Vernant is Professor of philosophy at the University Pierre Mendès France at Grenoble where he is director of the research group Philosophie, Langages & Cognition. His major research interests are logic and philosophy of logic, pragmatics of dialogue and praxiology. His main publications are *Introduction à la philosophie de la logique* (Mardaga, Bruxelles 1986), *La Philosophie mathématique de B. Russell* (Paris, Vrin, 1993), *Du Discours à l'action, études pragmatiques* (Paris, PUF, 1997), *Introduction à la logique standard* (Paris, Flammarion, 2001) and *Bertrand Russell* (Paris, Garnier-Flammarion, 2004). His electronic address is denis.vernant@upmf-grenoble.fr. For more information see his web page: `www.upmf-grenoble.fr/sh/persophilo/denisvernant/presentation.html`.

Lotfi A. Zadeh is Professor in the Graduate School and Director of Berkeley Initiative in Soft Computing (BISC). Since 1965, his research has been directed at the development of fuzzy set theory and fuzzy logic. Currently, his work is focused on computing with words and perceptions, and on development of a unified theory of uncertainty and a theory of hierarchical definability. His recent publications include "A new Direction in AI — Toward a Computational Theory of Perceptions," *AI Magazine 22(1)*: 73–84, 2001; and "Toward a Perception-Based Theory of Probabilistic Reasoning with Imprecise Probabilities," *Journal of Statistical Planning and Inference, 105*, 233–264, 2002. To appear: "Precisiated Natural Language (PNL)," *AI Magazine*; and "A Note on Web Intelligence, World Knowledge and Fuzzy Logic," *Data and Knowledge Engineering*. His web page is `http://www.cs.berkeley.edu/~zadeh`.

Chapter 1

INTRODUCTION*

Daniel Vanderveken

In contemporary philosophy as well as in human and cognitive sciences, language, thought and action are systematically related. One considers that the primary function of language is to enable human speakers not only to express and communicate their thoughts but also to act in the world. Thus speakers who communicate are viewed as intentional agents provided with rationality. By choosing to exchange certain words speakers first of all attempt to perform speech acts of different kinds (acts of utterance, acts of reference and predication, illocutionary and perlocutionary acts) in certain ways (literally or not). They also want to contribute to conversations whose goal is often to change rather than to describe the world they live in. So contemporary logic and philosophy of language study both thought and action. Underlying any philosophy of language there is a certain philosophy of mind and action.

The main purpose of this book is to present and discuss major hypotheses, issues and theories advanced today in the logical and analytic study of language, thought and action. One can find in the book major contributions by leading scholars of analytic philosophy, logic, formal semantics and artificial intelligence. Among fundamental issues discussed in the book let us mention the rationality and freedom of agents, theoretical and practical reasoning, the logical form of individual and collective attitudes and actions, the different kinds of action generation, the nature of cooperation and communication, the felicity conditions of

*I am very grateful to Springer's referee whose critical remarks have greatly helped to improve the book. I also wish to warmly thank my research assistant Florian Ferrand who, with so much care, has produced the camera ready final typescript and my colleague Geoffrey Vitale for his invaluable help in correcting the introduction. Most of all I want to express my gratitude to my wife Candida Jaci de Sousa Melo for her constant help and encouragement. Grants from the Social Sciences and Humanities Research Council of Canada and the Quebec Foundation for Research on Society and Culture have facilitated the collective work that underpins the publication of the present volume.

D. Vanderveken (ed.), Logic, Thought & Action, 1–24.

speech acts, the construction and conditions of adequacy of scientific theories, the structure of propositional contents and their truth conditions, illocutionary force, time, aspect and presupposition in meaning, the dialogical approach to logic and the structure of dialogues as well as formal methods needed in logic or artificial intelligence to account for choice, paradoxes, uncertainty and imprecision.

The book is divided into five parts. The first part, **Reason, Action and Communication**, contributes mainly to the general philosophy of language, mind and action, the second, **Experience, Truth and Reality in Science**, to the philosophy of science, the third, **Propositions, Thought and Meaning**, to the logic of language and formal semantics, the fourth, **Agency, Dialogue and Games**, to the logic of action, dialogues and language games and the last part, **Reasoning and Cognition in Logic and Artificial Intelligence**, to the role and formal methods of logic and computer science. Many authors participated in the *Decade on Language, Reason and Thought* which took place in June 1994 at the castle of Cerisy-la-Salle in France. Our dear and regretted colleague J-Nicolas Kaufmann to whom this book is dedicated was present and very active at that conference. We wish to pay him homage.

According to the Western conception of reason, the proper rationality of human agents basically rests on their capacity to weight on the scales of the balance of reason their different beliefs, reasons, desires, intentions and goals and having deliberated to select the best actions that will allow them to achieve their goals. The classical model of rationality goes back to Aristotle who claimed that deliberation is about means and not about ends. So, in the model underlying decision theory, human agents are supposed to have certain primary desires and well ordered preferences prior to making a deliberation and they reason on the basis of these desires and their beliefs about the state of the world in order to form other desires for means of coming to their ends. However it often happens that human agents have relatively inconsistent desires that cannot all be satisfied. Moreover their preferences are not always well ordered before deliberation. They often have to choose between conflicting desires in the process of deliberation. And finally there are the freedom and the weakness of the will. An agent who forms an intention after deliberation can revise or abandon the intention. Previous desires, beliefs and intentions of agents do not seem to cause their future actions. How can we account for such facts in a theory of rationality? Human agents are by nature social. They share forms of life, speak public languages, create social institutions and act together in the world. What kinds of speech acts do they attempt to perform in conversation? How can we explain in philosophy of mind their collective attitudes and actions and their communication abilities? The first part of the book, **Reason, Action and Communication**, contains a general philosophical discussion of these important questions.

In Chapter 2, **The Balance of Reason**, Dascal discusses the ideal of a perfectly reliable balance of reason, an ideal challenged by scepticism. He shows that the balance metaphor is compatible with two different conceptions of rationality which are both present in Western thought. The first conception expects the balance of reason to provide conclusive decisions in every rational deliberation. The second conception acknowledges the limits of human reason. It is clearly more appropriate for handling uncertainty, revision of intentions and more apt to face scepticism. Leibnitz, one of the most eminent rationalist philosophers, made a substantial contribution to both conceptions of rationality. Dascal discusses in detail his ideas. He shows how Leibniz came to grips with the balance metaphor. The state of equilibrium of the scales of a balance mirrors the *equilibrium of indifference* between the arguments for and the arguments against a belief, a decision or an action. Yet an indifference of that kind seems to model arbitrariness rather than rationality. Leibniz, as Dascal stresses, was well aware of the problem. He acknowledged that the balance of reason, when it is conceived as a *metric* and *digital* balance, lies open to the objection raised above, but he worked out another version of the balance of reason to circumvent this. We can conceive of a balance which permits us to directly compare the "values" of what is placed on the scales without reducing them to universal measuring units.

A major merit of Dascal's essay lies in the original response he gives to the new kind of scepticism that pervades the Post-modernist trend today. Developing Leibniz' insights, Dascal shows how Leibniz' revised metaphor of the balance of reason can apply even to what is imponderable and do justice to the idea that there are reasons (for believing, acting or deciding) which *incline* without *necessitating.* A new picture of reason emerges in which *hard rationality* represented by algorithms and *soft rationality* exemplified by the reasoning of lawyers can be seen as complementary rather than conflictual. Dascal considers foreground notions which are proper to the reasoning of lawyers (e.g. presumption, burden of proof) and shows that they anticipate Grice's theory of conversation and non monotonic reasoning studied nowadays in Artificial Intelligence.

In the third chapter, **Desire, Deliberation and Action**, Searle criticizes the classical conception of rationality underlying current analysis of practical reasoning and deliberation in philosophy of mind and in decision theory. It is wrong to require of rational agents a satisfiable set of desires. It is also wrong to think that an agent who prior to engaging in a deliberation already has certain primary beliefs and desires is thereby committed to other secondary desires or intentions. There could be no logic of practical reasoning stating valid principles of inference underlying such commitments of an agent. Searle denounces in detail the mistakes of this conception of practical syllogism. He first explains why desire differs radically from belief in both its logical and phenomenological features. He also briefly describes the nature of intentions and analyzes the relation between desire and action by discussing the nature of reasons for agents to act. In Dascal's chapter the digital and metric conception of the balance of reason was shown to be inadequate. Searle goes further and identifies the source of the trouble. That conception rests upon the faulty assumption that we can deal with choice, preference and desire without recognizing their intentional character.

In Searle's view, it is a mistake to suppose that the desire must always be the ground for the reason. An acknowledgement of the facts plus the agent's rationality can motivate the internal desire of an action. So the reason can also be the ground for the desire. Among desire independent reasons Searle considers previous commitments, obligations and duties of the agent. Searle carefully avoids the common mistake of assimilating an external reason to a physical cause. He argues that intentional causation is very different from physical causation. Prior beliefs, desires and intentions can be reasons for an action. However they do not really compel the agent to act. There is a certain gap in life between prior intentions and their execution just as there is a certain gap in the process of the deliberation between previous desires and beliefs and the formation of a prior intention. It is remarkable that Searle provides new and independent reasons for Dascal's idea (borrowed from Leibniz) of a desire which inclines without necessitating. Agents are free. They always have to act on reasons and intentions. So they can be weak. And their weakness of will or *akrasia* is not to be confused with self-deception. Searle's chapter ends up with an illuminating account of the formal resemblances and differences which exist between *weakness of the will* and *self-deception.*

In order to act together with success several rational agents sharing a common goal have to cooperate. It is now widely accepted that collective actions are more than the sum of individual actions of their agents. They require of agents a collective intention and a will to cooperate. But what is the very nature of cooperation in collective actions? How can agents share collective attitudes in general and collective intentions in particular? In the fourth chapter, **Two Basic Kinds of Cooperation**, Tuomela discusses these important questions for the philosophy of social sciences. According to him one must distinguish a full cooperation based on a shared collective goal (the "we" mode) and a weaker kind of cooperation that reduces to coordination (the "I" mode). While most current empirical studies concern simple coordination in the "I" mode, Tuomela emphasizes an analysis of full blown cooperation in the "we" mode. He also explains why shared collective goals tend to work better than shared private goals in most circumstances. Agents have to come to an agreement to solve many coordination problems. A logical lesson should be drawn here : there is no *non circular rational solution* to such problems. Mere private rationality will fail. The same remark can be made about coordination dilemmas . Developing an argument that goes back to Hume, Tuomela shows that only shared collective goals can reliably solve the game in a way satisfactory to all the participants.

Human agents use language in order to coordinate their actions in the world. They need to communicate their beliefs, desires and intentions in order to achieve shared collective goals. The basic units of meaning and communication are speech acts of the type called by Austin *illocutionary acts.* Unlike propositions, such acts have felicity rather than truth conditions. In Chapter 5, **Speech Acts and Illocutionary Logic**, Searle and Vanderveken analyze the logical form of illocutionary acts and their relations with other types of speech acts. Elementary illocutionary acts such as assertions, questions and promises consist of an illocutionary force and of a propositional content. Contrary to Frege and Austin whose notion of force was primitive, Searle and Vanderveken divide forces into several components (illocutionary point, mode of achievement, degree of strength, propositional content, preparatory and sincerity conditions). Rather than giving a simple list of actual forces, their speech act theory formulates a recursive definition of the set of all possible illocutionary forces. Moreover they rigorously define the conditions of successful and non defective performance of elementary illocutionary acts. Unlike Austin they distinguish between successful utterances which are defective (like promises which are insincere or that the speaker could not keep) and utterances which are not even successful (like promises to have done something in the past). They also analyze common illocutionary force markers such as verb mood and sentential type and propose a new declaratory analysis of performative utterances. Finally they show the importance of illocutionary logic for the purposes of an adequate general theory of meaning and for the foundations of universal grammar. Some illocutionary acts *strongly* or *weakly commit* the speaker *to* others. It is not possible to perform these illocutionary acts without *eo ipso* performing or being committed to other illocutionary acts. Thus commands contain orders and weakly commit the speaker to granting permission. One of the main objectives of illocutionary logic is to formulate the basic laws of illocutionary commitment. Searle and Vanderveken explain basic principles of illocutionary commitment. Later in ***Meaning and Speech Acts*** (1990-91) Vanderveken has used the resources of proof and model theories in order to formulate the laws of a general semantics containing illocutionary logic. Recently he has extended and generalized speech act theory so as to deal with discourse. In the special issue ***Searle With his Replies*** in the ***Revue internationale de philosophie*** (2001) he has also shown how to analyze the structure and dynamics of language games with a proper linguistic goal.

Verbal exchanges between speakers communicating with each other are standard cases of collective actions. They often consist in joint collective illocutionary acts like debates, consultations and negotiations that last during a certain interval of time in the conversation. In Chapter 6, **Comprehension, Communication and Minimal Rationality in the Tradition of Universal Grammar**, André Leclerc presents Arnauld and Nicole' theory of communication. Borrowing from neglected sources such as ***La grande perpétuité*** (1669-1672), Leclerc shows that Arnauld and Nicole were aware of the insufficiency of the *code model* of linguistic communication according to which the speaker codes his thoughts into sentences and the hearer decodes them in order to access the thoughts of the speaker. They fully realized the need to enrich this model with an *inferential model* of linguistic communication in order to account for the role of implicatures, insinuations and presuppositions. Leclerc provides evidence for the claim that Arnauld and Nicole anticipated Cherniak's principle of minimal rationality and Grice's maxim of quantity. He notices that they linked the maxim of quantity with the speakers' lack of logical omniscience. Leclerc does not only make an historical study which shows the roots of modern pragmatics, he also comments passages which can still teach us something important today. The treatment of metaphors is a case in point. Arnauld and Nicole interestingly explain why words could not acquire a metaphorical meaning in metaphors

Human reason fully manifests itself in scientific practice. Human agents formulate scientific theories in order to describe, explain and predict what is happening in the world they live in. According to empiricism scientific theories must be checked against the facts of our experience. We need to confirm or falsify them by observing the world. In order to be true, scientific statements must be empirically adequate. However the meaning of scientific terms and even the interpretation of observation sentences that are used for testing scientific statements are theory laden and depend on conventions which determine their use in a context of verification. There is a real construction of models in scientific theories. Observable phenomena are explained by reference to unobservable processes. Empirically equivalent scientific theories can differ in many aspects. How can we then relate **Experience, Truth and Reality in Science**? The second part of book discusses this fundamental question of the philosophy of science. It raises important issues for current empiricist, constructivist and realist views of science.

In Chapter 7, **Truth and Reference**, Lauener opposes to physical realism a pragmatic kind of relativism in the conception of truth and ontology. According to the received view, the question of meaning is prior to the question of truth. Before asking whether an assertive utterance is true or false, one has to understand the meaning of that utterance. Since meaning depends on sense and reference, one is led to think that *reference* precedes *truth.* Tarski's definition of truth reinforces the received view. Tarski equates truth with satisfaction by all sequences of objects of the domain. Quine however in ***Pursuit of Truth*** (1990) claims that truth precedes reference. Lauener criticizes Quine's claim and shows that reference plays a primordial role in the determination of truth conditions. According to him even the meaning of observation sentences is irreducible to stimulus meaning. Their use and interpretation in a context depend on both the senses and denotations of their terms which are relative to a given linguistic system and conceptual scheme. Scientific activity moreover requires rule governed intentional illocutionary acts such as assertions, conjectures, conventions and agreements that cannot be accounted for in an austere extensional ontology.

Lauener's objection to the priority of truth over reference leads to general conclusions which are independent of that issue. Quine upholds scientific realism and physicalism as far as truth is concerned, while he advocates relativism as regards to ontology. Lauener questions the compatibility of these two positions. Quine himself was fully aware of the problem. Lauener does not try to reconcile realism for truth with relativism for ontology. He advocates relativism in both cases, but the kind of relativism that he advocates has nothing to do with cultural or subjective relativism. Lauener does not so much challenge realism as he does the holistic view of science as a language-theory conglomerate conceived as a constantly evolving whole. Lauener is not opposed to *realism* but rather to *holism* in science and *universalism* in logic (the view of logic as a language as opposed to the view of logic as a calculus). Lauener does not really advocate a relativistic ontology. He rather advocates a *pluralistic ontology*: "Since a new domain of values for the variables is presupposed for each context I advocate a pluralistic conception of ontology in contrast to Quine who postulates a unique universe by requiring us to quantify uniformly over everything that exists ... according to my method of systematic relativization to contexts (of action), we create reality sectors by employing specific conceptual schemes through which we describe the world.". Lauener's argument for the recognition of *regional ontologies* is based on philosophical considerations.

It is very interesting to notice that in the next chapter Michel Ghins advances an *independent argument* supporting the claim that we need to circumscribe domains in science on the basis of purely scientific considerations. This spontaneous convergence between two chapters is worth stressing. In Chapter 8, **Empirical Versus Theoretical Existence and Truth**, Michel Ghins mainly argues in favour of a specific, selective and moderate version of scientific realism in accordance with the common use of the terms "existence" and "truth" in ordinary speech. The actual presence of an object in sensory perception and the permanence of some of its characteristics during an interval of time jointly constitute a sufficient condition, a "criterion", of existence of that object in scientific activity as well as in everyday experience. These features also ground the truth of statements about ordinary observable objects and of some physical laws connected to experience. So scientific statements can be accepted as true when they are inductively well established in a certain limited domain of experience.

Ghins illustrates his version of scientific realism by considering the two particular examples of electromagnetic and gravitational fields and crystalline spheres in ancient astronomy. As one might expect, his criterion supports the existence of the fields but not of the spheres. Herman Weyl had already compared the different observable manifestations of an electric field with different perceptions of an ordinary object and argued that if we see forces as corresponding to perceptions and different charges as corresponding to different positions of the observers, we are entitled to attribute objective reality to electric fields. Ghins takes advantage of Weyl's analogy and goes further. Sharing Kant's criterion of reality ("[reality is] that which is connected with perception according to laws"), Ghins shows that we have good grounds to consider as *laws* not only true physical statements like the three famous Newtonian laws but also mathematical laws of classical mechanics of point-like masses which *restrict* the domain in which Newton's laws are true. Such an extension of the coverage of the concept of *law* is by no means a trivial matter. It enables Ghins to answer Popper's objection according to which limiting a theory to a given domain would be tantamount to protecting it against adverse evidence. It also provides an independent support for Lauener's position on the question as to whether human knowledge is an "amorphous and unified language-theory" or as a constellation of separate theories, each endowed with its own language and its own ontology.

The scientist's preference for simpler theories is often seen as springing from *aesthetic* or *pragmatic* considerations which have nothing to with what reality is like. Ghins debunks this view and shows that *simplicity* is a reliable guide for those who want to know what there is. The opposite view leads to counter-intuitive consequence.

In Chapter 9, **Michel Ghins on the Empirical Versus the Theoretical**, Bas van Fraassen replies to Ghins' ideas on existence and truth in science. van Fraassen basically agrees with Ghins on the central role of experience, the need to reject the myth of the given and the hope for an empiricist philosophy of science. However as regards existence and truth van Fraassen considers that one must sharply separate questions of epistemology from questions of semantics and ontology. Sensory perception and invariance are not a necessary condition of existence. Ghins does not give a criterion of existence strictly speaking. He only offers a partial criterion of legitimacy for assertions of existence of certain objects of reference that we can observe. But perhaps there also exist in the world other sorts of entities which are "transient", "invisible" and "intangible". This is not an issue of semantics but of ontology. Is Ghins' epistemic principle right? van Fraassen does not give an answer to the question. Both Ghins and van Fraassen view reality from the standpoint of experience. According to Ghins, proponents of a scientific theory are committed to believing in the existence of all entities among those postulated that bear a certain relationship to what can be experienced. Directly observable ones are not privileged. Van Fraassen's empiricism is less moderate. Proponents of a scientific theory are only committed to believing in the existence of observable entities.

In contemporary philosophy of language, mind and action, propositions are not only senses of sentences provided with truth conditions. They are also contents of human conceptual thoughts like illocutionary acts (assertions, questions, promises) and attitudes (beliefs, desires, intentions). The third part of the book deals with **Propositions, Thought and Meaning**. The double nature of propositions imposes new criteria of material and formal adequacy on the logic of propositions and formal semantics. One can no longer identify so called strictly equivalent propositions having the same truth conditions. They are not the senses of synonymous sentences, just as they are not the contents of the same thoughts. Moreover human agents are not perfectly rational in thinking and speaking. They do not make all valid inferences. They can assert and believe necessarily false propositions. So we need very fine criteria of propositional identity in logic. It is important to take into account the creative as well as the restricted cognitive abilities of human agents. It is also important to consider tense and aspect as well as presupposition accommodation and assignment of scope in the understanding of truth conditions. For that purpose, we need a better explication of truth conditions with an account of aspect, tense and presupposition. We also need to take into account the illocutionary forces of utterances. Part three of the book contains logical contributions on the matter.

In Chapter 10, **Propositional Identity, Truth According to Predication and Strong Implication**, Daniel Vanderveken enriches the formal ontology of the theory of sense and denotation of Frege and Church. His main purpose is to formulate a natural logic of propositions that explains their double nature by taking into consideration the acts of reference and predication that speakers make in expressing propositions. According to his analysis, each proposition is composed of atomic propositions (each predicating a single attribute of objects of reference under concepts). Human agents do not know actual denotations of most propositional constituents. Various objects could fall under many concepts or could have certain properties in a given circumstance. So they also ignore in which possible circumstances atomic propositions are true. Most could be true in many different sets of possible circumstances given the various denotations that their attribute and concepts could have in the reality. For that reason atomic propositions have *possible* in addition to *actual* Carnapian *truth conditions*. For each proposition one can distinguish as many possible truth conditions as there are distinct sets of possible circumstances where that proposition would be true if its propositional constituents had such and such possible denotations in the reality. In understanding a proposition we just know that its truth in a circumstance is compatible with certain possible denotation assignments to its propositional constituents and incompatible with others. So logic has to distinguish propositions whose expression requires different acts of predication as well as those whose truth is not compatible with the same possible denotation assignments to their constituents. Consequently, not all necessarily false propositions have the same cognitive value. Some are *pure contradictions* that we *a priori* know to be false in apprehending their logical form. We cannot believe them. One can define the notion of *truth according to a speaker*, distinguish *subjective* from *objective possibilities* and formulate adequate principles of epistemic logic in predicative propositional logic.

As Vanderveken points out, the set of propositions is provided with a relation of *strong implication* that is much finer than strict implication. Strong implication is a relation of partial order which is paraconsistent, finite, decidable and *a priori* known. In the second part of the chapter Vanderveken proceeds to the predicative analysis of modal and temporal propositions of the logic of ramified time. He uses the resources of model theory and formulates a powerful axiomatic system. He also enumerates valid laws for propositional identity and strong implication. And he compares his logic with intensional and hyperintensional logics, the logic of analytic implication and that of relevance.

In predicating a property of an object a speaker expressing a propositional content can view the represented fact in different ways as a state, an event or an unfinished process. There are different ontological categories of fact. Having aspectualized the predication, the speaker has to insert the represented fact into his own time reference, which is distinct from the external time reference of the calendar. Verbal aspect and tense are fundamental to the understanding of truth conditions of elementary propositions. Their semantic analysis requires a logical calculus. In the late seventies Rohrer edited two collective books presenting a rigorous logical treatment of tense and aspect showing how one can represent in Montague grammar the temporal structure of verbs and how verbal meaning interacts with the meaning of tense forms and temporal adverbs. In Chapter 11, **Reasoning and Aspectual-Temporal Calculus**, Jean-Pierre Desclés analyzes aspect and time within the theoretical framework of *cognitive applicative grammar* which is an extension of Shaumyan's *Universal Applicative Grammar* incorporating combinatory logic and topology. Desclés analyses fundamental concepts of aspectuality: *state*, *event*, *process*, resultative state by means of topological notions. He uses open and closed intervals of instants for giving a semantic interpretation of aspectual concepts. Aspectual operators are obtained by an abstraction process from semantic interpretations. Curry's combinatory logic is used to build abstract aspectual operators.

Both the Montagovian and cognitive applicative approaches are rooted in Church's lambda calculus as regards logic and focus on intervals as regards semantics. Yet there are important differences between the two approaches. In "Universal Grammar" Montague interprets indirectly sentences of natural language via their translation into a formal *object language* of intensional logic for which he builds a truth conditional model-theoretical semantics. In his *Cognitive Applicative Grammar*, Desclés starts with defining a quasi-topological model of speech operations. Next he expresses model-theoretical concepts in terms of operators of combinatory logic. His formal language is not an object language related to natural language via translation. It is a meta-language which serves to describe natural language. Yet both approaches share a central concern for aspectual reasoning. In a natural deduction style Desclés formulates principles of valid inferences that enable us to derive from the sentence "This morning, the hunter killed the deer" conclusions like "Therefore the deer was killed this morning", "Now, the deer is dead" and "Yesterday, the deer was alive" An interesting feature of Desclés' approach lies in his concern for the *speaking act* as well as for the *dynamics of meaning*. He analyzes intricate connections between aspectual-temporal conditions and the learning of lexical predicates. His approach aims at shedding light on the interaction of language activities with other cognitive activities such as perception and action. Desclés also shows that applicative grammar can accommodate speech acts in its own way.

The problems raised by presupposition have been a challenge for logicians, linguists and philosophers of language for almost a century. The current notion of presupposition is ambiguous; among pieces of information which are not explicitly stated but taken for granted, one should distinguish between what is "induced" or "triggered" by lexical items or syntactic constructions, and what is "already given" but "not marked" because of *background knowledge*. In Chapter 12, **Presupposition, Projection and Transparency in Attitude Contexts**, Rob van der Sandt advocates an unified account of presuppositions which establishes a straightforward connection between the two kinds of phenomena. The central tenet of his *anaphoric theory of presupposition* is that one single process underlies both the process of anaphoric binding and the process of presupposition resolution. He treats all presuppositions as *anaphoric expressions* which are bound by some *previously established antecedent*. van der Sandt acknowledges that sometimes the so called antecedent of the presupposition is missing and has to be supplied by some kind of *accommodation*. But, contrary to others, he imposes a new constraint on accommodation. When no antecedent is in the offing, accommodation has to insert an identifiable object which can then function as antecedent for the presuppositional anaphor. Accommodation applies to discourse structures. This leads van der Sandt to work out a theory of presupposition in the framework of Kamp's Discourse Representation Theory. Using discourse representation structures he constructs a class of conditions that encode anaphoric material.

Accommodation is implemented by a projection algorithm. When it is applied to a modal sentence containing a definite description, that algorithm yields either the wide or the narrow scope reading of the description. This depends on the level at which the accommodation is made. Just as Desclés offered a dynamic conception of disambiguation, van der Sandt gives us a dynamic account of the contrasts between wide scope and narrow scope, or *de re* and *de dicto* readings. His projection mechanism has a greater explanatory power than the standard Russellian theory of descriptions. On Russell's account, there is no way to project a description in the consequent of a conditional to its antecedent. According to van der Sandt such a projection is possible. Are there truth value gaps in the case of presupposition failure? Kamp and Reyle' standard verification conditions for discourse representation side with Russell. van der Sandt revises verification conditions in such a way that no truth value is assigned when an presuppositional anaphor can neither be bound nor accommodated. This is a significant improvement in accordance with the Frege-Strawson theory. At the end, van der Sandt comes to grips with very difficult problems which arise when the pragmatic distinction between the first person and the third person interact with the semantic distinction between *de re* and *de dicto*. He shows how his theory can solve recalcitrant puzzles mentioned by Kripke and Heim for the presuppositional adverb "too."

The concept of assertion has played a crucial role in the development of contemporary logic. It took time for logicians and philosophers to clarify the role of assertion in formalization. In Chapter 13 **The Limits of a Logical Treatment of Assertion** Denis Vernant considers that any logical treatment of assertion is limited because the concept requires a pragmatic analysis. The first part of Vernant's contribution analyses Russell's account of assertion from the ***Principles of Mathematics*** to ***Principia Mathematica***, while emphasizing its characteristic aporias. Vernant carefully reconstructs successive stages of Russell's thought on the matter. The second part deals with the pragmatic treatment of assertion which began with Frege's ***Logische Untersuchungen*** and was to continue with Searle's definition of assertive speech acts and the formulation by Searle and Vanderveken of illocutionary logic. Vernant defends a solution which is based on Frege's account but which goes much beyond it. Assertion like judgment is well the acknowledgment of the truth of propositional content. Vernant shows that the new treatment of assertion as an *illocutionary act* (and not a mental state) removes Russell's aporias. Frege only dealt with propositional negation. However, there is another negation, called *illocutionary negation*, which applies to force. As Searle and Vanderveken pointed out, the point of an act of illocutionary denegation is to make it explicit that the speaker does not perform a certain illocutionary act. So one must distinguish between the assertion of the negation of a proposition and the illocutionary denegation of that assertion. Vernant shows that already in 1904 Russell had anticipated illocutionary negation in his treatment of what he called *denial.* Russell's insistence on denial as the expression of disbelief shows that he had understood the pragmatic complexity of assertion. At the end, Vernant criticizes current speech act theory for neglecting interactions and conversational exchanges between speakers and makes a plea for a *multi-agent speech act theory.* Speakers perform their assertions and other individual illocutionary acts with the intention of contributing to the conversation in which they participate. By pointing out the dialogical function of illocutionary acts Vernant shares actual concerns in a more general speech act theory adequate for dialogue analysis.

As Wittgenstein pointed out, meaning and use are inseparable. Human speakers are agents sharing forms of life whose language-games serve to allow them to act in the world. Their verbal and non verbal actions are internally related. Human agents first of all make voluntary movements of their own body. In oral speech they emit sounds. Their basic intentional actions generate others in various ways (causally, conventionally, simply, etc.) How do agents succeed to bring about facts in the world? What is the causal and temporal order prevailing in the world in which they act? As Belnap pointed out, the logic of action requires a theory of branching time with an open future as well as a theory of games involving histories that represent possible courses of history of the world. Such a theory is compatible with indeterminism. How can we formally account for the freedom of will and the intentionality, capacities and rationality of human agents? The fourth part of the book deals with **Agency, Dialogue and Games**. It is concerned with questions such as: What is the nature of agency? How can we explicate free choice, action in the present and in the future, mental causation, success and failure and action generation? What is the nature of basic actions? In language use, speakers make utterances, acts of reference and predication, they express propositional contents with forces and perform illocutionary acts which have perlocutionary effects on the audience? How do they succeed in doing all this? Is there an irreducible pragmatic aspect in predication and discourse? What is the logical structure of a dialogue? The first contribution by Paul Lorenzen to dialogical logic appeared more than fifty years ago. Since then different dialogical systems and related research programmes have been developed. Is there a general framework for the study of the various interactions between dialogue and logic? What kind of rationality do agents manifest in practising language-games? How can they reach outcomes given their knowledge and other attitudes?

In a recent book ***Facing the Future*** **Agents and Choices in Our Indeterminist World** (2000), Nuel Belnap and co-authors have outlined a *logic of agency* which accommodates both causality and indeterminism in a conception of ramified time where the set of moments of time is a tree-like frame. There is a single causal route to the past but there are multiple future routes. So agents are free: their actions are not determined. In the indeterminist theory of ramified time *moments* representing complete possible states of the actual world are instantaneously world-wide super-events. Because of the global nature of these causal relata (the instantaneous moments), there is a world-wide matter of action at a distance in the logic of agency with branching time. The theory remains non relativistic and commits us to an account of *action-outcomes* that makes them *instantaneously world-wide.* However it is clear that both our freedom and our actions are local matter. They are made up of events here now that have no effect on very distant regions of the universe. In Chapter 14, **Agents and Agency in Branching Space Times,** Nuel Belnap shows how to improve the logic of agency by using the theory of branching space-times which can account for *local indeterminism.* For that purpose the cosmological model proposed by Einstein and Minkowski is an invaluable source of insight. This model in which action at a distance is abandoned forces us to reconsider our conception of an *event.* As Belnap observes, "a causally ordered historical course of events can no longer be conceived as a linear order of *momentary super-events.* Instead, a history is a *relativistic space-time* that consists in a manifold of *point-events* bound together by a Minkowski-style causal ordering that allows that some pairs of points events are space-liked related". So the theory of branching space times better articulates better the indeterminist causal structure of the world. In that theory causal relata are point events which are limited in both time-like and space-like dimensions. Now indeterminism and free will are not global but local. Because the theory of branching space times is both indeterminist and relativist, it is a much better theoretical apparatus for the purpose of the logic of agency. As Belnap shows, logic can now more finely identify persisting agents and also describe their choices concerning the immediate future.

Belnap begins his chapter by explaining his basic ideas about choice and agency in branching time. Next he presents the theory of branching space-times. And then he considers how the two theories can be combined. He discusses new interesting postulates that the logic of agency could adopt as regards the nature of agents, their free choice and how they do things in branching space-times. Belnap's investigations could lead us to an important new *theory of games* in branching space times that would describe, as he says, "with utmost seriousness the causal structure of the players and the plays in a fashion that sharply separates (as von Neumann's theory does not) causal and epistemic considerations" One of Belnap's new postulate characterizes *causation* in branching space-times. Using the notion of transition between an initial event and a scattered outcome event together with the notion of causal loci, Belnap defines the notion of *joint responsibility* of two agents. He concedes that his account does not cover joint action which requires the additional concept of joint intention. Belnap's account of action in terms of causation does not consider at all the intentions of agents. The main objective of the next chapter is to take them into account.

In Chapter 15, **Attempt, Success and Action Generation,** Daniel Vanderveken presents a logic of agency where intentional actions are primary as in contemporary philosophy. In his view, any action that an agent performs unintentionally could in principle have been attempted. Moreover any unintentional action of an agent is generated by an intentional action of that agent. As strictly equivalent propositions are not the contents of the same attitudes, the logic of agency should distinguish intentional actions whose contents are different. For that purpose Vanderveken uses the resources of the predicative modal and temporal propositional logic presented in Chapter 9. His main purpose now is to enrich the logic of action thanks to a new account of *attempt* and *action generation.* Unlike prior intentions which are mental states, attempts are *mental actions* of a very specific kind that Vanderveken analyzes: they are *personal, intrinsically intentional, free* and also *successful.* (Whoever tries to make an attempt makes that attempt). Like intentions attempts have strong propositional content conditions. They are directed towards the present or the future, etc. Vanderveken explicates model theoretically these features within ramified time. As before, coinstantaneous moments are logically related in models by virtue of actions of agents at these moments. Now, moments of time and histories are also logically related by virtue of attempts of agents. Attempts have conditions of achievement. Human agents sometimes attempt to do impossible things. However they are rational and cannot attempt to do what they believe to be impossible. Thanks to his account of subjective possibilities, Vanderveken can deal with unachievable attempts. To each agent and moment there always corresponds in each model a non empty set of coinstantaneous moments which are *compatible according to that agent* with the achievement of his attempts at that moment.

He proceeds to a unified explication of attempt and action. In order that an agent *succeed* in doing things it is not enough that he try and that these things occur. It is also necessary that they occur *because* of his attempt. Vanderveken uses the counterfactual conditional in order to define intentional causation and intentional actions. He explicates how attempts can *succeed* or *fail*, which attempts are the *most basic actions* and *how they generate* all other actions. Not all unintended effects of intentional actions are contents of unintentional actions, only those that are historically contingent and that the agent could have intended. So many events which happen to us in our life (e.g. our mistakes) are not really actions. Vanderveken accounts for the *minimal rationality* of agents in explaining action generation. Agents cannot try to do things that they know to be impossible or necessary. Moreover agents have to minimally coordinate their knowledge and volition in trying to act in the world. He states the basic valid laws of his logic of action.

In the usual account of one-place predication where a general term serves to attribute a property to a particular of an independently given domain of objects, one takes for granted a conceptual framework which uses, among others, the metaphysics of substance and attribute, and which is, furthermore, dependent on the availability of individuated objects. In Chapter 16, **Pragmatic and Semiotic Prerequisites for Predication: A Dialogical Model**, Kuno Lorenz considers the prepropositional state where the task to utter a sentence and express a proposition is still to be achieved. He gives a rational reconstruction of the prerequisites for predication within a novel conceptual framework, a dialogical model, that is partly derived from ideas of Peirce and Wittgenstein. By relating the both pragmatic and semiotic approaches of Peirce and Wittgenstein to a dialogical methodology, Lorenz presents a sequence of nested dialogical constructions. His purpose is to lead us from modeling simple activity to modeling the growth of more complex activities up to elementary verbal utterances.

Lorenz uses *dialogue*, conceived as a generalized language-game, as a means of inquiry. Emulating Nelson Goodman's spirit he says that neither *particulars* nor *properties* exist out there. Lorenz argues that the contrast between *individuals* and *universals* is not something that we discover by observing the world. It emerges from a process of *objectivation* which is part of the acquisition of *action competence.* This process is best understood if we look at it from the perspective of the agent-patient opposition. The *agent* performs the *token* of an *action* and looks at it from the *I-perspective.* For him, action is a *means* to reach a goal. The *patient* recognizes an *action-type* and looks at it in a *You-perspective.* For him, action is an object among others. One has learned an action when one is able to go back and forth from one perspective to the other. This shift of perspective is exemplified in dialogue. Lorenz shows how the move of objectivation from action as a means to action as an object is accompanied by a split of the action into action particulars whose invariants may be treated as *kernels* of *universalia* and respective wholes which are *closures* of the actualization of *singularia.* Kernels and closure taken together (form and matter in the tradition) make up a *particular within a situation.* Hence particulars, for Kuno Lorenz, are the product of a *dialogical construction.* As he puts it, "particulars may be considered to be half thought and half action".

A major innovation of Lorenz lies in the role he gives to the *dialogical structure* of utterances. He distinguishes between two different functions in acts expressing elementary propositions: the *significative function* of *showing* and the *communicative function* of *saying*. Communication takes place between the two protagonists of the utterance. In his view, when a speaker makes an act of reference, he *shows* something to *a hearer*. Similarly when he predicates an attribute, he does that *for a hearer*. And even the ostensive function can involve a communicative component. Lorenz' account of predication is fine-grained. His conceptual framework enables him to distinguish between *part*, *whole*, *aspect* and *phase*. He give an account not only of familiar elementary propositions in which a universal is predicated of a particular but also of other propositions in which a particular is seen as a part of a whole. Lorenz' analysis of predication covers both the *class-membership predication* and the *mereological predication*. This is a remarkable advance.

The dialogical approach to logic and the theory of language games are part of the dynamic turn that logic took over the last thirty years. In Chapter 17, **On How to Be a Dialogician**, Shahid Rahman and Laurent Keiff present an **overview on recent development on dialogues and games**. Their aim is to present the main features of the dialogical approach to logic. The authors distinguish three main approaches following two targets: (1) the *constructivist approach* of Paul Lorenzen and Kuno Lorenz (1978) and (2) the *game-theoretical approach* of Jaakko Hintikka (1996) aim to study the dialogical (or argumentative) structure of logic. (3) The *argumentation theory approach* of Else Barth and Erik Krabbe (1982) is concerned with the logic and mathematics of dialogues and argumentation. It links dialogical logic with *informal logic* (Chaim Perelman, Stephen Toulmin). Now two very important lines of research attempt to combine the lines of the two groups: (4) the approach of Johan van Benthem (2001-04) aims to study interesting *interfaces between logic and games* as model for dynamic many-agent activities and (5) Henry Prakken, Gerard Vreeswijk (1999) and Arno Lodder stress the *argumentative structure of non-monotonic reasoning.* Rahman & Keiff describe main innovations of the dynamic approach from the standpoint of *dialogical logic.*

They give a new content to key logical notions. There are *illocutionary force symbols* in the object language of dialogical logic. The two players (proponent and opponent) perform illocutionary acts with various forces in contributing to possible dialogues of dialogic. The first utterance is an assertion by the proponent which fixes the *thesis* in question. That assertion is defective if the speaker cannot defend its propositional content so as to win the game. Other moves have the forces of attacks and defence. An *attack* is a demand for a new assertion. A *defence* is a response to an attack that justifies a previous assertion. The second utterance has to be an attack by the opponent. The third can be a defence or a counterattack of the proponent. And so on. In dialogical logic as in Frege's *Begriffschrift* force is part of meaning. Utterances serve to perform illocutions with different forces and conditional as well as categorical assertions. *Particle rules* determine how one can attack and defend formulas containing logical constants, whereas *structural rules* determine the general course of a dialogue. Dialogical logic can formulate different logical systems by changing only the set of structural rules while keeping the same particle rules. It can also formulate different logics by introducing new particles. Thus classical and intuitionistic logics differ dialogically by a single structural rule determining to which attacks one may respond. The dialogical approach to logic makes it simple to formulate new logics by a systematic variation and combination of structural and particle rules. Notice that the structural rules determine how to label formulas — number of the move, player (proponent or opponent), formula, name of move (attack or defence) — and how to operate with these labelled formulae.

The thesis advanced is *valid* when the proponent has a *formal winning strategy*: when he can succeed in defending that thesis against all possible allowed criticisms by the opponent. Rahman and Keiff show that the dialogical and classical notions of validity are equivalent under definite conditions. Like in illocutionary and paraconsistent logics, speakers can assert in *paraconsistent dialogic* incompatible propositions without asserting everything. Certain kinds of inconsistency are forbidden by dialogical logic. Like relevant logic *connexive dialogic* can discriminate trivially true conditionals from those where a determinate kind of meaning links the antecedent to the consequent. Each modal logic is distinguished by the characteristic properties of its accessibility relation between *possible worlds*. In dialogic accessibility relations are defined by structural rules specifying which *contexts* are accessible from a given context. Authors show the great expressive power of dialogic as a frame by presenting a dialogical treatment of *non normal logics* in which the law of necessitation does not hold. At the end, they advocate pluralism *versus* monism in logic.

In Chapter 18, **Some Games Logic Plays**, Pietarinen takes a game-theoretical look at the semantics of logic. Game-theoretical semantics has been studied from both logical and linguistic perspectives. Pietarinen shows that it may be pushed into new directions by exploiting the ressources of the theory of games. He focuses on issues that are of common interest for logical semantics and game theory. Among such topics

Pietarinen discusses concurrent *versus* sequential decisions, imperfect *versus* perfect and complete *versus* incomplete information. Furthermore, he draws comparisons between teams that communicate and teams that do not communicate, agents' short-term memory dysfunctions such as forgetting of actions and of previous information, screening and signalling, and partial and complete interpretations. Finally, Pietarinen addresses the relevance of these games to pragmatics and its precursory ideas in Peirce's pragmaticism.

The common reference point is provided by Independence-Friendly (IF) logics which were introduced by Hintikka in the early 1990s. In contrast to the traditional conception of logic, the flow of information from one logically active component to another in formulas of IF logics may be interrupted. This gives rise to *imperfect information* in semantic games. It is worth investigating, as Pietarinen does, in which senses game-theoretical approaches throw light on pragmatically constrained phenomena such as anaphora. Following Hintikka's idea that language derives much of its force from the actual content of strategies, Pietarinen extends the semantic game framework to hyper-extensive forms where one can speak about *strategies* themselves in the context of semantic games that are played in a move-by-move fashion. He further argues that Peirce's pragmatic and interactive study of assertions antedates not only the account of strategic meaning, but also Grice's programme on conversational aspects of logic.

The borderline between decision theory and game theory is one of the more lively areas of research today in the philosophy of action. Over the last ten years, the economist Robert Aumann renewed epistemic logic by his account of common knowledge and philosophers and logicians such as Cristina Bicchieri, Richard Jeffrey, Wlodeck Rabinowicz and Jordan Howard Sobel made decisive contributions to the analysis of *rational action.* In Chapter 19 **Backward Induction Without Tears?** Sobel focuses on a kind of game whose solution hinges on a pattern of reasoning which is well known in inductive logic : *backward induction.* The rules of the game under scrutiny are described in the following passage : "X and Y are at a table on which there are dollars coins. In round one, X can appropriate one coin, or two. Coins she appropriates are removed from the table to be delivered when the game is over. If she takes two, the game is over, she gets these, and Y gets nothing. If she takes just one, there is a second round in which Y chooses one coin or two. Depending on his choice there may be a third round in which it is X's turn to choose, and so on until a player takes two coins, or there is just one coin left and the player whose turn it is takes it."

Sobel distinguishes between weak and strong solutions to a game. A *weak solution* shows that players who satisfy certain conditions resolve a game somehow without explaining how. On the contrary, a *strong* solution shows *how* players reach the outcome. Conditions determine the level of rationality ascribed to the gameplayers. Game theorists disagree about rationality which should be granted to players even in a theory which is intended to reflect the behaviour of *idealized players.* Consider the backward-induction terminating game described above. The question arises whether ideally rational and informed players in that game satisfy a strong knowledge condition. Rabinowicz would give a negative answer. He claims that it is not reasonable to expect the players to be stubbornly confident in their beliefs and incorruptible in their dispositions to rational behaviour. On the contrary Sobel says that once we have granted that the gameplayers are *resiliently* rational, we should also admit that *past irrationality* would not exert a corrupting influence on present play. Even though he does not share Rabinowicz's view, Sobel wonders about the possibility of finding an intermediate solution which would be *less demanding* than his initial condition — the condition of knowledge compounded robustly forward of resilient rationality — but which nevertheless "would enable reasoning on X's part to her choice to take both coins and end the game" He argues that ideally rational and well informed players in the game would not have a strong solution to the game unless they satisfied demanding subjunctive conditions (involving counterfactual conditionals) which are not significantly different from "knowledge compounded robustly forward of resilient rationality". One can find in Sobel's contribution original and deep ideas on *ideal game-theoretic rationality.* Thus he investigates the consequences of holding *prescience* as being an ingredient of *game-theoretical rationality.*

Reasoning and computation play a fundamental role in mathematics and science. A primary purpose of logic is to state principles of valid inference and to formulate logical systems where as many logical truths as possible are provable by effective methods. The last part of the book, **Reasoning and Computation in Logic and Artificial Intelligence**, contains discussions on the matter. It is well known that material and strict implication, which are central notions for the very analysis of entailment and valid reasoning, lead to paradoxical laws in traditional logic. Among so-called paradoxes of implication there is, for example, the law that a contradiction implies any sentence whatsoever. Do inconsistent theories really commit their proponents to asserting everything? This part of the book presents paraconsistent and relevant logics which advocate like intuitionist logic rival conceptions of implication and of valid reasoning. It also discusses important issues for artificial intelligence. Human agents take decisions and act in situations where they have an imperfect knowledge of what is happening and they do many things while relying on imprecise perceptions. They are not certain of data and they can revise their conclusions. Which new methods should logic and artificial intelligence use in order to deal with uncertainty and imprecision in computing data?

Developing insights due to Vasilev and Jáskowski, da Costa and Asenjo invented paraconsistent logic. In Chapter 20, **On the Usefulness of Paraconsistent Logic**, Newton da Costa, Jean-Yves Béziau and Otavio Bueno examine intuitive motivations to develop a paraconsistent logic. These motivations are formally developed using semantic methods where in particular, bivaluations and truth-tables are used to characterize paraconsistent logic. The authors then discuss the way in which paraconsistent logic, as opposed to classical logic, demarcates inconsistency from triviality. (A theory is *trivial* when every sentence in the theory's language is a theorem.) They also examine why in paraconsistent logic one cannot infer everything from a contradiction.

As a result, paraconsistent logic opens up the possibility of investigating the domain of what is inconsistent but not trivial. Why is it desirable to rescue inconsistent theories from the wreck? The reason is that *in practice* we live with inconsistent theories. From 1870 to 1895 Cantor derived important theorems of set theory from two quite obvious principles: the postulate of extensionality and the postulate of comprehension. Yet around 1902, Zermelo and Russell discovered a hidden inconsistency in the second principle. However when the shaky foundations of set theory were brought to light, mathematicians and logicians did not abandon the whole body of set theory. They decided instead to search for a way of *correcting* the faulty postulate and found several solutions (Russell's theory of types, Zermelo's separation axiom etc...) This historical fact shows that an inconsistent theory can be useful, that working mathematicians do not derive anything whatever from an inconsistency and that we need a logic if we want to continue to use reasoning during the span of time which elapses after the discovery of an inconsistency and before the discovery of a solution which removes the inconsistency. Inventors of paraconsistent logic intended to provide such a logic. da Costa, Béziau and Bueno briefly consider applications of paraconsistent logic to various domains. In mathematics they consider the formulation of set theory, in artificial intelligence the construction of expert systems, and in philosophy theories of belief change and rationality. With these motivations and applications in hand, the usefulness and legitimacy of paraconsistent logic become hard to deny.

According to relevance logic what is unsettling about so-called paradoxes of implication is that in each of them the antecedent seems irrelevant to the consequent. Following ideas of precursors such as Ackermann and Anderson & Belnap, relevance logicians tend to reject laws that commit fallacies of relevance. The most basic system of relevance logic is the system B+ that Paul Gochet, Pascal Gribomont and Didier Rosetto consider in Chapter 21, **Algorithms for Relevant Logic**. Their main purpose is to investigate whether the connection method can be extended to that basic system of relevance logic. A *connection proof* proceeds like a refutation constructed by tableaux or sequents. It starts with the denial of the formula to be proven and attempts to establish by applying reduction rules which stepwise decompose the initial formula that such a denial leads to a contradiction. The connection method which has been recently extended to modal and intuitionistic logic is much more efficient than the sequent calculi and tableau method. So it is very useful to extend it to other non classical logics especially to those used in artificial intelligence. Gochet, Gribomont and Rosetto begin their chapter by presenting the basic axiomatic system B+ of relevant logic. They also briefly present Bloesch's tableau method for B+ and next adapt Wallen's connection method to the system B+. The authors give a decision procedure which provides finite models for any satisfiable formula of system B+. They also prove the soundness and the completeness of their extension. This is an important logical result.

Extensional languages with a pure denotational semantics are of a very limited interest in cognitive science. Intensional object languages are needed in artificial intelligence as well as in philosophical logic and semantics to deal with thoughts of agents who are often uncertain. However, many natural intensional properties existing in artificial and natural languages are hard to compute in the algorithmic way. In Chapter 22, **Logic, Randomness and Cognition**, Michel de Rougemont shows that randomized algorithms are necessary to represent well intensions and to verify some specific relations in computer science. There are two main intensional aspects to take into consideration in artificial intelligence namely the complexity and the reliability of data. When data are uncertain, the advantage of randomized algorithms is very clear according to de Rougemont for both the uncertainty and complexity can then be improved in the computation. Rougemont concentrates on the reliability of queries in order to illustrate this advantage. This important contribution to "exact philosophy" fits in with the previous chapter in which complexity issues were also raised.

Perceptions play a key role in human recognition, attitudes and action. In Chapter 23, **Computing with Numbers to Computing**

with Words — From Manipulation of Measurements to Manipulation of Perceptions, Lofti Zadeh provides the foundations of a computational theory of perception based on the methodology of computing with words. There is a deep-seated tradition in computer science of striving for progression from perceptions to measurements, and from the use of words to the use of numbers. Why and when, then, should we compute with words and perceptions? As Zadeh points out, there is no other option when precision is desired but the needed information is not available. Moreover when precision is not needed, the tolerance for imprecision can be exploited to achieve tractability, robustness, simplicity and low solution cost. Notice that human agents have a remarkable capability for performing a wide variety of actions without any need for measurements and computations. In carrying out actions like parking a car and driving, we employ perceptions — rather than measurements — of distance, direction, speed, count, likelihood and intent.

Because of the bounded ability of sensory organs to resolve detail, perceptions are intrinsically imprecise. In Zadeh's view, perceived values of attributes are *fuzzy* and *granular* — a granule being a clump of values drawn together by indistinguishability, similarity, proximity or functionality. In this perspective, a natural language is a useful system for describing perceptions. In Zadeh's methodology, computation with perceptions amounts to computing with words and sentences drawn from natural language labelling and describing perceptions. Computing with words and perceptions provides a basis for an important generalization of probability theory. Zadeh's point of departure is the assumption that subjective probabilities are, basically, perceptions of likelihood. A key consequence of this assumption is that subjective probabilities are f-granular rather than numerical, as they are assumed to be in the standard bivalent logic of probability theory. In the final analysis, Zadeh's theory could open the door to adding to any measurement-based theory the capability to operate on perception-based information.

Fuzzy logic, even more than relevant and paraconsistent logics, had to overcome deep-seated prejudices and hostility. Nowadays the hostility has vanished and the merits of fuzzy logic have been widely recognized. Like other non-standard logics, fuzzy logic brings together concern for logic, for thought and for action.

I

REASON, ACTION AND COMMUNICATION

Chapter 2

THE BALANCE OF REASON*

Marcelo Dascal
Tel Aviv University

> *If we had a balance of reasons, where the arguments presented in favor and against the case were weighed precisely and the verdict could be pronounced in favor of the most inclined scale... [we would have] a more valuable art than that miraculous science of producing gold.*
>
> —*Gottfried W. Leibniz*

1.

Western conceptions of rationality have been dominated by one image: that of the balance. According to this image, human rationality rests essentially on our capacity of **weighing**. Animals react instinctively and emotively to their environment and to their impulses. Humans, on the contrary, are able to escape from the influence of immediate stimuli (external or internal) thanks to their capacity to control their actions on the basis of a comparative evaluation of their different beliefs, motives, desires, values, and goals. Such an evaluation consists in **weighing** them on the scales of the Balance of Reason.[1] A rational belief is reached by carefully weighing data, evidence, and justifications; a rational prefer-

*A version of this paper was published in Spanish in O. Nudler (ed.), *La Racionalidad: Su Poder y sus Límites.* Buenos Aires: Paidós, 1996, pp. 363-381. I thank Oscar Nudler and the publisher for granting me permission to use that edition as the basis for the present version. I would also like to thank Catherine Wilson and an anonymous referee for their comments on the earlier version of this paper.

[1]In contemporary English, it would be more natural to use "Scales of Reason" instead of "Balance of Reason", which strongly suggests equilibrium. I will however preserve the latter

D. Vanderveken (ed.), Logic, Thought & Action, 27–47.

ence is based on a choice of goals that have value or weight; a rational decision is the one that opts for the best means to achieve a goal, after weighing the alternatives; a rational action consists in applying a rational decision without falling prey to the weight of non-rational factors (when this happens, it is customary to attribute the failure to the weakness of the will — *akrasia* — rather than to the weakness of Reason). Ideally, in a rational human being the Balance of Reason is the engine that activates and controls all beliefs, preferences, decisions, and actions.

This image of rationality is as dominant in the 17th century, when Leibniz hails it as the most valuable and desirable achievement of man (see the motto above), as it is in the 20th century, when Rescher, 1988(82), expressing a view shared by most contemporary theories of rationality, claims that:

> The aim of the cognitive project is to secure the *best achievable overall balance* between information and misinformation. ... [T]he best epistemic policy is clearly one that optimizes the overall balance of information, minimizing the sum total of errors...

It is through this image that domains as diverse as justice, theology, economy, politics, ethics, and even art are conceptualized and thereby connected to their underlying rational engine.[2]

In the wake of the work of Mary Hesse in the philosophy of science, of Martin Heidegger in metaphysics, of George Lakoff and his associates in linguistics and cognitive science, and of many others, we now know that one should not underestimate the cognitive importance of metaphors and images. They can no longer be conceived of as mere rhetorical ornaments, easily disposable, but rather as means through which we organize our conceptual and linguistic schemata and perform creative intellectual work.[3] Some of these metaphors deserve to be called "root metaphors", due to their dominant philosophical role. The scales/balance metaphor

---

phrase, which was currently used (with the meaning I assign to it) in the 17th and 18th centuries (see, for instance, Samuel Clarke's quotation in section V).

[2]Here are some illuminating quotes to this effect: "There is no action without will, but there is will without action. If all will were to break out into open action man would perish, since there would be no rational balance or moderating reason" (Swedenborg). "Poetic Justice, with her lifted scale, / Where, in nice balance, truth with gold / she weighs, / And solid pudding against empty praise" (Pope). After posting an earlier version of this article in my web-site, I received a message from an Australian colleague, where he says: "I was wandering around the www and found your very interesting paper on the metaphor of balance in our thinking about reasoning and rationality. Now that you've highlighted the issue, I couldn't help but be struck by the extent to which the metaphor of balance infuses our thinking about rational deliberation in the *Reason!* Project" (Tim van Gelder). For information on this project, see `http://www.philosophy.unimelb.edu.au/reason/`.

[3]The literature on metaphor has increased dramatically in the last quarter of the twentieth century. For good surveys and discussions of this literature, see Kitay (1987), Gibbs (1994) and Barcelona (2000), as well as the collection of essays edited by Ortony (1979). Recent

is certainly one of these root metaphors, and it deserves careful analysis.[4] In this paper, I undertake to bring to the fore some of the effects of this metaphor upon the conceptualization of rationality in Western philosophical thought.

I will first try to show how the main problems of epistemology correspond to the technical problems involved in creating and operating a perfectly reliable balance — an ideal challenged by Skepticism. The balance metaphor, it will be further argued, is compatible with two different conceptions of rationality, both present in Western thought. One of them, here dubbed 'hard rationality', expects the balance to provide unquestionable, conclusive decisions in every matter submitted to Reason. The other, here dubbed 'soft rationality', acknowledges the limitations of the former, and considers the balance of reason to be valuable even when it is only able to provide less than conclusive — and therefore questionable — decisions. Whereas the former conception equates rationality with certainty, and is vulnerable to skeptical doubt, the latter is appropriate for handling uncertainty and, by mitigating the claims of Reason, more apt to face the skeptical challenge. Leibniz, who contributed substantially to the development of both views of rationality, will, as usual, occupy a prominent place in my reflections.

2.

It should come as no surprise that Leibniz, the most deeply rationalist of the rationalist philosophers, is the one who paid close attention to the importance of the image of the balance for the conception of rationality. In a virtually unknown text,[5] to which the quote used as motto also belongs, he elaborates:

> Just as in weighing it is necessary to pay attention that all the weights are put into place, to check that they are not in excess, to check that they are not adulterated by other metals nor heavier or lighter than they

work on the essential role of metaphor includes, among others, (Hesse (1966), Lakoff (1987), Lakoff and Johnson (1980, 1999), Lakoff and Turner (1989). For the import of Heidegger's contribution to the topic, see Rorty (1989).

[4]The expression 'root metaphor' was coined by Stephen Pepper (1935), whose early recognition of the philosophical import of metaphor grants him also a position in the pantheon of metaphor champions of the twentieth century (see Pepper 1928, 1935, 1961). Among other root metaphors, one could mention the conceptualizations of thought in terms of vision and of ideas and meanings as mental content — both predominant in Western thought for many centuries. For a criticism of the former and of its epistemological implications, see Rorty (1979); for an analysis of the communicative effect of the latter, see Reddy (1979). I have analyzed two other root metaphors in Dascal (1991 and 1996).

[5]*Brief Commentaries on the Judge of Controversies or the Balance of Reason and Norm of the Text* (A, 6, 1, 548-559). This text was written in Latin, presumably between 1669 and 1671. A translation and commentary of this text is included in AC.

> should, to verify the balance's correct position, with the arms equidistant, the scales with equal weights, etc.; so too in this rational Balance attention must be paid to the propositions as to the weights, to the balance as to their connection, and no unexamined weight or proposition is to be admitted. Just as one is to estimate the gravity of the weights, so too [one should measure] the truth of a proposition; just as the gravity of the weights measures the gravity of the things to be weighed, so too the truth of the propositions adduced in the proof measures the truth of the principal proposition of the question under discussion; just as one must take care that no weight be omitted or added, so too one is to take care that nothing unfavorable or favorable to the topic examined be omitted or that the same thing, expressed in different words, be repeated. The mechanism of the Balance is similar to the connection of the propositions; just as one scale should not be lighter than the other, so too if one of two premises is weaker than the other, the conclusion must follow from the weaker one; just as the arms must be linked to each other by the beam, so too from pure particulars nothing follows, for they are sand without lime; just as the arms must be at equal distances from the yoke, so too the place of the proposition must be such that the middle term be equidistant from the major and the minor, which is achieved by observing an exact and eternal Sorites.[6]

In this text, Leibniz — with his usual acumen — singles out the main tasks rationality, conceived within the framework of the balance metaphor, has to face:

1 How to *calibrate* the balance?

2 How to ensure the *reliability* of the weights?

3 How to establish a suitable weighing *procedure*?

The calibration problem has to do, on the one hand, with the *mechanism* of the balance: that the scales are equidistant from the yoke, that they do not differ in weight, etc. Without a perfect mechanism, the balance wouldn't be able to fulfill its mission, for it would not be *neutral* vis-à-vis that which it is supposed to weigh. The Balance of Reason itself should not lean *a priori* towards one or another reason. But in order to ensure its neutrality one should also avoid the undesirable influence of other *causes* on its functioning. Just as a balance may be imperceptibly affected by a magnetic or gravitational field acting differentially on one of its scales, so too socio-historical or psychological pressures (e.g., current prejudices, traditions, political interests, passions, limitations of attention or memory, unconscious desires) may surreptitiously take the place

[6] *Brief Commentaries*, # 65.

of reasons. No doubt factors such as these are those that often end up determining our beliefs, preferences, decisions, and actions. But when that happens, the result cannot be called *rational*. For a rational human being is supposed to protect his Balance from such causal influences which are alien to rationality. Apriorism, anti-historicism, anti-sociologism, anti-psychologism — in short, anti-*contextualism* — are examples of the efforts to build up the protection in question. Whether they have successfully *insulated* the Scales of Reason is a controversial matter (cf. Dascal 1990).

The problem of *reliability* of the weights is, in the particular case examined by Leibniz, that of the *truthfulness* of the propositions taken as reasons (or premises of an argument). An adulterated weight corresponds to a piece of "information" or "data" which have not passed the tests required for them to be considered part of our "knowledge". There is no use for a perfect Balance if what we weigh with it is of doubtful value. A rational human needs, therefore, a criterion of knowledge that ensures the reliability of the information upon which she bases her rational deliberations. The centuries-old search for a satisfactory concept of "evidence" and related concepts looms large in the effort to elaborate such a criterion. That such a search continues today (see, for example, Gil 1993) is proof enough that the issue is far from settled.

The problem of the weighing *procedure* consists in determining the rules of *method* that ensure the valid *extension* of our knowledge. A satisfactory theory of *reasoning* is the cornerstone of such a procedure. In the above quote, Leibniz envisages such a theory as consisting mainly of deductive *logic*, which he instantiates by the classical theory of syllogisms. However, in the light of the well-known limitations of deduction as a means of expanding knowledge, other forms of logic have been considered — by himself as well as by others. For instance, inductive logic, probabalistic logic, juridical logic and, more generally, the entire set of procedures Leibniz subsumed under the label *ars inveniendi*, which includes, among other things, the *Topica* and *Dialectica*, as well as a gamut of semiotic "helps" for the proper conduct of reasoning (cf. Dascal 1978) and his hitherto overlooked art of conducting and resolving controversies by means other than strictly formal ones. It is this ensemble of reasoning procedures that Leibniz sought to incorporate in a broadened conception of logic, which he viewed as corresponding to a "softer reason" or "softer procedure" (*blandior tractandi ratio*; C, 34 — see Dascal 2001), insofar as it went beyond strict formal deduction. Needless to say, in spite of the progress made in some of these fields, the task is still far from completion. The difficulties range from the psychological fact that our "natural reasoning" often deviates from the norms of correct reasoning (so that we fall short of being Ideal Reasoners), through the problems in establishing such norms when they go beyond those of formal logic, up to the reluctance in acknowledging the need to do so in order to account for a wide range of ways of extending our knowledge that cannot be handled by formal logic alone.

3.

A substantial portion of the well-known skeptical critique of rationality — ancient, modern, or contemporary — consists in raising doubts about the possibility of accomplishing satisfactorily the three tasks singled out by Leibniz. The skeptics attempt to show the impossibility of certifying that the mechanism of the rational balance functions perfectly, the impossibility of determining the value of the weights, and the inevitable errors involved in every procedure of rational decision. Many of Sextus Empiricus's tropes, as well as many of the arguments of Montaigne, of Bayle and of the post-moderns, belong to one or another of these kinds of criticism.

Besides the specific difficulties pertaining to each of the three tasks, the skeptics have also raised problems shared by them. One example is the well-known "problem of the criterion" (cf. Popkin 1979: 15, 51, 71, 141, etc.), which hinges on the need for an additional criterion or rule — i.e., of *another* Balance — for determining the calibration, the reliability, and the correctness of the procedures of the Balance of Reason — in short, on the fact that the Balance is incapable of grounding itself. The following passage, taken from Hobbes's *Dialogue between a Philosopher and a Student of the Common Law of England*, illustrates well this kind of problem:

> Lawyer: The manner of punishment in all crimes whatsoever, is to be determined by the common-law. That is to say, if then the judgment must be according to the statute; if it be not specified by the statute, then the custom in such cases is to be followed: but if the case be new I know not why the judge may not determine it according to reason.
>
> Philosopher: But according to whose reason? If you mean natural reason of this or that judge authorized by the King to have cognizance of the cause, there being as many several reasons as there are several men, the punishment of all crimes will be uncertain, and none of them ever grow up to make a custom. Therefore a punishment certain can never be assigned, if it have its beginning from the natural reasons of deputed judges... (Hobbes [1740]: 121-122).[7]

If accepted, this criticism can lead to the admission that the choice of rationality as a "form of life" is not, ultimately, open to rational justification (Popper).

Another example of skeptical critique addressed to all three tasks is the observation that a multiplicity and variety (historical, cultural, individual) of methods or criteria lay claim to be *the* correct ones. The lack of agreement among scientists or philosophers regarding such claims and how to adjudicate them suggests a relativism that seems to destroy the alleged universality of the Balance of Reason.[8]

Finally, another source of skepticism vis-à-vis the Balance of Reason is the problem of interpretation: even when one applies universally accepted methods, the data used as well as the results of the "weighing" always require interpretation. But the latter involves a non-eliminable amount of indeterminacy, because it depends upon the context (historical, social, or psychological) of the interpreter, upon the theoretical framework embedded in the balance used itself, and upon the interpretive practices employed. If — adapting a phrase employed by Quine (1969) to a somewhat different kind of indeterminacy — there is no "fact of the matter" capable of eliminating such an indeterminacy, then, regardless of how accurate is the Balance, its use will be always infected by relativity.

The strategies employed by the defenders of Reason against its skeptical detractors are also well-known. The *tu quoque* argument, already employed by Aristotle, attempts to show that the skeptic himself in fact employs the Balance of Reason in order to criticize it, a fact that demonstrates its universality and reliability (since even its declared enemies rely upon it).

Another familiar strategy — which I have called 'insulation' (Dascal 1990) — consists in admitting the validity of the skeptical critique, while denying that it affects *all* the uses of Reason: there is at least some "pure" domain of rationality where the Balance of Reason is entirely protected from skepticism; it is in this privileged domain that the three tasks of grounding the Balance would be satisfactorily performed. In his reply to Hobbes's criticism, Leibniz alludes to this possibility:

> Thomas Hobbes thus mocks those who appeal to right reason, [arguing that] by the name of right reason they understand their own [reason],

[8]Hobbes, incidentally, doesn't consider the diversity of "natural reasons" argued for in the above quote as leading necessarily to relativism. To the Lawyer's distressful question, "If the natural reason neither of the King, nor of any[one] else, be able to prescribe a punishment, how can there be any lawful punishment at all?", the philosopher replies: "Why not? For I think that in this very difference between the rational faculties of particular men, lieth the true and perfect reason that maketh every punishment certain" (Hobbes [1740]: 122).

> so that in fact they appeal to themselves. But those who object in this way have not, so far, understood what I have in mind. In the first place, it is not clear that it is impossible to choose right reason as a judge, at least in some questions, examples of which follow.[9]

Gassendi's "mitigated skepticism" and Kant's "transcendental idealism" instantiate different implementations of the insulating strategy. Descartes's strategy, even though he too "insulates" *one* proposition which he considers immune to skeptical doubt and employs it both as a criterion of calibration and as a paradigmatic example of truthfulness and of a procedure of evaluation of reasonings, does not properly belong to this family of strategies, since he believes that it is possible to extend the Balance (or what he labels "natural light"), once calibrated by the *Cogito*, to virtually *all* domains.

4.

Leibniz, I believe, is the first Western philosopher who develops a new type of strategy to combat skepticism and to ground the Balance of Reason. Like Gassendi and Mersenne, he does not believe in the objectivity of Descartes's natural light, which can always be contaminated by subjectivism. But, whereas Gassendi's solution consists in assigning to the controlled use of "experience" a role in cognition and Mersenne's, in enhancing the role of mathematics, Leibniz — without overlooking these two elements — emphasizes rather the need for a rigorous formalization of reasoning (see Dascal 1978: 212-214). In order to be reliable, the Balance of Reason must be based on a rigorous *filum Ariadnes*, accessible to all, where errors are easily detectable as in arithmetic; and such a thread is nothing but the logical structure of reasoning, expressed in a precise and transparent notation.

Leibniz's critique of what Yvon Belaval (1960) described as Descartes's "intuitionism", leads him to develop a research programme which, beginning with the *De Arte Combinatoria* and evolving through many formulations of a logical calculus, reaches its apex in the idea of a *Characteristica Universalis*. The aim is to formalize the methods of reasoning and of representation of knowledge, so as to cover areas other than mathematics and logic, such as jurisprudence, physics, engineering, metaphysics, ethics, politics, and theology. If we had an adequate notation for representing all types of knowledge and a rigorous calculus for the manipulation of these representations, all questions would be solved by calculation and all mistakes would be easily detectable and

[9] *Brief Commentaries*, ## 55-56.

correctable as mere errors of calculation. Thus equipped, the Balance of Reason would permit us to resolve all disputes and would function universally and perfectly.

This is Leibniz's "maximalist" project — as Gil (1985) proposes to call it. Leibniz's enthusiasm in describing it is contagious, and has inspired, among other works, Frege's *Begriffschrift*.[10] This project is connected with a considerable portion of Leibniz's semiotics, which contributes not only to the task of devising the perfect notation, but also to the first of the tasks incumbent on whoever wants to improve the Balance of Reason: to overcome psychological limitations and other forms of interference. This is what I have called the "psychotechnical function" of symbol systems: abbreviations, synoptic tables, "naturally expressive" notations, mnemonic methods, etc. are designed to overcome the deficiencies of our attention and memory, thereby allowing for a considerable expansion of the Balance's scope of application. The various types of "indices" Leibniz proposes to compile, at the end of the *Brief Commentaries* (# 70), are an example of this semiotic improvement of the Balance. In other paragraphs of the same text (notably # 58) Leibniz refers explicitly to the maximalist project of the *Characteristica Universalis*, which would permit the entirely formal resolution of some controversies, especially juridical ones.

5.

But can this maximalist project really overcome all the difficulties and ensure the universal efficacy of the Balance of Reason? What should we do as long as we do not have the means to formalize *all* the areas of knowledge and action? And what should we do if there are areas which do not permit — by their very nature — formalization? Before tackling these difficulties, there is another problem, even more fundamental, to be addressed.

[10]Here is one example of Leibniz's enthusiasm. In a letter to Princess Elizabeth (1678), after listing Descartes's mistakes, Leibniz says: "All of this could give some people a bad opinion of the certainty of our knowledge in general. For, one can say, with so many able men unable to avoid a trap, what can I hope for, I, who am nothing compared to them? Nevertheless, we must not lose our courage. There is a way of avoiding error... In brief, it is to construct arguments only in proper form [*in forma*]... Any rigorous demonstration that does not omit anything necessary for the force of reasoning is of this kind... In order to determine the formalism that would do no less in metaphysics, physics, and morals, than calculation does in mathematics, that would even give us degrees of probability when we can only reason probabilistically, I would have to relate here the thoughts I have on a new characteristic, something that would take too long.... I dare not say what would follow from this for the perfection of the sciences — it would appear incredible. The only thing I will say here is that... all reasoning in demonstrative or probable matters will demand no more skill than a calculation in algebra does" (A, 2, 1, 437-438; translation in A&G, 239-240).

Let us suppose that there is no field of knowledge or action whose nature forbids formalization. Let us assume also that we have at our disposal the perfect Universal Characteristic. Now, the tasks, difficulties, and solutions so far mentioned — including the innovative one proposed by Leibniz — refer either to the functioning of the Balance or to the need to establish its proper foundations. They do not question the efficacy of the Balance as an instrument of decision, once such problems are satisfactorily solved. That is to say, the Ideal Balance would *always* lead us to the solution of any question. Furthermore, it is usually assumed that the Ideal Balance provides *the* rational solution which is endowed with the status of a *necessary* conclusion of the weighing procedure.

Nevertheless, Pyrrhonism, beyond its critique of the functioning and grounding of the Balance, has developed a more radical critique: even if the Balance were to function perfectly, it would not allow us to decide *anything*, because it would remain in *equilibrium*. This is the well-known skeptical doctrine of *isostheneia*. Such an equilibrium is reached by employing the very same Ideal Balance in order to oppose reasons of equal weight to the reasons that support any given conclusion. In this kind of critique, the skeptic makes full and conscious use of the *tu quoque*, with the aim of showing not that the Balance cannot exist, but that — were it to exist — it would be useless for the purpose of providing rational decisions. But, if it is the case that the most perfect Ideal Balance of Reason could not permit one to decide, either we are condemned to paralysis (like Buridan's Ass) or else our decisions are, from the point of view of Reason, arbitrary, i.e., irrational.

In a sense, it is this radical critique that characterizes the post-modern version of skepticism. For it emphasizes the intrinsic insufficiency or under-determination of Reason, whence it follows its uselessness, the arbitrariness of its decisions, and the purely political (Foucault) or honorific (Rorty) character of the appeal to terms such as "Reason", "Science", "Method", and "Truth".

When Samuel Clarke repeatedly appeals to the notion of "freedom out of indifference", which requires a mysterious capacity of the agent to act even when there are no *reasons* for choosing a course of action, he is in fact admitting the limitation of Reason and the arbitrariness of action:

> A Balance is no Agent, but is merely passive and acted upon by the Weights; so that when the Weights are equal, there is nothing to move it. But Intelligent beings are Agents; not passive, in being moved by Motives, as a Balance is by Weights; but they have Active Powers and do move Themselves, sometimes upon the View of strong Motives, some-

> times upon weak ones, and sometimes where things are absolutely indifferent.[11]

What Clarke does not realize perhaps is the consequence of this admission for the status of the Newtonian science he defends, whose results he considers absolute.

The same problem arises in the moral sphere with those who — like Ruth Barcan-Marcus — affirm that the existence of genuine moral dilemmas does not entail the inconsistency of moral principles. It only shows their insufficiency for the determination of the choice of a particular course of action. According to her, it is not the principles that are to be blamed (nor, we might add, the Balance of Reason). It is the world that sometimes defeats us.[12]

6.

An extreme rationalist like Leibniz cannot accept such a defeat. For it would mean accepting the irrationality of the world, i.e., the incompetence of its creator. Ultimately, this would amount to acknowledging the triumph not only of the skeptics, but also of the gnostics. Furthermore, it would mean admitting — as the modern tradition on the whole has done (cf. Unger 1975) — the schizophrenic character of the human being, split into a Reason and a Will that more often than not are not in harmonious relation, and dominated more by the latter than by the former — a situation that would provide further proof of divine imperfection.[13]

It is well-known that Leibniz, in his metaphysics, rejects altogether the idea of a complete equivalence of alternatives: just as there are no two individual substances which share all their properties, being different only numerically (*solo numero*), so too there are no two possible worlds equivalent in their degrees of perfection. God, who is able to weigh the totality of reasons, has always a sufficient reason for his choice of the most perfect world to be created. But what we are concerned with here is the Human Balance, not the Divine one. Hence, Leibniz's metaphysics is of no avail to us.

The crucial question for a rationalist is whether the Balance of *Human Reason* has the means to avoid non-arbitrarily the catastrophic conse-

[11] Samuel Clarke, Fourth letter to Leibniz (GP 7, 381). Leibniz's reply will be discussed below.

[12] These claims were put forth by Ruth Barcan-Marcus in her lecture "More about consistency of principles and moral dilemmas", delivered at a Cérisy-la-Salle colloquium (June 1994) on rationality, organized by Jacques Poulain and Daniel Vanderveken.

[13] Cf. Swedenborg's quote, in note 2.

quences of the equilibrium of indifference. Is there a Balance of Human Reason which, in this respect, mirrors — even though modestly and imperfectly — the absolutely rational Divine one? Obviously, Leibniz's answer must be an emphatic "Yes!". Nevertheless, paradoxically, this "Yes!" entails a significant modification in his anti-skeptic strategy. The maximalist algorithmic model, which was the core of this strategy, can no longer be considered the only and exclusive paradigm of rationality.

If not metaphysics, ethics — in so far as it is concerned with human action — might perhaps provide the clue. In his reply to Clarke's argument quoted above, Leibniz says:

> ... motives do not act on the mind as the weights act on a balance; it is the mind that acts by virtue of the motives, which are its dispositions to act. [...] the motives include *all* the dispositions the mind may have in order to act voluntarily, since they include not only the reasons, but also the *inclinations* which come from the passions or from other previous impressions. So that if the mind would prefer the weak inclination over the strong one, it would act against itself, and otherwise than it is disposed to act (GP 7, 392).

Rather than a strict dichotomy passive/active or a complete split between the Will and the Intellect, as Clarke seems to assume, Leibniz, in conformity with his overarching principle of continuity, includes — rather than excludes — the passions among the motives for action. In this way, he places them along a single scale, where the relative weights of the passions can be compared with those of reasons in the determination of human choices.[14] I have italicized two key words in the passage quoted, which indicate, on the one hand, the fact that — for Leibniz — the 'calculus of motives' that leads us to action must always be *global* and, on the other, that this calculus takes into account that which *inclines* us to act (without *forcing* us to do so). The result of this calculus, then, is itself an inclination.

Leibniz agrees with Locke that a person should be able to control his passions so as to avoid their forcing one to act (*Nouveaux Essais* II.21.53; GP V, 186), and also accepts that the decisive consideration

[14]Leibniz studied carefully the controversy between Hobbes and Bramhall, whose central topic was the issue of freedom and necessity. He appended to the *Théodicée* an account of this controversy, under the title "Reflexions sur l'ouvrage que M. Hobbes a publié en Anglois, de la Liberté, la Necessité et du Hazard" (GP VI, 389-399). Leibniz sided with Hobbes in claiming that the notion of 'free will' cannot mean that we are able to presently determine our will. Our present will, he says, is a function of our reasons and dispositions. Nevertheless, he points out, we can have some influence — albeit "obliquely" — upon our future will, by looking for and shaping new reasons and dispositions. For a collection of essays on the controversy, see Dascal and Fritz, eds. (2001). Marras (2001) suggests an analysis of the argumentative structure of Leibniz's comments on the controversy. A selection of the texts of the controversy has been recently published by V. Chappell (1999).

for this purpose is to take into account not only the present moment or the present life, but also eternal happiness. "Were everything limited to the present moment — he says — there would be no reasons to refuse the pleasure that presents itself to us" (*Nouveaux Essais* II.21.58; GP V, 187). Nevertheless, whereas for Locke, if there were nothing to hope for beyond the grave, one would be entitled to conclude: "*let us eat and drink*, let us enjoy what we delight in, *for tomorrow we shall die*" (*Essay* II.21.55), Leibniz — in conformity with his principle of uniformity — argues that, even within this life it is possible to establish an order of preferences of the different (terrestrial) goods that would establish the superiority of some of them over others, "even though the obligation [to choose the former] would not be then so strong nor so decisive" (*Nouveaux Essais* II.21.54; GP V, 186). As rational human beings we cannot overlook the fact that a present perfection (and pleasures are perfections, for him) may lead to greater imperfections, for our lives unfold in time, rather than in eternity.

Accordingly, similarly to God, we have a criterion for our choices, namely, to maximize the total amount of perfection we can achieve in life. Unlike God's, however, *our* calculus of perfections cannot be "decisive" or "demonstrative" since, unlike Him, we cannot but rely on "confused perceptions" along with those (relatively few, alas!) bits of clear and distinct knowledge we manage to achieve.[15] Unlike Him, we need a Balance of Reason, with the help of which we can, albeit only approximately and non conclusively, guide rationally our lives rationally.[16]

7.

The ethical need for such a Balance, which is due to our epistemic limitations, only emphasizes its epistemic need for the achievement of knowledge in most fields. Both needs must, of course, be translated into the development of adequate epistemic means to operate rationally within the framework of human limitations.

Already in the *Brief Commentaries*, when he mentions a method that would permit one to reach "moral certainty or practical infallibility"

---

[15] *Nouveaux Essais* II.21.54; GP V, 186. Leibniz is here referring to his well-known classification of types of knowledge, which he introduces elsewhere in the *Nouveaux Essais* (II.30; GP V, 236-244).

[16] In fact, our reasoning activity — be it approximative or strictly deductive — does not correspond to any similar "activity" of God. For "God does not reason, strictly speaking, by using time as we do in order to move from one truth to another: however, since he understands at once all the truths and all their connections, he knows all the consequences, and he contains eminently in himself all the reasonings we can make — and this is why his wisdom is perfect" (GP VI, 399).

(# 37), Leibniz is suggesting an alternative model, presumably complementary to the algorithmic one, for improving and implementing the Balance of Reason. But the *Brief Commentaries* is still impregnated with elements belonging to the algorithmic model. It mentions a "true Logic or form of proceeding which is perfectly exact and rigorous" (# 61). The errors of judges are compared with errors of calculation (# 58). The *metric* function of the Balance is stressed: the truth of the premises *measures* that of the conclusion, just as the gravity of the weights *measures* that of the thing weighed (# 64); the arguments in favor and against are said to be "rigorously quantified" (# 62), and those men who are patient and diligent are said to "be in all questions practically as infallible as a *calculator* or a *measurer* are" (# 65). It would seem that Leibniz here anticipates the modern digital balances we now have.

But the excessive fixation on this paradigm of a Universal and Rigorous Metric is easy prey for the earlier mentioned skeptical arguments. The digital balance does not exhibit with perceptible evidence the weighing mechanism that yields its "conclusions". It depends on the theories — themselves in need of "weighing" — which govern its mechanism.[17] The multiplication of logics and the fragmentation of mathematics would force us to devise a "super-logic" or a "super-mathematics", were we to wish to evaluate the respective merits of each form of logic or of mathematics in order to choose the one most appropriate for governing the mechanism of the Balance — the problem of the criterion would strike again with full force. The practical (if not principled) impossibility of reducing all concepts to their atomic components introduces an element of tentativeness and arbitrariness in any notation we may invent. And the extrapolation of the algorithmic model to all fields of knowledge and to all kinds of issues risks rendering it a purely abstract schema, leaving unsolved the thorny problem of granting it an interpretation in each particular field of application.[18] In view of these facts, wouldn't those who argue that — as Leibniz himself puts it — this Balance, "abstractly taken, is a useless idea, empty, inefficient, and remote from real life" (*Brief Commentaries,* # 54) be right? Shouldn't the very demand of

[17]When one weighs what is "evident", on the contrary, one does not get involved in such a circularity: "...in the problems which are immediately evident to the senses, there is no need for a judge of controversies other than the senses themselves" (*Brief Commentaries*, # 57).

[18]In other words, unlike Leibniz's Universal Characteristic, which is conceived as a *semantic* representation, using *interpreted* symbols, we would land upon the notion of a purely *syntactic* calculus, the burden of endowing it with a semantic interpretation being left to someone other than formal logicians. This is an example of what Yehoshua Bar-Hillel (1970) appropriately called "the logicians' treason".

algorithmic perfection be blamed for leaving us without an instrument of decision-making in most of the real problems we face?

A balance, however, need not be digital, i.e., it need not have exclusively a metric function and a metric mode of operation. A more complete balance has also what I would call a "dialectical" function. It permits us to confront and compare the "values" of what is placed on its scales directly, i.e., without reducing them to universal measuring units: "Let the right to explain to the other his own reasons be given to everyone"; let "each of the parties listen to the reasoning of the other, along with the judges" (*Brief Commentaries*, # 63). No doubt the "judge of controversies" must follow "the thread of true Logic" and he should not deviate from the "eternal Law of reasoning" (*Brief Commentaries*, # 63). But this is not enough for satisfactorily fulfilling his duty. For he also must be capable of distinguishing what is relevant from what is irrelevant, of separating what is merely verbal from what is essential, of eliminating redundancies, of filling the gaps, of ordering and evaluating the reasons offered by both parties (de Olaso 1990: 117). All of these tasks, which precede the possibility of applying logical form in the process of decision-making, require capabilities of evaluation and interpretation which are irreducible to formalization.[19] Furthermore, the strict application to controversies (and to many other practical matters) of the requirement of full formalization would soon lead to absurdities, as Leibniz himself points out:

> For if we wanted to carry through a formal disputation, several days would be spent on a syllogism, and where would the audience and the other opponents be by then? The large number of prosyllogisms, moreover, would compose a real labyrinth from which we could not escape without a protocol, to say nothing of the great understanding and unusual acuteness needed to carry a demonstration back to its primary sources and fundamental truths on the spur of the moment. It is thus a human perversity to use logical form only where it can be of little help and must soon be stopped...[20]

[19]It is not by coincidence that in the texts that deal with the "judge of controversies" — like the *Brief Commentaries* — Leibniz also discusses at length hermeneutics. For other remarks on hermeneutics, see Leibniz's *Nova methodus discendae docendaeque jurisprudentiae* of 1667 (A, 6, 1, 337-338); see also "On the interpretation, foundations, application, and system of laws" (in AC).

[20]Letter to Gabriel Wagner, 1696 (L, 466; translated in AC). Leibniz himself attempted at least once to provide a (partial) syllogistic reduction of his controversy with Denis Papin on the problem of the *perpetuum mobile*. He boasted to have at least reached thereby an agreement with Papin about what was at issue: "We carried the matter beyond the twelfth prosyllogism, and from the time we began this, complaints ceased, and we understood each other, to the advantage of both sides" (L, 467). Needless to say, Papin himself did not accept Leibniz's reduction, nor — for that matter — did he agree to his "understanding" of the issue. For an analysis of the Leibniz-Papin controversy, see Freudenthal (2000).

What is required of a Balance of Reason capable of being applied efficiently beyond those few domains where the algorithmic model is viable, is the sensitivity to all that which is — according to this model — *im*ponderable. A balance endowed with this kind of sensitivity will certainly not be able to produce in all cases absolute, i.e. demonstrable or calculable certainties.[21] Hence, it will be a balance that *inclines without necessitating*. It will be a balance capable of operating not only within the realm of the necessary, but also within that of the contingent.[22]

Without abandoning his efforts to develop the algorithmic model, Leibniz — aware of its insufficiency for establishing the universality of rationality[23] — has undertaken to develop also another, non-algorithmic model of rationality. Admittedly, it will be needed only where strict demonstration, which is applicable to "necessary matters where eternal truths occur", is not possible; that is to say, the alternative model will be appealed to "in contingent matters where the most probable must be chosen". According to Leibniz, the application of such a model raises two problems:

> The first concerns presumption, that is, when and how one has the right to shift the demonstration from oneself to someone else; the second concerns the degrees of probability, how to weigh and evaluate considerations which do not constitute a perfect demonstration but run counter to each other (*indicantia* and *contraindicantia*, the medics call them), and to reach a decision. For the common saying is true enough — *rationes*

[21] Even Rescher, whose account of rationality seeks to devise an epistemic policy based on the optimization (i.e., calculability) of cost effectiveness, admits that such a policy "will have to tolerate errors and inconsistencies, being such that an inconsistent family of contentions will occasionally (though no doubt rarely) manage to slip through the net" (Rescher 1988: 82). I take this to mean that, at least in such cases, the maximalist Balance will have to be complemented by some other way of "weighing", if it is not to accept arbitrariness or, in Ruth Barcan-Marcus's terms, defeat.

[22] The phrase *incliner sans necessiter*, in Leibniz's mature metaphysics, refers to the realm of contingency as well as to that of ethics. As far as I know, it appears for the first time in the *Discours de Métaphysique* of 1685 (cf. # 30, for instance), written at the time he was in the Harz mining region, designing pumps based on very slight deviations from the equilibrium point. The notion of inclination withou necessitation becomes the fundamental piece of Leibniz's defence against the charges of determinism or "spinozism" that had been often levelled against him. It appears also in his fifth letter to Clarke, where it is explicitly linked to the image of the balance: "It is true that Reasons perform in the mind of the sage, and Motives in any mind whatsoever, that which corresponds to the effect of the weights on a balance. It is objected that this notion leads to necessity and fatality. But this is said without proof... A motive inclines without necessitating, i.e., without imposing an absolute necessity" (GP 7, 389-390).

[23] "There is never *indifference of equilibrium*, i.e. [a situation] where eveything is perfectly equal on one side and the other, without there being more inclination towards one side.... It would have been a big defect, or rather a manifest absurdity, if it were otherwise, even in men down here, if they were able to act without an inclining reason" (*Essais de Théodicée* # 46, GP 6, 128).

> *non esse numerandas sed ponderandas* [reasons are not to be counted but to be weighed]. But no one has yet devised the scales, though no one has come closer to doing so than the jurists.[24]

Certainly there are many more problems to be solved. In fact, ever since Aristotle pointed out the need for a Dialectics which should be called into action when Logic reaches its limits, little has been done to work out the details of this complementary side of Reason. Leibniz gives here and elsewhere valuable hints, some of which he developed in considerable detail. He mentions the jurists as those who have contributed more than anyone else to this enterprise, suggesting that much can be learned from them in this respect.[25] Part of what one can learn from the jurists is no doubt the role of such notions as *burden of proof* and *presumption*, which Leibniz singles out as especially important.[26] He refers to the need to develop a calculus of probabilities as a part of what has to be done.[27] He suggests that hermeneutics — i.e., a theory of interpretation — is also an essential component of this other side of rationality.[28] Finally, he not only engages in a "dialectical" construction of knowledge through his vast correspondence and multiple polemics, but also undertakes to provide a theory of controversies which should account for the rationality of such an activity. And, of course, this is not an exhaustive list.[29]

[24]Letter to Gabriel Wagner of 1696 (L, 467).

[25]In a paper called "For a Balance of Jurisprudence Regarding the Degrees of Proofs and Probabilities", written around 1676 (C 210-214; translated in AC), Leibniz says: "[J]ust as the Mathematicians have excelled in the practice of logic, i.e. the art of reason in necessary propositions, so too the jurists have practiced it better than anybody else in contingent matters". Leibniz's studies of juridical logic deserve careful attention.

[26]The latter, it should be recalled, is the heart of what is nowadays called "non-monotonic logic" or "default reasoning". It lies also at the heart of the "logic of conversation" due to Grice, which became one of the cornerstones of current pragmatic theory (cf. Dascal 1983, 2003).

[27]And he did indeed contribute extensively to developing the calculus of probabilities. The extent of this contribution has been highlighted by recent research. See, for example, the texts published by Parmentier [P] and by Mora Charles (1992). It is not clear whether the calculus of probabilities really belongs to the non-algorithmic model of rationality I am talking about here. For, in so far as it is a "calculus", it belongs to the algorithmic model. See the end of the passage quoted in note 10, where Leibniz clearly includes probabilistic reasoning within his dream of the Universal Characteristic. It must be said, then, that at least the use of probabilities is not typical — *pace* Fernando Gil — of the minimalist program offered as an alternative to the maximalist, algorithmic one. On the other hand, in so far as probabilities "incline without necessitating", they belong to the minimalist program, just as the logic of presumption does.

[28]See note 19.

[29]For all the points mentioned in this paragraph, see the texts collected in AC, as well as the Introduction to that volume. For the special epistemological and historical importance I attach to controversies in general and to scientific controversies in particular, see Dascal (1998, 2000).

What is shared by all these methods is their modest character. The conclusions they permit us to reach, which are not obtained in a strictly deductive form, are provisional and likely to be revised without leading to contradiction. Nevertheless, they are sufficient to incline the Balance of Reason, i.e. to provide rational justification even in the absence of necessitating proof.

It is remarkable that, next to the well-known 'hard' rationalist, there is 'another' Leibniz, a 'soft' rationalist, so far hardly noticed. In this other side of Leibniz's thought one can find, I think, the basis for a strategy of defense of rationality which is in a better position to cope with rationality's tougher critics, past and present. For, whereas the pretentiousness and arrogance of the traditional conception of an unfailing decisive and apodictically ruling Reason can hardly be sustained in the light of the skeptics' attacks, a more modest rationality, which cannot be blamed for not providing certainty but nevertheless provides justified inclination toward one of the scales, stands a good chance of not having to surrender to the skeptics.

As every image or metaphor, the 'Balance of Reason' allows for several interpretations. We have seen how one of these interpretations — the one I have called 'metric' or 'algorithmic' — leads to a 'hard' (another metaphor, of course) conception of Reason, while the other — the one I dubbed 'dialectical' — leads to a 'soft' conception of rationality. The fact that the second interpretation has been found, along with the first one, in the work of an uncompromising rationalist such as Leibniz, suggests that the two views of rationality are indeed complementary rather than competing with each other. Once revised as suggested here, the image of the balance regains vitality and may be further used by those who are persuaded that, unless it is somehow softened along the lines discussed here, rationality will hardly be able to secure its position.

References

Barcelona A. (ed.) (2000). *Metaphor and Metonymy at the Crossroads: A Cognitive Perspective.* Berlin-New York: Mouton de Gruyter.

Bar-Hillel Y. (1970). "The Logicians' Treason". In Y. Bar-Hillel, *Logic, Language, and Method.* Tel Aviv: Sifriat Hapoalim, pp. 112–118 [Hebrew].

Belaval Y. (1960). *Leibniz critique de Descartes.* Paris: Gallimard.

Chappell V (ed.). (1999). *Hobbes and Bramhall on Liberty and Necessity.* Cambridge: Cambridge University Press.

Chuang Tzu [= BW]. *Basic Writings.* Translated by B. Watson. New York: Columbia University Press, 1964.

Dascal M. (1978). *La sémiologie de Leibniz.* Paris: Aubier-Montaigne.
— (1983). *Pragmatics and the Philosophy of Mind,* vol. 1. Amsterdam: John Benjamins.
— (1990). "La arrogancia de la Razón".*Isegoría* 2: 75–103.
— (1991). "The Ecology of Cultural Space". In M. Dascal (ed.), *Cultural Relativism and Philosophy: North and Latin American Perspectives.* Leiden: E. J. Brill, pp. 279–295.
— (1996). "The Beyond Enterprise". In J. Stewart (ed.), *Beyond the Symbol Model: Reflections on the Representational Nature of Language.* Albany, New York: State University of New York Press, pp. 303–334.
— (1998). "The Study of Controversies and the Theory and History of Science". *Science in Context* 11(2): 147–154.
— (2000). "Epistemology and Controversies". In Tian Yu Cao (ed.), *Philosophy of Science [Proceedings of the Twentieth World Congress of Philosophy, vol. 10].* Philadelphia: Philosophers Index Inc., pp. 159–192.
— (2001). "Nihil sine ratione → Blandior ratio" ('Nothing without a reason → A softer reason'). In *Nihil sine ratione* (Proceedings of the VII. Internationaler Leibniz-Kongress), Poser H. (ed.) Berlin: Leibniz Gesellschaft, Volume I, pp. 276–280.
— (2003). *Interpretation and Understanding.* Amsterdam: John Benjamins.
Dascal M. and Fritz G (eds.). (2001. *The Hobbes-Bramhall Controversy* (= Technical Report 1, Research Group "Controverses dans la République des Lettres). Tel Aviv (available from e-mail: controil@post.tau.ac.il.
Freudenthal G. (2000). *Perpetuum-mobile — the Leibniz-Papin Controversy.* Berlin: Max-Planck–Institut für Wissenschaftsgeschichte (Preprint 127).
Gil F. (1985). "Leibniz et la charge de la preuve". *Revue de Synthèse* 118–119: 157–173.
Gil F. (1993). *Traité de l'Evidence.* Paris: Millon.
Hesse M. (1966). *Models and Analogies in Science.* Notre Dame, Indiana: Notre Dame University Press.
Hobbes T. [1740]. *Dialogue Between a Philosopher and a Student of the Common Law of England.* In *The English Works of Thomas Hobbes* (ed. W. Molesworth), vol. 6. London: John Bohn, pp. 3–160.
Kitay E.F. (1987). *Metaphor: Its Cognitive Force and Linguistic Structure.* Oxford: Clarendon Press.
Lakoff G. (1987). *Women, Fire, and Dangerous Things: What Categories Reveal about the Mind.* Chicago: The University of Chicago Press.

Lakoff G. and Johnson M. (1980). *Metaphors we Live By.* Chicago: The University of Chicago Press.

Lakoff G. and Johnson M. (1999). *Philosophy in the Flesh: The Embodied Mind and its Challenge to Western Philosophy.* New York: Basic Books.

Lakoff G. and Turner M. (1989). *More than Cool Reason: A Field Guide to Poetic Metaphor.* Chicago: The University of Chicago Press.

Leibniz G.W. [= A]. *Sämtliche Schriften und Briefe.* Berlin: Deutsche Akademie der Wissenschaften [1923 –].

— [= AC]. *The Art of Controversies and Other Writings on Dialectics and Logic.* Ed. M. Dascal *et al.* (in preparation).

— [= A&G]. *Philosophical Essays.* Ed. R. Ariew and D. Garber. Indianapolis: Hackett [1989].

— [= C]. *Fragments et opuscules inédits.* Ed. L. Couturat. Hildesheim: Olms [reprinted 1966].

— [= GP]. *Die Philosophischen Schriften von G.W. Leibniz.* Ed. by C.I. Gerhardt. Hildesheim: Olms [reprinted 1965].

— [= L]. *Philosophical Papers and Letters*, 2nd ed. Ed. by L.E. Loemker. Dordrecht: Reidel [1969].

— [= P]. *Textes sur les probabilités.* Ed. M. Parmentier. Paris: Vrin [1994].

Locke J. 1690 [1961]. *An Essay Concerning Human Understanding*, ed. J. Yolton. London / New York: Everyman's Library.

Marras C. (2001). "The Reception of the Hobbes-Bramhall Controversy According to Leibniz's 'Réflexions sur l'ouvrage que M. Hobbes a publié en anglois, de la liberté, de la necessité et du hazard". In Dascal and Fritz (eds.).

de Mora Charles M.S. (1992). "Quelques jeux de hazard selon Leibniz (manuscrits inédits)". *Historia Mathematica* 19: 125–157.

de Olaso E. (1990). "Sobre la filosofía leibniziana de las controversias". In F. Gil (ed.), *Scientific and Philosophical Controversies.* Lisboa: Fragmentos, pp. 115–130.

Ortony A. (ed.) (1979). *Metaphor and Thought.* Cambridge: Cambridge University Press.

Pepper S. (1928). "Philosophy and Metaphor". *Journal of Philosophy* 25: 130–132.

— (1935). "The Root Metaphor Theory of Metaphysics". *Journal of Philosophy* 32: 365–374.

— (1942). *World Hypotheses.* Berkeley: University of California Press.

Popkin R.H. (1979). *The History of Scepticism from Erasmus to Spinoza.* Berkeley: University of California Press.

Quine W.V.O. (1969). *Ontological Relativity and Other Essays.* New York: Columbia University Press.

Reddy M. (1979). "The Conduit Metaphor". In Ortony (ed.), pp. 284–324.

Rescher N.*Rationality: A Philosophical Inquiry into the Nature and the Rationale of Reason.* Oxford: Clarendon Press.

Rorty R. (1979). *Philosophy and the Mirror of Nature.* Princeton: Princeton University Press.

Rorty R. (1989). "Philosophy as Science, as Metaphor and as Politics". In A. Cohen and M. Dascal (eds.), *The Institution of Philosophy: A Discipline in Crisis?* La Salle, Illinois: Open Court, pp. 13–33.

Unger R.M. (1975). *Knowledge and Politics.* New York: The Free Press.

Chapter 3

DESIRE, DELIBERATION AND ACTION*

John R. Searle
University of California, Berkeley

1. Introduction

This article is an interim report on my struggles to try to understand some of the logical features of desires as they relate to rationality and to actions. I begin with a discussion of practical reasoning and what is sometimes called "the practical syllogism".

Discussions that I have seen of practical reason and the practical syllogism usually — not always — exhibit a certain conception of rationality that I think is probably mistaken and, if so, profoundly so. I remember this conception of rationality from the economic theory I learned as an undergraduate, and it is also exhibited by many versions of decision theory. If it is as widespread as I think, it is worth looking at very closely. I hope I am not being tendentious in labelling it, "the classical conception". According to this conception human rationality is at least partly constituted by the precepts of practical reason. On this conception we human beings are supposedly given a set of desires prior to engaging in practical reasoning, and on the basis of these *primary* desires, we reason from our beliefs about how the world is to form *secondary* desires[1] (or intentions, or on some views, actions themselves) about how to satisfy

*The first version of this article was delivered at a conference on Practical Reason in Stanford some years ago. Later versions were delivered at a conference on Aristotle at the University of Rochester and at other "Practical Reason" conferences at the University of Dayton, and at the University at Santa Clara. I have also discussed these matters in seminars in Berkeley and Rutgers. In the course of these many discussions I have benefited from so many criticisms that I cannot possibly acknowledge all of my indebtedness here. Some of the people whose criticisms I especially remember are: Michael Bratman, John Etchemendy, Bernard Williams, Ernest LePore, Dagmar Searle, Deborah Modrak, Barbara Horan, Brian McLaughlin, and Thomas Nagel.

[1]As far as I know the first person to use the terminology of "primary" and "secondary" to describe this distinction was Thomas Nagel (1970).

D. Vanderveken (ed.), Logic, Thought & Action, 49–78.

our primary desires; on this view we reason from our "ends", which are given by the primary desires, and our beliefs about the "means", to form desires for the means. Thus for example a paradigm of practical reasoning would be a case of a man who has a primary desire to go to Paris, a set of beliefs about the means to get to Paris and who then reasons to form a secondary desire, e.g., to buy a plane ticket.

Implicit and sometimes explicit in this classical conception are a set of constraints which rationality is supposed to place on human desires.

1 Rationality requires that the set of *desires* be *consistent.* There will of course be conflicts of the sort where the satisfaction of one desire frustrates another, but a rational agent cannot simultaneously both want that p and want that not p.[2]

2 Rationality requires that the *preferences* of an agent be *well ordered* prior to engaging in deliberation. Since practical reasoning typically involves the allocation of scarce resources (for example, money) among competing ends, it is essential that a rational agent have a well ordered ranking of his or her preferences.

3 Rationality requires that an agent who has the appropriate combination of beliefs and desires is thereby *committed* to certain secondary desires (or intentions, etc.) and the aim of a deductive logic of practical reason is to state the principles according to which these can be logically derived from the primary desires and the beliefs.

I think all these principles are false. And they are not harmlessly false, in the way that idealizations in the sciences give us literally false but importantly true idealized models (e.g. frictionless systems) but they are importantly false, and treating them as true has given us a misconception of the real nature of practical reason. I think many people would concede that they are literally false, but would maintain that it doesn't really matter, because what we are trying to construct is not a mere description, but a *model* of rational behavior, and it doesn't matter if the model is not literally true as along as the it gives us insight into the phenomena. I think, on the contrary, that in many ways the classical conception prevents us from getting certain important insights. I will not try to provide an alternative model but will try to state some of the facts which, I believe, should constrain any such model. However, the

[2]For example, Elster (1983), p.4 "Beliefs and desires can hardly be reasons for action unless they are consistent. They must not involve logical, conceptual or pragmatic contradictions."

investigation has opened up a whole lot of other subjects and I fear is now much longer than I ever intended it to be. I begin by discussing the possibility of a deductive logic of practical reason.

* * * * * * * * * * *

Practical reason, we are sometimes told, is reasoning about what to do; just as theoretical reason is reasoning about what to believe. But if this is so, it ought to seem puzzling to us that we do not have a generally accepted account of the deductive logical structure of practical reason in a way that we apparently do for deductive theoretical reason. After all, the processes by which we figure out how to best achieve our goals seem to be just as rational as the processes by which we figure out the implications of our various beliefs, so why do we seem to have such a powerful logic for the one and not for the other?

To see what the problem is, let us review how it is apparently solved for theoretical reason. We need to distinguish questions of logical relations from questions of philosophical psychology. Great advances in deductive logic were made when, in the nineteenth century, Frege separated questions of philosophical psychology (the "laws of thought") from those of logical relations. After Frege it has seemed that if you get the logical relations right the philosophical psychology should be relatively easy. For example once we understand the relations of logical consequence between propositions then many of the corresponding questions about belief seem fairly simple. If I know that the premises `all men are mortal' and `Socrates is a man' jointly entail the conclusion `Socrates is mortal' then I already know that someone who *believes* those premises is *committed* to that conclusion; that someone who *knows* the premises to be true is *justified* in *inferring* the truth of the conclusion, etc. There seems in short to be a fairly tight set of parallels within theoretical reason between such "logical" notions as premise, conclusion, and logical consequence on the one hand and such "psychological" notions as belief, commitment, and inference on the other. The reason for this tight set of parallels is simply that the psychological states have propositional contents; and they therefore inherit certain features of the logical relations between the propositions. An assertion that p, for example, has the same truth conditions as a belief that p, and therefore, the assertion and the belief have the same logical consequences. The tacit principle that has worked so well in assertoric logic is that if you get the logical relations right, then most of the philosophical psychology will take care of itself.

Now, supposing we accept this distinction between the logical relations and the philosophical psychology, how is it all supposed to work for practical reason? What are the logical relations in practical reason and how do they bear on the philosophical psychology? Some of the questions about logical relations would be: What is the formal logical structure of practical argument? In particular, can we get a definition of formal validity for practical reason in the way that we can for deductive "theoretical" reason? Does practical logic use the same or does it require different rules of inference than assertoric logic? The questions about the philosophical psychology of deliberation would concern the character of the intentional states in practical reasoning, their relation to the logical structure of deliberation, and their relations to action. Some of the questions in this set are: what sorts of intentional states figure in deliberation and what are the relations between them? What sorts of things can be reasons for action – beliefs, desires, motives, pro attitudes, obligations? What is the nature of motivation and how does deliberation actually motivate, (lead to, or cause) action? How is weakness of will possible?

2. Three patterns of practical reason

To begin with let us consider some attempts to state a formal logical structure of practical reason. I will confine the discussion to so-called means-ends reasoning, since most authors on the subject think that all, or at least most, practical reason is deliberating about means to achieve ends. Oddly enough it is not at all easy or uncontroversial to state the formal structure of means-ends reasoning, and there is no general agreement on what it is. In the philosophical literature there is a bewildering variety of formal models of such reasoning, and even fundamental disagreements over what its special elements are supposed to be – are they desires, intentions, fiats, imperatives, norms, noemata or what?[3] Many philosophers speak rather glibly about the belief-desire model of explanation and deliberation, but what exactly is the structure of this model supposed to be? Several philosophers [4] have suggested the following as the correct model:

[3]For a good survey of the literature, see Aune (1977).

[4]E.g., Kenny (1975). This form is not always stated explicitly as being about beliefs and desires.

Kenny gives the following example:

I am to be in London at 4 p.m.

If I take the 2:30, I will be in London at 4 p.m.

So, I'll take the 2:30.

I hope that it does not misrepresent his views to put the example explicitly in the form of an inference concerning beliefs, desires, and intentions.

I want to achieve (end) E.

I believe if I do (means) M I will achieve E.

Therefore I want to do M.

We can represent this schematically as:

DES (I achieve E)

BEL (If I do M I will achieve E)

Therefore DES (I do M)

But it seems this could not be right because premises of this form simply do not commit one to having the corresponding desire (much less the corresponding intention). To see this remember that a lot of the E's one can think of are quite trivial and many M's are ridiculous. For instance I want this subway to be less crowded and I believe that if I kill all the other passengers it will be less crowded. Of course one *might* form a homicidal desire on a crowded subway, but it seems absurd to claim that rationality *commits* me to a desire to kill just on the basis of my other beliefs and desires. The most that this pattern could account for would be *possible* motivations for forming a desire. Someone who has the appropriate beliefs and desires has a possible motive for desiring M. But there is no *commitment* to such a desire.

It is sometimes said that this pattern fails because there is no entailment relationship between the propositional contents of the premises and the conclusion. Indeed, if we just look at the propositional contents, the inference is guilty of the fallacy of affirming the consequent. Some philosophers think the standard form of practical reason is to be found in cases where the means is a necessary condition of achieving the end. Thus, they endorse the following (or variations on it):

DES (I achieve end E.)

BEL (The only way to achieve E is by means M.)(sometimes stated as "M is a necessary condition of E", or "to achieve E, I must do M")

Therefore, DES (I do M.)

But, again, if you think about this in terms of real life examples it seems quite out of the question as a general account of practical reason. In general there are lots of means, many of them ridiculous, to achieve any end; and in the rare case where there is only one means, it may be so absurd as to be out of the question altogether. Suppose that you have any end you care to name: you want to go to Paris, become rich, or marry a Republican. Well in the Paris case, for example, there are lots of ways to go. You could walk, swim, take a plane, ship, kayak, or rocket; you could tunnel through the earth or go via the moon or the North Pole. In very rare cases there may be only one means to an end. As far as I know there is no quick way to get rid of flu symptoms short of death. Therefore, on the above model, if I seriously want (will, intend) to get rid of my flu symptoms really fast I should commit suicide. This model, like the first one, has very little application. As an account of a general structure of practical reason it is a nonstarter.[5]

In the first of these examples there was no entailment relation between the propositional contents of the premisses and the conclusion; but in the second there was. The fact that entailment relations do not generate a commitment to a secondary desire reveals an important contrast between the logic of beliefs alone and the logic of belief-desire combinations. If I believe p and believe (if p then q), then I am committed to a belief in the truth of q. But if I want p and believe that (if p then q), I am not committed to wanting q. Now why is there this difference? When we understand that, we will go a long way toward understanding why there is no plausible logic of practical reason.

Let's try again to construct a formal logical model of practical reason. Generally when you have a desire, intention or goal you seek not just *any* means; nor do you search for the only means; you seek the best means (as Aristotle says you seek the "best or easiest" means.) And if you are rational, when there isn't any good or at least reasonable means you give up on the goal altogether. Furthermore, you don't just have a goal, but where rational you appraise and select your own goals in the light of... well, what? We will have to come back to this point later. But in the meantime suppose you have seriously selected a goal and appraised it as reasonable. Suppose you seriously want to go to Paris, i.e. you have "made up your mind", and you try to figure the best way to get there and conclude that it is by plane. Is there a plausible formal model of the logic of means-ends reasoning for such a case?

[5] Aune (1977), who sees that the first model is inadequate for reasons similar to those I have suggested, nonetheless fails to see that the same sorts of objections seem to apply to the second model.

In such a case, the form of the argument seems to be:

DES (I go to Paris.)

BEL (the best way, all things considered, is to go by plane.)

Therefore DES (I go by plane.)

If we separate out the questions of logical relations from the questions of philosophical psychology – as I have been urging – we see that from a logical point of view this argument, as it stands, is enthymematic. In order to be formally valid it would require an extra premise of the form:

DES (If I go to Paris I go by the best way, all things considered.)

If we add this premise, the argument is valid by the standards of classical logic. Let P = I go to Paris, Q = I go by the best way, and R = I go by plane, then its form is:

$$\begin{array}{l} P \\ P \longrightarrow Q \\ Q \longleftrightarrow R \\ \hline R \end{array}$$

And though the argument is not truth preserving because two of its premises and its conclusion don't have truth-values; this doesn't really matter since the argument is satisfaction preserving, and truth is just a special case of satisfaction. Truth is satisfaction of representations with the word-to-world direction of fit.

I have tried to make a sympathetic attempt to find a formal logical model of the traditional conception of means-ends reasoning, the conception that goes back to Aristotle, and this is the best that I can come up with. I have also tried give a statement of its formal structure which seems to me an improvement on other versions I have seen. But I think it is still hopelessly inadequate. Once again, as in the earlier examples, it seems the logical relations don't map onto the philosophical psychology in the right way. It is by no means obvious that a rational person who has all those premisses must have or even be committed to having a desire to go by plane. Furthermore, to make it plausible, we had to introduce a fishy sounding premise, about wanting to do things "by the best way all things considered". And indeed it looks as if any attempt to state formally the structure of a practical argument of this sort would in general require such a premise, but it is not at all clear what it means. What is meant by "the best way", and what is meant by "all things considered"? Notice furthermore that such premises have no analogue in standard cases of theoretical reason. When one reasons from one's belief that all men are mortal and that Socrates is a man to the conclusion that Socrates is mortal, one does not need any premise about what is the best thing to believe all things considered.

After various unsuccessful tries I have reluctantly come to the conclusion that it is impossible to get a formal logic of practical reason which is adequate to the facts of the philosophical psychology. To show why this is so, I now turn to the discussion of the nature of desire.

3. The structure of desire

In order to understand the weaknesses in my revised logic for practical reasoning, and in order to understand the general obstacles to a formal logic of practical reasoning, we have to explore some general features of desire and especially explore the differences between desires and beliefs. To save time and space I am simply going to assume that the general account given of desires, beliefs, intentions, etc. in *Intentionality*[6] is correct. Specifically I am going to assume that contrary to the surface grammar of sentences about desire, desires all have whole propositions as intentional contents (thus "I want your car" means something like "I want that I have your car"); that desires have the world-to-mind direction of fit; whereas beliefs have the mind-to-world direction of fit; and that desires do not have the restrictions on intentional contents that intentions have. Intentions must be about future or present actions of the agent and must have causal self referentiality built into their intentional content. Desires have no such causal condition, and they can be about anything, past or present. Furthermore, I am going to assume that the usual accounts of the *de re/de dicto* distinction are hopelessly muddled as is the view that desires are intensional-with-an-s. The *de re/de dicto* distinction is properly construed as a distinction between different kinds of sentences about desires, not between different kind of desires. The claim that all desires, beliefs, etc. are in general intensional is just false. *Sentences about* desires, beliefs, etc. are in general intensional. Desires and beliefs themselves are not in general intensional, though in a few oddball cases they can be.[7]

---

[6]Searle (1983).

[7]For a discussion of these points about intensionality-with-an-s and the *de re/de dicto* distinction, see Searle (1983) chapters 7 and 8.

Where a state of affairs is desired in order to satisfy some other desire, it is best to remember that each desire is part of a larger desire. If I want to go to my office to get my paycheck, there is indeed a desire whose content is simply: I want that (I go to my office). But it is part of a larger desire whose content is: I want that (I get my paycheck by way of going to my office). This feature is shared by intentions. If I intend to do a in order to do b, then I have a complex intention whose form is I intend (I do b by means of doing a). I will say more about this point later.

The first feature to notice about desiring (wanting, wishing, etc.), in which it differs from belief is that it is possible for an agent consistently and knowingly to want that p and want that not p in a way that it is not possible for him consistently and knowingly to believe that p and believe that not p. And this claim is stronger than the claim that an agent can consistently have desires which are impossible of simultaneous satisfaction because of features he doesn't know about. For example, Oedipus can want to marry a woman under the description "my fiancee" and want not to marry any woman under the description "my mother" even though in fact one woman satisfies both descriptions. But I am claiming that he can consistently both want to marry Jocasta and want not to marry Jocasta, under the same description. The standard cases of this are cases where he has certain reasons for wanting to marry her and reasons for not wanting to. For example, he might want to marry her–because, say, he finds her beautiful and intelligent, and simultaneously not want to marry her–because, say, she snores and cracks her knuckles. Such cases are common, but it is also important to point out that a person might find the same features simultaneously desirable and undesirable. He might find her beauty and intelligence exasperating as well as attractive and he might find her snoring and knuckle-cracking habits endearing as well as repulsive. (Imagine that he thinks to himself: "It is wonderful that she is so beautiful and intelligent, but at the same time it is a bit tiresome; her sitting there being beautiful and intelligent all day long. And it is exasperating to hear her snoring and cracking her knuckles, but at the same time there is something endearing about it. It is so human"). Such is the human condition.

In order to understand this point and its consequences for practical reason we need to probe a bit deeper. It is customary, and I think largely correct to distinguish, as the classical conception does, between primary and secondary or derived desires. It is literally true to say to my travel agent, "I want to buy a plane ticket." But I have no lust, yearning, yen or passion for plane tickets–they are just "means" to "ends". A desire which is primary relative to one desire may be secondary relative to another. My desire to go to Paris is primary relative to my desire to buy a plane ticket, secondary relative to my desire to visit the Louvre. The primary/secondary desire distinction will then always be relative to some structure whereby a desire is motivated by another. This is precisely the picture that is incorporated in the classical conception of practical reason. In such cases, as I just noted, the complete specification of the secondary desire makes reference to the primary desire. I don't just want to buy a ticket, I want to buy a ticket in order to go to Paris.

Once we understand the character of secondary desires we can see that there are at least two ways in which fully rational agents can form conflicting desires. First as noted earlier, an agent can simply have conflicting inclinations. But secondly he can form conflicting desires from consistent sets of primary desires together with beliefs about the best means of satisfying them. Consider the example of the man who reasons that he wants to go to Paris by plane. Such a man has a secondary desire to go by plane motivated by a desire to go to Paris together with a belief that the best way to go is by plane. But the same man might have constructed a practical inference as follows: I don't want to do anything that makes me nauseated and terrified, but going anywhere by plane makes me nauseated and terrified, therefore I don't want to go anywhere by plane, therefore I don't want to go to Paris by plane. It is easy enough to state this according to the pattern of practical reasoning I suggested above: all things considered the best way to satisfy my desire to avoid nausea and terror is *not* to go to Paris by plane. Since this can be stated as a piece of practical reasoning, it seems that *the same person using two independent chains of practical reason, can rationally form inconsistent secondary desires from a consistent set of his actual beliefs and a consistent set of primary desires.* A consistent set of "premises" will generate inconsistent secondary desires as "conclusions". This is not a paradoxical or incidental feature of reasoning from beliefs and desires, but rather, it is a consequence of certain essential differences between practical and theoretical reason.

Let's probe these differences further: in general it is impossible to have any set of desires, even a consistent set of primary desires, without having or at least being rationally motivated to having inconsistent desires. Or, to put this point a bit more precisely: if you take the set of a person's desires and beliefs at any given point in his life, and work out what secondary desires can be rationally motivated from his primary desires, assuming the truth of his beliefs, you will find inconsistent desires. I don't know how to demonstrate this, but any number of examples can be used to illustrate it. Consider the example of going to Paris by plane. Even if planes do not make me nauseated and terrified, still I don't want to spend the money; I don't want to sit in airplanes; I don't want to eat airplane food; I don't want to stand in line at airports; I don't want to sit next to people who smoke, are too fat, or who put their elbows where I am trying to put my elbow. And indeed, I don't want to do a whole host of other things that are the price, both literally and figuratively, of satisfying my desire to go to Paris by plane. The same form of reasoning that can lead me to form a desire to go to Paris by plane can also lead me to form a desire *not* to go to Paris by plane. A possible answer to this, implicit in at least some of the literature, is to invoke the notion of preference. I prefer going to Paris by plane and being uncomfortable to not going to Paris by plane and being comfortable. But this answer, though acceptable as far as it goes, mistakenly implies that the preferences are given *prior* to practical reasoning; whereas it seems to me they are often the product of practical reasoning. Ordered preferences are typical products of practical reason, and hence they cannot be treated as its universal presupposition. Just as it is a mistake to suppose that a rational person must have a consistent set of desires, so it is a mistake to suppose that rational persons must have a rank ordering of (combinations of) their desires prior to deliberation.

This points to the following conclusion: Even if we confine our discussion of practical reasoning to means-ends cases, it turns out that practical reason essentially involves the adjudication of conflicting desires (and other sorts of conflicting reasons) in a way that theoretical reason does not essentially involve the adjudication of conflicting beliefs. That is why in our attempt to give a plausible account of the classical conception of practical inference we needed crucially a step about wanting to go by "the best way, all things considered." Such a step is characteristic of any rational reconstruction of a process of means-ends reason, because "best" just means the one which best reconciles all of the conflicting desires that bear on the case. However, this also has the consequence that the formalization of the classical conception I gave is essentially a trivialization of the problem, because the hard part has not been analyzed: How do we arrive at the conclusion that such and such is "the best way to do something all things considered" and how do we reconcile the inconsistent conclusions of competing sets of such valid derivations?

If all one had to go on were the classical conception of reasoning about means to ends, then in order to reach a conclusion of the argument that could form the basis of action one would have to go through a whole set of other such chains of inference and then find some way to settle the issue between the conflicting desires. *The classical conception works on the correct principle that any means to a desirable end is desirable at least to the extent that it does lead to the end. But the problem is that in real life any means may be and generally will be undesirable on all sorts of other grounds and the model has no way of showing how these conflicts are adjudicated.*

The matter is immediately seen to be worse when we consider another feature of desires, which we already noticed in passing. A person who believes that p and that (if p then q) is committed to the truth of q; but a person who desires that p and believes that (if p then q) is not committed to desiring that q. If you believe the premises you are committed to a belief that q at least to the extent that, first, you cannot believe that not q without contradiction; and second, you cannot, in consistency, acknowledge that you believe that p and that (if p then q) while denying that you have the belief that q.

Of course you are not committed in the sense that you must actually have *formed* the belief that q. You might believe that p and that (if p then q) without having thought any more about it. (Someone might believe that 29 is an odd number and that it is not evenly divisible by 3, 5, 7, or 9 and that any number satisfying these conditions is prime, without ever having actually drawn the conclusion, i.e. formed the belief, that it is prime.) But these conditions simply do not hold for combinations of belief and desire. You can want that p and believe that (if p then q) without being committed to wanting that q. For example, there is nothing *logically* wrong with a couple who want to have sexual intercourse and who believe that if they do she will get pregnant but who do not want her to get pregnant.

We can summarize these points about desire and the distinction between desire and belief as follows: Desires have two special features which make it impossible to have a formal logic of practical reason parallel to our supposed formal logic of theoretical reason. The first feature we might label "the necessity of inconsistency." Any rational being in real life is bound to have inconsistent desires. The second, we might label "the non-detachability of desire." Sets of beliefs and desires as "premises" do not necessarily commit the agent to having corresponding desire as "conclusion" even in cases where the propositional contents of the premises entail the propositional content of the conclusion. These two theses together account for the fact that there is in the philosophical literature no remotely plausible account of a deductive logical structure of practical reason. In my view, they explain why it is impossible to give such an account if by "practical reason", we mean the structure of reasoning from desires for "ends" and beliefs about "means" to desires and intentions regarding the means, and if by "deductive logical structure", we mean anything like our existing deductive logical systems.

The moral is: As near as I can tell, the search for a formal deductive logical structure of practical reason is misguided. Such models either have little or no application, or if they are fixed up to apply to real life it can only be by trivializing the essential feature of practical deliberation: the reconciliation of conflicting desires (and conflicting reasons for action generally) and the formation of rational desires on the basis of the reconciliation. We can always construct a deductive model of any piece of reasoning at all; but where an essential feature of the reasoning contains both p and not p, as in I want that p and I want that not p; deductive logic is unilluminating, because it cannot cope with such inconsistencies. The models either have to pretend that the inconsistencies do not exist or they have to pretend that they have been resolved ("by the best way all things considered"). The first route is taken by the models I criticized at the beginning, the second route is taken by my revised version. The possibility, indeed the inevitability, of contradictory desires renders the classical conception unilluminating as a model of the structure of deliberation. Furthermore even if you do fudge to the extent of trivializing the problem you still do not get a commitment to a desire, as the conclusion of the argument. *Modus ponens* simply doesn't work for desire/belief combinations to produce a commitment to desiring the conclusion.

4. Explanation of the difference between desire and belief

Now why should there be these differences? What is it about the philosophical psychology of desire that makes it so logically unlike belief? Well, any answer to that has to be tautological, and so, disappointing, but here goes anyhow:

First, let us remind ourselves of the general structure of intentional states. The structure is S (p) where "S" marks the psychological mode, and "p" marks the propositional content, the content which determines the conditions of satisfaction. This structure is common to both beliefs and desires. Where the mode is belief, the propositional content represents a certain state of affairs as actually existing. But where the mode is desire, the prepositional content does not function to represent an actual state of affairs, but rather a *desired* state of affairs, which may be actual, non-existent, possible, impossible or what have you. And the propositional content represents the state of affairs under the aspects that the agent finds desirable.

Both desires and beliefs have propositional contents, both have a direction of fit, both represent their conditions of satisfaction, and both represent their conditions of satisfaction under certain aspects. So, what is the difference that accounts for the different logical properties between desires and beliefs? The difference derives from the different directions of fit. The job of beliefs is to represent how things are. To the extent that the belief does this or fails to do it, it will be true or false respectively. The job of desires is not to represent how things are, but how we would like them to be. And desires can succeed in representing how we would like things to be, even if things don't turn out to be the way we would like them to be. There is nothing wrong with unsatisfied desires, *qua* desires, whereas there is something wrong with unsatisfied beliefs, *qua* beliefs, namely: they are false. They fail in their job of representing how things are. Desires succeeds in their job of representing how we would like things to be even in cases where things are not the way we would like them to be, i.e. even in cases where their conditions of success are not met. Roughly speaking, when my belief is false, it is the belief that is at fault. When my desire is unsatisfied it is the world that is at fault.

The two points, inconsistency and nondetachabality, both derive from this underlying feature of desires; desires are inclinations towards states of affairs (possible, actual or impossible) under aspects. There is no necessary irrationality involved in the fact that one can be inclined and disinclined to the same state of affairs under the same aspect; and the fact that one is inclined to a state of affairs under an aspect together with knowledge about the consequences of the existence of that state of affairs does not guarantee that where rational one will be inclined to those consequences.

But if you try to state parallel points about belief it doesn't work. Beliefs are convictions that states of affairs exist under aspects. But one cannot rationally be convinced both that a state of affairs exists and does not exist under the same aspect. And the fact that one is convinced of the existence of a state of affairs under an aspect together with knowledge about the consequences of the existence of that state of affairs does guarantee that where rational one will be convinced of (or at least committed to) those consequences.

These features of desire are characteristic of other sorts of representations with the world-to-word direction of fit. The features of inconsistency and nondetachability apply to needs and obligations as well as desires. I can consistently have inconsistent needs and obligations and I do not necessarily need the consequences of my needs, nor am I obligated to achieve the consequences of my obligations.[8]

In objecting to this account, one might say, "Look, when I believe something, what I believe is that it is true. So, if I believe something and know that it can't be true unless something else is true, then my belief and knowledge must commit me to the truth of that other thing as well. But now why isn't it the same for desire? When I want something what I want is that something should happen or be the case, but if I know that it can't happen or be the case unless something else happens or is the case then surely I must be committed to wanting that something else." But the analogy breaks down. If I want to drill your tooth to fill your cavity and I know that drilling the tooth will cause pain it simply does not follow that I am in any way committed to causing pain, much less committed to wanting to cause pain. And the proof of this distinction is quite simple: if I fail to cause pain one of my beliefs is thereby false, but none of my desires is thereby unsatisfied.

When I want something, I want it only under certain aspects. "Yes, but when I believe something I believe it only under certain aspects as well. Sentences about belief are just as opaque as sentences about desire." Yes, but there is this difference: When something is desired under certain aspects it is, in general, the aspects that make it desirable. Indeed the relation between the aspects and the reasons for desiring are quite different from the case of belief, since *the specification of the reasons for desiring something is, in general, already a specification of the content of the desire*; but the specification of the evidence on the basis of which I hold a belief is not in general itself part of the specification of the belief. The reasons for believing stand in a different relation to the propositions believed than the contents of reasons for wanting do to the proposition which is the content of the desire, because in general the statements of the reasons for wanting state part of what one wants. If

[8]However, one cannot consistently have inconsistent intentions or issue inconsistent directives, even though they also have the world-to-word direction of fit. Why not? I am not sure but I think the reason is that they are designed to cause actions and so cannot achieve their function if we allow for an agent consistently to have inconsistent intentions or issue inconsistent orders. It's okay–up to a point–for a speaker to say reflectively "I both wish you would go and wish you would stay". But he is irrational if he says simultaneously "Go!" and "Stay!", and you are equally irrational if you form the simultaneous intentions to go and to stay. There cannot consistently be inconsistent intentions and orders, because intentions and orders are designed to cause actions, and there cannot be inconsistent actions. For the same reason both intentions and directives imply a belief that the action is possible, but it is not possible to carry out the conjunction of inconsistent actions. Desires and obligations have no such condition.

one wants something for a reason then that reason is part of the content of one's desire.

For example, if I want it to rain in order to make my garden grow, then I both want that it should rain and that my garden should grow. If I believe it will rain and I believe that the rain will make my garden grow, then I both believe that it will rain and that my garden will grow. But there is still a crucial difference. If I want it to rain *in order to* make my garden grow, then my reason for wanting it to rain is part of the whole content of the entire complex desire. My reason for believing both that it will rain and that the rain will make my garden grow, on the other hand, have to do with a lot of evidence about meteorology, the reliability of newspaper weather predictions, and the function of moisture in producing plant growth. All of these considerations count as evidence for the truth of my belief, but they are not themselves the content of that belief. But in the case of my desire, the role of reasons is not at all like that of evidence, for the reasons state the aspects under which the phenomenon in question is desired. The reasons, in short, are part of the content of the complex desire.

In sum: beliefs have the mind-to-world direction of fit. Their job is to represent how things are. Desires have the world-to-mind direction of fit. Their job is not to represent how things are, but how we would like them to be. It is the notion of "how things are" that blocks the simple possibility of consciously held contradictory beliefs, and that requires a commitment to the consequences of one's beliefs, but there is no such block and no such requirement when it is a question of to how we would like things to be. In spite of certain formal similarities, then belief is really radically unlike desire in both its logical and its phenomenological features.

For these reasons, it is misleading to think of theoretical reason as reasoning about what to believe in the way that we think of practical reason as reasoning about what to do. What one should believe is dependent on what is the case. Theoretical reasoning, therefore, is only derivatively about what to believe. It is primarily about what is the case — what must be the case given certain premises. Furthermore, we can now see that it is misleading to think even that there is a "logic" of theoretical reason. There is just logic — which deals with logical relations among, e.g., propositions. Logic tells us more about the rational structure of theoretical reason than it does about the rational structure of practical reason, because there is a close connection between the rational constraints on belief and the logical relations between propositions. This connection derives from the fact that, to repeat, beliefs are meant to be true. But there is no such close connection between the structure of desire and the structure of logic. All the facts in the world cannot commit me to having an inclination, if I just don't feel like having it. And I both can and do have conflicting inclinations even after all the facts are in.

If we could fully appreciate these points, I think we could see that many decision theoretical models of human rationality are really quite crazy. It is, for example, a standard consequence of Bayesian decision theory that if I value two ends differently, then if I am rational, there must be some mathematical odds at which I would be willing to bet one outcome against the other. If, for example, I value my life and I value a dime, there must be some odds at which I would bet my life against a dime. Now, I want to say: there aren't any odds at which I would bet my life against a dime, and I believe there is nothing irrational in my refusal to do so. And even if there were some such odds, there aren't any odds at which I would bet my child's life against a dime. And this is not because either of our lives has "infinite value." I think nobody's life has infinite value. It is rather because the scale on which I value my life is not the same scale on which I value a dime. And this is true even though there are points where the scales intersect.

The answers that one gets to these sorts of objections reveal a very deep misconception about the nature of choice, preference, desire, and rationality. The standard answers are always to treat choices, preferences, and desires, as if they lacked intentionality, and hence as if *statements* of choice and preference were always fully extensional. So the standard answer to this sort of objection is to say something like the following: I am, as a rational agent, willing to bet my life against a dime, if the odds are sufficiently favorable, because I am willing to do things that are *extensionally equivalent* to that. Thus, for example, if somebody offered me a thousand dollars to drive him to the airport, I would accept the offer immediately. Now, it will be possible to divide the trip to the airport in small enough units so that for any given unit, I will be accepting a dime for driving that unit. I will also be increasing the odds of my own death over the odds that I would have had if I had stayed at home. So, it seems to follow that there are some odds at which I am betting my life against a dime during those instants. But this answer is a mistake. From the fact that I desire to be in state a, and the fact that I know that a occurs only if b occurs, it simply does not follow that I am logically committed to desiring to be in state b.[9](In this case the

[9]A similar mistake afflicts the traditional doctrine of revealed preference in economics. As Sen (1973) points out, the traditional theorists try to treat preferences purely extensionally, purely in terms of the overt behavior of the agent. But the system cannot be made to work

fact that it is rational for me to buy the whole package does not imply that rationality requires that I be willing to buy each bit of the package separately.)

5. Internal and external reasons

I now want to turn to closely related question concerning the relation of desire and action. In its simplest and crudest form: Can there be reasons for action for an agent which do not appeal to some antecedently existing desire of the agent. This question is supposed to have received a negative answer from Hume and, on one possible interpretation, a positive answer from Kant. It has surfaced recently in a slightly different form in a problem by Bernard Williams.[10] Can there be "external reasons" for an agent to act as well as "internal reasons"? Internal reasons appeal to the agent's existing "subjective motivational set", external reasons do not. Williams, in the spirit of Hume, claims that there are no external reasons, that external reasons statements are always false.

The heart of the argument appears to be that a reason for an agent must be capable of motivating the agent to act, and it is impossible that reason could rationally motivate an agent to act unless it made some appeal to an existing motivation. It is hard to know how to take this claim without some further analysis of motivation and of the relations between motivation and action. If we accept the traditional conception according to which all motivation is a form of desire, then the thesis that there are no external reasons follows immediately as a trivial logical consequence. On that conception no reason could be a reason for an agent to deliberate from unless it already appealed to some antecedently existing desire of the agent. If practical reason is reasoning from desires and beliefs then, trivially, you can't do practical reasoning from beliefs without also having desires. But the problem is that there are still lots of cases of reasons for action which do not appeal to the classical model.

Someone might reject the classical conception and still insist that there are no external reasons (this I take it is Williams's position). But how exactly is the argument then supposed to go? It looks as if the argument might turn out to be equally trivial, even independently of the classical conception. Why? Well the premise that both sides agree to is:

if it is treated purely extensionally. The theorist has to smuggle the mental representations back in tacitly, otherwise the system collapses. If revealed preferences are supposed to be revealed entirely in terms of behavior, then Buridan's ass would simply reveal a preference for starvation.

[10]Williams, B. (1981).

> Premise 1. For a reason to be a reason for an agent to act it must be capable of motivating that agent.

And the conclusion is:

> For a belief (or statement) to function as a reason, given only its content and rationality alone, it must appeal to a pre-existing motivation of the agent.

But now what is the extra premise that is supposed to get us to that conclusion? It must be something like the following:

> Premise 2. The only way a belief (or statement) could rationally motivate an agent is by appealing to a pre-existing motivation of the agent.

But if this is the argument it is really no advance on the classical conception. It just substitutes the more general notion of "motivation" for the original notion of "desire".

A slightly different way to see this same point is the following. The interest of the claim that there are no external reasons depends crucially on the account of motivation, for it has to give an account of motivation which could show how it is impossible for beliefs plus rational processes to provide a rational ground for motivations. The whole issue hangs on that. If motivations are always taken to be desires or desire like impulses and desires in turn are construed on the classical conception then the falsity of the externalist position follows trivially. If on the other hand they are not just desires so construed then what are they? Williams is untypically vague on the subject.[11] But suppose somebody claims to be motivated entirely by reflection on Kant's Categorical Imperative. What do we say? That in his motivational set he must have had a "disposition of evaluation" to act on the Categorical Imperative? Or that his reasoning processes could not have been purely rational? Without an independent account of motivation we have no natural way of knowing what it means to reject the classical conception but still insist on the internalist position. Furthermore, even construed just as an item of ordinary English, the word "motivation" is a source of some confusion because it tends to be ambiguous between, roughly speaking, "reason" and "urge", i.e. between a rational basis and a felt inclination.

[11] He says, all the elements of the subjective motivational set can be characterised "formally" as desires, but cautions that this may make us forget that the set "can contain such things as dispositions of evaluation, patterns of emotional reaction, personal loyalties, and various projects, as they may be abstractly called, embodying commitments of the agent." Williams (1981) p. 105.

At times Williams seems not to be providing an argument for the internalist position, but rather shifting the onus to the externalist. So construed, his aim is not to show that it is impossible that there should be external reasons but that no argument is forthcoming to show that they are in fact possible. Let us now accept that onus. My aim will be to show that the claim that all reasons for action must appeal to a pre-existing motivation has consequences which we know independently to be absurd. If we suppose that the agent must have an antecedently existing motivation to which a reason appeals before rationality can have any hold on him at all then the following imaginary speech by me ought to make a kind of sense that it does not seem to make. Suppose you have loaned me a thousand dollars and I have promised to pay you back on such and such a date. Suppose the date arrives and you remind me of my obligation and demand payment. Suppose I then say,

"I agree that I made a promise, and I agree that promises create obligations. I agree that the promise was an undertaking that you counted on and that none of the ways in which promises can be defeasible or invalid apply to this case. I agree indeed that this particular promise created an overriding obligation, that it does not conflict with any other obligations or other sorts of reason that I have, and that I have no reason against doing the thing I promised to do. But the point is that I find nothing in my pre-existing subjective motivational set in favor either of keeping promises in general or this promise in particular or in doing the thing that I promised to do. Therefore I have no reason for paying you back the thousand dollars, absolutely no reason at all. The point is not that I don't have enough of a reason or that there might be other more powerful reasons on the other side, but rather that I have *no reason whatever* for doing what I promised to do. Therefore the obligation to do the action is not a reason for me to do it. I have read Professor Williams on the topic and am assured that unless a reason can appeal to my pre-existing motivational set then there is no reason for me to do it. No doubt you will have various nasty things to say about me, but you cannot show that there is any inconsistency, irrationality, illogicality or unreasonableness in my denial that there is any reason for me to pay you back."

Why is this speech absurd? The short answer is because obligations are reasons for action; and to *recognize something, as an obligation is already to recognize that it is a reason for an action quite independently of what antecedently existing motivation one has.* But how, asks Williams, how could such a reason motivate? To answer that question we now need to turn to the relation between reasons and desires, and between reasons and intentional states generally. I want to work up to the problem of the relation of reason to desire and action by starting with the problem of the relation between reason and belief and acceptance.

As we have had occasion to remark over and over, there are many important differences between reasons for acting and reasons for believing or accepting, between practical and theoretical reason, but the word "reason" means the same in both occurrences and we will find it easier to see what is wrong with the first speech if we see what is wrong with the following imagined speech:

"I recognize that you have given me evidence in favor of the truth of the proposition that p; indeed overwhelming evidence. I also agree that there is no evidence against the proposition. But what I don't see is how you have given me any reason for *accepting* p or *believing* it. In order that your evidence should be a reason for me to accept or believe it, it would have to appeal to something in my pre-existing motivational set. But there is nothing in my motivational set that your evidence appeals to, therefore there is no reason for me to believe this proposition. The evidence, after all, is about the subject matter of the proposition. And what has that to do with my mental state of believing or my behavior of accepting?"

Now, what exactly is wrong with this speech? Just as the first speech treated the relation of obligation to reasons for acting as purely external, so the second treats the relation of evidence to reasons for belief as purely external. "Well why isn't it external — after all believing a proposition is one thing; recognizing something as evidence for the truth of a proposition is something else? You can easily have one without the other."

The relation is not external because beliefs are a certain type of intentional state which renders them subject to the constraints of rationality. Specifically beliefs are a type of intentional states whose purpose is to represent how things are. To recognize something as evidence is already to acknowledge a reason — and hence a motivation — for accepting it or believing it. If you think of beliefs and evidence as just a lot of neutral phenomena then it will seem mysterious that there is any essential connection. But of course they are not neutral objects. There are a series of internal relations between belief, evidence and truth: belief is belief in truth, and evidence is evidence for truth. To have evidence and to know that it is evidence is *eo ipso* to have grounds or reasons for belief. Notice, furthermore, that if I form a belief on the basis of overwhelming evidence, I do not require in addition a general desire to believe propositions which are supported by overwhelming evidence. Such desires derive from the recognition of the way evidence grounds belief; but the converse is not the case, that is, the grounding relation between evidence and belief does not derive from any such desire. (What would it be like *not* to have that desire?)

My aim here is not to try to give general account of the evidentiary relation and of the grounding relation between evidence and belief, but simply to remind us of their existence. I think it is obvious that rationality can motivate belief, I now want to show how it can motivate desire. The point of the digression into theoretical reason is to help us to remove the disguise from disguised nonsense by showing how it is like obvious nonsense. Could there ever be reasons which provide a rational grounding for psychological states and events which do not appeal to antecedently existing motivation? Obviously. Belief on the basis of evidence is a case in point. The evidence provides a motivation for me to believe the proposition; to recognize it as evidence is to acknowledge the motivation.

To repeat, there are many differences between theoretical and practical reason and many differences between beliefs and desires, but it seems to me no more mysterious in principle that a set of "external" reasons can motivate a desire than that they can motivate a belief. A person who has acknowledged that such and such is overwhelming evidence for p, or who has acknowledged that p follows from premises whose truth he accepts, has already acknowledged that he has a reason, and thereby a motivation, for believing p. Of course rationality alone doesn't guarantee that he will actually go ahead and believe that p. All that rationality plus the set of propositions can provide are grounds or reasons for accepting or believing p. Now similarly a person who has acknowledged that he has an overriding obligation to do something has already acknowledged that he has grounds or reasons, and thereby a motivation for doing it. Of course such considerations alone do not guarantee that he will actually have a desire to act, will actually feel motivated to act or will actually go ahead and do the act. All that rationality can provide are *grounds* or *reasons* for acting and thereby grounds or reason for desiring to act. The recognition of reasons can, by rational processes, cause both beliefs and desires, but rationality alone doesn't guarantee that they actually will cause anything.

It is perhaps important to emphasize this last point since Williams seems to think that the external reasons theorist is committed to the view that the recognition of an external reason plus rationality must *guarantee* the creation of an internal motivation. He says, "the external reasons statement itself will have to be taken as roughly equivalent to, or at least entailing, the claim that if the agent rationally deliberated, then, whatever motivations he originally had, he would come to be motivated."[12] But the theorist need not maintain anything as strong as this. The claim is rather that rational deliberation *can* be the ground of a desire where none existed before, not that it *must* be.

Well even supposing it works for evidence and belief, how does it work for obligations and desires? It is not my aim to try to give general account of obligations any more than it was to give an account of evidence. But I do want to remind us of some of their logical relations:

1. X has an obligation to do a.

entails

2. X has a reason to do a.

And

3. X recognizes that he has an obligation to do a

entails

4. X recognizes that he has a reason to do a.

which entails

5. X recognizes that he has a motivation for doing a.

Now let's suppose that X actually satisfies 1- 5, then X has rational grounds for wanting to do a, for intending to do a, and for doing a.

Of course, to repeat, none of this entails that he actually will want to do it, that he will intend to do, or that he will do it. Any more than in the theoretical case the parallel set of relations about evidence and belief, or deductive validity and acceptance, entails that the agent actually will believe what he acknowledges he has overwhelming evidence for or will accept that for which he recognizes he has a deductive proof. In both cases rationality plus intentional contents provide rational grounds for further intentional contents; they don't causally guarantee their occurrence. Notice furthermore that we do not require an extra premise stating that the agent wants to fulfill his obligations in the practical case any more than we needed a premise stating that the agent wants to believe propositions for which there is overwhelming evidence in the theoretical case.

Well, how does it work when the agent forms these further rationally based intentional contents? There are at least three possibilities. First, he might on the basis of deliberation form a desire to do it, and then

[12]Williams (1981), p. 109.

on the basis of the desire form a prior intention, and then carry out the prior intention by doing it. Second, he might skip the first stage and just form a prior intention and then act on that intention. Third, he might just haul off and do it. (Aristotle: straightway he acts) The form of the intentionality is revealing because it shows in every case that the desire or intention plays the role of a secondary desire based on an obligation and not on a primary desire. The reference to the obligation becomes part of the content of the secondary desire or intention, even though there need never have been a primary desire. Thus, where he forms a secondary desire based on the recognition of the obligation, the form of the desire is:

DES (I do it in order to carry out my obligation)

These three cases meet the counterfactual test for derived desires and intentions: If I hadn't been under an obligation I would not have wanted / intended to do it. All three cases involve the formation of a rationally motivated desire or intention.

There remains a puzzle. Why is it so easy to see how beliefs can be rationally motivated, but the idea that desires might be rationally motivated encounters terrific resistance. Well there are several reasons. Most importantly there is the difference in direction of fit, which I mentioned earlier. Belief is related to truth and therefore to objectivity in a way that is not shared by desire. Secondly with desire it is very easy to make the mistake I noted earlier of thinking that if a desire *can be* rationally motivated by some consideration then that consideration *must in every case* actually cause the desire. But thirdly, I think the classical conception has a powerful hold on our imagination and it emerges in this case as follows:

It is part of our concept of desire and of voluntary action that everything one does voluntarily at some level of desire, one desire to do. If I did it voluntarily then my voluntary action was an expression or manifestation of a desire to do it, by definition. This is consistent with a fact that I may have wanted to do something else more but was prevented from doing that other thing (compulsion) or I might not have been prevented but still did something other than what I most wanted to do (akrasia). But all intentional actions are, as such, expressions of some level of desire. So it appears that unless I had an *antecedent* desire, there is no way that rational deliberation from the facts could motivate me.

That, I believe, is the crucial mistake. The mistake is to suppose that the desire must always be the ground of the reason and never the reason the ground of the desire. If you think of desire as a primal juice, (squirted out by the id, say) then it will seem not merely mysterious but impossible that the recognition of any phenomenon by itself could rationally motivate a desire. But then it should seem mysterious and impossible that the recognition of truth should rationally motivate acceptance, or the recognition of overwhelming evidence should rationally motivate belief.

The basic idea I am trying to get across in this section can be summarized as follows: Even though all actions are expressions of desire, there can nonetheless be external reasons for actions, reasons which do not appeal to an antecedently existing desire or other desire like motivation, because an acknowledgement of the facts plus rationality alone *can* motivate the internal desire. In such cases the external reason (e.g. I made a promise) can provide the grounds for the internal motivation (e.g. I want to do it because I promised to do it.) Our failure to see this derives from a misconception of the nature of rationality.

6. Weakness of will

Sometimes, indeed all too frequently, it happens that one goes through a process of deliberation, makes a considered decision, thereby forms a firm and unconditional intention to do an act and when the moment arrives, one does not do the act. Now if the relation between deliberation and intention is both causal and rational or logical, that is if the rational processes cause intentions, and if intentions in turn cause actions by intentional causation, then how could there ever be genuine cases of weakness of will? In the last section we answered the first half of the question. Rational processes can provide the grounds for forming desires and intentions even though in many cases the agent might not actually form the desire or the intention. In this section we address the second half: How could there be cases where an agent forms an all out inclusive, unconditional intention to do something, where nothing prevents him from doing that thing, and still does not do that thing? Amazingly, many philosophers think that such a thing is impossible and they have advanced ingenious arguments to show that it is impossible, that the apparent cases are really cases of something else. Alas, it is not only possible but all too common. Here for example is an absolutely common sort of case: A student forms a firm and unconditional intention to work on his term paper Tuesday evening, nothing prevents him from working on it, but when midnight comes, it turns out that he has spent the evening watching television and drinking beer. Such cases, as any teacher can attest, are quite common. Indeed we ought to insist as conditions of adequacy on our account of akrasia that it allow for the fact that akrasia is very common in real life and involves no logical errors.

Well, how could such cases be possible? Let us turn the question around and ask, why would anyone doubt or even be puzzled by their possibility since in real life they are so common? I think the basic mistake, and it is a mistake that has characterized analytic philosophy of action for almost forty years, is to misconstrue the relationships between the antecedents of an action and the performance of an action. There is in analytic philosophy a tradition that runs from Hare (195x) through Davidson (1980), according to which pure cases of weakness of will never really occur. It is logically impossible that they should occur. On Hare's account it turns out that if the agent acts contrary to his professed moral conviction, that shows that he really did not have the moral conviction that he claimed to have. On Davidson's account, if it turns out that the agent acts contrary to his intentions, then he really did not have an unconditional intention to perform the action. Both Hare and Davidson hold variations of the basic idea that someone who makes an all-out evaluative judgement in favor of doing something, must do that thing (unless, of course, he is prevented, etc.), and consequently, if the action is not performed then it follows that there was no all-out evaluative judgement, but on Davidson's account only "prima facie" or conditional value judgement.

To these analyses, we can make the obvious objection that you can make any kind of evaluative judgement you like and still not act on that judgement. That is, the problem with these philosophers' analyses of akrasia is that it neglects the fact that the antecedents of action can be of any sort that you like, moral commitments, firm evaluations, fully formed unconditional intentions, etc. And all the same it is logically possible that the agent might voluntarily fail to act in accordance with the content of those antecedents. There is simply a gulf between the antecedents of the action, whatever they are, and the actual performance of the action, and this is what makes akrasia possible. The only way that the Hare-Davidson tradition can avoid this objection is to build the notion of acting intentionally into the notion of evaluation. But then the analyses avoid falsity at the prize of circularity. It now becomes a definition of having a certain sort of antecedent of an action that one acts on in accord with that antecedent. The problem of akrasia to repeat, is that, any type of antecedent whatever, provided that it is described in a non-question-begging way, that is provided in such a way that it does not trivially entail the performance of the action, is such that it is always possible for a fully conscious rational agent to have the relevant type of antecedent, (e.g., the relevant moral judgement, unconditional intention, anything you like) and still not act in accordance with the content of that antecedent. Furthermore, this is not a rare occurrence. It happens all the time to anybody who has ever tried to lose weight or give up smoking.

In its crudest form the mistake we are making, which makes it seem puzzling that there can be akrasia, derives from a mistaken conception of causation. We think that the antecedents of the action produce the action according to simple causal models, and that therefore if the action was not produced there must have been something wrong with the causes. If, for example, we think of causation on the model of billiard balls hitting billiard balls or gear levers activating gear wheels, then it just seems impossible that we should have the causes without the effects. If intentions cause behavior and the intention was present and the agent did not undertake the intended action, it can only be because some other cause interfered, or it was not the type of intention we thought it was or some such.

But intentional causation is in certain important respects unlike billiard ball causation. Both are cases of causation, but in the case of desires and intentions, once the causes are present they still do not compel the agent to act; the agent has to *act on* the reasons or on his intention. In the case of voluntary action there is a certain amount of slack between the process of deliberation and the formation of an intention, and again there is slack between the intention and the actual undertaking.

As usual, where intentionality is concerned, it is best to think of the cases of akrasia from the first person point of view. Well, what is it like for me to form an intention and then not act on it? In such cases am I always prevented from acting on it, compelled by causes, conscious or unconscious, to act contrary to my intentions? Of course not. Well, does it always turn out in such cases that the intention was somehow defective, conditional, or inappropriate, that it was not an all out, unconditional, no holds barred intention, but only a prima facie, conditional intention? Once again, of course not. It is possible, as we all know, for an intention to be as strong and unconditional as you like, for nothing to interfere, and still the action does not get done.

To see how akrasia occurs we have to remind ourselves how actions proceed in the normal, non akrasia cases. When I form an intention I still have to act on the intention that I have formed. I can't just sit back and wait to see the action happen, in the way that in the case of the billiard balls I can just sit back and wait to see what happens. But from a first person point of view, the only view that really matters here, actions are not just things that happen, they are not just events that occur, rather from the first person point of view they are *done*; they are, for example, undertaken, initiated or launched. Making up your mind is not enough; you still have to do it. It is in this slack between intention and action that we find the possibility, indeed the inevitability of at least some cases of weakness of will. Because of the inevitability of conflicting desires, for most premeditated actions there will also be the possibility of conflicting desires, desires not to do the thing one has made up one's mind to do.

What would it be like if akrasia were genuinely impossible? Imagine a world in which once a person had formed an unconditional intention to perform an action, (and had satisfied any other antecedent conditions you care to name, such as forming an all out value judgment in favor of performing it, issuing a moral injunction to himself to perform it, etc.) the action then followed by causal necessity unless some other cause overcame the causal power of the intention or unless the intention grew weak and lost its power to cause action. What this fantasy asks us to imagine is a world in which intentions have a connection to actions on the model of levers moving other levers and billiard balls hitting billiard balls. But if that were how the world worked in fact, we would not have to *act on* our intentions we could, so to speak, wait for them to act by themselves. We could sit back and see how things turned out. But we can't do that, we always have to act.

Akrasia in short is but a symptom of a certain kind of freedom, and we will understand it better if we explore that freedom further. On a certain classical conception of decision making we, from time to time, reach a "choice point" a point at which we are presented with a range of options from which we can — or sometimes must — choose. Against that conception I want to propose that at any normal conscious waking moment in our lives we are presented with an indefinite, indeed strictly speaking infinite range of choices. We are always at a choice point and the choices are infinite. At this moment I can wriggle my toes, move my left hand, my right hand, or set out for Timbuctoo. Any conscious act, any intention in action, contains the possibility of not performing that act, but of performing some other act. All but a tiny handful of these options will be out of the question as fruitless, undesirable or even ridiculous. But among the range of possibilities will be a handful we would actually like to do, e.g., have another drink, go to bed, go for a walk or simply quit work and read an escapist novel.

Now the way in which akrasia characteristically arises is this: as a result of deliberation we form an intention. But when the moment comes there are an indefinite range of choices open to us and several of those choices are attractive or motivated on other grounds. For many of the actions we do for a reason, there are conflicting reasons for doing not that action but something else. Sometimes we act on those reasons and not on our original intention. The solution to the problem of akrasia is as simple as that.

It might seem puzzling then that we ever act on our best judgement with all these conflicting demands made on us. But it is not so puzzling if we remind ourselves why we have deliberation and prior intentions at all. A large part of the point of these is to regulate our behavior. Sane behavior is not just a bundle of spontaneous acts, each motivated by the considerations of the moment, rather we bring order and enable ourselves to satisfy more of our long range goals by the formation of prior intentions through deliberation.

It is common to draw an analogy between akrasia and self deception, and there are indeed certain similarities. A characteristic form of akrasia is that of duty versus desire, just as a characteristic form of self deception is reasons versus desire. For example the lover deceives himself that his beloved is faithful to him in the teeth of the evidence to the contrary, because he wants desperately to believe in her faithfulness. But there are also certain crucial differences, mostly having to do with direction of fit. The akrasiak can let everything lie right on the surface. He can say to himself, "Yes I know I shouldn't be smoking another cigarette and I have made a firm resolve to stop but all the same I do want one very much; and so, against my better judgement, I am going to have one." But the self deceiver cannot say to himself, "Yes I know that the proposition I believe is certainly false, but I want very much to believe it; and so, against my better judgement, I am going to go on believing it." Such a view is not self deception, it is simply irrational and perhaps even incoherent. In order to satisfy the desire to believe what one knows to be false the agent must suppress the knowledge. "Akrasia" is the name of a certain type of conflict between intentional states, where the wrong side wins. "Self deception" is not so much the name of a type of conflict at all but rather it is a form of conflict avoidance by suppression of the unwelcome side. The name of a form of concealment of what would be a conflict, indeed an inconsistency, if the conflict were allowed to come to the surface. The form of the conflict is:

> I have overwhelming evidence that p (or even perhaps, I know that p) but I wish very much to believe that not p.

But that conflict cannot be won by desire if it emerges in that form. The conflict itself requires suppression if desire is to win, and that is why it is a case of self deception. Akrasia is a form of conflict but not a form of logical inconsistency or irrationality. Self deception is a way of concealing what would be a form of inconsistency or irrationality if it were allowed to surface. For these reasons self deception logically requires the notion of the unconscious; akrasia does not. Akrasia is often supplemented by self deception as a way of removing the conflict: e.g. The smoker says to himself: "Smoking isn't really so bad for me, and besides the claim that it causes cancer has never been proved".

To summarize these differences: akrasia and self deception are not really similar in structure. Akrasia has the typical form:

> It is best to do A but I am voluntarily and intentionally doing B.

There is no logical absurdity or inconsistency here at all, though there is a conflict between inconsistent desires.

Self deception has the typical form:

> Conscious: I believe not p
>
> Unconscious: I have overwhelming evidence that p and want very much to believe that not p.

References

Aune B. (1977). *Reason and Action.* Dordrecht, Holland: D. Reidel Publishing Company.

Elster J. (1983). Sour Grapes, *Studies in the Subversion of Rationality.* Cambridge University Press, Cambridge, England.

Nagel T. (1970). *The Possibility of Altruism.* Clarendon Press, Oxford.

Kenny A. (1975). *Will, Freedom and Power.* London, England: Basil Blackwell.

Davidson D. (1980). *Actions and Events.* "How is Weaknessof Will Possible?", pp.21–42. Oxford University Press. Oxford and New York.

Searle J. (1983). *Intentionality: An Essay in the Philosophy of Mind,* Cambridge, England: Cambridge University Press.

Sen A.K. (1973). "Behavior and the Concept of Preference", reprinted in Elster, John, ed. *Rational Choice*, pp.60–81, New York University Press, New York, 1986.

Williams B. (1981). "Internal and External Reasons" in *Moral Luck, Philosophical Papers 1973–1980*, pp. 101–113.

Chapter 4

TWO BASIC KINDS OF COOPERATION

Raimo Tuomela
University of Helsinki

1. Introduction

This paper will discuss the broad topic of cooperation from a conceptual and philosophical point of view. Its basic problem will be to answer the question "What is cooperation?". It will be seen that when viewed *teleologically*, from the point of view of the goals involved, two different basic kinds of cooperation exist. What I will call *full cooperation* is intuitively cooperation in something like the dictionary sense. For instance, Collins Cobuild Dictionary defines 'cooperate' thus: "1. If people cooperate, they work or act together for a purpose. 2. If you cooperate, you help willingly when they ask you for your help." My basic idea is the related one that cooperation is many-person activity based on a *shared collective goal*, understood in a strong sense. Accordingly, cooperation in the present full sense is acting together as members of a group (however temporary) to achieve a shared collective goal. I will speak of the resulting kind of cooperation as *we-mode cooperation.*

The other kind of cooperation is cooperation where no collective goal is involved but which rather is based on reciprocity (exchange) and compatible, possibly type-identical individual goals (state-goals or action-goals). This may also be called cooperation in the sense of *coordination* and the term to be used in this paper is *I-mode cooperation.* Typically, cooperation as spoken of in the context of a Prisoner's Dilemma game, for instance, is I-mode cooperation. No shared collective goal is involved, at best similar private goals.

Cooperation is a very broad and many-sided topic, and I can only deal with a few questions in this paper and properly argue only for some

D. Vanderveken (ed.), Logic, Thought & Action, 79–107.

of the theses expressed.[1] Accordingly, this paper concentrates on a) a general discussion of (full) cooperation, b) a more detailed discussion of cooperative joint action — a central kind of cooperation, c) arguments for the presence of a shared collective goal in full cooperation, and d) a brief discussion of I-mode cooperation.

A shared collective goal is a goal satisfying the so-called Collectivity Condition. According to this principle, if one or more agents satisfy the goal (intention content), then necessarily, on non-contingent, "quasi-conceptual" grounds it is satisfied for all participants. There "quasi-conceptual" grounds involve that it is satisfied in part on the basis of the participants' acceptance of the goal as a collective goal, one applying to the collective in question, the participants being collectively committed to the goal. This entails that the collective goal is shared in a we-mode sense, because it is the participants goal accepted for the purposes and use of the group such that the participants are collectively committed to the goal.[2]

It is a commonplace that human beings are social and are disposed to cooperate. We have learned from biology and ethology that such factors as "kin-altruism" and "reciprocal altruism" can ground cooperative behavior in animals. In the case of human beings, we think somewhat similarly but in more general terms that people are social. This sociality is a many-faceted thing, which involves at least that people on the whole need, and enjoy, the company of other human beings. This kind of dependence can be intrinsic (sociality as an irreducible basic want or

[1] See Tuomela (2000) for a detailed account. In the present paper I draw on the account of cooperation presented in the aforementioned book. (Cf. also Bratman, 1992, Tuomela, 1993, and Tuomela and Tuomela, 2004, for cooperative joint action.)

[2] The following is a more precise analysis of the notion of a (shared) collective goal (cf. Tuomela, 2000, Chapter 2):

(*ICG*) G is an intended *collective goal* of some persons $A_1, \ldots, A_m$ forming a collective g in a situation S if and only if G is a state or (collective) action such that

1) each member of g has G as his goal in S, entailing that he intends to contribute (at least if "needed"), together with the others – as specified by the mutually believed presupposition of the shared goal G – to the realization of G;
2) part of a member's reason for a), viz. for his having G as his goal is that there is a mutual belief among them to the effect that a);
3) it is true on "quasi-conceptual" grounds that G is satisfied for a member A_i of g if and only if it is satisfied for every member of g; and this is mutually believed in g. (Collectivity Condition with mutual belief).

Here is an analysis of what the we-modeness of a collective goal can be taken to amount to, given collective commitment (Miller and Tuomela, 2001, Tuomela, 2002a, 2002b):

Goal P of an agent X is in the we-mode relative to group g (or X has P in the we-mode) if and only if, X is functioning qua a member of g, X intends or wants to satisfy (or participate in the satisfaction of) P at least in part for g (viz. for the use of g).

need) or instrumental (related to features like self-respect, honor, pride, etc. or to various things that they want to achieve but cannot alone achieve). Leaving a debate about this to another occasion, let me in any case mention one central feature — assumed fact — relevant to our present concerns. It is that people in their thinking and acting tend to take into account what others think and do. Thus others' approval and disapproval of one's ways of thinking and acting form an important motivational element.[3] All this induces an element of conformity and cooperativity (at least "harmony") into human life in a social context. Thus it can be claimed that even if human beings often have different goals, they are still on many occasions disposed to behave cooperatively at least with respect to their kin and, perhaps by a kind of analogical extension, with respect to their friends and other close group members and possibly more generally with all human beings (we could term this "friendship" or, more broadly, "we-ness" cooperation). Furthermore, people often have different but suitably interlocking goals, and then they tend to cooperate with strangers in terms of reciprocal exchange in business and related contexts ("exchange cooperation"). While a general, biologically based disposition to cooperate can perhaps be seen to exist in human beings, it is not that easy to specify under what conditions people actually cooperate rather than defect, act competitively, selfishly, or even aggressively. These latter kinds of behavior are all in their different ways opposite to cooperation, and people seem also to be disposed to such behavior. It is surely of interest to investigate deeper the nature of cooperation and especially the conditions and circumstances that make it feasible for people to cooperate, but in this paper I cannot go deeper into that.

It can also be argued that it is a *necessary* feature of human beings qua thinkers and agents (conceived in terms of the "conceptual framework of agency") that they are social and at least to some extent cooperative beings. At least this sociality assumption is a general *presupposition* underlying any person's thinking and action on the whole, although in actual practice this presumption may be retracted on particular occasions. A central argument for this brand of the sociality view goes in

[3]While it is not the purpose of this paper to review the empirical findings concerning people's motivation to cooperate, let me still mention the recent research reported in Tyler and Blader (2000). These authors studied the effect of the (believed) status of one's group and of their own (believed) status within their group on cooperation. They found out that both factors are central motivating elements, the first one being somewhat more central. While the (material) gains and resources obtainable from the achievement of group goals also matter, issues of status may still be even more important. These empirical findings fit well the emphasis on we-mode cooperation in this paper.

terms of the assumption that human beings conceived as thinking and acting persons necessarily are language users. As language necessarily is based on shared meanings and shared uses, we arrive at the sociality view (or at least its presupposition version) of human beings.

My approach to cooperation is based on a philosophical theory of social action. It is argued that cooperative acting together forms the core of full cooperative action. The term 'we-mode cooperation' will be used for this kind of cooperation — or more precisely for cooperation based on a shared collective goal in the we-mode. This view of full cooperation as *we-mode cooperation*, to be properly argued for in Section IV, will be called the *collective goal theory* of full cooperation. As already mentioned, there is also *I-mode cooperation*, which is based on the participants private or I-mode preferences and goals. Both kinds of cooperation are important and worthwhile objects of study in their own ways. While most current empirical studies concern cooperation in the I-mode sense, I here emphasize full cooperation. In game theory there are in a sense parallel developments. Thus, cooperative game theory can to some extent be connected and related to my theory of we-mode cooperation, while non-cooperative game theory deals with I-mode cooperation.[4]

2. Cooperation and joint action

It is a platitude that cooperation is collective activity: we speak of two or more agents cooperating in order to achieve their ends or their shared collective end. One always cooperates in some collective context, yet we can speak of a single agent's action being cooperative as long as it is somehow based on a collective end (or some kind of "jointness", such as processual jointness in activity) to be achieved by it. I will speak of this end as a shared collective goal – a state goal or process goal. In addition, we can also speak of somebody's being cooperative in the sense of his having a cooperative or willing attitude towards some collectively endorsed goal or activity and towards the participants in such activity.

We intuitively think that some social action is cooperative while some is not. For instance, carrying a table jointly or singing a duet together seem to be unproblematic paradigm cases of cooperation whereas quarreling is non-cooperative. What about playing a game of tennis? Is walking in a crowded street cooperative if the people intend to avoid

[4]One of my claims, discussed in Tuomela (2000), Chapter 7, is, however, that current game theory is not really capable of giving an adequate account of cooperation based on a collective goal. This is basically due to the fact that the conceptual basis of game theory is too meager. It lacks proper means for presenting intentions, commitments, and norms, and all these features belong to a theory of full cooperation.

bumping into each other? How about each of us lighting candles in the evening of Independence Day? A philosopher wishes to know in more detail what is involved in examples such as these. What kinds of elements are or must be involved in cooperation, and how weak can cooperation be?

Acting together involves sociality in the relatively strong sense that such action must be based on joint intention or shared collective goal. This makes any case of acting together cooperative at least to the extent that the persons are collectively committed to making true a certain state of affairs in a "harmonious" way (that can find expression also in their relevant practical inferences). One can distinguish between different kinds of acting together. In the strongest sense of acting together we require intentional acting on a joint, agreed-upon plan. Examples of such joint action are jointly singing a duet, playing tennis, building a house. This kind of joint action is collective social action in its most central sense, acting as a team. The joint action has a cooperative element in that the participants are jointly committed to acting together and to relying on the other participants' performing their parts. Cases of joint action with an inbuilt element of conflict, such as in playing a game of tennis, are to a considerable degree cooperative — in contrast to cases of pure conflict such as being involved in a fight for one's life. While playing tennis still at least typically is plan-based acting together, there is weaker kind of acting together based on mutual belief or mere shared belief that also can represent cooperation (cf. below).

Let us now consider plan-based cooperative joint action: The participants have formed a joint plan for a joint action; the plan is taken to involve a relevant joint intention, entailing for each participant the intention to perform her part of the joint action (think e.g. of two agents deciding to prepare a meal together or of their forming the plan to repair the roofs of their houses). Each participant is assumed to believe (and rely on the fact) that the various conditions of the success of the joint action will be fulfilled at least with some probability, and she must also believe that this is mutually believed by the participants. In general, the performance of a joint action can be regarded as agreement-based if the plan has been accepted by the participants and if they have communicated their acceptances appropriately to the others so that a joint commitment to perform the joint action has come about. This shared plan to perform a joint action gives a kind of cooperative base for a joint action and both a "quasi-moral" and an epistemic basis for the participants to trust that the others will perform their parts and will not let them down.

As said, people can also cooperate in the full sense involving acting towards satisfying a shared collective goal also without plan-based joint action. For instance, people may cooperate to keep the streets clean. They share the collective goal of keeping the town tidy but need not have formed a joint plan concerning the matter at hand. It suffices that they accept the collective goal and act, possibly independently but in harmony with the others, towards this goal, which they believe to be generally shared among the inhabitants. This is still a weak kind of cooperative joint action.

It is important to notice that the participants' preference structures in a joint action can be perfectly cooperative (cf. carrying a table) in the sense of being highly correlated (or "correspondent") or they can be to some extent opposed (cf. chess, selling and buying). This is a feature of cooperation in the sense of preference correlation; it will be emphasized when the underlying motivation and rationality conditions of cooperation are discussed. We can speak of *given* preferences concerning the joint or collective activity (or, more precisely, concerning the end(s) and/or the means-process) in question. These are the preferences the participants have before action and before their (possibly) having considered the situation of cooperation in strategic terms, and they contrast with *final* preferences to be defined below. Such given preferences may (or may not) be based on the agents' needs and interests, and they may be rational or "considered" preferences. Furthermore, they may reflect objective payoffs such as money or other quantifiable and transferable objective goods.

Depending on the case at hand, the given preferences may be either "natural" or "institutional" and culture-dependent. This last distinction need not be regarded as a dichotomy, and it is not a very clear one either. It corresponds roughly to a similar distinction concerning joint action – a joint action may be physical (e.g., carrying a table jointly) or it may involve a conventional or normative element such as transfer of property rights. As to the latter kind of cases, think e.g. of toasting a national victory together (based on a social convention), making a business deal (legal transfer of rights), getting married (legal creation of rights and duties).[5]

When cooperating in a normatively defined situation the participants accept, or are assumed to accept, the goals, tasks, and parts defined by the norms (or, more generally, by some kind of normative authority). The correspondence between the outcome-preferences in question thus

[5]Bowles (1998) contains interesting empirical and theoretical points related the influence of markets and other economic institutions on preferences

will be at least in part normatively determined. Normative determination is taken to mean that the correspondence is defined by means of an agreement, or a social norm (either a so-called rule-norm or a so-called proper social norm), or a normative mutual expectation (perhaps based on an authority's directive).

Actual cooperative activity will take place on the basis of the participants' *final* preferences, which so to speak by definition take all the relevant considerations in the situation of cooperation into account. One may participate in cooperation either willingly, *viz.*, with a cooperative attitude, or only reluctantly. Part of what is meant by a cooperative attitude is that a participant with such an attitude is supposedly disposed to transform his relevant situational preferences in a cooperative way, e.g., to transform his given preferences into final ones so as to take the participants' joint reward in the situation into account. This joint reward can concern both the collective end or ends in question and the means-activities related to its achievement. A standard example of a person acting for a cooperative attitude would be a moral person "loving one's neighbor as himself" and acting accordingly.

Whether a singular example should correctly be classified as we-mode cooperation or I-mode cooperation depends on what kinds of goals and satisfaction modes the participants have in their minds. Determining this may often be difficult and will rely on observing people's actions (especially helping behavior) and asking them relevant questions about their goals. Thus, people can cooperate to keep the streets clean, and this can be either we-mode or I-mode cooperation depending on whether a shared collective goal or only shared I-mode goals are involved.

As said, we-mode cooperation amounts to cooperation towards a shared collective goal (in the we-mode). This kind of cooperation does require acting together, but not a joint plan or an agreement to act together. What I will call the Basic Thesis of Cooperation says the following: Two or more actors cooperate in the full sense if and only if they share a collective (or joint) goal (in the we-mode) and act together to achieve the goal. This can be taken to involve the following subtheses: 1) full cooperation entails acting together towards a shared collective goal, viz. we-mode cooperation, and conversely, 2) we-mode cooperation exhausts full cooperation. Both 1) and 2) will be discussed and defended later in the paper. Before that I will discuss cooperative joint action in more detail, in view of the centrality of this kind of cooperation.

Every intentionally performed joint action in the *plan-based* sense of joint action is full cooperation, even when some conflict is involved. This is due to the interdependent and (to some degree) harmonious acting to satisfy the joint plan. In plan-based joint acting, the joint plan, qua involving a collective goal, makes the joint action (say painting a house together) a full-blown one, and acting for the reason that there is the plan (or agreement) to be satisfied or fulfilled makes the joint action intentional. The jointness of the action (and the cooperative attitude that may but need not be involved towards others) serves to make the activity cooperative.

In cooperative joint action the participants' relevant preferences concerning their part performances can be strongly correlated (cf. painting a house together) or even to a large extent contrary and conflict-involving (cf. playing tennis), but there must be some kind of joint plan (or agreement) to engage in cooperative joint action or joint action with cooperative elements, and thus there must be a cooperative base at least in this sense. Every plan-based joint action, accordingly, is cooperative to some degree (or in a certain sense). Cooperation in its fullest possible sense is cooperation in the present sense of full joint action and it thus involves a shared collective goal (however, all acting together yields full cooperation, we-mode cooperation, even if perhaps not the fullest kind of cooperation – cf. the willingness requirement for fullest cooperation below).

A joint action based on a shared collective goal involves cooperation, since when trying to achieve the collective goal together each participant suitably harmoniously coordinates his performance of his part of the joint action with the others' performances of theirs. Here helping others may occur, but no direct helping is strictly required in normal situations. Nevertheless, one's part-performance often involves behavioral interaction with the others. For instance, if you and I are jointly carrying a table upstairs we have to be responsive to each other's bodily movements. I must adjust my movements to yours and try to see to it that I am not hindering your part-performance. This kind of responsiveness entails cooperation and, one may say, helping in a *weak* sense. Helping in the strong sense — which is not required normally — entails helping the other person directly to perform his part of the joint action (cf. I perform a part of your task). The first-mentioned weak sense of cooperation (and helping) relates to what one does, so to speak, within the limits of one's performing one's own, unconditionally required part, whereas cooperation and helping in this stronger sense is concerned with crossing those boundaries. We can speak of helping in a strong sense as activity contributing to the satisfaction of another agent's want, need, goal, or part-performance, thus incurring costs not related directly to one's goals, tasks, and preferences. Obviously, if a participant regards another participant as helpful and cooperative, he has good reason to trust the other one in the course of the joint action.[6]

There is of course helping involved in the course of the joint activity itself as well. Cooperative agents are assumed to help each other at least in the weak sense in the various phases of the performance of the joint action in question and to help in the strong sense in the case of full cooperation if help is strictly needed (e.g. in some kind of breakdown case) and if it is rational for them to give it. However, when a participant acts with a cooperative attitude, with relevant willingness, he is supposed to be helpful in the strong sense (cf. below).

The extent to which helping is required depends on the action context. In the case of, say, carrying a table jointly it is under normal circumstances rational to help (at least weakly help) to make the joint action successful concerning all the aspects of joint action, including helping the other participants to perform their parts. Consider, on the other hand, a game (rule-governed joint activity) in which the interests of the participants' (qua participants) are opposite — except for the very playing of the game itself. Playing poker would constitute such a joint action, since in poker it is not rational to help the other players to win by trying to lose as much money as possible. One surely can take part in playing poker and cooperate without that.

As these examples suggest, in a joint action situation the participants' preferences (interests) may 1) correlate perfectly, 2) be positively, although not perfectly, correlated, 3) be negatively correlated, or 4) be fully opposed. The preferences may be in part "built" into the action in question. Thus they may be "suggested" or "generated" by the physical nature of the joint task involved in the joint action (cf. lifting a table), or they may be determined or generated by suitable rules (cf. singing a song together), or they may be due to more or less purely personal or social psychological factors.

[6]Let me mention here that recent empirical studies suggest that the kind of trust that disposes to cooperation is more likely to develop in cases where there are no contracts or, rather, the contracts that exist are not complete and fully explicit (see Bowles, 1998, for references and for discussion). Fully specified explicit contracts and agreements — as well as the "anonymity" and lack of personal contact typically accompanying them – thus are enemies of cooperativity. (In the language of Tuomela, 1995, we can say somewhat loosely that belief-based or "s-notions" tend to be friends of cooperation while rule-based or "r-notions" tend to be its enemies.) An obvious point to make about helping is that while cooperation entails the requirement of mutual helping in general, the converse is not true, for mutual helping can take place without a shared collective goal. Cf. Van Vugt et al., 2000, Chapter 1, for a view of cooperation as mutual helping of a suitable kind: "Cooperation is a type of helping that can be distinguished from other forms of helping in (1) *the number of people who profit*, (2) *the common interdependence*, (3) *the duration of help*, and (4) *the nature of the helping act*".

We may speak of a perfectly cooperative action type in the case of 1) and of a cooperative context involving opposed preferences or interests in the case of 4). Cases 2) and 3) are mixed cases. A central feature of perfectly (and "almost perfectly") cooperative joint actions is that the participants' preferences in them are strongly correlated. A second relevant aspect is that the participants must act because of a cooperative attitude and willingly (unreluctantly) perform relevant cooperative actions. In fact, we can say on conceptual grounds that any joint action, no matter whether involving highly correlated preferences or not, can be performed cooperatively or non-cooperatively in the present "adverbial" sense.

Cooperative joint actions are performed in situations where the participants, qua "normally rational" persons, have certain expectations and preferences about the joint activity in question. First, the performance of a perfectly cooperative joint action can be expected under favorable conditions to give each of the participating agents (at least in the case of an optimal choice of rational agents) a better result (reward or utility) than they can attain without cooperation. This so-called individual rationality (or reward) condition is the content of a (rational) mutual expectation involved in every situation of rational cooperation prior to action; it obviously also serves to motivate the agents to perform cooperative joint actions. If all participants are expected to gain from joint action, then there is a collective gain. There may be collective gain compared with acting alone without individual gain in the case of all participants. Whether it is depends on the fairness of cooperation, viz. how fairly to outcome of cooperation is divided among the participants.[7] In some situations people can cooperate without expecting to gain themselves as long as they expect their group will gain.

Obviously, there are also many presuppositions in cooperative joint action that the participants must take for granted. If the participants are to cooperate by performing an action jointly, they must not only have the joint intention to perform it but they must also understand the nature of the action at least in the sense of being disposed to have the right beliefs about it. In particular they must know whether the action is perfectly cooperative in its nature (cf. painting the house) or not (cf. poker) and in the latter case in what sense the preferences or interests of the participants are irreconcilably in conflict.

In the case of a perfectly cooperative action situation – a situation of joint action with no conflict – help, in the sense of actions strictly

[7] Cf. Rawls (1993), Moulin (1995), and Roemer (1996) for these ideas.

contributing to other participants' performing their parts well, is always in accordance with the participants' preferences. Furthermore, at least in some clear-cut cases of perfectly cooperative actions, the more the participants actually help each other, the more successful the joint action will be, other things being equal. (It must be assumed in the case of rational agents that the costs of helping actions do not exceed the gains accruing from them.) We can say that in the case of genuinely cooperative action the participants tend to "stand or fall together". This is strictly true in cases where the participants' preferences qua a participants) are strongly positively correlated, which explains the possibility of rationally helping the other participants (provided the cost condition is satisfied).

In a joint action type involving strictly opposed preferences, in contrast, there can be only the joint action "base" or "bottom." In the case of plan-based joint action the joint bottom is – or is in part – a plan to perform the action. An example of a joint action in which the participants' interests ("within-the-action preferences") are in conflict with each other is organized fighting, such as boxing; another one is the game of tennis. Here I am speaking only about plan-based joint actions and saying that they all involve a cooperative element and that making and fulfilling such an agreement is cooperative activity. However, beyond that base there is no way to help other participants with their part performances (given that the participants act on their part-related preferences built into the structure of the action). In these kinds of actions a participant can at best – and, indeed, ought to when needed – contribute to the preconditions of the other participants' part-performances. (Note that joint actions with opposite preferences do not include taking part in an unorganized fight for one's life, a situation of full conflict without any cooperative element; this is not a joint action at all.)

To be a little more specific, let us consider the game of chess with players A and B. In this 1) both prefer playing to not playing; 2) both prefer the disjunction "A wins or B wins or there is a stalemate" to its negation (viz., both prefer playing a complete game to an incomplete one); 3) there are opposed preferences only concerning who wins (or, more generally, concerning the individual disjuncts). Thus A, qua player, paradigmatically prefers his winning to a stalemate and a stalemate over B's winning; and B prefers his winning to a stalemate and a stalemate over A winning. Here 1) and 2) concern the cooperative bottom or base that I have spoken about, while 3) relates to the issue of preference correlation. In organized fights such as boxing matches all of 1), 2), and 3) at least paradigmatically hold, whereas in an unorganized fight 1) and 2) in general fail, or at least may fail to hold.

An example of a mixed joint action type which is neither perfectly cooperative nor one with strictly opposed preferences (viz., a zero-sum situation) is cooperative joint action based on exchange – cf. selling and buying, where making the exchange can be taken to be the plan-based "bottom".

3. Classifying cooperative joint action

Given the above discussion, I will next present a more systematic and more detailed classification of the basic types of cooperative joint action, and discuss cooperativeness as an attitude or strategy.

A joint action type can be (perfectly) *cooperative* (one with perfectly correlated preferences) or one with (strictly) *opposed preferences* or it can be "mixed". Let us list the components that we have discerned earlier in this paper. First there is a participant's (subjectively and objectively required) *part-action.* Next there are the extra actions that a joint action, X, may require. Let us call a *required extra action* Z. The performance of Z has to be divided among the participants. In addition, there may be actions that contribute to X but are not necessary for its successful performance. Let us call these actions *unrequired extra actions.* In finer treatment one would have to make an explicit distinction between (mind-independently) objective, mutually believed, and plainly subjectively believed "requiredness".

What is preference correlation? My index measuring the degree of correspondence (*corr*) of utilities (*corr* is an index of covariance standardized by the sum of the variances of utilities, and it can vary between -1 and +1): At least as a first approximation it can be said that the agents' preferences are perfectly cooperative if and only if correspondence is maximal (= 1); they are to some degree cooperative if and only if *corr* > 0; and they are maximally conflicting if and only if *corr* = -1). In general, a given amount of covariance between the preferences of the participants can be achieved by means of several different patterns of interdependence (viz., different patterns of control over their own and their partner's action as well as of purely interactive control) between their actions.[8]

Which of these components are present in perfectly cooperative joint actions (joint actions with highly correlated preferences) and which in joint actions with opposed preferences? In the case of both kinds of joint action types, part-actions must be involved, and extra actions of

[8]My technical approach to the problem of correspondence and the components of social control was presented in Tuomela, 1985, and it is discussed at length in Tuomela, 2000, Chapters 8 and 9.

both kinds may also be present (depending on the agent-external circumstances of action and, in the case of unrequired extras, on their relative cost and the agents' attitude to them). In the case of joint action with highly correlated preferences, contribution to other participants' performances of their parts is rational relative to a participant's preferences (utilities) and the principle of acting on one's preferences, given that the cost related to those helping actions is less than the gains expected to accrue from them. Or we can say that — disregarding the cost related to helping actions — it is conducive to the satisfaction of anyone's part-related preferences in question to help the others. Note that required extra actions are also included in a successful performance of X, although there may be a dispute among the participants about how to divide them into parts (and here normative considerations can come in). As a joint action is at stake, the participants have the collective responsibility to do what the joint action in question requires. The notion of helping in the case of perfectly cooperative action concerns helping other participants not only to perform their parts but also to do it *well.*

Considering perfectly cooperative actions, those in which the participants' interests are to a large extent shared, the following helping activities may in general be involved: i) contributing to the coming into existence of a precondition of another person's part, ii) contributing to or participating in the performance of the part itself (in the latter case the other person's part is actually performed as a joint action), iii) contributing by counteracting negative interference affecting the other person's part-performance. A perfectly cooperative joint action allows for both required and unrequired extra actions; in the case of rational action the costs of the unrequired extras must be smaller than the expected gains generated by their performance. (In the case of required helping actions there will not be much room for cost calculations, at least as long as their cost does not exceed the specific gain from the joint performance of X.) In a joint action type with opposed preferences helping actions only of kind i) and iii) can occur.

The central point about cooperative joint action situations is thus that one can help others to perform their parts (indeed, performing them well) and can also make other contributions increasing the joint utility accruing from X. In a cooperative joint action everybody is assumed to contribute positively to the joint utility, which in fair cases is expected to be divided among the participants in a way respecting their contributions. In game-theoretical terminology, games with coinciding or nearly coinciding interests roughly correspond to this notion of a cooperative action type, given that there is an agreement to play the game.

In the case of joint action types involving opposing interests — qua their being joint action types — the participants must perform their parts plus the required extra actions (contributory actions related to the bringing about and maintaining of joint action opportunities), and they may also perform some unrequired extra actions. As already noted, the essential difference between actions with perfectly correlated preferences and actions with opposed preferences is that in the case of the latter it is not in accordance with a participant's part-related preferences to help the others with their performances of their parts beyond what is required for the joint action to come about. That is, this is not possible relative to such a participant's preferences (utilities), assuming that he qua participant acts on his preferences. Competitive joint actions are examples of joint actions in which the parts are in conflict — here the preferences (or utilities) of the participants (part-performers) are at least to some extent antagonistic, and helping another participant to do well in his part-performance will reduce one's gain (payoff, utility) from the joint action. In game theory, zero-sum games and some mixed-interest games represent these cases.[9]

Next we distinguish between *cooperatively* and *non-cooperatively* performed joint actions (viz., action tokens). All joint action types (irrespective of how the part-preferences correlate) can be performed either cooperatively (viz., out of a *cooperative attitude* or *action strategy*) or non-cooperatively. A person having a cooperative attitude towards a joint action X must be disposed to reason and act in ways contributing to X. He must thus be disposed willingly to perform his part of X and must also willingly accept a share — reasonable for him relative to his capacities and skills — of a required extra action, Z; and he must be willing to perform unrequired extra actions related to X — as long as they are not too costly related to the gains accruing. As the performance of Z is the agents' collective responsibility, we can see that the notion of a cooperative attitude has a normative aspect. I would like to emphasize that willingness in the sense meant here is assumed to be an action-disposition, which can be based on various kinds of motivation, so that willingness need not involve any particular desire or emotion (such

[9]It may be noted that in game theory the distinction between cooperative and noncooperative games is different than my distinction between perfectly cooperative (or common-interest) joint action and that with opposed preferences. Game theorists call a game cooperative if the players are allowed to communicate and make binding or enforceable agreements; otherwise it is noncooperative. The plan-based joint actions under consideration in the present chapter will often be noncooperative in the game-theorist's terminology, for the agreements or plans involved are often are not strictly binding in the game-theoretical sense. Agreement-based games with "strictly identical interests" will represent perfectly cooperative joint action situations in my sense (cf. Harsanyi, 1977).

as intrinsic desire to help or enthusiasm to cooperate) towards the situation. In typical cases, cooperativeness no doubt will involve goodwill not only towards X (which I invariably require) but also goodwill and faithfulness towards the other participants and acting in part for their personal sake. While tendency to faithfulness in a moral or quasi-moral sense is typically involved in cooperativeness in the case of cooperation between friends, it may not be there in the case of businesslike cooperative transactions or in coerced cooperation.

A person with a non-cooperative (or uncooperative) attitude towards X, however, is disposed to freeride and to minimize his contributions to X (and thus his related costs). He will reluctantly perform his part of X, but he must perform it, for otherwise he could not be a participant in X at all. He will be reluctant to accept any extra tasks required. (Often an "Italian strike" — in which the activities are performed slowly and exactly according to all the regulations — occurring as part of larger joint action is an example of a joint activity performed with a non-cooperative attitude.) He will not perform any unrequired extra tasks. The presence of a cooperative attitude may clearly affect the effectiveness and speed of performance and the general "social atmosphere" as well as the amount of "we-feeling" in the group.

There are broader issues of cooperation in society overall for which the present account is relevant. Thus, we can make a connection to Rawls' theory of political philosophy in which society is viewed "as a fair system of social cooperation between free and equal persons viewed as fully cooperating members of society over a complete life".[10] In his idealized system justice as fairness is a shared common end (in my terminology: a shared collective goal in the we-mode) which serves to make society-wide institutional cooperation full cooperation in the sense of the present theory. In addition to Rawlsian kind of liberal theory, communitarian and republican accounts of society rely on shared collective goals and cooperation (cf. e.g. material and structural equality, freedom as non-domination). Therefore, the present account should help to clarify and even fortify these kinds of theories.[11] Society consists largely of social institutions (and social relationships of a less structured kind), and social institutions overall – or in many cases – rely on cooperation. For example, they provide and build on collectively rational solutions to collective action dilemmas, and such solutions require the existence of cooperative action patterns.

[10]Cf. Rawls (1993), p. 9; also cf. pp. 300-301.

[11]These matters are discussed e.g. in Pettit (1997); see also the discussion in Tuomela (2000), Chapter 13 and the points made in Tuomela, 1995, Chapter 10.

Our discussion leads us to the following simple, self-explaining diagram:

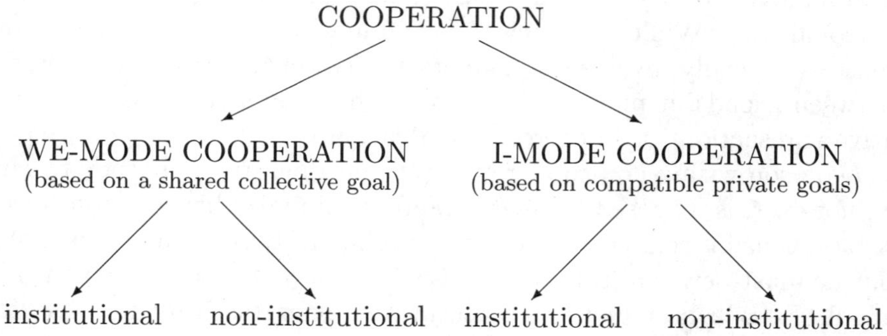

4. Full cooperation and shared collective goal

Full cooperation necessarily involves a collective goal in the strong sense satisfying the Collectivity Condition (*CC*) mentioned in Section I. This condition says that it is criterion of a collective goal that, necessarily due to participants collective acceptance with collective commitment of the goal as their collective goal, if it is fulfilled for one of the participants it is fulfilled for every other participant as well.

Above I have discussed the collective goal theory of full cooperation mainly in terms of cooperative plan-based joint action, the flagship of cooperation. But as will soon be argued in detail, while all full cooperation requires a shared collective goal, there is full cooperation which is not based on joint action in the plan-based sense. We can speak of acting together in a more general sense as follows — supposing, for simplicity's sake, that there are only two participant, A and B. Suppose both of them intend to participate in a joint action such as cleaning a back yard. Also suppose that they believe that the other one will participate (or probably participate) and that the other one believes similarly. Now, if A and B in part base their participation intentions on this belief and start acting, we get a case of acting together in a general sense. This sense does not even require full mutual belief about participation and still less is it based on a jointly formed plan to act jointly.[12]

[12]Here is the most general formulation of acting together for the two-person case given in Tuomela and Bonnevier-Tuomela (1997) and Tuomela (2000):
(*AT*) You and I *intentionally act together* in performing X if and only if

According to the Basic Thesis of Cooperation, 1) full cooperation entails acting together towards a shared collective goal, viz. we-mode cooperation, and conversely 2) we-mode cooperation (assumed to entail acting together) exhausts full cooperation. These claims will be discussed below. Thesis 1) actually contains the following two claims: a) Cooperation in the full sense involves a collective goal (end, purpose) and b) a collective goal entails that the participants must act together to achieve the goal. The kind of acting together that is required here need not literally concern the means-actions by means of which the goal in question is reached. As claimed, the participants are collectively committed to the goal and the activity realizing this collective commitment is collective seeing to it that the goal is achieved. The participants may use various "tools" or "instruments" for reaching the goal. Thus, they may hire agents to perform relevant means-actions or decide that a cer-

---

1) X is a collective action type, viz,. an "achievement-whole" divided into A's and B's parts, although not necessarily on the basis of an agreement or even a social norm;

2) (a) I intend us to perform X together, and I perform my part of X (or participate in the performance of X) in accordance with and partly because of this intention;

   (b) you intend us to perform X together, and you perform your part of X (or participate in the performance of X) in accordance with and partly because of this intention;

3) (a) I believe that you will do your part of X or participate in the performance of X at least with some likelihood;

   (b) you believe that I will do my part of X or participate in the performance of X at least with some likelihood;

4) 2) in part because of 3).

This analysis can be understood without much explanation (for further clarification see Tuomela, 2000). The central elements are belief-responsive intentions with collective, "collective action type" content, and a participant's beliefs about the other's participation. I am suggesting that the clauses of (AT) give a rather good idea of what should minimally be presupposed to be understood of acting together by the participants in question. As intention to act together basically is already involved in clause 2), that notion does not here need further explication. Using the "shared plan" and "acting as a group" locutions, (AT) can be said to express the most rudimentary idea of acting together (and as a group) in order to realize a *de facto* shared plan (involving a shared we-mode collective goal) although the existence of the plan is based only on the participants' beliefs (which need not be mutual beliefs and which need not have rational grounds). Obviously, more must be required to capturing non-rudimentary acting together in the we-mode. For one thing, clause 2) must in those cases be rendered in the following form applying we-thoughts to the participants as intentional subjects of the thoughts:

2*) (a) We intend to perform X together, and I perform my part of X (or participate in the performance of X) in accordance with and (partly) because of this intention;

   (b) We intend to perform X together, and you perform your part of X (or participate in the performance of X) in accordance with and (partly) because of this intention;

Clause 2*) is taken to entail 2), but the converse holds only in we-mode contexts.

tain group member will be the one actually to bring about the goal. Typically, however, the participants perform the means-actions together (cf. painting a house together). Cooperation here consists of the collective seeing to it that the goal is reached. This collective seeing to it that the goal is achieved clearly is a form of acting together (cf. the clauses of the analysis (AT) in note 12). (Note that acting together requires an intention to act together and hence an intended collective goal in our above sense. Thus the notion of an intended collective goal and that of acting together intentionally are conceptually intertwined. Thus not only is 1) b) true but so is its converse.)

The argument just discussed shows that we-mode cooperation (defined as cooperation towards a shared we-mode goal or we-mode goal) entails acting together. Thus, we-mode cooperation is entailed by full cooperation, and conversely, according to the Basic Thesis of Cooperation. This Basic Thesis, consisting of the subtheses 1)a), 1)b), and 2), has already been discussed by means of a variety of considerations and supporting examples. Subthesis 1)b will not be further discussed below, but 1)a) will be argued for. As to 2), the variety of examples to be presented below and elsewhere which do involve acting together but not necessarily plan-based acting together speak in its favor and the dictionary definitions of cooperation (to be cited below) rather directly support my thesis (also cf. my argumentation for 1)a) below). Some amount of stipulation must be involved, however, if subthesis 2) is to be regarded as true, for its antecedent e.g. does not require the presence of a cooperative attitude of full cooperation.

Recall our example of keeping the streets of a town tidy. We may ask what is cooperative about such a situation. My basic but somewhat circular answer to this is that the presence of a shared collective goal requires *cooperative* collective action for its satisfaction. This "package" makes the interaction situation cooperative and is seen to involve the following three central components or dimensions in the case of full cooperation (not necessarily plan-based cooperative joint action): 1) correlated preferences (and hence the possibility of helping), 2) a cooperative attitude (which need be only "actionally" understood), and 3) collective commitment to a jointness-feature or collectivity feature. Collective commitment to a collective goal is of course just a required kind of collectivity feature.

Basically, my account of cooperation with acting together in a wider than the plan-based sense just relaxes some of the assumptions made earlier. The differences between plan-based joint action and the other kinds of acting together are mainly epistemic and doxastic, as there is more uncertainty about the others' participation and about such things as the precise part-structure or ("contribution-structure") of the collective action in question. An example fitting the generalized account – but which is not a case of plan-based joint action would be the following. In this example a sign urges the passengers to cooperate by refraining from the use of the Victoria Station. This example involves an institutionally determined collective goal. Analogously, London people could accept the non-institutional collective goal of cleaning some parts of the city or of fighting an armada of giant bats, etc. No plan making (agreements) is needed for this kind of collective cooperative action.

The main thesis to be defended below is the subthesis 1)a) viz. that all full cooperative activity relies on a collective or joint goal towards which the participants in cooperation contribute, understanding here that in some cases the collective goal can be the very cooperative action in question. Often the participants help — or can help — each other in their performances, and often they are also in some ways dependent on each other (over and above sharing a collective goal as the basis for cooperation). However, there can be cooperation without these features. When minimally rational such cooperation also involves the participants' expectation that cooperative activity be rewarding to them individually at least when things go well, as compared with separate action or action not directed towards a collective goal.

The thesis that all full cooperation involves a shared collective goal in the we-mode contains the following possibilities: a) cooperation as plan-based joint action, thus based on a joint intention (possibly only a joint "intention in action"); b) cooperation as collective action based on, and directed towards, a collective *state*-goal (such as the collective action of keeping the house clean by collecting the litter and by trying not to litter); c) cooperation as collective action based on, and directed towards, a collective *action*-goal; the action here can be a different action or it can be the cooperative collective action itself. Cases b) and c) need not involve plan-based joint action. In all these cases merely personal goals may be involved. For instance, in case c) the cooperative action will often be a means towards the participants' (further) private goals.

Let us now consider the crucial problem of why a collective goal is needed for full cooperation. I will now advance some arguments for the presence of a collective goal (in support of subthesis 1)a) of the Basic Thesis of Cooperation. We need such arguments in view of the fact that most treatments of cooperation operate without the notion of a collective goal more or less in my sense.

Here are my arguments for the presence of a we-mode collective goal in cooperation:

1) (a) The *conceptual* argument is simply that the very notion of full cooperation depends on a collective goal in about the sense of my analysis of the notion of a we-mode collective goal.

   (b) Linguistic evidence also supports this view, even if one should be cautious about dictionaries for purposes of conceptual analysis. Thus, for example, according to Collins' dictionary, to cooperate is to act together for a purpose, and, according to Webster's dictionary, to cooperate is to work with another or others to a common end. These dictionary definitions of course support subthesis 2) of the Basic Thesis of Cooperation as well. Note, however, that an expression such as 'common end' still can be taken to cover also shared I-mode goals, resulting in I-mode cooperation. However, this is no problem for my argument as long as genuine cases of we-mode cooperation exist.

2) One may view cooperation from a *group's point of view*: the group has achieving something as its goal and intentionally acts to achieve it; and, conversely, when the group acts intentionally there must be a goal of some kind involved. Here the group is treated as an agent. I will start by considering simple unstructured (or, one may also say, "egalitarian") groups. Viewed from the participating members' point of view, the group action, X, is their cooperative collective action or their joint action. More precisely, it must be collective action of the general acting together kind coordinated by means of the group's goal, say G. Basically this is because a group can act and have attitudes only via its members' actions and attitudes involving that the group members function as group members.[13] This is just the intuitive idea in we-mode cooperation: cooperation qua group members where the group might be only a spontaneously formed, fleeting group. From the group members' point of view their reaching or failing to reach the goal means necessarily "standing or falling together".
   Consider now the following point. If a) the group members have the same goal G, b) collectively accept to achieve this goal for their collective use, and c) are collectively committed to achieving G and act so as to realize this commitment, then (by definition) the goal G is a we-mode collective goal and one satisfying the Collec-

[13]See Chapters 5-7 of Tuomela (1995) and Tuomela (2002b).

tivity Condition (CC).[14] The conditions a)-c) can be regarded as satisfied in the case of full cooperation involving collective or joint activity (which upon analysis turns out to be of the acting together kind). The members' collective commitment to G is central here in showing that a shared we-mode goal and not only an I-mode goal must be present. Consider an example: I am mowing the lawn and you are planting flowers in a garden. This could be I-mode cooperation with different private goals (to have the lawn mowed and the flowers planted), it could be I-mode cooperation towards a shared I-mode goal (cleaning up the garden), or it could be a case of we-mode cooperation. In the first case we separately perform our activities and are free to change our minds about our tasks without the other's criticism. In the last case we collectively accept to clean up the garden and are accordingly collectively committed to doing so. Our collective action of cleaning up the garden is a cooperative acting together. The upshot of the present argument is that the adoption of the group perspective warrants the claim that there must be cases of we-mode cooperation. The categorical premise for the necessity argument is that a group perspective not only is often adopted but must frequently be adopted by such social group beings as human beings basically are.

The case of normatively structured groups is somewhat different, but it is worth considering briefly, as it also involves another, weak kind of cooperation. In the case of structured groups there are special members or perhaps hired representatives for decision making and acting. I have spoken of "operative" members here (recall note 13). The basic idea is that an autonomous structured group performs an action X just in case its operative members (for action) jointly or collectively perform actions such that X becomes generated, the non-operative members being obligated to tacitly accept what the operative members do. The operative members' activity can on this occasion be regarded as cooperation (even if they as group members may do much else which is not cooperation in this sense). Tacit acceptance, when it indeed occurs, represents actual or potential action with a cooperative element — the non-operative members should at least refrain from interfering with what the operative members do and purport to do. Obviously, the

[14]This is essentially what thesis (CCG) of Section VI of Chapter 2 of Tuomela (2000) states. It is proved there that the Collectivity Condition is fulfilled given a)-c).

non-operative members are then not required to share the collective goal in question.

3) Related closely to the second argument, we have the following *normative argument*: Actual life abounds with cases of cooperation in which the participants take themselves to be collectively committed to cooperative action and accordingly tend to think partly in normative ways such as "I will participate because I quasi-morally ought to do my part of our joint project". This collective commitment, which is stronger than aggregated private commitment, indicates the presence of a collective goal and g-cooperation, for the norm is one related to the group context in question. A special example of this is joint action based on *agreement* making. An agreement ties the group members normatively together. Agreement, when wholeheartedly endorsed, is a we-mode notion, and the fulfillment of the agreement is a shared we-mode goal.

4) *Instrumental* argument: Shared collective goals tend to work or function better, e.g. for achieving coordination and stability than shared private goals (and shared compatible private goals). In some cases — such as in the case of games of "pure coordination" — coordination cannot optimally be achieved without the participants sharing a collective goal.[15] The fact that shared collective goals tend to offer better coordination and better goal-achievement than private goals is in part explained by the fact that the participants are *epistemically* in a better situation. The participants are *collectively committed* and may be more strongly personally committed to the collective end than in the case of shared private goals. This offers collective persistence, which is more likely to lead to the achievement of the goal than in the shared I-mode goal case. Collective commitment to the goal in question helps the participants to trust each other especially when they cannot effectively monitor each other's part-performances.
A special case of the superiority (and in some cases of the necessity) of shared collective goals over shared private goals is provided by pure coordination situations.[16] To see why, consider the familiar case where two agents, A and B, wish to meet each other (at least in the thin sense of arriving at the same place). They can meet in

[15]Various aspects of this argument are considered in many chapters of Tuomela (2000), see especially Chapters 2-4, 6, 9, 11, and 12.
[16]This is the notion used by Lewis (1969), p.14. For a detailed examination of the philosophical problems involved in this kind of case, see Tuomela (2002c).

two ways, by going to the railway station (s_1, s_2, respectively) or by going to the church (c_1, c_2). These actions (viz. the pairs s_1, s_2 and c_1, c_2) achieve coordination and lead to the satisfaction of their goals. The two other action pairs do not satisfy their goals. There is no conflict in this situation. Both agents are assumed to act individually ("privately") rationally to successfully achieve their goals. However, there is a coordination problem here because A will go to the church (respectively station) given that B will go to the church, and B will go to the church given that A will, and so on. There is no noncircular rational solution to be obtained unless the agents somehow "agree" to go to a certain one of these places. In other words, they *must* base their acting on a shared goal if they are to achieve coordination rationally. This shared goal need only be one in the I-mode, but, because of the collective commitment involved, a we-mode goal can be argued to result in more effective and stable action. Accordingly, a detailed thesis says that in many cases (and one can even say "normally") a shared collective goal in the we-mode is required for a rational stable solution to a coordination situation with a coordination problem (several equally good alternatives to be coordinated on), while in a coordination situation with one best outcome shared I-mode goals suffice.

5. Summary analysis of we-mode cooperation

In this section I will analytically sum up full cooperation for the general case where plan-based cooperative joint action need not be present but only acting together in the more general sense. In my account below I will be concerned with action situations involving action alternatives for all the participants. It is also assumed that the participants can interact behaviorally. In interaction the performances of individual actions result in collective or joint outcomes in the game-theoretical sense. (Understood in its minimal sense the joint outcome consists of what the participants do in a situation of interaction, irrespective of consequences.)

Using this kind of framework one can define different notions of cooperative situation and cooperation that are increasingly weaker as to the assumptions of rationality (rewardingness) and correlation of preferences.[17] Below I will only consider the most general of these notions. This notion requires neither commonality of interest nor rewardingness.

[17] See Tuomela (2000), Chapter 4.

We then get this summary analysis of the notion of a (potential) situation of cooperation:

(*COS*) S is a *we-mode cooperative situation* if and only if

1) the participants share a collective goal (state or action satisfying the Collectivity Condition), believed by them to be realizable by one of the outcomes in S or by collectively performed joint actions (viz. in at least a weak sense of acting together) leading to such an outcome, and are willing to act together towards its achievement by means of the actions available to them in S;

2) the participants have a mutual belief to the effect that 1.

When this "dispositional" situation is realized we get cooperation:

(*COA*) The participants in S *cooperate* with each other in the *we-mode* if and only if

1) S is a we-mode cooperative situation;

2) based on their final preferences concerning outcomes from their part-performances or contributions, assumed to correlate positively, the participants willingly perform actions believed by them to contribute to the collective goal S is assumed to involve;

3) the participants have a mutual belief to the effect that 1 and 2.

The preference correlation assumption in clause 2 of (*COA*) concerns the part actions or shares by the agents by which they contribute to the achievement of their shared collective goal. Note that as there need be no "joint action bottom" based on a shared plan of action here, we must understand the preference correlations in a wide sense. While in the treament of plan-based joint action preference correlations were meant to be concerned with "within-action" outcome preferences (e.g. in painting a house or playing chess the part actions within the agreed upon joint action). In contrast, in the present context where no joint action bottom is assumed the outcome correlations concern simply the outcomes of the part or share performances, be these shares preassigned or not. As there is no clear technical framework here for the actual computation of preference correlations, the problem must be solved *in casu*. Another point to be made here is that the preference correlations are assumed to concern final rather than given preferences. Final preferences can in general be taken to correlate to a higher degree than given preferences in the context of cooperation, because the assumed willingness to perform part actions in (*COA*) tends to prevent conflicts between part performances. This kind of compatibility of part performances translates into the requirement of positive (here rather: non-negative) preference correlation.

This analysis states concisely what it is for the dispositional notion of a full we-mode cooperative situation to become rationally manifested. Note that the assumption of willing performance entails that the contributions will be intentionally performed. However, while one cannot unintentionally cooperate, one can be mistaken about one's beliefs. In this sense (*COA*) (and similar analyses) deals with *subjective* cooperation, without success requirement. For instance, to have a grotesque example, two persons are supposed to have the goal to paint a school building green. A person can in this sense subjectively cooperate with another person who, being color blind without knowing it, paints his side of the house taken by him to be the school building with a wrong color (say red instead of the meant green). He might even have been mistaken about the house in question. So I allow for mistakes concerning both the identification of the goal and about the various belief related to cooperation. An "objectivist" might here say that our agent did not cooperate but only tried to, while the other agent with the correct beliefs did cooperate. Contrary to this, (*COA*) emphasizes the mental conditions and allows for a subjective idea of cooperation (the agent certainly believed he was cooperating).

The newly introduced (*COS*) and (*COA*) are central notions as they serve to exhaust potential and actual we-mode cooperation. They can also be applied to institutional contexts.[18]

6. Cooperation in the individual mode

This paper has concentrated on full cooperation, viz. we-mode cooperation defined as acting together towards a shared collective goal. In contrast, I-mode cooperation, or cooperation as coordination, is based only on I-mode goals. Such I-mode cooperation then involves interaction with compatible private goals and with the intention of satisfying one's goal by means-actions which do not conflict with others' attempts to achieve their goals. Somewhat more precisely I require of such "compatible coaction" (as I have also called it): 1) the presence of compatible goals, viz. goals which can be satisfied in the situation without the kind of conflict preventing the others to satisfy their goals; 2) the presence of intention to avoid satisfying those goals by means-actions which strongly conflict with others' attempts to reach their goals; 3) the actors are dependent in that action situation on each other's action, viz., they have

[18]For institutional applications, see Tuomela (2000), Chapter 6.

to take the others' actions into account in attempting to achieve their goals in an optimal way; 4) the goals may (but need not) be of the same type; they may also be shared; 5) the agents must have beliefs about the participants' goals being compatible and about their intending to avoid satisfying them by means-actions conflicting with the others' goals (or at least they must think the latter kind of "cooperativeness" will probably exist); 6) each agent intends to achieve his goal and believes he can do it in that context at least with some probability without coming in conflict with the other persons' attempts to satisfy their goals.[19]

We thus arrive at the following schema explicating the most general notion of I-mode cooperation:

(*COI*) Agents A_1 and A_2 *cooperate in the I-mode* in a situation S relative to their I-mode goals G_1 and G_2 if and only if

1) their respective primary goals (viz., action-goals) in S, *i.e.* types of states or actions, G_1 and G_2, which relate to the same field of action dependence in S, are compatible in the sense of being satisfiable without making it impossible for the other agent to satisfy her goal;

2) (a) A_1 intends to achieve G_1 without means-actions conflicting with A_2's attempts to satisfy his goal and believing that he can achieve it at least with some probability in that context although his relevant G_1-related actions are dependent on A_2's relevant G_2-related actions, and he acts successfully so as to achieve G_1; and

   (b) analogously for A_2:

3) (a) A_1 believes that 1) and 2), and

   (b) analogously for A_2.

I-mode goal in (*COI*) means private goal, one which does not satisfy the Collectivity Condition. More exactly, goal G of an agent is in the *I-mode* relative to group g if and only if he is not functioning (t least fully) qua a member of g, and he privately intends or wants to satisfy G for himself.

(*COI*) can be argued to be in conflict which the mentioned Collectivity Condition assuming that the former entails that a participant's goal can be achieved by acting alone (but with the social and other environment not preventing goal-satisfaction) rather than by acting together. (An

[19]See Tuomela and Bonnevier-Tuomela (1997), and especially Tuomela and Tuomela (2004) for a detailed discussion of I-mode cooperation.

analogue of this condition with a "contingently" true equivalence may still hold true, where the equivalence is not true on the ground that the participants accept it as true.) (*COI*) by itself is rather weak but it does exclude cases of conflict. Although it deals with cases of dependence, the participants are required not to engage in conflict-involving behavior on purpose.

Recall from Section IV that a coordination problem can be solved in terms of individual commitments to the same goal. That is, we get a solution when the participants have succeeded in coordinating successfully (viz. selected one of the action pairs leading to coordination). Accordingly, (*COI*) entails that the coordination problem in question has been coordinatively (and thus cooperatively) solved. Conflict can be introduced into the situation by assuming that the agents have different preferences concerning where to meet, although they still prefer to meet rather than not. Here the conflict is not disturbingly big, and this modified coordination case (a "Battle of the Sexes" situation) qualifies as I-mode cooperation in the sense explicated by (*CO*). Competitive cases and "zero-sum" cases with strongly conflicting means actions do not belong to I-mode cooperation in our present sense. (*COI*) can accordingly be regarded as a general schema for I-mode cooperation. Within it we have the central case of a shared divided goal in which G_1 and G_2 are identical as types of goals (viz. $G_1 = G_2$).

Cooperating in the sense of I-mode cooperation can help agents to resolve collective action dilemmas such as the Prisoner's dilemma in a collectively rewarding way (in Pareto's sense). However, as I have argued, acting for *collective* reasons (e.g. towards a shared we-mode collective goal) will in general be needed here. For instance, Hume's farmer's dilemma and similar "centipede" type of dilemmas applying e.g. to conditional promising and various cases of reciprocal revenge-taking can at least under some conditions be resolved only when acting for a collective reason (not necessarily a we-mode reason, however). In the farmer's dilemma the farmers' crops will have to be harvested at consecutive times and harvesting requires the other's help. Why should the first one, having received help, bother to help? The collective reason needed for the solution of a centipede situation can be either an I-mode reason or a we-mode reason.[20]

---

[20]Here is the famous quotation from Hume on the farmer's dilemma:

> "Your corn is ripe today; mine will be so tomorrow. 'Tis profitable for us both that I shou'd labour with you today, and that you shou'd aid me tomorrow. I have no kindness for you, and know that you have as little for me. I will not, therefore, take any pains on your account; and should

References

Bratman M. (1992). "Shared Cooperative Activity", *The Philosophical Review 101*, 327–341.

Bowles S. (1998). "Endogenous Preferences: The Cultural Consequences of Markets and other Economic Institutions", *Journal of Economic Literature XXXVI*, 75–111.

Harsanyi J. (1977). *Rational Behavior and Bargaining Equilibrium in Games and Social Situations.* Cambridge, Mass.: Cambridge University Press.

Hume D. (1965) (orig. 1740). *A Treatise of Human Nature*, L.A. Selby-Bigge (ed.). Oxford: The Clarendon Press.

Lewis D. (1969). *Convention, A Philosophical Study.* Cambridge, Mass: Harvard University Press.

Moulin H. (1995). *Cooperative Microeconomics: A Game-Theoretic Introduction.* Princeton: Princeton University Press.

Pettit P. (1997). *Republicanism: A Theory of Freedom and Government.* Oxford: Oxford University Press.

Rawls J. (1993). *Political Liberalism.* New York: Columbia University Press.

Roemer J. (1996). *Theories of Distributive Justice.* Cambridge, Mass.: Harvard University Press.

Tuomela R. (1985a). "The Components of Social Control", *Quality and Quantity 19*, 1–51.

Tuomela R. (1993). "What is Cooperation?", *Erkenntnis 38*, 87–101.

Tuomela R. (1995). *The Importance of Us: A Philosophical Study of Basic Social Notions.* Stanford Series in Philosophy, Stanford University Press

Tuomela R. (2000). *Cooperation: A Philosophical Study.* Philosophical Studies Series, Dordrecht and Boston: Kluwer Academic Publishers.

Tuomela R. (2002a). *The Philosophy of Social Practices: A Collective Acceptance View.* Cambridge: Cambridge University Press.

Tuomela R. (2002b). "The We-mode and the I-mode", in F. Schmitt (ed.), *Socializing Metaphysics: The Nature of Social Reality* (2003). Lanham, Md: Rowman and Littlefield, pp. 93–127.

I labour with you on my account, I know I shou'd be disappointed, and that I shou'd in vain depend upon your gratitude. Here then I leave you to labour alone: You treat me in the same manner. The seasons change; and both of us lose our harvests for want of mutual confidence and security." (David Hume, 1965, III, part II; sec. v.)

See the detailed discussion of the centipede in Tuomela (2000), Chapter 11.

Tuomela R. (2002c). "Joint Intention and Commitment", in Meggle G. (ed.), *Social Facts & Collective Intentionality*, in German Library of Sciences, *Philosophical Research, 1*: 385–418. Frankfurt: Dr. Hänsel-Hohenhausen AG.

Tuomela R. and Bonnevier-Tuomela M. (1997). "From Social Imitation to Teamwork", in Holmstrom-Hintikka, G. and Tuomela R. (eds.), *Contemporary Action Theory, Vol. II: Social Action.* Dordrecht and Boston: Kluwer Academic Publishers, pp. 1–47.

Tuomela R. and Tuomela M. (2004), "Cooperation and Trust in Group Context", ms, available at `www.valt.helsinki.fi/staff/tuomela`.

Tyler T. and Blader S. (2000). *Cooperation in Groups.* Philadelphia: Psychology Press.

Van Vugt M., Snyder M., Tyler T. and Biel A. (2000). *Cooperation in Modern Society.* London: Routledge.

Chapter 5

SPEECH ACTS AND ILLOCUTIONARY LOGIC*

John R. Searle[1] and Daniel Vanderveken[2]

[1] *University of California, Berkeley*

[2] *University of Québec, Trois-Rivières*

1. Illocutionary acts and illocutionary logic.

The minimal units of human communication are speech acts of a type called *illocutionary acts.*[1] Some examples of these are statements, questions, commands, promises, and apologies. Whenever a speaker utters a sentence in an appropriate context with certain intentions, he performs one or more illocutionary acts. In general an illocutionary act consists of an illocutionary force F and a propositional content P. For example, the two utterances "You will leave the room" and "Leave the room!" have the same propositional content, namely that you will leave the room; but characteristically the first of these has the illocutionary force of a prediction and the second has the illocutionary force of an order. Similarly, the two utterances "Are you going to the movies?" and "When will you see John?" both characteristically have the illocutionary force of questions but have different propositional contents. Illocutionary logic is the

---

*This paper is Chapter 1 Introduction to the Theory of Speech Acts of John Searle & Daniel Vanderveken *Foundations of Illocutionary Logic* (Cambridge University Press, 1985). We thank Cambridge University Press for granting permission to republish that chapter in the present volume. The theory which follows is based on and is a development of the theory expressed in J. R. Searle, *Speech Acts* (Cambridge University Press, 1969), and *Expression and Meaning* (Cambridge University Press, 1979). There is a proof and model-theoretical formalization of a unified intensional and illocutionary logic in D. Vanderveken, *Meaning and Speech Acts,* Volume 1 *Principles of Language Use*, Volume 2 *Formal Semantics of Success and Satisfaction* (Cambridge University Press, 1990-91).

[1]The term is due to J. L. Austin, *How to Do Things with Words* (Oxford: Clarendon Press, 1962).

D. Vanderveken (ed.), Logic, Thought & Action, 109–132.

logical theory of illocutionary acts. Its main objective is to formalize the logical properties of illocutionary forces. Illocutionary forces are realized in the syntax of actual natural languages in a variety of ways, e.g. mood, punctuation, word-order, intonation contour, and stress, among others; and it is a task for empirical linguistics to study such devices as they function in actual languages. The task of illocutionary logic, on the other hand, is to study the entire range of possible illocutionary forces however these may be realized in particular natural languages. In principle it studies all possible illocutionary forces of utterances in any possible language, and not merely the actual realization of these possibilities in actual speech acts in actual languages. Just as propositional logic studies the properties of all truth functions (e.g. conjunction, material implication, negation) without worrying about the various ways that these are realized in the syntax of English ("and", "but", and "moreover", to mention just a few for conjunction), so illocutionary logic studies the properties of illocutionary forces (e.g. assertion, conjecture, promise) without worrying about the various ways that these are realized in the syntax of English ("assert", "state", "claim", and the indicative mood, to mention just a few for assertion) and without worrying whether these features translate into other languages. No matter whether and how an illocutionary act is performed, it has a certain logical form which determines its conditions of success and relates it to other speech acts. We will try to characterize that form independently of the various forms of expression that may exist in actual natural languages for the expression of the act. However, though the results of our investigation are in general independent of empirical linguistic facts, the method of the investigation will require us to pay close attention to the facts of natural languages, and the results should help us to analyze actual performative verbs and other illocutionary force indicating devices of natural languages. In Chapter 9 we will apply our results to the analysis of English illocutionary verbs.

Any element of a natural language which can be literally used to indicate that an utterance of a sentence containing that element has a certain illocutionary force or range of illocutionary forces we will call an *illocutionary force indicating device.* Some examples of illocutionary force indicating devices are word order and mood as in: (i) "Will you leave the room?", (2) "You, leave the room!", (3) "You will leave the room", (4) "If only you would leave the room!" In each of these examples, there is some syntactical feature which, given the rest of the sentence and a certain context of utterance, expresses an illocutionary force F, and some syntactical feature p which, given the rest of the sentence and a context of utterance, expresses a propositional content P.

From the point of view of the theory of speech acts, then, the general form of such simple sentences, which express elementary illocutionary acts of form $F(P)$, is $f(p)$. We will call these elementary sentences.

A special class of elementary sentences are the *performative sentences.* These consist of a performative verb used in the first person present tense of the indicative mood with an appropriate complement clause. In uttering a performative sentence a speaker performs the illocutionary act with the illocutionary force named by the performative verb by way of representing himself as performing that act. Some examples of performative sentences (with the performative verbs italicized) are: (5) "I *promise* that I will come tomorrow", (6) "I *apologize* for what I have done", (7) "I *order* you to report to the commanding officer", (8) "I *admit* that I committed the crime." There has been a great deal of philosophical controversy concerning the proper analysis of performative sentences. The two most widely held views are: First, that the performative element in the sentence functions simply as an illocutionary force indicating device on all fours with other devices, such as word order. On this view an utterance of a sentence such as (5) consists simply in the making of a promise. Secondly, that all utterances of performative sentences are statements, and thus for example in utterances of (5), a speaker makes a promise only by way of making a true statement to the effect that he promises. On the first view, performative utterances such as (5) do not have truth values; on the second view they do. In this paper we will try a third approach, according to which performative utterances are *declarations* whose propositional content is that the speaker performs the illocutionary act named by the performative verb. On this account, the illocutionary force of a performative sentence is always that of a declaration, and then, derivatively, the utterance has the additional force named by the performative verb. Since the defining trait of a declaration[2] is that it actually brings about the state of affairs represented by its propositional content, and since the propositional content of a performative utterance is that the speaker performs a certain sort of illocutionary act, the successful declaration that a speaker performs that act will always constitute its performance.

Not all illocutionary acts are of the simple $F(P)$ form. More complex cases we will call *complex illocutionary acts* and the sentences used to express them *complex sentences.* Complex sentences are composed of simple sentences using connectives that we will call *illocutionary connectives.* For example, the connectives of conjunction ("and", "but")

[2]See J. R. Searle, 'A taxonomy of illocutionary acts', in *Expression and Meaning*, pp. 1-29.

enable speakers to conjoin different illocutionary acts in one utterance. In general, the utterance of a sentence which is the conjunction of two sentences constitutes the performance of the two illocutionary acts expressed by the two sentences. Thus in a certain context by uttering (9) "I will go to his house, but will he be there?", a speaker both makes an assertion and asks a question. This conjunction of two illocutionary acts constitutes the performance of a complex illocutionary act whose logical form is $(F_1(P_1) \& F_2(P_2))$. The illocutionary connective of conjunction is "success functional" in the sense that the successful performance of a complex illocutionary act of form $(F_1(P_1) \& F_2(P_2))$ is a function of the successful performances of its constituents. Not every pair of sentences will grammatically admit every illocutionary connective. For example, the following conjunction is syntactically ill formed in English: (10) "When did John come and I order you to leave the room?"

Another type of complex illocutionary act involves the negation of the illocutionary force, and we will call these acts of *illocutionary denegation.* It is essential to distinguish between acts of illocutionary denegation and illocutionary acts with a negative propositional content, between, for example, (11) "I do not promise to come" and (12) "I promise not to come." The utterance of (11) is typically an act of illocutionary denegation and it is of form $\neg F(P)$. The utterance of (12) by contrast is an illocutionary act with a negative propositional content and it is of the form $F(\sim P)$. We can say generally that an act of illocutionary denegation is one whose aim is to make it explicit that the speaker does not perform a certain illocutionary act.

The fact that illocutionary denegation is not success functional is shown by the fact that the non-performance of an illocutionary act is not the same as the performance of its illocutionary denegation; for example, from the fact that I did not make a promise, it does not follow that I declined or refused to make a promise. And the usual asymmetry between the first person present and other occurrences of performative verbs reveals the same phenomenon. A person's silence may be sufficient for somebody to say truly of him (13) "He did not promise." But a person's silence is not the same as the overt act of saying (14) "I do not promise." Most acts of illocutionary denegation are performed in English by way of negating a performative verb as in (11) but some, very few, verbs are explicit performatives for illocutionary denegation. "Permit" is the denegation of both "forbid" and "prohibit"; "refuse" is frequently used as the denegation of "accept" and "disclaim" as the denegation of "claim".

The conditionals "if" and "if... then" are also used as illocutionary connectives. A conditional speech act is a speech act which is performed on a certain condition; its characteristic forms of expression therefore are sentences of the form "If p then $f(q)$" and "If $p, f(q)$". Some examples are: (15) "If he comes, stay with me!", (16) "If it rains, I promise you I'll take my umbrella." It is essential to distinguish between a conditional speech act and a speech act whose propositional content is a conditional. In a conditional speech act expressed by a sentence of the form "If p then $f(q)$" the speech act expressed by "$f(q)$" is performed on condition p. Syntactically the "if" clause modifies the illocutionary force indicating device. This form is quite distinct from that of the speech act performed by an utterance of a sentence of the form "f(if p then q)" whose propositional content is conditional, for in this case an illocutionary act of force F is categorically performed. Thus, for example, in a bet on a conditional of the form (17) "I bet you five dollars that if a presidential candidate gets a majority of the electoral votes he will win" one either wins or loses five dollars depending on the truth or falsity of the conditional proposition (provided all the presuppositions hold). On the other hand, in a conditional bet of the form (18) "If Carter is the next Democratic candidate, I bet you five dollars that the Republicans will win", there is a winner or a loser only if Carter is the next Democratic candidate. The logical form of (18) is $P \rightarrow F(Q)$. This conditional is not truth-functional, for from the fact that Carter does not run for the presidency, it does not follow that every speaker performs a conditional bet of the form (18). Part of the task of illocutionary logic is to analyze illocutionary denegation and illocutionary conditionals.

In carrying out the general project of illocutionary logic some of the main questions we will attempt to answer are: (1) What are the components of illocutionary force and what are the necessary and sufficient conditions for the successful performance of elementary illocutionary acts? How can the conditions of success of complex illocutionary acts be defined in terms of the conditions of success of their constituent parts? (2) What is the logical structure of the set of all illocutionary forces? Is there a recursive definition of this set, i.e. can all illocutionary forces be obtained from a few primitive forces by applying certain operations and, if so, how? (3) What are the logical relations between the various types of illocutions? In particular, under which conditions does the successful performance of one illocutionary act commit the speaker to another illocutionary act?

A theory of the foundations of illocutionary logic capable of answering these questions should be able to characterize a set of logical laws governing illocutionary forces. Thus, for example, there are laws of distribution of illocutionary forces with respect to truth-functions, e.g. if a speaker succeeds in asserting a conjunction of two propositions (P and

Q) then he succeeds both in asserting P and in asserting Q. Furthermore, such a theory should explain the relations between illocutionary forces and intensionality, modalities, time, presuppositions, and indexicality. It should also explain the reasons why the utterances of certain sentences of natural language constitute self-defeating illocutionary acts. Self-defeating illocutionary acts have self-contradictory conditions of success and are thus odd semantically.[3] Some examples of sentences expressing self-defeating illocutions are: (19) "I promise you not to keep this promise", (20) "I assert that I do not make any assertion", (21) "Disobey this order!"

A theory of illocutionary logic of the sort we are describing is essentially a theory of illocutionary commitment as determined by illocutionary force. The single most important question it must answer is simply this: Given that a speaker in a certain context of utterance performs a successful illocutionary act of a certain form, what other illocutions does the performance of that act commit him to? To take the simplest sort of example, a speaker who warns a hearer that he is in danger is committed to the assertion that he is in danger. A speaker who denies a proposition P is committed to the denegation of an assertion that P. And, as is obvious from even these examples, we will need to distinguish between the overt performance of an illocutionary act and an illocutionary commitment. The overt performance of one illocutionary act may involve the speaker in a commitment to another illocution, even though that commitment does not involve a commitment to an overt performance of that illocution. Thus, for example, if I order you to leave the room I am committed to granting you permission to leave the room even though I have not performed an overt act of granting you permission and have not committed myself to performing any such overt act. Among other things, a logical theory of illocutionary acts will enable us to construct a formal semantics for the illocutionary force indicating devices of natural language.

Illocutionary logic is part of the overall project of logic, linguistics, and the philosophy of language for at least the following two reasons:

[3]For further discussion of self-defeating illocutionary acts, see D. Vanderveken: 'Illocutionary Logic and Self-Defeating Speech Acts', in Searle *et al.* (eds.), *Speech-Act Theory and Pragmatics* (Dordrecht, Netherlands: D. Reidel, 1980).

1.1 Illocutionary force is a component of meaning.

Part of the meaning of an elementary sentence is that its literal utterance in a given context constitutes the performance or attempted performance of an illocutionary act of a particular illocutionary force. Thus, for example, it is part.of the meaning of the English sentence, (22) "Is it raining?", that its successful literal and serious utterance constitutes the asking of a question as to whether it is raining. Every complete sentence, even a one-word sentence, has some indicator of illocutionary force; therefore, no semantical theory of language is complete without an illocutionary component. A materially adequate semantics of a natural language must recursively assign illocutionary acts (elementary or complex) to each sentence for each possible context of utterance. It is not sufficient for it simply to assign propositions or truth conditions to sentences. In order to assign illocutionary acts to sentences an illocutionary logic would need first to provide a semantic analysis of illocutionary verbs and other illocutionary force indicating devices found in actual natural languages. In the sense that it provides an analysis of the illocutionary aspects of sentence meaning, illocutionary logic is part of a theory of meaning.[4]

1.2 An adequate illocutionary logic is essential to an adequate universal grammar (in Montague's sense of 'universal grammar').[5]

Since illocutionary forces and propositions are two components of the meanings of elementary sentences, the ideal language of a universal grammar must contain logical constants and operators capable of generating names for all possible illocutionary forces of utterances. Any sentence in any natural language should be translatable into sentences of the ideal language of universal grammar, and those sentences must reflect the illocutionary potentiality of the natural language sentences. Up to the present time universal grammar has been mostly concerned with propositions, but it also needs to include an account of illocution-

[4]For further discussion, see D. Vanderveken, 'Pragmatique, sémantique et force illocutoire', *Philosophica*, vol. 27, no. I, 1981.

[5]See R. Montague (1970), "Universal Grammar", *Theoria 36.* The general semantics for natural language developed in D. Vanderveken *Formal Semantics of Success and Satisfaction* Volume 2 of *Meaning and Speech Acts* (Cambridge University Press, 1991) is a generalization and extension of Montague Grammar. Its ideal object language has richer expressive powers than that of Montague. It can express illocutionary forces as well as propositions.

ary forces, and therefore, it goes beyond the boundaries of intensional logic as traditionally conceived.

2. Illocutionary acts and other types of speech acts.

In order to prepare the way for a formalization of the theory of illocutionary acts we need first to clarify the relations between an illocutionary act and certain types of speech acts, specifically utterance acts, propositional acts, indirect speech acts, perlocutionary acts and conversations.

Just as the sentences used to perform elementary speech acts have the form $f(p)$, where f is the indicator of illocutionary force and p expresses the propositional content, so we can say that the illocutionary act itself has the logical form $F(P)$, where the capital F stands for the illocutionary force, and P for the propositional content. The distinction between illocutionary force and propositional content, as was suggested by our earlier remarks, is motivated by the fact that their identity conditions are different: the same propositional content can occur with different illocutionary forces and the same force can occur with different propositional contents. The character of the whole illocutionary act is entirely determined by the nature of its illocutionary force and propositional content. This distinction also motivates the introduction of another speech act notion, that of the propositional act.

In the performance of an illocutionary act the speaker performs the subsidiary act of expressing the propositional content and this act we will call the *propositional act.* A propositional act is an abstraction from the total illocutionary act in the sense that the speaker cannot simply express a proposition and do nothing more. The performance of the propositional act always occurs as part of the performance of the total illocutionary act. Syntactically this fact is reflected in natural languages by the fact that "that" clauses, the characteristic form of isolating the propositional content, cannot stand alone; they do not make complete sentences. One can say "I promise that I will leave the room", but one cannot say simply "That I will leave the room".

Some, but not many, types of illocutionary forces permit a content that does not consist of an entire proposition but only a reference, as in an utterance of "Hurrah for the Raiders!" Such an utterance does not have the form $F(P)$ but rather $F(u)$ where u is some entity of the universe of discourse. And some permit an utterance consisting only of an illocutionary force and no propositional content, e.g. "Hurrah", "Ouch", and "Damn". These utterances simply have the form F. With these very few sorts of exception, all illocutionary acts have a propositional content and hence (with such exceptions) all performances of illocutionary acts are performances of propositional acts.

Illocutionary acts are performed by the utterance of expressions, and this fact motivates the introduction of yet another speech act notion, that of the *utterance act*: an utterance act consists simply in the utterance of an expression. One can perform the same illocutionary act in the performance of two different utterance acts, as, for example, when one says either "It's raining" in English or "Il pleut" in French; or even in the same language, when, for example, one uses synonymous sentences, as one may say either "John loves Mary" or "Mary is loved by John" to perform the same illocutionary act. Furthermore, an utterance act can be performed without performing an illocutionary act, as, for example, when one simply mouths words without saying anything. And finally, the same utterance act type can occur in the performance of different illocutionary acts. For example, if Bill says "I am hungry" and John says "I am hungry", in the two token utterances the same utterance act type is performed but two different illocutionary acts are performed, since the reference and hence the proposition is different in the two cases.

This account of the general form of the illocutionary act and the relation of its performance to that of propositional and utterance acts can be summarized as follows. In the utterance of a sentence of the form $f(p)$ the speaker performs an utterance act. If the utterance is in certain ways appropriate he will have expressed the proposition that P (which proposition is a function of the meaning of p), and he will thereby have performed a propositional act. If certain further conditions are satisfied he will have expressed that proposition with the illocutionary force F (which force is a function of the meaning of f) and he will thereby have expressed an illocutionary act of the form $F(P)$. Furthermore, if the conditions of success of that act obtain, he will thereby have successfully performed that act.

Often speakers perform one illocutionary act implicitly by way of performing another illocutionary act explicitly. The explicitly performed act is used to convey another speech act; and the speaker relies on background knowledge and mental capacities that he shares with the hearer in order to achieve understanding. So, for example, if someone on the street says to you, "Do you know the way to the Palace Hotel?", it would be in most contexts inappropriate to respond simply "yes" or "no", because the speaker is doing more than just *asking a question* about your knowledge: he is *requesting* that you tell him the way to the hotel. Similarly, if a man says to you, "Sir, you are standing on my foot", the chances are he is doing more than describing your location: he is requesting you to get off his foot. In these cases two speech acts are involved: the non-literal primary speech act ("Tell me the way to the Palace Hotel!", "Get off my foot!") is performed indirectly by way of performing a literal secondary speech act ("Do you know the way to the Palace Hotel?"; "Sir, you are standing on my foot"). Such implicit acts are called *indirect speech acts.*[6] The speaker may convey indirectly a different illocutionary force or propositional content from what is directly expressed; hence in one utterance act he may perform one or more non-literal indirect illocutionary acts.

Just as indirect speech acts are quite pervasive in real life, so in real life illocutionary acts seldom occur alone but rather occur as parts of conversations or larger stretches of discourse. Traditional linguistics tends to construe a speaker's linguistic competence as a matter of his ability to produce and understand sentences; and traditional speech act theory tends to construe each illocutionary-act as an isolated unit. But we will not get an adequate account of linguistic competence or of speech acts until we can describe the speaker's ability to produce and understand utterances (i.e. to perform and understand illocutionary acts) In *ordered speech act sequences* that constitute arguments, discussions, buying and selling, exchanging letters, making jokes, etc. For terminological convenience we will call these ordered sequences simply *conversations.* The key to understanding the structure of conversations is to see that each illocutionary act creates the possibility of a finite and usually quite limited set of appropriate illocutionary acts as replies. Sometimes the appropriate illocutionary act reply is very tightly constrained by the act that precedes it, as in question and answer sequences; and sometimes it is more open, as in casual conversations that move from one topic to another. But the principle remains that just as a move in a game creates and restricts the range of appropriate countermoves so each illocutionary act in a conversation creates and constrains the range of appropriate illocutionary responses.

When an illocutionary act is successfully and nondefectively performed there will always be an effect produced in the hearer, the effect of understanding the utterance. But in addition to the illocutionary effect of understanding, utterances normally produce, and are often intended to produce, further effects on the feelings, attitudes, and subsequent behavior of the hearers. These effects are called *perlocutionary effects*[7] and the acts of producing them are called *perlocutionary acts.* For ex-

[6]J. R. Searle, "Indirect Speech Acts", in *Expression and Meaning*, pp. 30-57; and H. P. Grice, "Logic and conversation", in P. Cole and J. L. Morgan (eds.). *Syntax and Semantics*, vol. 3, *Speech Acts* (New York: Academic Press, 1975).

[7]Following Austin, *How to Do Things with Words.*

ample, by making a statement (illocutionary) a speaker may convince or persuade (perlocutionary) his audience, by making a promise (illocutionary) he may reassure or create expectations (perlocutionary) in his audience. Perlocutionary effects may be achieved intentionally, as, for example, when one gets one's hearer to do something by asking him to do it, or unintentionally, as when one annoys or exasperates one's audience without intending to do so.

Perlocutionary acts, unlike illocutionary acts, are not essentially linguistic, for it is possible to achieve perlocutionary effects without performing any speech act at all. Since illocutionary acts have to do with understanding they are conventionalizable. It is in general possible to have a linguistic convention that determines that such and such an utterance counts as the performance of an illocutionary act. But since perlocutionary acts have to do with subsequent effects, this is not possible for them. There could not be any convention to the effect that such and such an utterance counts as convincing you, or persuading you, or annoying you, or exasperating you, or amusing you. And that is why none of these perlocutionary verbs has a performative use. There could not, for example, be a performative expression "I hereby persuade you", because there is no way that a conventional performance can guarantee that you are persuaded, whereas there are performative expressions of the form "I hereby state" or "I hereby inform you", because there can be conventions whereby such and such counts as a statement or counts as informing you. It is essential to keep this distinction clear in what follows, for we will be investigating speech acts proper that is, illocutionary acts. Perlocutionary acts will figure only incidentally in our discussions.

3. The seven components of illocutionary force.

The study of illocutionary logic is mainly the study of the illocutionary forces of utterances. We therefore need to analyze the notion of illocutionary force into its component elements. On our analysis there are seven interrelated components of illocutionary force, and in this section we will provide an informal explanation and definition of these seven components and of the ways in which they are interrelated. The formalization will be presented in subsequent chapters.

One way to understand the notion of an illocutionary act is in terms of the notion of the conditions of its successful and non-defective performance. Illocutionary acts, like all human acts, can succeed or fail. An act of excommunication, for example, can be successful only if the speaker has the institutional power to excommunicate someone by his utterance. Otherwise, it is a complete failure. Just as any adequate talk of propositions involves the pair of concepts truth and falsity, so any adequate talk of speech acts (and of acts in general) involves the pair of concepts success and failure. And even when they succeed, illocutionary acts are subject to various faults and defects, such as insincerity or failure of presuppositions. We therefore have the following three possibilities: a speech act may be unsuccessful, it may be successful but defective, and it may be successful and nondefective. For example, if one of us now attempts to excommunicate the other by saying "I hereby excommunicate you" the speech act will be totally unsuccessful. The various conditions necessary for such an utterance to be a successful excommunication do not obtain. But if one of us now makes a statement for which he has hopelessly insufficient evidence or warrant, he might succeed in making the statement; however, it would be defective, because of his lack of evidence. In such a case the speech act is successful but defective. Austin's distinction between "felicitous" and "infelicitous" speech acts fails to distinguish between those speech acts which are successful but defective and those which are not even successful, and for this reason we do not use his terminology,, but instead use the terminology of *Speech Acts.*[8] In the ideal case, a speech act is both successful and nondefective, and for each illocutionary force the components of that illocutionary force serve to determine under what conditions that type of speech act is both *successful* and *nondefective*, at least as far as its illocutionary force is concerned. In this section we will present the seven components in a way which will make clear how they determine the conditions of successful and nondefective performance of illocutions.

3.1 Illocutionary point.

Each type of illocution has a point or purpose which is internal to its being an act of that type. The point of statements and descriptions is to tell people how things are, the point of promises and vows is to commit the speaker to doing something, the point of orders and commands is to try to get people to do things, and so on. Each of these points or purposes we will call the *illocutionary point* of the corresponding act. By saying that the illocutionary point is internal to the type of illocutionary act, we mean simply that a successful performance of an act of that type necessarily achieves that purpose and it achieves it in virtue of being an act of that type. It could not be a successful act of that type if it did not achieve that purpose. In real life a person may have atl sorts of other

[8]Searle, *Speech Acts* (1969).

purposes and aims; e.g. in making a promise, he may want to reassure his hearer, keep the conversation going, or try to appear to be clever, and none of these is part of the essence of promising. But when he makes a promise he necessarily commits himself to doing something. Other aims are up to him, none of them is internal to the fact that the utterance is a promise; but if he successfully performs the act of making a promise then he necessarily commits himself to doing something, because that is the illocutionary point of the illocutionary act of promising.

In general we can say that the illocutionary point of a type of illocutionary act is that purpose which is essential to its being an act of that type. This has the consequence that if the act is successful the point is achieved. Some characteristic illocutionary points are the following: The illocutionary point of a promise to do act A is to commit the speaker to doing A. The illocutionary point of an apology for having done act A is to express the speaker's sorrow or regret for having done A. The illocutionary point of issuing a declaration that P (e.g. a declaration of war) is to bring about the state of affairs that P represents.

Illocutionary point is only one component of illocutionary force, but it is by far the most important component. That it is not the only component is shown by the fact that different illocutionary forces can have the same illocutionary point, as in the pairs assertion/testimony, order/request and promise/vow. In each pair both illocutionary forces have the same point but differ in other respects. The other elements of illocutionary force are further specifications and modifications of the illocutionary point or they are consequences of the illocutionary point, but the basic component of illocutionary force is illocutionary point.

In the performance of an act of form $F(P)$ the illocutionary point is distinct from the propositional content, but it is achieved only as part of a total speech act in which the propositional content is expressed with the illocutionary point. We will say therefore that the *illocutionary point is achieved on the propositional content.* A speaker can be committed to an illocutionary point that he does not explicitly achieve. Thus, for example, if he promises to carry out a future course of action he is committed to the illocutionary point of the assertion that he will carry out that course of action, even though he may not have explicitly asserted that he will do it.

3.2 Degree of strength of the illocutionary point.

Different illocutionary acts often achieve the same illocutionary point with different degrees of strength. For example, if I *request* someone to do something my attempt to get him to do it is less strong than if I *insist* that he do it. If I *suggest* that something is the case the degree of strength of my representation that it is the case is less than if I *solemnly swear* that it is the case. If I *express regret* for having done something my utterance has a lesser degree of strength than if I *humbly apologize* for having done it. For each type of illocutionary force F whose illocutionary point requires that it be achieved with a certain degree of strength, we will call that degree of strength the *characteristic degree of strength* of illocutionary point of F. There are different sources of different degrees of strength. For example, both pleading and ordering are stronger than requesting, but the greater strength of pleading derives from the intensity of the desire expressed, while the greater strength of ordering derives from the fact that the speaker uses a position of power or authority that he has over the hearer.

3.3 Mode of achievement.

Some, but not all, illocutionary acts require a special way or special set of conditions under which their illocutionary point has to be achieved in the performance of the speech act. For example, a speaker who issues a command from a position of authority does more than someone who makes a request. Both utterances have the same illocutionary point, but the command achieves that illocutionary point by way of invoking the position of authority of the speaker. In order that the utterance be a successful command the speaker must not only be in a position of authority; he must be using or invoking his authority in issuing the utterance. Analogously a person who makes a statement in his capacity as a witness in a court trial does not merely make a statement, but he *testifies*, and his status as a witness is what makes his utterance count as testimony. These features which distinguish respectively commanding and testifying from requesting and asserting we will call *modes of achievement* of their illocutionary points. When an illocutionary force F requires a special mode of achievement of its point we will call that mode the *characteristic mode of achievement* of illocutionary point of F. Sometimes degree of strength and mode of achievement are interdependent. For example, the characteristic mode of achievement of a command will give it a greater characteristic degree of strength of illocutionary point than that of a request.

3.4 Propositional content conditions.

We have seen that the form of most illocutionary acts is $F(P)$. In many cases the type of force F will impose certain conditions on what can be in the propositional content P. For example, if a speaker makes a promise, the content of the promise must be that the speaker will perform some future course of action. One cannot promise that someone else will do something (though one can promise to *see to it* that he does it) and one cannot promise to have clone something in the past. Similarly if a speaker apologizes for something it must be for something that he has done or is otherwise responsible for. A speaker cannot successfully apologize for the law of *modus ponens* or the elliptical orbit of the planets, for example. Such conditions on the propositional content which are imposed by the illocutionary force we will call *propositional content conditions.* These conditions obviously have syntactic consequences: sentences such as "I order you to have eaten beans last week" are linguistically odd.

3.5 Preparatory conditions.

For most types of illocutionary acts, the act can be both successful and nondefective only if certain other conditions obtain. For example, a promise might be successfully made and so have achieved its illocutionary point but it would still be defective if the thing the speaker promised to do was not in the hearer's interest and the hearer did not want him to do it. In making a promise the speaker presupposes that he can do the promised act and that it is in the hearer's interest to do it. Similarly if a speaker apologizes he presupposes that the thing he apologizes for is bad or reprehensible. Such conditions which are necessary for the successful and nondefective performance of an illocutionary act we call *preparatory conditions.* In the performance of a speech act the speaker *presupposes* the satisfaction of all the preparatory conditions. But this does not imply that preparatory conditions are psychological states of the speaker, rather they are certain sorts of states of affairs that have to obtain in order that the act be successful and non-defective. Speakers and hearers internalize the rules that determine preparatory conditions and thus the rules are reflected in the psychology of speakers/hearers. But the states of affairs specified by the rules need not themselves be psychological.

Preparatory conditions determine a class of presuppositions peculiar to illocutionary force. But there is another class of presuppositions peculiar to propositional content. To take some famous examples, the assertion that the King of France is bald presupposes that there exists a King of France; and the question whether you have stopped beating your wife presupposes both that you have a wife and that you have been beating her. Regardless of which of the various philosophical accounts one accepts of these sorts of presuppositions, one needs to distinguish them from those that derive from illocutionary forces. The same propositional presuppositions can occur with different illocutionary forces, as, for example, one can both ask whether and one can assert that Jones has stopped beating his wife.

As we noted earlier a speech act can be successfully, though defectively, performed when certain preparatory conditions are unsatisfied. Even in such cases, the presupposition of the preparatory conditions is internal to the performance of the illocutionary act, as is shown by the fact that it is paradoxical to perform the act and deny that one of the preparatory conditions is satisfied. One cannot, for example, consistently make a promise while denying that one is able to do the act promised.

Many preparatory conditions are determined by illocutionary point. For example, all acts whose point is to get the hearer to do something – orders, requests, commands, etc. – have as a preparatory condition that the hearer is able to do the act directed. But some preparatory conditions are peculiar to certain illocutionary forces. For example, a promise differs from a threat in that the act promised must be for the hearer's benefit. Preparatory conditions and mode of achievement are connected in that normally certain preparatory conditions must obtain in order that an illocutionary act can be performed with its characteristic mode of achievement. For example, a speaker must satisfy the preparatory condition of being in a position of authority before he can non-defectively issue an utterance with the mode of achievement of a command.

3.6 Sincerity conditions.

Whenever one performs an illocutionary act with a propositional content one expresses a certain psychological state with that same content. Thus when one makes a statement one expresses a belief, when one makes a promise one expresses an intention, when one issues a command one expresses a desire or want. The propositional content of the illocutionary act is in general identical with the propositional content of the expressed psychological state.

It is always possible to express a psychological state that one does not have, and that is how sincerity and insincerity in speech acts are distinguished. An insincere speech act is one in which the speaker performs a speech act and thereby expresses a psychological state even though he does not have that state. Thus an insincere statement (a lie) is one where the speaker does not believe what he says, an insincere apology is one where the speaker does not have the sorrow he expresses, an insincere promise is one where the speaker does not in fact intend to do the things he promises to do. An insincere speech act is defective but not necessarily unsuccessful. A lie, for example, can be a successful assertion. Nevertheless, successful performances of illocutionary acts necessarily involve the expression of the psychological state specified by the sincerity conditions of that type of act.

The fact that the expression of the psychological state is internal to the performance of the illocution is shown by the fact that it is paradoxical to perform an illocution and to deny simultaneously that one has the corresponding psychological state. Thus, one cannot say "I promise to come but I do not intend to come", "I order you to leave but I don't want you to leave", "I apologize but I am not sorry", etc. And this incidentally explains Moore's paradox that one cannot say consistently "It is raining but I don't believe that it is raining" even though the proposition that it is raining is consistent with the proposition that I do not believe that it is raining. The reason for this is that when one performs the speech act one necessarily expresses the sincerity condition, and thus to conjoin the performance of the speech act with the denial of the sincerity condition would be to express and to deny the presence of one and the same psychological state.

Just as the performance of an illocution can commit the speaker to an illocution that he has not performed, so the expression of a psychological state in the performance of an illocution can commit him to having a state he has not expressed. Thus, for example, a speaker who expresses a belief that P and a belief that if P then Q is committed to having the belief that Q, The expression of a state commits the speaker to having that state; and one can be committed to having a state without actually having it.

The verb "express", by the way, is notoriously ambiguous. In one sense a speaker is said to express propositions and in another to express his feelings and attitudes such as fear, belief, or desire. In this discussion of the sincerity conditions of speech acts we are using it in this second sense, which should not be confused with the first. Both senses of "express" are used throughout this book and we believe the contexts will make it clear in each case which sense is intended.

3.7 Degree of strength of the sincerity conditions.

Just as the same illocutionary point can be achieved with different degrees of strength, so the same psychological state can be expressed with different degrees of strength. The speaker who makes a request expresses the desire that the hearer do the act requested; but if he *begs*, *beseeches*, or *implores*, he expresses a stronger desire than if he merely requests. Often, but not always, the degree of strength of the sincerity conditions and the degree of strength of the illocutionary point vary directly, as in the above examples. But an order, for example, has a greater degree of strength of its illocutionary point than a request, even though it need not have a greater degree of strength of its expressed psychological state. The greater degree of strength of the illocutionary point of ordering derives from the mode of achievement. The person who gives an order must invoke his position of power or authority over the hearer in issuing the order.

In cases where illocutionary force requires that the psychological state be expressed with a degree of strength, we will call that degree of strength *the characteristic degree of strength* of the sincerity condition.

4. Definitions of illocutionary force and related notions.[10]

4.1 Definition of the notion of illocutionary force.

Our discussion so far of the components of illocutionary force enables us to define the notion of illocutionary force as follows: An illocutionary force is uniquely determined once its illocutionary point, its preparatory conditions, the mode of achievement of its illocutionary point, the degree of strength of its illocutionary point, its propositional content conditions, its sincerity conditions, and the degree of strength of its sincerity conditions are specified. So two illocutionary forces F_1 and F_2 are identical when they are the same with respect to these seven features. To illustrate these points, here are a few examples of illocutionary forces that differ in (at least) one aspect from the illocutionary force of assertion. The illocutionary force of the testimony of a witness differs from assertion in that a speaker who testifies acts in his status as a witness when he represents a state of affairs as actual. (This is a special mode of achievement that is specific to testimony.) The illocutionary force of a conjecture differs from assertion in that the speaker who conjectures commits himself to the truth of the propositional content with a weaker degree of strength than the degree of commitment to truth of an assertion. The illocutionary force of a prediction differs from assertion in that it has a special condition on the propositional content. The

[10] These definitions are in Vanderveken, "Illocutionary Logic and Self-Defeating Speech Acts".

propositional content of a prediction must be future with respect to the time of the utterance. The illocutionary force of reminding (that P) differs from assertion only in that it has the additional preparatory condition that the hearer once knew and might have forgotten the truth of the propositional content. The illocutionary force of complaining differs from assertion in that it has the additional sincerity condition that the speaker is dissatisfied with the state of affairs represented by the propositional content.[11]

4.2 Definition of a successful and nondefective performance of an elementary illocutionary act.

Whether or not an utterance has a certain force is a matter of the illocutionary intentions of the speaker, but whether or not an illocutionary act with that force is successfully and nondefectively performed involves a good deal more than just his intentions; it involves a set of further conditions which must be satisfied. Prominent among these conditions are those that have to do with achieving what Austin called "illocutionary uptake".[12] The conditions for correctly understanding an utterance normally involve such diverse things as that the hearer must be awake, must share a common language with the speaker, must be paying attention, etc. Since these conditions for understanding are of little theoretical interest in a theory of speech acts, we will simply henceforth assume that they are satisfied when the utterance is made; and we will concentrate on the speaker and on how his utterance satisfies the other conditions on successful and nondefective performance. The seven features of illocutionary force that we have specified reduce to four different types of necessary and sufficient conditions for the successful and nondefective performance of an elementary illocution. Assuming that all the conditions necessary and sufficient for hearer understanding are satisfied

[11] Additional note of the editor. Searle and Vanderveken formulate in chapter 3 of *Foundations of Illocutionary Logic* the following recursive definition of the set of all possible illocutionary forces on the basis of their analysis of the notion of illocutionary force into components. According to them, there are five and only five basic illocutionary points: the assertive, commissive, directive, declaratory and expressive illocutionary points. So there are five and only *five primitive illocutionary forces* of utterances in the logical structure of language. These are the simplest possible illocutionary forces with a given illocutionary point: they have that illocutionary point, no special mode of achievement of that point, neutral degrees of strength and only the propositional content, preparatory and sincerity conditions which are determined by their point. All other illocutionary forces are derived from these five primitive illocutionary forces by a finite number of applications of operations which consist in adding new components or in increasing or decreasing the degrees of strength.

[12] *How to Do Things with Words.*

when the utterance is made, an illocutionary act of the form $F(P)$ is successfully and nondefectively performed in a context of utterance iff:

1) The speaker succeeds in achieving in that context the illocutionary point of F on the proposition P with the required characteristic mode of achievement and degree of strength of illocutionary point of F.

2) He expresses the proposition P, and that proposition satisfies the propositional content conditions imposed by F.

3) The preparatory conditions of the illocution and the propositional presuppositions obtain in the world of the utterance, and the speaker presupposes that they obtain.

4) He expresses and possesses the psychological state determined by F with the characteristic degree of strength of the sincerity conditions of F.

For example, in the performance of a particular utterance act, a speaker succeeds in issuing a nondefective command to the hearer iff:

1) The point of his utterance is to attempt to get the hearer to do an act A. (illocutionary point). This attempt is made by invoking his position of authority over the hearer (mode of achievement), and with a strong degree of strength of illocutionary point (degree of strength).

2) He expresses the proposition that the hearer will perform a future act A. (propositional content condition).

3) He presupposes both that he is in a position of authority over the hearer with regard to A. and that the hearer is able to do A. He also presupposes all of the propositional presuppositions if there are any. And all his presuppositions, both illocutionary and propositional, in fact obtain (preparatory conditions and propositional presuppositions).

4) He expresses and actually has a desire that the hearer do A (sincerity condition) with a medium degree of strength (degree of strength).

As we remarked earlier, a speech act can be *successful* though *defective.* A speaker might, actually succeed in making a statement or a promise even though he made a mess of it in various ways. He might, for example, not have enough evidence for his statement or his promise might be insincere. An ideal speech act is one which is both successful and nondefective. Nondefectiveness implies success, but not conversely. In our view there are only two ways that an act can be successfully performed though still be defective. First, some of the preparatory conditions might not obtain and yet the act might still be performed. This possibility holds only for some, but not all, preparatory conditions. Second, the sincerity conditions might not obtain, i.e. the act can be successfully performed even though it be insincere.

4.3 Definition of illocutionary commitment.

The idea behind the notion of illocutionary commitment is simply this: sometimes by performing one illocutionary act a speaker can be committed to another illocution. This occurs both in cases where the performance of one act by a speaker is *eo ipso* a performance of the other and in cases where the performance of the one is not a performance of the other and does not involve the speaker in a commitment to its explicit performance. For example, if a speaker issues an order to a hearer to do act A he is committed to granting him permission to do A. Why? Because when he issues the order he satisfies certain conditions on issuing the permission. There is no way he can consistently issue the order and deny the permission. And the kind of consistency involved is not the consistency of sets of truth conditions of propositions, but illocutionary consistency or compatibility of conditions of success. In many cases illocutionary commitments are trivially obvious. For example, a report commits the speaker to an assertion because a report just is a species of assertion, an assertion about the past or the present. A report differs from an assertion in general only by having a special propositional content condition. Similarly, a speech act of reminding a hearer that P commits the speaker to the assertion that P because reminding that P is a species of assertion that P made with the preparatory condition the hearer once knew and might have forgotten that P. Thus reminding differs from assertion only by having a special additional preparatory condition. In such cases, which we will call *strong* illocutionary commitments, an illocutionary act $F_1(P)$ commits the speaker to an illocutionary act $F_2(Q)$ because it is not possible to perform $F_1(P)$ in a context of utterance without also performing $F_1(Q)$.

But there are also cases, which we will call *weak* illocutionary commitments, where the speaker is committed to an illocutionary act $F(P)$ by way of performing certain illocutionary acts $F_1(P_1),\ldots, F_n(P_n)$ although he does not perform $F(P)$ and is not committed to its performance. Thus a speaker can be committed to an illocution without explicitly achieving the illocutionary point of that illocution, and similarly he can be committed to an illocution without explicitly expressing the propositional content or without expressing the psychological state mentioned in the sincerity conditions. For example, if he asserts that all men are mortal and that Socrates is a man, he is committed to the assertion that Socrates is mortal; even though he has not explicitly represented as actual the state of affairs that Socrates is mortal, nor expressed the proposition representing that state of affairs, nor expressed a belief in the existence of that state of affairs.

As a general definition we can say that *an illocutionary act of the form* $F_1(P_1)$ *commits the speaker to an illocutionary act* $F_2(P_2)$ iff in the successful performance of $F_1(P_1)$:

1) The speaker achieves (strong) or is committed (weak) to the illocutionary point of F_2 on P_2 with the required mode of achievement and degree of strength of F_2.

2) He is committed to all of the preparatory conditions of $F_2(P_2)$ and to the propositional presuppositions.

3) He commits himself to having the psychological state specified by the sincerity conditions of $F_2(P_2)$ with the required degree of strength.

4) P_2 satisfies the propositional content of F_2 with respect to the context of utterance.

Both strong and weak illocutionary commitments satisfy this definition. Thus, for example, a speaker who asserts that all men are mortal and that Socrates is mortal is committed to the illocutionary point of the assertion that Socrates is mortal and similarly he is committed to having the belief that Socrates is a man. A report commits the speaker to an assertion because a report is simply an assertion about the past or the present. Giving testimony commits the speaker to an assertion because to testify is simply to assert in one's status as a witness. A complaint about P commits the speaker to an assertion that P because to complain that P just is to assert that P while expressing dissatisfaction with the state of affairs represented by the proportional content. A *speaker is committed to an illocution* $F(P)$ in a context of utterance iff he successfully performs in that context a speech act which commits him to $F(P)$. Thus, for example, a speaker who successfully testifies, reports, or complains that P is committed to an assertion that P.

4.4 Definition of a literal performance.

A speaker performs *literally* an illocutionary act $F(P)$ in a context of utterance when he performs $F(P)$ in that context by uttering a sentence which expresses literally that force and content in that context. Thus, for example, a speaker who requests someone to leave the room by uttering in an appropriate context the sentence "Please leave the room" performs a literal request. Many speech acts are not performed literally but rather are performed by way of metaphor, irony, hints, insinuation, etc. Two classes of speech acts which are not expressed literally in an utterance are of special interest to us: First, there are speech acts $F_1(P)$ performed by way of performing a stronger illocutionary act $F_2(Q)$. In such cases the conditions of success of $F_1(P)$ are conditions of success of $F_2(Q)$, and $F_2(Q)$ strongly commits the speaker to $F_1(P)$). For example, begging commits the speaker to requesting. Second, as we noted earlier, there are indirect speech acts $F_1(P)$ performed by way of performing another illocutionary act $F_2(Q)$ that does not commit the speaker to them. In such cases, all the conditions of success of $F_2(Q)$ are satisfied, but the speaker conveys $F_2(Q)$ by relying on features of the context as well as on understanding of the rules of speech acts and of the principles of conversation to enable the hearer to recognize the intention to convey $F_2(Q)$ in the utterance of a sentence that literally expresses $F_1(P)$.[13]

4.5 Definitions of illocutionary compatibility.

Attempts to perform several illocutionary acts in the same context can break down because of various sorts of inconsistency. For example, if a speaker attempts to perform an illocutionary act and its denegation (if he says for example "Please leave the room!" and "I am not asking you to leave the room") his speech act will be unsuccessful because of illocutionary inconsistency. The denegation of an illocutionary act is incompatible with that act because the aim of an act of illocutionary denegation of form $\neg F(P)$ is to make it explicit that the speaker does not perform $F(P)$. We will say that a set of illocutionary acts is *simultaneously performable* iff it is possible for a speaker to perform simultaneously all illocutionary acts belonging to it in the same context of utterance. Two illocutionary acts are *relatively incompatible* iff any set of illocutionary acts that contains both of them is not simultaneously performable. Otherwise they are relatively compatible.

[13]For further discussion see Searle, "Indirect Speech Acts", and D. Vanderveken, "What is an Illocutionary Force?', in M. Dascal (ed.), *Dialogue: An Interdisciplinary Study* (Amsterdam: Benjamins, 1985).

Two possible contexts of utterance are *relatively compatible* when the union of the two sets of illocutionary acts that are performed in them is simultaneously performable, i.e. when it is possible to perform simultaneously in the context of an utterance all illocutionary acts that are performed in them. If two contexts of utterance are relatively compatible, no illocutionary act performed in one is incompatible with any illocutionary act performed in the other.

Chapter 6

COMMUNICATION, LINGUISTIC UNDERSTANDING AND MINIMAL RATIONALITY IN THE TRADITION OF UNIVERSAL GRAMMAR

André Leclerc
Federal University of Paraíba/Brazil

1. Arnaud and Nicole's theory of communication

My aim in this study is to reconstruct the main aspects of Port-Royal's theory of communication and linguistic understanding, mainly spontaneous (non-inferential) linguistic understanding, the true basis, as we shall see, of any hermeneutic practice established on rationality principles. My point is simply that this theory, this model of linguistic communication, is one of the most original, complete and subtle ever produced in the history of linguistics and philosophy of language, because it takes into account, in the total sense communicated in any verbal interaction, both coded elements and inferred elements, as well as elements that are neither coded nor inferred. I shall argue moreover that these rationality principles that govern our hermeneutic practices are clearly principles of *minimal* rationality.[1]

Sperber & Wilson [1986/1989] placed the *Grammaire générale et raisonnée* [1660; henceforth the *Grammaire*] of Port-Royal, together with Aristotle's *De interpretatione*, among the greatest classical representatives of the theories of verbal communication that follow the "code model". To communicate, according to that model, is to codify ideas, thoughts or some piece of information that are then interpreted or "recaptured" by the hearer.

[1]See C. Cherniak, *Minimal Rationality*, Cambridge (MA), M.I.T. Press, "A Bradford Book," 1986.

D. Vanderveken (ed.), Logic, Thought & Action, 133–150.

Sperber & Wilson's appraisal is fair indeed as far as it is restricted to the *Grammaire*, which tries to account for the aspects of linguistic communication that depend on a code or any sign system apt to serve as a public human language.

However, *La logique ou l'Art de penser* [1662; henceforth the *Logique*] goes much farther. In this work, Arnauld & Nicole show clearly the insufficiency of the code model. Later works, like the *Grande Perpétuité* [1669-1672], confirm this insufficiency. Grammatical knowledge is not sufficient, neither for a full comprehension of a whole text, nor for a full understanding of a single utterance, which involves more than a simple cognitive device for decoding or recovering the ideas expressed by the speaker in the context of utterance. The full understanding of an utterance must integrate, most of the time, ideas that are not conventionally signified but nevertheless communicated, either by the facial expressions or the tone of voice, or by giving clues enabling the hearer to make the inferences required to recover parts of the speaker's meaning.

In Arnauld & Nicole's view, linguistic understanding is based on fundamental capacities and abilities, like what they call "*sentiment*," or that "imperfect penetration of the speaker's mind," that may look strange to us at first sight — but it shouldn't be so — and that seem to have been neglected in the historiography of Port-Royal's logico-linguistic theories. We shall see that the normal use of language presupposes, in this view, a lot of extra-linguistic knowledges, and a great number of tacit, "secrete conventions," governing our linguistic exchanges and hermeneutic practices. These tacit conventions allow the hearer to choose the more likely among various interpretations, and they include "maxims of rationality" (Dominicy [1984], chap. 3). These are based on an "embodied" Reason, limited in its resources and capacities, and sometimes strongly affected by the passions. From a logical point of view, the two more basic principles of rationality (the "condition of consistency" and the "condition of inference"), as they are formulated by Arnauld & Nicole, are clearly principles of minimal rationality in Cherniak's sense. The *Messieurs de Port-Royal* made use of these principles to criticize interestingly interpretations based on abusive practices of attribution of propositional attitudes.

2. The Utterance: Signifying and Communicating

The model of linguistic communication proposed by the *Grammaire* seems to reduce to a double metaphor: language is the *expression* of thought, and the linguistic expression of (conceptual) thought is destined to produce a corresponding *impression* in the hearer's mind. To say something and to mean it is to try to produce (intentionally), by the use of conventional means, a certain impression in the hearer's mind.

The world is composed of things, thought of ideas, and discourse of words; words signify ideas that represent things. In the framework of the *ideational* theories of language (to borrow Morris's [1938] and Alston's [1964] word), ideas are what fulfil the role of word meanings and, as we shall see, they are structured entities. We can reconstruct functionally in a very simple way what it is to signify and understand something in this framework: to signify is to apply a function or operation **F** to an idea **i** to produce the sign of this idea (**Si**); to understand is to apply the inverse function $\mathbf{F}^{-1}$ to **Si** to recover, that is, to conceive in one's turn, the idea **i** expressed by the speaker.[2] This model elaborated in the small *Grammaire* gives an account of this part of the total sense that is *conventionally* communicated. This is why Sperber & Wilson classified it, rightly, under the heading "code model". However, the *Grammaire* exposes just one part of Port-Royal's conception of communication. The aim of the *Grammaire* was to determine the universal constraints or the constitutive rules that any system of signs has to satisfy to be a human language. General or Universal Grammar is the study of the universal conditions that any system of signs must satisfy to *completely* represent human thought, and to communicate it *efficiently* in discourse. Any language must have expressive powers sufficient to signify, distinctly, all the possible objects (the idea of which) we can conceive, and also the forms or manners of our thoughts, whose main form is judgement (the *Grammaire* still mentions wish, command, interrogation and strong emotion). Communication, too, imposes its own constraints on the systems of signs we use everyday: these systems must allow their users to communicate efficiently, that is, easily, clearly, in a brief and elegant way. The logical theory of judgement and the famous "theory of ideas" were the foundations of classical Universal Grammar. These theories provided the explanations for the universal constraints just mentioned. So the *Grammaire* was interested mainly in the universal constraints on any code (public language) for the expression and communication of our thoughts. But almost all our utterances carry much more than conventionally signified ideas.

[2]See S. Auroux, *La Sémiotique des Encyclopédistes*, Paris, Payot, 1979. For more details, see also, by the same author, *La Logique des idées*, Paris/Montréal, Vrin/Bellarmin, 1993; M. Dominicy, *La Naissance de la grammaire moderne*, Brussels, Pierre Mardaga, 1984, and J.-C. Pariente, *L'Analyse du langage à Port-Royal*, Paris, Ed. Minuit, 1985. These are certainly among the best books ever written on the subject.

In the *Logique* [chap. xiv, First Part], Arnauld & Nicole completed the model introducing a new concept, that of *accessory ideas* [*idées accessoires*] (the *Grammaire* already used "added meanings" [*significations ajoutées*] for verbal inflexions). This concept made it possible to explain a lot of ideas added to the principal (literal) meaning of an utterance just in the context of use (more precisely, added to the "perceptual idea" — *idée adventice* —, auditive or visual, of the token of the sentence used). The theory of accessory ideas is the most important auxiliary theory in the research programme of Universal Grammar, and universal grammarians of the next century will use it regularly to explain a huge variety of linguistic phenomena, such as: "connotations" (those affective values associated to words by a linguistic community), the working of demonstratives, verbal morphology, or topics like synonymy, tropes, what has been called the "genius of the language", problems of translation, and all that is communicated by the tone of voice, gestures, facial expressions, etc.

In the ideational theory of language, for instance in the work of the Encyclopedist N. Beauzée, the total meaning of a word, say a verb, can be decomposed, in virtue of the rules and conventions of the language, in two parts: the *objective meaning*, and the *formal meaning*. The objective meaning, the idea of the thing, is the idea (in the case of a verb) of the attribute associated to the word by convention; this idea can be modified by various accessory ideas of the type "connotation." "You're lying!," for instance, seems to be a much stronger and offensive assertion than "You're telling me what you know to be false". The formal meaning can be divided in *specific meaning* of a verb (the meaning of a verb *qua* verb), and in *accidental meanings* (formal accessory meanings or ideas associated conventionally to the morphemes that "cosignify" mood, tense, person, number, etc.), that modify the specific meaning of the verb. Hence, ideas as meanings are structured entities, with accessory ideas modifying either the principal objective meaning or the specific meaning of the word. These accessory ideas are cosignified, associated to the word by the formal rules or conventions of the language. They belong to semantics when they modify the objective meaning, or to morpho-syntax when they modify the specific meaning. But there is still another kind of accessory ideas belonging to pragmatics that the speaker adds to the principal objective meaning *only* in the context of utterance.

> There is still a last type of accessory ideas added by the speaker to the principal meaning coded in the discourse, but only at the time of the utterance, by the tone of voice, facial expression or gestures, & by other natural signs that attach to our words infinitely many ideas, which

> diversify, change, decrease, increase the meaning, adding to it the image of the emotions, judgements, opinions of the speaker. [*Logique*, I, xiv, p. 95; my translation].

Sometimes, the tone of voice, Arnauld & Nicole say, can signify as much as the words used: "There is a voice to teach, a voice to please and a voice to reprimand" (*ibid.*). The accessory ideas of the type "connotation" associated to words by a common use, as far as they are effectively registered in the lexicons, pertain to semantics: they are signified by the words *and* by the speaker, being added to the objective meaning, which is associated by convention to the acoustic or visual image of words. But the other accessory ideas added only at the time of the utterance are expressed *only by the speaker*. Nonetheless, they are added to the ideas conventionally signified to compose a unique "impression" in the mind of the hearer. In that sense, they refer to pragmatics as the theory of speaker-meaning. As we can already see, the "total impression" that our utterances produce in the mind of the hearers, spills over in different ways (or goes far beyond) the principal or linguistic meaning conventionally attached to words. In other words, for Arnauld & Nicole, speaker-meaning always spills over word-meaning and sentence-meaning. But this excess on the side of speaker-meaning is not always, at least in this case, *inferred* by the hearer. The ideas expressed by facial expressions or intonation, for instance, and the ideas conventionally signified by words are received *at the same time* and together they contribute to form a single total impression in the mind. They come along with the coded message but are not part of the code itself; they are expressed by signs that Arnauld & Nicole call "natural signs," signs analogous to the symptoms of the internal states of the organism.[3] The code model and the inferential model are present in the common works of Arnauld & Nicole, as we shall see, but they add an interesting ingredient neither inferred nor coded: these "natural signs" coming along with the coded message. The *Grammaire* already treated interjections like "voices more natural than artificial" ("*des voix plus naturelles qu'artificielles*"). These "natural signs" seem to be quite similar to those by which we recognise fear, pain or eagerness in animals or in human communication. The Cartesians always recognised to the beasts a certain form of sentiment, a capacity to feel hunger, thirst, pain, etc., that is, internal states depending upon the working of the machine, and that cannot be referred to

[3]These signs, of course, are not so "natural". We all know that a smile, for example, does not have the same "value" in all cultures. But I think it is clear enough what Arnauld & Nicole have in mind.

the soul.[4] Here the link between the internal state of the "machine" and the "natural sign" of this state is clearly causal. So it seems that what is expressed by these "natural signs" is not intentionally communicated.

Are we always betrayed by our intonation or facial expression, by these signs of our internal states? Of course not. There is what we call self-control. Actors and swindlers are quite good in using intentionally and convincingly intonation and facial expressions. They play with them, and Descartes shows how this is possible in *Les Passions de l'âme* [§ 50]. The result is sometimes impressive: they look so sincere! One can also adopt a stoic attitude and facing up moral or physical pain. But in all these cases training is required. Most of the time, however, we do not have a full control over all the aspects of our verbal interactions. We all know what it means "to have the voice broken by the emotion", or raised by wrath. But if intonation and facial expressions really are "natural signs," most of the time independent of the will, are we really communicating when we just involuntarily let them be known to the hearer? The answer, of course, is "yes", and I think it is the right way to interpret Port-Royal's theory of communication. Human communication is, for the most part, an intentional, rule-governed activity that requires control to be successful, but not every part of this extremely complicated activity needs to be under full control.[5] Condillac, almost one century later, expressed nicely this idea when he said that "before we know how to communicate, we communicated without knowing how to do it" ("*Nous avons communiqué sans le savoir avant de savoir communiquer*"). Be that as it may, Arnauld & Nicole integrate these "natural signs" in the total sense communicated and, moreover, they recognise and attribute to normal speaker-hearers a capacity to penetrate, however imperfectly, the mind of the speaker, a capacity to "read", as it were, his states of mind or humours. (More on this later).

How these accessory ideas excited by the tone of voice, facial expressions or demonstrative gestures, add up to the meaning of the sentence used in a context of utterance to modify it and to compose a unique impression in the hearer's mind? Well, Arnauld & Nicole do not really explain the mechanism; they just describe it. Perhaps, the clearest case is that of the demonstratives. The linguistic (conventional) meaning of "this" is "the confuse idea of a thing present at the place of the utter-

[4]See the famous letter from Descartes to the Marquis of Newcastle, 23 of November 1646.

[5]This is a complicated matter. For instance, I speak English and Portuguese with an accent. When I speak, I do not usually pay attention to the accent I have. When I speak English with a French accent, am I communicating that I am a French native speaker? I would say "no", but how to separate what is unintentionally communicated from what is not communicated at all?

ance"; this is the idea *signified* by the word "this". But each utterance of the word "this", accompanied most often by an act of pointing at the present thing, *excites* in the hearer's mind various (accessory) ideas of the thing in question, which determine the *content* of the word "this" in the context. Thus, an utterance of "this" pointing to an emerald will excite in the hearer's mind the accessory ideas of a green, translucent and very hard thing, these ideas adding up, in the context, to the confuse idea signified by "this." However, Port-Royal's theory adopted the point of view of the ideas signified and communicated by the speaker using a demonstrative and does not capture quite well the main logical function of a demonstrative (identifying a referent) and the way it achieves this function. This is, of course, highly controversial: are complex demonstratives quantificational phrases or devices of direct reference? I personally prefer the second approach. Be that as it may, Port-Royal's theory does not stress the important difference between descriptive identification and demonstrative identification, the latter being logically much stronger than the first, so stronger, indeed, that the speakers intention, or "whats in the head," in many cases, does not play any decisive role in the determination of the referent. But here I cannot go farther than that on that topic.

As to the "natural signs" (intonation and facial expressions), the mechanism seems to be the following: during a verbal interaction, a speaker produces a set of signs, either conventional or linguistic, or "natural," as symptoms of various states of mind (humours, disdain, approval, distrust, etc.). The *simultaneous perception* of these natural and conventional signs produces the perceptual ideas (*idées adventices*) of these signs (the acoustic and visual images of them). To the acoustic images are associated, on the one hand, the meanings (these structured entities composed of one principal idea and various accessory ideas for mood, tense, etc.), and on the other hand, accessory ideas excited by the tone of voice. To the visual images of facial expressions and gestures, are associated the accessory ideas, the thoughts, preferences, dispositions and emotions of the speaker. Verbal behaviour is something quite complex, and in the perception of something complex the attention is naturally directed to the most salient or relevant features of the complex. In a verbal interaction, the attention is mainly directed at the conventional signs emitted by the speaker's voice; but the hearer's mind receives also the impressions of the circumstances in which the conventional signs have been emitted. The "natural signs" are perceived "obliquely" by the hearer's mind. The simultaneous perception of conventional and "natural" signs forms a single total impression in the mind, with the ideas as meanings conventionally associated to words at its centre, and these can be (and indeed most of the time are) attenuated, enhanced, modified or determined in many ways by the accessory ideas provided by the circumstances in the context of utterance. Arnauld & Nicole tell us, in effect, that "...the mind does not consider only the ideas expressed [conventionally by the words], it goes through all the ideas that are joined to the former, above all when it realises that it is the speaker's intention to lead the hearer to conceive them at his turn." [*Grande Perpétuité*, Vol. III, Book 1, chap. iii, p. 692; my translation].

If simple signified ideas compose complex ones (judgements, interrogation, wishes, etc.) according to the clear and fixed relationships of *determination* and *explication* [6], the relationship between principal ideas and accessory ideas are very diversified and flexible. Arnauld & Nicole simply say that many of these accessory ideas are joined by the speaker only, and contribute to form a unique total impression in the mind of the hearer.

So we communicate much more by our utterances than we signify linguistically. Furthermore, Arnauld & Nicole constantly insist on the fact that, most of the time, we leave a lot to supply in our discourses; in other words, beyond what is literally said, there is a great deal of implication, insinuation, suggestion or presupposition. As they say, "there are more judgements in the mind than in the words." There are different reasons explaining why it is so. First, there is "this tendency of men to abridge their discourse," responsible for the fact that we do not waste our time saying things when we realise they are already well-known. There is also this remarkable capacity to guess what's happening in the mind of the speaker, a capacity to determine approximately the humours, states of mind and many things that we believe to be authorized to suppose well-known by those who are listening to us. This implicit part of the message, of course, has to be inferred by the hearer. (I'll be back soon on this topic). There is still this enormous difference between the rapidity of the conceiving, thinking and reasoning of our mind, and the slowness of the linguistic expression of our ideas, judgements and reasonings. Sometimes, what happens in the mind in one stroke or instant has to be wrapped in a long sequence of words. This is why we often signify just a half of what we want to say, letting to the hearer the task of supplying the ellipses, to catch the allusions, understatements, presuppositions, etc., in such a way that our discourses run faster, enabling our trains of words (almost) to follow the rapid train of our thoughts. Finally, both thought and language have a kind of deficiency or consti-

[6]On the concepts of determination and explication, see M. Dominicy's excellent book, *La Naissance de la grammaire moderne*, Brussels, Pierre Mardaga, 1984, chap. 2, 4 and 5.

tutive limitation: thought cannot represent a thing in its full extension and under all its aspects, but only under a small number of aspects, because our mind is not great enough to grasp them all at once. But our language too has its limitations: it does not represent our thought in all its extension, it does not fully do justice to its richness. There are much more objects in the world and ideas in the mind than words to denote and signify them. This is why there are, in our discourses, so many presuppositions, allusions, ellipses, suggestions, insinuations and other implicit ways of communicating our trains of thoughts. The number of our perceptions, ideas and judgements much exceeds the number of linguistic means we have at our disposal to express them; the systems of signs that our human languages are, consequently, are efficacious and viable as far as they make it possible to do many things with few resources. The same economy prevails in our use of signs in communication.

The communication model proposed by Arnauld & Nicole contains also a set of maxims of rationality that M. Dominicy patiently extracted from their common works. I shall consider a few maxims later. Before that, I would like to examine certain fundamental capacities on which the normal use of language runs, according to Arnauld & Nicole, and especially the spontaneous understanding of discourse. But let us keep in mind the enormous, tacit knowledge required to simply speak a language and that determines the meaning of the words used by a speaker in a given context. The traditions concerning the use of certain words can be, in this respect, extremely important. The Church, for example, which is "the master of its language" ("*qui est maîtresse de son langage*"), as Arnauld says, could determine and fix, by an explicit declaration, the sense of an ambiguous expression in such a way that this expression ceases completely to be ambiguous "in the use of the Church." In the language of criminal gangs, "a trigger" might signify a professional killer, not the part of a gun. Without any knowledge of this kind, linguistic communication would be just a web of misunderstandings.

3. Spontaneous understanding of discourse

As Pariente [1985] pointed out, at Port-Royal, grammatical art is neither for animals nor for angels. Only beings provided with Reason, this "unlimited capacity of adapted innovation," (as Pariente says), are able to use language normally as we do all the time. This excludes animals, according to Descartes, Arnauld and the Cartesians. Nevertheless, human communication seems to retain, as we saw, something from animal communication, namely, these "natural" signs, more or less independent of the will, symptoms of the internal states of the organism. However, if, like the angels according to tradition, "human beings could see immediately what happens in the mind and the heart of one another, they would not talk at all and the words would become useless." (*Grande Perpétuité*, Vol. I, Book ix, p. 989). The main use of our words is to communicate our thoughts when we have the right to suppose they are not known by the hearer we are talking to. This predicament, the necessity for us to use words to communicate our thoughts, prove, if it were necessary, that we are not angels...

But it is clear that we wouldn't talk at all in the same way if the mind of the others would be completely closed or opaque to us, if we were unable to see what their dispositions, desires, preferences or intentions are. As a matter of fact, we choose our words, rule our discourse according to the thoughts and humours we spontaneously ascribe (sometimes mistakenly, of course) to the hearers. Only someone very badly educated would make a very severe or pinpointed criticism to someone else burying his wife or his child. This capacity by which we "penetrate imperfectly" the speaker-hearer's mind to grasp partially its content is placed, by Arnauld & Nicole, among the most important corner-stones of human language:

> We cannot reflect, however little, on the nature of human language, without recognising that it is entirely founded on this imperfect penetration of the mind of the others. And this is why, in talking, there are so many things we do not express. (*Grande Perpétuité*, Vol. 2, Book I, p. 81; my translation).

We do not express them, as we have seen, precisely because we suppose them already known. This is also why we frequently say just the half of what we want to say because we perceive that the hearer has already understood, and that we often answer in advance to what we "read" in the mind of the speaker-hearer. So communication, as it is thought of at Port-Royal, is not just an affair of coding and decoding ideas or thoughts, it is also a matter of "recognising intentions," as it is in the pragmatics *à la* Grice. (This capacity to penetrate (imperfectly) the mind of others was attributed to the Christ, but to a degree and a potency much superior to ours. And given that He was obliged, like us, to express His thoughts by words, He also had the capacity to determine, in advance, the exact impression that His utterances would make in the hearer's mind.)

There is still another fundamental capacity on which, according to Arnauld & Nicole, the spontaneous understanding of discourse is based, and by which we determine sentence-meaning and word-meaning, and grasp nuances, sometimes very subtle, among expressions reputed synonymous. This capacity, they call it *sentiment*. Most people do not judge and assess the meaning of words through abstract and complicated reflections. But most competent speakers do not commit mistakes when they use expressions having different but very similar meanings. However, they are often unable to mark explicitly these slight differences. So how do they succeed? "It is by a simple view of the mind, an impression they feel...", say Arnauld & Nicole. This way to spontaneously assess the meaning of words by sentiment is however not exceptional at all. On the contrary:

> This is the way human beings assess almost all the variety of the things in this world. We recognise in one stroke that two very resembling persons are nonetheless different, without paying attention to details, to what is in the face of one that is not in the face of the other. The impression marks all this in the mind, without revealing distinctly the particular differences. (*Grande Perpétuité*, Vol. 2, Book 1, p. 990.)

This way to assess differences is not only the more common and universal, it is also, in Arnauld & Nicole's opinion, "the surest, the finest, and the subtlest." We are effectively able to recognise in one instant subtle differences that are based on minute details. We recognise that we are talking with Joe, and not with his twin Joey, by a characteristic gesture, a facial expression, the manner of walking, etc. In the same way, "there are, between the terms a thousand imperceptible differences that the mind feels, and that it cannot explain but with great difficulties. There are such differences that it feels, but would not be able to mark the meaning and precise idea." (*Grande Perpétuité*, Vol. 2, Book II, chap. 1, p. 122; my translation). This sentiment, that can be viewed as a kind of speaker's intuition *à la* Chomsky, really is a corner-stone of the normal use of language, since not only it enables us to recognise "in one stroke the finest differences between expressions better than all the rules in the world," but also "the rules themselves are true only whenever they are conformed to this sentiment" (*Grande Perpétuité*, Vol. 2, Book II, chap. 1, p. 122; my translation). What happens when the sentiment of one group of speakers conflicts with that of another group speaking the same language? The *grammairiens philosophes*, most of the time, designated a group of reference serving as an "ideal speaker-hearer", for example the "best authors," the Court, or the "healthier part of it" (*sa partie la plus saine*).

So it is this sentiment, this spontaneous comprehension, or this "natural impression" made by words in the mind of the hearer, which is the basis of any reflected interpretation, the starting point of any hermeneutic work.[7] In their numerous controversies, Arnauld & Nicole condemned the methods of their opponents that neglected the sentiment or this "natural impression," showing how they try to repress it under a lot of subtleties or hidden senses that only a stubborn hermeneutic work could discover.

This sentiment enables us to distinguish also, in one stroke, metaphorical expressions from literal ones. Arnauld & Nicole often called "natural sense" the literal sense, or the sense an expression has through its (first) *impositio*; it is also, usually, the most current and common use of an expression, the one that comes first to mind. The mind of the speaker-hearer "naturally" and spontaneously expects the literal sense, and one needs reasons to take a word in a sense other than its literal sense. Of course, we do not need any particular reason to take a word according to its first, "natural sense." In other words, one does not have to justify what is normal or standard.

> ... although almost every word that we use in any language sometimes can be taken in its literal sense, sometimes in a metaphorical sense, there is, nonetheless, this difference between the first or proper sense and the metaphorical sense, that one does not need a particular proof showing that a word should be explained according to its proper sense; it is enough that there is no particular reason obliging us to take it in another sense. (*Grande Perpétuité*, Vol. 2, Book IV, chap. xiii, p. 355; my translation).

We do not need reasons to take a term in its "natural sense," but such reasons are required to take it in a metaphorical sense, and the mere absence of these reasons is a clear indication that the only correct interpretation is the literal one. On the contrary, when there are such reasons, the "maxim of metaphorisation" (Dominicy [1984], pp. 116-117) recommends searching for a metaphorical interpretation, both intelligible and charitable. It is interesting to note that in Arnauld & Nicole's theory of metaphor, there is no such thing as a mysterious "change of word-meaning," as in the semantic theory of metaphor. In a metaphorical use, we get, so to speak, two ideas for one word, and this is possible because the words do not loose their first or "natural sense." The word "lion," for instance, in the sentence "This knight is a lion," does not mysteriously change its meaning, in a given context of use, to signify

[7]Compare with Tyler Burge, "Comprehension and Interpretation," in L.E. Hahn, *The Philosophy of Donald Davidson*, "The Library of Living Philosophers." Vol. XXVII, Chicago, Open Court, 1999, pp. 229-250. According to Burge, comprehension is immediate, non-inferential, linked to the knowledge of an idiom, while interpretation would be always inferential. See especially pp. 236 and 237.

suddenly courage, bravery or strength; otherwise, the knight couldn't be *seen as a lion.*

4. Rationality and interpretation

The normal use of language is a rational goal-directed activity whose main end is communication. In the common works of Arnauld & Nicole, the normal use of language, spontaneous linguistic comprehension, and interpretation (thought of as a hermeneutic — inferential and therefore non-spontaneous — activity), are clearly based on a *presumption of rationality.* Arnauld & Nicole gave different versions of this presumption. Here are a few:

> ... one ordinarily supposes that one is talking to people that deviate from reason the less they can. (*Grande Perpétuité*, Vol. 1, Book 1, chap. vii, p. 538; my translation).

> ... the principle of all the knowledge one can extract from men's writings or from the relationship we have with them by the use of words, is that they talk reasonably, and that they do not conceal, in their words, senses or ideas that these expressions are unable to impress in the mind, whenever the one who pronounces or writes them should have seen that they cannot produce these senses or ideas. Otherwise, there is no rule or measure to be taken on men's discourse. (*Grande Perpétuité*, Vol. 2, Book III, chap. viii, p. 242; my translation).

> Men talk to be understood, and whoever talks in a way that cannot be understood, and that is not appropriated to produce in the hearer's mind the idea of his own thoughts, is talking without reason or judgement. (*Grande Perpétuité*, Vol. 1, p. 369; my translation).

> As it is only by an extraordinary reversion of nature that men would come to think and talk against reason (*bon sens*) all the consequences one can draw from what they say are established on the fact that one never supposes gratuitously that they deviate from the lights of common sense, particularly in the simplest things, that are, so to speak, uncovered to the mind's eyes. (*Grande Perpétuité*, Vol. 2, VI, viii, pp. 544-545; my translation).

That's enough to show that it is not an insignificant digression, but a recurrent theme in the common works of Arnauld & Nicole. Without this "presumption of rationality," there would be "no rule or measure to be taken on men's discourse."

The maxims of rationality discovered by Dominicy in the works of the great Arnauld do function only when we're supposing that the speaker-hearer is rational. But to what extent do we have to be rational just to use language normally?

At Port-Royal, the *Messieurs* have a conception of reason much less triumphant than that of the Cartesians. The *Logique* complains that "common sense is not a quality so common as one thinks," and that "the most ridiculous stupidities always find minds to which they are proportioned." (*Logique*, First Discourse). Arnauld & Nicole, constantly, direct our attention to the very narrow limits of our mind, and that the limits of our cognitive capacities are also limits for sciences in general. So a lot of questions about metaphysics, mathematics, or theology are declared futile, particularly those concerning the infinite: Are there various "infinites" with different extensions or just one? God could have created a body infinite in extension? Etc. Our finite mind cannot understand the infinite (*Logique*, IV, xi); when one tries, one gets lost, obfuscated and stunned, simply at loss. When Arnauld & Nicole transcribe and reformulate, in their *Logique* (IV, xi) Descartes' rules V and VI (from the *Regulae ad directionem ingenii*), rules that prescribe considering things in their natural order, beginning with the most general and simple (Rule V), and to divide each genus in all its species, each whole in all its parts, etc. (Rule VI), they take care to add the pragmatic, restrictive clause: "as far as it is possible," precisely to take into account the cognitive limits of our mind, and also the enormous waste of time that would represent, most of the time, a strict application of these rules for a very small benefit. A genus might have so many species, for example, that it would be practically impossible to enumerate all of them; or may be it could be much more practical to consider one species before considering all the genus in detail. Moreover, our weak intelligence is constantly obfuscated by our passions and prejudices. The *Logique* enumerates various "fallacies of self-love, self-interest and passion," (III, xx). Envy, vanity, stubbornness and prejudice often are the motors of argumentation. From this, the *Messieurs* draw the conclusion that, although we are all *capable* of following our reason, most of the time we don't.

> One knows that, although it is easy to judge most of the things according to reason, it would be misleading not to complete that knowledge by another, which is that one does not always follow our reason, or rather that one rarely follows it, because of a lot of secret inclinations or profoundly rooted prejudices that ordinarily vanquish the most obvious proofs, there having many people for whom the authority of the ones they most respect is an invincible reason. (*Grande Perpétuité*, Vol. 2, Preface).

If we can be still considered as rational animals, it is, above all, because we are "capable of reason," not because we are rational most of the time.

These severe limitations of our mind, and the passions that go through it all the time, make it possible that one contradicts oneself regularly, either in speech or in thought. "A speaker is not supposed to contradict himself/ herself" (Dominicy's maxim of non-contradiction); there is a corresponding principle of charity that recommends not suspecting the speakers of committing extravagant errors without a very good reason. This is what happens regularly in the cases of slips of the tongue or in faulty uses; in spite of the faulty use, we recognise immediately the speaker's intention. Arnauld & Nicole give the example of a member of the Paris Parliament who claimed: "The Cardinal Mazarin here has his hemispheres"; of course, everybody understood that he wanted to say "emissaries." So one always must research, as far as possible, for any curious and apparently contradictory utterance, an interpretation that avoids the attribution of silly or contradictory beliefs to the speaker.

> There is no criticism more common than that of self-contradiction; but this criticism does not suppose that one is accusing the addressee of having had two directly contradictory beliefs all at once. One supposes only that he said things that are effectively contradictory, although in saying them he didn't see the contradiction, or that deviating the words from their natural meaning, perhaps he did ally in his thought what is contradictory in the expression. (*Grande Perpétuité*, Vol. 1, Book VII, pp. 780-781; my translation).

Thus we suppose that the speakers/hearers are consistent. But this consistency is *minimal* and does not suppose the logical capacity to eliminate all the contradictions there are in the set of our beliefs.

> ...one must bear in mind that it is not a proof absolutely certain that people do not hold two opinions at the same time that these are effectively contradictory. Because the stubbornness and the smallness of the human mind easily provide ways to ally these contradictions, mostly in abstract matters, of which one has only confuse ideas. (*Grande Perpétuité*, Vol. 3, Book V, pp. 926-927; my translation).

The same predicament holds for our logical capacity to draw consequences from our beliefs. Given the smallness of our mind, we cannot calculate and perceive but a very small number of valid consequences of our beliefs. So one cannot attribute to a writer or a speaker an opinion that would be just a remote consequence of an opinion explicitly held.

> ...one must distinguish very clearly between the consequences and the dogmas explicitly held; because one cannot conclude that those who hold an opinion hold also all its consequences, if these consequences are not correctly drawn. But even when the consequences are correctly drawn, one does not have the right to attribute these consequences to them, if they do not hold the consequences independently, and if it does not seem that they did perceive and accept them. (*Grande Perpétuité*, Vol. 1, Book II, vi, p. 349; my translation).

This minimal condition of inference plays an important argumentative role in the controversies in which Arnauld & Nicole get involved.

They use it regularly to contest certain hermeneutic practices of their opponents (Protestants) that drew, from the writings of the Fathers of the Church, consequences that seem favourable to their point of view on the Eucharist, or to criticize proceedings utilised in the Inquisition trials where someone is accused and condemned for holding an opinion (never explicitly held) that directly offends an important dogma, just because this opinion was a remote consequence of another opinion explicitly professed. These practices are very intolerant in that they overestimate the cognitive capacities of human beings. An inferred opinion drawn from a passage can be attributed to an author only if he (she) expresses it in other passages or in other texts.

However limited as they might be, our inferential capacities are extremely important for the interpretation of utterances, particularly when the speaker-meaning differs from the sentence-meaning. Thus, any speaker is supposed to provide the strongest information possible on the theme under discussion, this information implying pragmatically weaker information. (See Dominicy, p. 119). If the Emperor is also Archduke, by calling him "Emperor," his highest title, one is not denying that he is also Archduke. And when a hearer is searching for an acceptable interpretation of a metaphorical utterance, she (he) has to reason in accordance with the maxims of non-contradiction and intelligibility (which prescribe to any speaker, respectively, to avoid contradictions, and to talk in such a way that she (he) can be understood).

5. Conclusion

The theory of communication that can be found in the common works of Arnauld & Nicole proposes an original combination of the code model and the inferential model. I hope I have shown its complexity. The aspects of communication linked to the code constitute the core of the total sense communicated. But the code model, as we saw, is far from sufficient for the reconstruction of this theory. The recognition of the speaker's intentions, guided by various "secret conventions" and contextual indications supply the lacuna. But something is still missing. In Arnauld & Nicole's theory, the coded and inferred elements of the total message can be modified or strongly influenced by another part of the message, by facial expressions and tones of voice, which are considered as "natural signs." Most of the time, these signs are not under a full, voluntary control, because we usually do not pay attention to them when we're talking. Finally, in Arnauld & Nicole's theory, communication is based on fundamental background capacities and abilities, on a lot of tacit conventions, shared knowledge and presuppositions, and on logical abilities obeying principles of minimal rationality. Arnauld & Nicole's theory really deserves, in my opinion, the title of "Integrated Theory of Communication."

What makes the history of the sciences of language interesting for us is that we constantly meet in this history good theoreticians facing more or less the same problems as ours. Their answers to these problems, the writings they have left, are still talking to us. As long as we ignore their contributions, we cannot have a clear idea of what have been done in the last centuries. Sperber & Wilson, in their excellent and rightly famous book on relevance, claim that:

> To distinguish meaning and communication, to accept that something can be communicated without having been signified by the communicator or by his (her) behaviour, constitutes a first essential step, that radically takes us away from the traditional approach of communication and from most of the modern approaches.[8]

Part of my intention was to show that this "first essential step" has been taken resolutely more than three centuries ago.[9]

References

Alston W. (1964). *The Philosophy of Language.* Prentice-Hall.

Arnauld A. & Lancelot C. (1660). *Grammaire générale et raisonnée*, ed. by H. Brekle, reprinted from third edition of 1676, Stuttgart-Bad Cannstatt, Friedrich Frommann, 1966.

Arnauld A. & Nicole P. (1662). *La logique, l'Art de penser*, ed. by P. Clair & F. Girbal. Paris: P.U.F. 1965.

Arnauld A. & Nicole P. (1669-1672). *La grande perpétuité de la foi de l'église catholique sur l'eucharistie*, ed. by l'Abbé M***, Paris, Imprimerie de Migne, chez l'éditeur rue d'Ambroise, Hors la barrière d'Enfer, 1841.

Auroux S. (1993). *La logique des idées*, Paris/Montréal: Vrin/Bellarmin.

— (1979). *La sémiotique des encyclopédistes*, Paris: Payot.

Burge T. (1999). "Comprehension and Interpretation," in L.E. Hahn, *The Philosophy of Donald Davidson.* The Library of Living Philosophers, Vol. XXVII, Chicago: Open Court.

Cherniak C. (1986). *Minimal Rationality*, Cambridge (MA): MIT Press.

Descartes R. (1649). *Les passions de l'âme.* Paris: Gallimard. 1969.

---

[8] I am quoting and translating from the French edition of *Relevance* [1986], the only edition at my disposal now. D. Wilson & D. Sperber, *La Pertinence*, Paris, Minuit Editions, 1989, pp. 92-93.

[9] I am very grateful to my colleagues and friends Rubem Mendes Oliveira and Maria Leonor Maia dos Santos, who read this paper and suggested many improvements to me.

Dominicy M. (1984). *La naissance de la grammaire moderne.* Brussels: Pierre Mardaga.

Morris C. (1938). *Foundations of the Theory of Signs*, in the *Encyclopedia of Unified Science, 1*, no. 2. Chicago: Chicago University Press.

Pariente J.-C. (1985). *L'analyse du langage à Port-Royal.* Six études logico-grammaticales. Paris: Minuit.

Sperber D. & Wilson D. (1986/1989). *La pertinence.* Communication et cognition, Paris: Minuit.

II

EXPERIENCE, TRUTH AND REALITY IN SCIENCE

Chapter 7

TRUTH AND REFERENCE*

Henri Lauener[†]
University of Bern

To current naturalistic views on philosophy I oppose a pragmatically relativized version of transcendental philosophy. Quine's system, as a paradigmatic case, forms a subtle and solidly woven fabric of theses which seem difficult to attack from within in spite of certain apparent tensions. As I am not prepared to concede all the semantic indeterminacies it involves I object to its founding principles and reject naturalism as a general approach. Shunning any notion of absolute (external) truth, I replace the doctrine of physical realism by a distinctive kind of relativism which is not to be confused with so-called cultural or subjective relativism. On the basis of an entirely different conception of language, I consider Quine's claim that truth precedes reference as an error due to his particular brand of holism and to his one-sidedly behavioristic method. Questions concerning truth are so central in philosophy that it should not be introduced, at the outset, as a pretheoretic notion relying on such a vague criterion as that of assenting to sentences. I doubt that he can be right when he asserts that what objects there are according to a theory is indifferent to the truth of observation sentences, for the meaning, i.e. the intension <u>and</u> the extension of the terms occurring in sentences used for testing that theory, depends on it insofar as their truth, in accord with Tarski's definition, requires the existence of empirically discoverable objects satisfying the respective open sentences. Therefore, holophrastically conceived observation sentences, held true merely on account of their stimulus meaning, cannot do the job since

*This article appeared in the issue of the *Revue internationale de philosophie* devoted to "Quine with his replies" 1997, vol. 51, pp. 557–566. We thank both the editor of the Revue internationale de philosophie, Prof. Michel Meyer, and Herr Michael Frauchiger acting on behalf of the Lauener Stiftung, for granting us permission to re-publish this paper to which Quine replied in the issue mentioned (*Ibid.* pp. 581–582)

D. Vanderveken (ed.), Logic, Thought & Action, 153–161.

they do not properly belong to the language in which the theory has been couched.

Jaakko Hintikka has distinguished two radically contrasting approaches to language which he labels the universalistic view and the view of language as a calculus.[1] According to the first, language is a universal medium which we cannot contemplate from an external vantage point in order to examine its relation to the world. As a partisan of the second approach I do not consider language an amorphous, constantly evolving whole; I rather hold that we create a great number of distinct linguistic systems which we use as instruments for various purposes. Facing the fact that there are different uses of expressions, I lay much weight on the possibility of interpreting or reinterpreting token systems in the way this is done in model theory.

According to my method of systematic relativization to contexts (of action), we create reality sectors by employing specific conceptual schemes through which we describe the world. Since a new domain of values for the variables is presupposed for each context I advocate a pluralistic conception of ontology in contrast to Quine who postulates a unique universe by requiring us to quantify uniformly over everything that exists. The divergence comes from the different forms of holism we countenance. In "Two Dogmas of Empiricism", he favors an extreme sort of holism claiming that our global theory of the world is confronted with the tribunal of experience as a whole and affirming later that only sentences at the periphery have an empirical content of their own. This is so because their (stimulus) meaning — as in the case of all occasion sentences — is constituted by the fact that assent to them is directly prompted in presence of adequate sensory stimuli. Doubts have been raised whether the Duhem-thesis is really compatible with the claim that the meaning of observation sentences does not depend on the theory. With my contextual holism no such difficulty occurs. Rejecting the view of a constantly evolving, unified language-theory, I claim that the intension of all the terms is determined by the axioms of the specific theory with which we operate in a given context, and that consequently the meaning of the observation sentences must depend on that theory, too. Whereas the naturalist aims at theories which are supposed to explain the causal relation between semantic facts and utterances as physical tokens, I stress the normative aspect of semantics. Equating talk about meaning with talk about rules, I consider that intensions and extensions

[1]Cf. "Is Truth Ineffable?", in *Les formes actuelles du vrai*, Palermo, 1988, and "Quine as a Member of the Tradition of the Universality of Language", in R. Barrett and R. Gibson (eds.), *Perspectives on Quine*, Cambridge (Massachusetts), 1990.

are fixed by the totality of the rules which prescribe the correct use of the expressions. My method has the advantage that it permits us to distinguish language from theory and to separate the distinctive contributions to the truth conditions made by the language and by matters of fact. If linguistic rules alone are involved we have analytically true statements. Contrary to what Quine affirms, it is not the semantic distinction between analytic and synthetic which is a matter of degree, but the psychological faculty of an individual to comprehend a language. In order to understand exactly the theoretical terms of quantum mechanics, for instance, one must thoroughly master the whole theory including its integral parts of logic and mathematics. Philosophers tend to overestimate the capacities of a layman when they suggest that he can grasp the precise meaning of terms like 'electron', 'nucleus', 'spin' etc. The fact that they are able to utter some true sentences about particles is not sufficient, since full understanding requires potential knowledge of all the truths on the matter — an ideal which can be approximately realized only by professionals.

Thus, considered from my transcendental point of view, the very possibility of expressions having meaning depends on conventions, i.e. on a community of users agreeing on a set of rules which determine their use in a context. Insofar as the intensions of the theoretical terms are implicitly defined by the system of empirical laws and their extensions fixed by the intended model, their meaning cannot remain the same in the event of (even a slight) theory change. For, when we give up a theory, replacing it by a new one, the conditions under which sentences can be rightly asserted have been altered altogether so that the two theories must be considered, strictly speaking, as semantically incommensurable. Of course, it is possible, by means of ascent to a meta-language, to speak about the words and to ascertain that they have a similar meaning due to some similarities of the axioms in which they appear, but this does not amount to making them synonymous, since synonymy, according to the present view, must be an intralinguistic property (if it occurs at all). Moreover, we have no obvious guarantee that terms like 'electron' have the same extension in successive physical theories, contrary to what Hilary Putnam has suggested, because we must secure that equal methods of empirical identification have been applied, before we can assert the extensional identity of two terms used in different contexts. In view of the fact that linguistic individuals cannot be taken to be (identical with) physical tokens, as I have argued elsewhere,[2] I insist on the different nature of scientific theories dealing with empirical objects and meta-theories talking about abstract linguistic entities. I wonder how Quine with his vision of a unique language-theory manages to keep together such a mixed bag of disorderly things which is fatally threatened — it would seem — by paradox. Therefore, I prefer to resort to my notion of a limited context which allows us to avoid the dismissal of useful semantic distinctions as merely gradual. The resulting concept of meaning diverges radically from any concept developed along naturalistic lines. One important consequence touches on the sentences which are to serve for testing theories. Since a language used in daily matters is subjected to semantic rules differing from those of a language used in science, a sentence taken from the first cannot have a proper function within a scientific test procedure. For this reason the so-called protocol sentences of early logical positivism or Quine's observation sentences are of no avail when we are confronted with the problem of testing scientific theories. The fact that we may describe, on a meta-level, the words 'water' and 'H_2O' as denoting roughly the same substance does not entitle us to declare them synonymous, for they do not belong to the same linguistic systems; as 'water' is not an appropriate chemical term it should be banished from the language used in the context of chemistry. This explains why I reject sentences like 'Water is H_2O' as semantically incoherent while I maintain that metalinguistic statements such as, "'Water' in ordinary language denotes roughly the same substance as 'H_2O' in chemical terminology", do make sense.

Every philosopher must assume the consequences of his fundamental options. In my case the price to pay is a relativized conception of truth and ontology. Quine, for his part, has become more and more insistent in defending his position of physicalist realism. Yet, somehow surprisingly, he favors, at the same time, an extreme form of ontological relativity, originating in the fact that we cannot determine a speaker's referential intentions from his linguistic behavior and that, therefore, reference remains empirically inscrutable. But then how is it possible that indeterminacy of reference does not undermine physicalism, the doctrine which assumes the posits of our overall world theory?

Quine argues that alternative theories obtained by means of proxy functions are structurally identical with our physical theory down to the observation sentences, through which it gains its empirical content,

[2] Cf. "Speaking about Language: On the Nature of Linguistic Individuals" in A. P. Martinich and M. White (eds.), *Certainty and Surface*, New York, Edwin Mellen Press, 1992, pp. 117-134.

and that, therefore, they must have the same cognitive import: "The structure of our theory of the world will remain undisturbed, for the observation sentences are conditioned holophrastically to stimulation, irrespective of any reshuffling of objective reference. Nothing detectable has happened. Save the structure and you save all." The position is clearly stated in the quotation: Since the observation sentences which establish the contact with sensory stimuli are our only access to the one and absolute reality and since they are accepted as true without any regard to referential matters, there can be no objective, i.e. physical, facts about reference. As Quine, on the other hand, concedes that we must ascribe denotations to the terms in order to understand a language he admits a derivative kind of semantic facts in the form of referents which are assumed relative to a background language taken "at face value". But as such ascriptions are posterior to our attributions of a truth value to sentences we must eventually grant the precedence of truth over reference. It seems very questionable to me whether ontological relativity so conceived is compatible with Quine's professed realism according to which our physical theory must count as true (pending further information). For if no empirical evidence in favor of the objects posited by the physical theory — inclusively those of physiology, as nerve endings and stimuli etc. — can be adduced, how is it possible to affirm absolutely its (external) truth to the detriment of empirically equivalent[3] rivals with platonistic, pythagorean or other ontologies?

As I consider any attempt to establish an absolute correspondence relation between theory and reality (neutrally given through sense experience) as hopeless, I am not prepared to accord to physicalism the status of a true theory about the world. It is at best a recommendation to adopt a physicalist ontology on the ground of practical reasons partaking of scientific methodology.

Renouncing realism with its dubious notion of external truth and opting for a form of relativism which turns truth into a strictly internal matter, akin to model theoretic treatment, does not prevent me from clinging to an empiricist attitude appropriate to scientific method. Relativity after all is a concept familiar to physicists. Quine's most implausible theses — especially the ones concerning semantic indeterminacy — have their origin in his deep-rooted conviction that scientific language must be purged of intensionality. Flight from intensions is one of his well-known slogans. As it is impossible to treat intentionality by

[3]Two theories are empirically equivalent if they have corresponding predicates interrelated in the same way and if their corresponding observation sentences are identically conditioned to sensory stimulations, irrespective of the differing kinds of objects they are talking about.

purely extensional means he adopts a strategy according to which intentions along with propositional attitudes have to be reduced to a strictly descriptive treatment in line with the methods of natural science, particularly of behavioristic psychology which is the seemingly best candidate for being incorporated into our global physical theory. Yet, I doubt that this will do because we cannot be content with simply describing past intentional acts. Ordinary life as well as science requires innovation, i.e. decisions of all sorts in order to achieve tasks which are not predictable by means of a physical theory. My main objection to a naturalistic view on semantics resides in the fact that interpretation presupposes intentional acts to the effect that the members of a community speaking a specific language will agree on a set of rules prescribing the use of the expressions and that it cannot, therefore, be considered as a merely descriptive matter to be handled exclusively with extensional tools. Accepting (conventionally fixed) rules, choosing, conforming, asserting etc. are typical kinds of actions without which language would not exist. If they were absent we could not speak but only produce noises. Consequently there can be no hope for integrating semantics into a physicalist doctrine with its corresponding semantic facts, as long as nobody has succeeded in reducing talk about intensions to talk about extensions.

The realist, being committed to the posits of his overall theory, must assume the existence of physical objects and cannot, consequently, justify his claim that it is (externally) true by taking recourse to the pretheoretical truth of some unanalyzed sentences. As we need determinate categories of objects, i.e. definite domains of values for the variables, in order to fix the truth conditions, reference has to play a primordial role. This is the reason why I propose to relativize the concept of truth to contexts in which we operate with specific linguistic systems and theories. According to my normative viewpoint, semantic questions are settled by the fact that we have accepted the rules which determine the intension and the denotation of the terms for a particular language. By relativizing ontology to a given theory we gain the advantage that referential relations become determinate. Thus the analogy with relativity in physics, where position and velocity are determinable relative to an inertial frame, works well while it fails for Quine since relativized reference in his sense cannot be behavioristically determined. The weakness of the naturalist's position resides in his ignoring the trivial fact that posits presuppose positing and that intentional acts cannot be accounted for in an austerely extensional language as he wants to have it. I conclude then that, notwithstanding his claims to the contrary, reference does matter and that a scientific theory cannot be properly identified without assuming an intended domain whose individuals must satisfy certain open sentences in order to make corresponding closed sentences true. At any rate, a philosopher should not resort to the notion of a posit if his very doctrine renders intentional acts of positing unintelligible.

According to my special brand of transcendental philosophy, concerned with the optimal conditions for the elaboration of reliable science, we employ specific linguistic systems in order to structure the world about which we acquire knowledge by describing and explaining it with help of various sorts of theories. We create what I call reality sectors by imposing different conceptual schemes on the raw material provided by sensory experience. The imposition of linguistic forms is a precondition for the possibility of individuating objects and for specifying the ontology to which a given theory is committed. Electrons *qua* electrons do not exist absolutely but only relative to a context in which we use quantum mechanics. Through the selection of a language appropriate to the intended purpose we create a relative *a priori* such that the truth of certain sentences will be determined by the semantic rules alone. Thus we exclude the very possibility for analytic statements to be refuted by empirical facts (internal to the operating theory) as long as we stick to the same conceptual framework. Such a relativized concept of analyticity depending on linguistic rules explicitly stated in a context has nothing to do with the old absolutistic notion inspired by Kant and rightly dismissed in "Two Dogmas of Empiricism". It is perfectly compatible with the conviction that no statement is immune to revision. If a conceptual scheme proves to be inadequate for some practical reason we give it up and replace it by a new one. Two tokenwise identical sentences may occur in both contexts, but with different meanings since the connections within the semantic network have been altered, and it can even happen that the one is analytic while the other is synthetic according to the respective sets of accepted rules. As I do not believe that we can do with the continually moving mass of a total language-theory, I insist on the necessity of introducing stability points in our conceptual apparatus by stipulating that certain statements must be held true without regard to empirical matters in the reality sector created by the context. By recommending such a procedure I do justice to the widespread intuition that there are statements whose truth is elucidated without recourse to empirical considerations. Contrary to Quine, I do not rate classical logic as objectively true because we have integrated it into our overall scientific theory. I rather estimate that, being free to make alternative choices, we can use any system that fits best our practical needs. One may prefer, for philosophical reasons, intuitionistic to classical logic and consequently deny the truth of 'p ν -p'. In doing so he does not, however, enter into an objectively decidable conflict with his rival who claims the truth of the same (token) sentence. No logical contradiction can arise between the competing positions since the rules fixing the correct use of the connective, i.e. the axioms determining its intension, are not the same. For this reason the meaning of 'ν' must be different and it should not, therefore, be affirmed that the systems have a common stock of truths. Insofar as no statement of the one can be expressed in the language of the other, they have to be considered incommensurable.

My transcendental method requires a uniform treatment of truth by means of modeltheoretic procedures. The difference between mathematics and empirical theories resides in the distinct nature of the denizens who populate their respective ontologies. The domain of the former consists of abstract entities created by the fact that we use a mathematical theory through which they are precisely definable, whereas in the domains of the latter we have physical objects whose existence must be ascertained by way of experimental procedures. Accordingly existential claims like '$(\exists x)$(x is a pentagon)' are analytically true in the context of Euclidean geometry since it follows logically from the axioms that there must be at least one individual which satisfies the predicate 'pentagon'.[4] On the other hand, synthetic statements like '$(\exists x)$(x is an electron)' are true in the context of quantum mechanics because physicists have been able to fix traces of such particles on photographic plates placed in cloud chambers in order to confirm the theory. It may be finally remarked that the theses of a system of logic (whose axioms and rules of deduction determine the intension of the logical constants) remain true under any interpretation of the descriptive terms and that they are, therefore, extensionally indistinguishable.

I hope that the reasons I have invoked in the present paper will convince the reader that truth without reference does not make sense. In accord with my transcendental method, I propose to apply the semantic predicate 'true' only to sentences seen from within a theory, complete with its posited ontology, which we use in a specific context. It seems to me that, in view of the unsurmountable difficulties with which (scientific) realism is confronted, we have no other choice than to banish any notion of external truth. Superseded theories cannot be deemed false in an absolute sense; they only have a more or less extensive range of more

[4]For a platonist believing in the absolute existence of mathematical objects who endeavours to render clear the informal notion of 'arithmetically true' by resorting to a formal system, there is a problem: as such systems are incomplete according to Gödel's theorem there will always be truths not captured by them. For me the problem does not arise because, from the start, I limit domains to objects specifiable within an (axiomatized) theory itself specified by an explicit set of rules.

or less precise applications. Newtonian mechanics, for instance, still works satisfactorily for a limited class of phenomena which we may call classical phenomena. Since it fails in cases where great distances and high velocities are involved, we must use for such domains Einstein's more efficient theory which, in turn, is not to be termed (externally or absolutely) 'true' any more than its predecessors.

In way of conclusion I remark that 'truth' is ultimately to be considered as an evaluative term designed to assess sentences with regard to their practical reliability. We can decide objectively whether a sentence used in a given context is true or not only after having accepted rules which fix convenient standards. Thus everything will finally rest on certain agreements about norms which I call conventions and which cannot be captured within a purely descriptive scientific theory, but must rather be discussed and settled by deliberation on a metalevel.

Chapter 8

EMPIRICAL VERSUS THEORETICAL EXISTENCE AND TRUTH*

Michel Ghins
Université Catholique de Louvain

Abstract On the basis of an analysis of everyday experience and practice, criteria of legitimate assertions of existence and truth are offered. A specific thing, like a newspaper, can be asserted to exist if it has some invariant characteristics and is present in actual perception. A statement, like "This newspaper is black and white", can be accepted as true if it is well-established in some empirical domain. Each of these criteria provides a sufficient condition for acceptance of existence and truth, respectively, at the empirical level. Following Hermann Weyl, it is argued that they can be extended to the scientific theoretical level to support a selective and moderate version of scientific realism according to which entities like the electromagnetic and gravitational fields, but not crystalline spheres or some topological manifolds, can legitimately be asserted to exist.

Keywords: Existence, truth, scientific realism, constructive empiricism, underdetermination.

1. Existence and truth in ordinary experiencce

Everyday experience and our sensory presence to ordinary objects are the *starting point* of any knowledge. On this, I agree with logical and constructive empiricists (and also with Aristotle and Aquinas). An examination of the use of the terms "existence" and "truth" in the context of everyday experience must reveal the criteria of their legitimate ap-

*This paper has been published, followed by Prof. Bas van Fraassen's comments, in *Foundations of Physics* (Vol. 30, No. 10, 2000) in a special issue dedicated to Prof. Maria Luisa Dalla Chiara. I would like to thank Kluwer for granting permission to republish this paper in the present volume. I am also grateful to Giovanni Boniolo, Harvey R. Brown, Benito Muller, Mauricio Suárez, Bas van Fraassen for their stimulating comments and suggestions.

D. Vanderveken (ed.), Logic, Thought & Action, 163–174.

plications. I am allowed to say that this newspaper in front of me, for example, exists when I am visually acquainted with it. Some empiricists tried to analyse statements like "This newspaper exists" in terms of more elementary statements about "immediate" *sense data.* I do not intend to discuss this question. Let me just point out that the "myth of the given" has been widely criticized, including within the empiricist tradition, by Quine (1953) and van Fraassen (1980) among others, and I deem these criticisms successful.

Why am I entitled to assert the existence of this newspaper? In the first place because I see it, and *not* because I am making some kind of (justified?) *inference* from *data* (or, more accurately, from *statements* on *data).* Actual *presence* in sensory perception is the first condition for the legitimacy of an affirmation of existence. But it is not sufficient on its own. A second condition is the permanence or *invariance*[1] for some time of some characteristics of the perceived object. These two conditions of presence and invariance constitute *jointly* the sufficient condition, the *criterion*[2], of existence that will be used.

An affirmation of existence goes usually beyond actual presence. Its acceptance calls, at least implicitly, for other possible experiences, by myself or other people. When I say that this newspaper exists, I also implicitly say that I will be able, for some (even very short) time, to perceive its shape, its colour, its texture, etc. In other words, I assert the permanence in time of some properties of the object. The Cartesian notion of "punctual" (durationless), vanishing existence seems unintelligible and, moreover, does not seem to find any correlate in ordinary experience.

But an affirmation of existence also calls for possible perceptions by other observers, at different spatiotemporal locations. Nobody doubts (except perhaps lunatics and some — very rare — philosophers...[3]) that, in the usual contexts, several people can see the *same*, unique, object, even if their visions differ, precisely because these perceptions also have constant, invariant, aspects, and because the observed variations show a systematic character. This point is stressed by phenomenologists when they say that the object presents itself through a variety of profiles

[1] A connection between objectivity and reality on the one hand, and invariance on the other has been discussed by phenomenologists, espoused by Einstein and Weyl and revived more recently by Michael Friedman (1983, p. 321).

[2] I do not propose a *definition* of existence. In accordance with a philosophical tradition which goes back to Aristotle and includes Aquinas, Kant and Carnap, I think that existence is not a property. Moreover, I want to leave open the *possibility* of the existence of metasensible entities even if they may be cognitively inaccessible to us.

[3] Even Sextus Empiricus, the "sceptic", did not put the existence of ordinary objects into question, unlike the radical sceptic *fabricated* by Descartes.

(*abschattungen*), potentially infinite in number. These profiles are not *sense data* but different perceptions of the same object.

The two conditions of presence and invariance are jointly sufficient for the legitimacy of the assertions of existence about ordinary observable objects. But these conditions do not exhaust the meaning (or, rather, the meanings) of the term "existence".

We often associate to the existence of an object some idea of independence with respect to our desires, our language, our actual perceptions, etc.[4]: a real object is something that imposes itself upon us and opposes resistance to our actions. A real thing is something that can hurt us. It is generally accepted that an actually perceived tree (to take over a famous example due to Hans Reichenbach) will continue to exist when nobody looks at it, and existed for some time before. The affirmation of its existence or independent reality rests upon the possibility of perceiving it at arbitrarily chosen times. If the independent existence of perceived objects is admitted, then the existence of *perceivable* objects must also be accepted. Thus, the acceptance of an assertion of existence implies also the acceptance of counterfactuals like: "If I looked at the tree now, then I would see it".

It seems thus reasonable to admit that any affirmation of existence based on presence and invariance, can legitimately be extended to moments and circumstances that go beyond the actual presence in order to include merely possible presence as well. Assertions about the reality of specific things or objects, even confined to ordinary, sensory experience, reach beyond actual perception to include possible observations, and, to that extent, they run the risk of falsification. Empirical everyday assertions about observable objects are not immune from error. The statement of the independent existence of specific objects carries an anticipating power with respect to possible perceptions (Nelson Goodman (1955) speaks of *projectibility*). If I say that this newspaper in front of me exists, I also predict that myself, and others, will see it in the (even very short) future. And this also implies that I would see this newspaper if I was located elsewhere, etc. But the newspaper could also disappear (by burning), and this would simply falsify the affirmation of its existence.

It must be stressed that the independent reality discussed here pertains to ordinary perceived objects. This does not commit us to the existence *an sich* or "for God", or in any other way, of ordinary objects. We do not have any reason, at least on the basis of the foregoing argu-

[4]See for example Putnam (1981).

mentation, to believe that ordinary objects exist, in themselves (whatever this may mean) in the way they are perceived (remember the 17th century debate about primary and secondary qualities).

2. Existence and truth in theoretical science

Now, I want to argue, if we accept the existence (or reality) of observable everyday objects, and the truth of some statements about them, there is no reason not to accept that *some* theoretical entities exist and that *some* physical laws are true. My whole argumentation rests on a parallelism on certain crucial respects between everyday assertions and scientific assertions.

The problem is to apply the criterion of acceptance of existence to *some* (not all) unobservable theoretical entities, postulated by scientific theories accepted today. Kant, in his *Critique of Pure Reason*, formulates the following criterion of reality : "that which is connected with perception according to laws" *is real* (A231; B284)[5]. This passage is quoted by Hermann Weyl in his *The Philosophy of Mathematics and Natural Science* (1963, p. 122). I will limit my discussion to physical, mathematized, theories. But it is in physics, where the most "abstract" and seemingly most remote from experience entities occur, that the issue of realism is considered most controversial[6].

Hermann Weyl takes the example of the electric field. We have the following law:

$$\bar{F}(P) = \bar{E}(P)q$$

The force $\bar{F}$, experienced by a charge q at a point P is proportional to the size of the charge. Weyl sees an analogy between the different perceptions of an ordinary object and the different observable manifestations of the electric field. Forces correspond to perceptions[7], and different charges correspond to different positions of the observers. An objective reality can thus be attributed to an electric field[8] when it is experienced by means of actual forces (condition of presence) and when a systematic variation of some properties together with the invariance of other properties is ascertained within the sequence of perceptions, namely the

[5] Kant specifies that the laws in question are *empirical,* thus not *apodictic,* laws.
[6] A realist like Ernan McMullin (1984) for example does not want to commit himself to the existence of theoretical entities introduced by mathematical physics, like electrons and fields.
[7] For the sake of argument, Weyl takes forces to be observable.
[8] Actually, the real object is the electromagnetic field which is covariant under the Lorentz group. Here Weyl restricts himself to one reference frame in which there is no magnetic component.

forces in the present case (condition of invariance). For example, the direction of the force remains constant whereas its strength varies in a systematic way in function of the electric charge. The mathematical expression above gives a precise formulation of this variation.

The electric field $\bar{E}$, occurs in other laws too (optical, among others). This reinforces the credibility of its existence, in the same manner that a larger variety of perceptions (shape, colour, texture, smell, etc.) of a newspaper gives more weight to the assertion of its existence.

After having defined a criterion of existence we must now give a criterion of *truth*. Any statement about ordinary observable objects can be accepted as true on the basis of simple experiences. The truth of the statement "This newspaper is black and white" is grounded on actual, effective perceptions. I say: effective perceptions, in the plural. Even at the level which is closest to sensory experience, truth rests on several perceptions, even if these perceptions are relatively few in number, belonging to a given domain (visual, tactile, acoustic, etc.) We already have here a sort of "induction"[9]. I do not mean here an inference or an inductive *argument* (which is a set of statements). It is a kind of fact (call it a *metafact*, if you wish) that human beings make assertions on the basis of their perceptions and that the latter are put forward as grounds for the truth of their assertions. Moreover, as we saw above, what holds for existence also holds for truth: both reach beyond the strict framework of actual perceptions. Assertions can then go beyond the individual sphere and enter the domain of intersubjectivity. The affirmation "This newspaper is black and white" includes an invitation, as it were, to verify it yourself. This assertion implies, at least implicitly, a series of counterfactuals like: if you came closer, you would continue seeing a newspaper, etc.

The criterion of truth that comes out from these (admittedly too brief) considerations is this: a statement "inductively" well-established in a certain domain of perceptions can be accepted -if only provisionally- as true. Acceptance of the truth of a statement does not commit oneself to the belief in its absolute or unrevisable truth. It does not commit oneself to the belief in the corresponding actualisation of an independent fact, existing somehow "in itself", either.

According to this criterion, we can also accept as true numerous physical statements (commonly referred to as "laws"), although their connection with experience is less direct (we briefly discuss the question of

[9] I will use the term "induction", with quotation marks, to mean the, somewhat mysterious, connection between statements and observations which would require further analysis ("Induction" is not abduction in Peirce's sense.)

underdetermination below). The mathematical laws of classical mechanics of pointlike masses, for example, are "inductively" well established in an empirical domain limited by what David Speiser (1990) called *negligibility relations*[10]. They circumscribe the domain of truth of the famous three Newton's laws[11]:

$$m\bar{v} = \bar{k}$$

$$\bar{F} = m\bar{a}$$

$$\bar{F_A} = -\bar{F_R}$$

The domain of truth of these laws is bounded by the following *negligibility relations*:

$$v \ll c$$

$$\int pdq \gg h/2\pi$$

$$Gm/c^2 \ll R$$

These relations[12] restrict the truth of Newton's laws to velocities which are small with respect to the velocity of light, to actions which are large relatively to Planck's constant and to weak gravitational fields (if the force is given by Newton's law of gravitation). They must be considered as an integral part of classical mechanics because we know that outside this domain we must use special relativity, general relativity or quantum mechanics. These relations permit also to endow the notion of approximate truth with a precise meaning. A physical law is not approximately true when it is a more or less faithful image of some "reality", but when measurement results in a certain domain do not deviate from the predictions more than some antecedently specified value.

Even within this limited empirical domain, classical mechanics (including the three negligibility relations) is still able to predict new facts, in empirical fields that have not been investigated yet, and, consequently, this theory runs the risk of being falsified. If that happens, new negligibility relations will be introduced. This provides an answer to Popper's objection according to which limiting a theory to a given domain would be tantamount to protect it against any adverse evidence, any possible falsification, which would be the very negation of the scientific enterprise.

[10]Krajewski (1977, p. 11) speaks of *limit conditions*. But this expression can lead to a confusion with *boundary conditions*.

[11]These laws hold for point mechanics only: for rigid bodies, fluids and elastic bodies respectively, other sets of laws must be used.

[12]v is the velocity of a moving body, c the velocity of light, p the linear momentum, q the position, h Planck's constant, G the gravitational constant, M the mass of the source, R the distance from the source.

We must acknowledge that logical positivists were essentially right when they claimed that it is extremely unlikely that a well-established theory in a certain domain would be falsified in that same domain. It is totally pointless to indefinitely repeat Galileo's experiments on inclined planes (except perhaps for students in physics...), in the same way that it would be a waste of time to keep looking for hours at a newspaper to make sure it is still really there. On the other hand, well-established statements and theories remain conjectural, in the sense that they could be abandoned some day, in the face of other, different, experiences. Previous experience does not warrant, with necessity, that future experience will be similar to it. It is quite possible that bodies suddenly begin to move in a quite surprising way according to new "laws", although we all consider this extremely unlikely.

3. Illustrative applications of the criteria

I would like now to illustrate the specific, selective and moderate, version of scientific realism I advocate with two particular examples, the gravitational field and the crystalline spheres of ancient astronomy.

Are we entitled to accept that the sun is surrounded by a real gravitational field[13] **g**, the Schwarzschild solution of Einstein's field equations? The metric **g** represents, according to the general theory of relativity, both the gravitational and the inertial field. The metric determines the geodesics, which are at the same time the straightest (along which parallely transported tangent vectors stay parallel) and the extremal — *longest* [14]- paths. These paths are not only the trajectories of free (submitted to the sole gravitational field) particles, but are also the extremal paths marked by the behaviour of metrical devices (clocks): the affine and the metrical structures coincide.

The distinction between geodesics and other paths is an objective feature, i.e. invariant under the wide group of continuous transformations[15]. Physically, this means that *any* observer, whatever its state of motion, who explores a sufficiently large region of spacetime will detect the presence of the gravitational field. Although the affine connection can be cancelled out *locally* by means of an adequate (and non-linear) transformation of the coordinates, there is no way to annihilate a non-

[13]One often refers to **g** as the field. But the field is actually represented by the affine connection Γ and **g** is the gravitational potential. The components of the affine connection are expressed in terms of derivatives of the components of the metric.

[14]They correspond to a maximal proper time.

[15]It is possible to give a "coordinate-free" formulation of the metric **g**. See for example Friedman (1983).

uniform gravitational field in any finite region of spacetime[16]. Particles, and generally bodies, moving along paths which are *not* geodesics will experience some deformations (just as water in Newton's rotating bucket) or will manifest other phenomena which may be called "inertial manifestations" (We want to avoid the expression "inertial effects" which would seem to imply some sort of causality. Only functional, mathematical correlations are at stake here). We can then apply the criterion of existence to the field **g**. It is indeed connected with observable manifestations in agreement with "inductively" well-established mathematical laws, like the law for geodesics[17]. A spacetime can then be considered to exist to the exact extent that we can accept the existence of the metrical field in some region (notice that this argument doesn't warrant the belief in the existence of a *continuum* of points).

Let us now examine an objection that can be directed against our realistic conception of the metrical-gravitational field. This objection relies on the possibility of formulating equivalent descriptions of the same observations. If another ("non-normal", according to Reichenbach's terminology) definition of congruence[18] is used, we can account for the same inertial phenomena by using another metrical structure. In that case, however, the inertial and metrical structures do not coincide anymore. The trajectories of free particles are no longer geodesics and extremal paths are no longer the straightest lines. Everybody, including constructive empiricists like van Fraassen, will prefer the "model" corresponding to the normal definition of congruence, since it is simpler and more integrated. The antirealist will nonetheless refrain from attributing a reality to the simplest model which was chosen for purely practical reasons.

What could the realist reply to this powerful objection? Notice first that equivalent descriptions can also be provided at the level of everyday experience. Let us suppose, to take over one of van Fraassen's examples (1980, p. 19), that we observe the following phenomena: cheese disappears, scratches are heard in the wall, hair is found on the floor, etc. From these phenomena, the existence of a mouse is established. Since mice are observable entities, the assertion of the existence of a mouse on the basis of experience, according to van Fraassen, is legitimate: at the level of phenomena, empirical adequacy is tantamount to truth. With a little luck, one could observe the mouse "directly", but this would only enlarge the amount of empirical evidence in favour of the existence of

[16]For more details on this see Ghins and Budden (2001).

[17]Details can be found in standard textbooks, like Sklar (1974). See also Ghins (1990).

[18]This would involve an, at least partial, change, in the empirical meaning of the word "congruence", but I leave this (already widely discussed) issue aside here.

a mouse and would not bring in any evidence of a different sort. I am *not* claiming here that the existence of the mouse *explains* the presence of some observations and that this explanatory role gives support to the assertion of existence[19]. On the contrary, the observations, and the more numerous and various they are the better, give support to the assertion of the existence of the mouse, irrespective of the (perhaps disputable) explanatory value of the fact of its existence *per se.*

But we could give the following equivalent description as well. Instead of saying that there exists only one mouse, we can posit the existence of different entities which we may call $mouse_1$, $mouse_2$, $mouse_3$, etc. $Mouse_1$ eats cheese, $mouse_2$ scratches the door, $mouse_3$ looses hair, etc. Someone may point out that you may see the mouse eat cheese *and* loose hair at the same time. But then you may say that $mouse_4$ performs both tasks. This seems highly artificial and Ockham would find this unacceptable, but it is logically and empirically admissible. In fact, we are not obliged to suppose that the *same* mouse perdures in time and performs these various actions. We can formulate an empirically equivalent description that resorts to a plurality of mice.

If the constructive empiricist accepts that the (unique) mouse exists and that empirical propositions about this mouse are true, it seems that he or she has no reason to deny that a unique entity, the gravitational field, is at the same time responsible for both metrical and inertial phenomena, instead of assuming that there is an affine connection, associated to inertial phenomena, and a metric, associated to metrical devices. Moreover, a spacetime geodesic is surprisingly easy to visualize: just let a pen drop on the floor, it will follow an inertial and extremal path in spacetime. Admittedly, we can, as we saw, restrict the existence of a tree to the times when it is actually perceived. (We can also go further and posit the existence of a plurality of entities : $tree_1$, $tree_2$, $tree_3$, etc.). But if the existence of the same tree during some time is conceded, even when it is not actually perceived, then it seems that we have no reason to deny that the gravitational and metric field is real. The assertion of the existence of this field is supported, in an immediate and natural way, by its observable manifestations at arbitrarily chosen spacetime locations and the truth of this assertion implies the truth of counterfactuals like: if I put a (free) particle at this or this place, it would follow this or this path.

As far as crystalline — transparent and hard — spheres of ancient astronomy are concerned, we must first notice that, although not visi-

[19]For a critique of the vindication of scientific realism by means of the 'no-miracle argument', i.e. inference to the best explanation of the success of science, see Ghins (2002).

ble, they were not unobservable in principle since a possible cosmonaut would hit on them (if they existed). But their hardness was not represented by a mathematical parameter connected to specific observations by means of invariant well-established mathematical laws. Crystalline spheres, even at the time of Ptolemy[20], failed to comply with our criterion of existence. It could be perhaps retorted that, according to our criterion, the circles and epicycles of Ptolemaic astronomy could be legitimately asserted to be real. But the parameters characterizing those circles (radius and angular velocity) were not connected to observations in a clear-cut and well-established manner (Gardner 1983, p. 207-210). On the other hand, the geometrical trajectory — resulting from various possible combinations of circles the existence of which is indefensible (underdetermination is present here) — actually followed by a planet can be asserted to exist. The geometrical trajectory empirically marked by a planet has a different status from a purely geometrical figure, since it is not like a circle drawn on a blackboard. The latter exists spatially "all of a piece", so to speak, whereas the former is successively and temporally actualised by the planet and could be said perhaps to exist "all of a piece" in spacetime.

It can be objected further that Ptolemaic astronomy is false anyway and has been replaced by Keplerian, Newtonian and finally (up to this date) by Einsteinian astronomy. To this I reply with the well-known approximation arguments. Suppose I say that this soccer ball, as I perceive it, is round. Somebody could challenge that contention by pointing out that the soccer ball is not exactly spherical but is closer to an ellipsoid, or that its surface is slightly irregular. Would this falsify my previous assertion ? In a sense, yes, if the ball is not *exactly* round. But, on the other hand the truth of the assertion "This ball is round" is acceptable in most circumstances. I am ready to concede that truth is relative to some precision requests that are usually left vague in ordinary contexts[21].

4. Conclusion

Let me conclude with a few remarks. First, as we saw, the acceptance of the existence of an entity does not involve any commitment to the belief in its existence "in itself" with the properties we attribute to them. The acceptance of the truth of a statement does not commit one

[20]Whether Ptolemy himself believed in solid spheres or not is a controversial issue, but Adrastus of Aphrodisias and Theon of Smyrna did (Duhem 1969). However, we are not concerned here with what they actually believed but rather with what they were *entitled* to believe.
[21]This notion of approximate truth is also discussed in Ghins (1992).

to the belief in the existence "in nature" of some state of things that makes the statement true, nor to his unrevisable and absolute truth. To that extent, the realism advocated in this paper is *moderate*.

Second, we are not allowed to accept the truth of all statements of currently accepted scientific theories. For example, we cannot accept the truth of statements about the global topology of the universe since empirically indistinguishable but non-isomorphic spacetimes with distinct topologies occur in the framework of Robertson-Walker cosmology (Glymour (1977) and Malament (1977)). (It may be pointed out however that current "standard" cosmology is far from being well-established). Thus, the brand of scientific realism we favour is *selective*.

Third, in the same way as objectivity comes in degrees, reality also comes in degrees and is proportional to the size of the invariance group. The same line of reasoning was followed by John Locke (1690) when he defended that primary (geometrical) qualities belong to the things themselves, while secondary qualities (colour, texture, smell, etc.) depend on the interaction of the objects with our sensory organs. Geometrical qualities (shapes) are, according to Locke (1690, Book II, §19), invariant under a larger variation of observational conditions than secondary qualities (like the colour of porphyry). I do not want to maintain that any property or quality, be it mathematical or not, belong to the things "in themselves" and to endow some privilege to mathematical properties on this respect. I only wish to say that a physical, theoretical, mathematically represented entity is more real than another to the extent that it is more invariant. For example, in the theory of general relativity, the scalar curvature of spacetime is more objective, and real, than the velocity of a particle in some reference system.

References

Duhem P. (1969). *To save the Phenomena. An Essay on the Idea of Physical Theory from Plato to Galileo* translated by Doland, E. and Maschler, C. Chicago University Press.

Friedman M. (1983). *Foundations of Space-Time Theories. Relativistic Physics and Philosophy of Science.* Princeton University Press.

Gardner M. (1983). "Realism and Instrumentalism in Pre-Newtonian Astronomy" in Earman, J. (Ed.), *Minnesota Studies in the Philosophy of Science. Vol. X. Testing Scientific Theories.* Minneapolis: University of Minnesota Press.

Ghins M. (1990). *L'inertie et l'espace-temps absolu de Newton à Einstein. Une analyse philosophique.* Bruxelles: Académie Royale de Belgique.

Ghins M. (1992). "Scientific Realism and Invariance" in *Rationality in Epistemology, Philosophical Issues 2.* Ridgeview.

Ghins M. & Budden T. (2001). "The Principle of Equivalence", *Studies in History and Philosophy of Modern Physics.* Vol. 32B, 1. March 2001, 33–51.

Ghins M. (2002). "Putnam's No-Miracle Argument: a Critique" in Clarke S. and Lyons T.D. *Recent Themes in the Philosophy of Science. Australasian Studies in the Philosophy of Science, 17.* Dordrecht: Kluwer.

Glymour C. (1977). "Indistinguishable Spacetimes and the Fundamental Group" in Earman, J. Glymour, C. Stachel, J. (Eds.) *Minnesota Studies in the Philosophy of Science. Vol. VIII. Foundations of Space-Time Theories.* Minneapolis: University of Minnesota Press.

Goodman N. (1955). *Fact, Fiction and Forecast.* Harvard University Press.

Krajewski W. (1977). *Correspondence Principle and Growth of Science.* Dordrecht: D. Reidel.

Locke J. (1690). *An Essay concerning Human Understanding..*

McMullin E. (1984). "A Case for Scientific Realism" in Leplin J. (ed.), *Scientific Realism.* Berkeley: University of California Press.

Malament D. (1977). "Observationally Indistinguishable Space-times" in *Minnesota Studies in the Philosophy of Science. Vol. VIII. Foundations of Space-Time Theories.* Minneapolis: University of Minnesota Press.

Putnam H. (1981). *Reason, Truth and History.* Cambridge University Press.

Quine W.V.O. (1953). *From a Logical Point of View.* Harvard University Press.

Sklar L. (1974). *Space, Time and Space-Time.* Berkeley: University of California Press.

Speiser D. (1990). "Truth and Reality in the Exact Sciences" in *Rivista Trimestrale di Analisi e Critica. Nuova Civiltà delle Macchine 8, 4.* (Special issue: *La verità nella scienza*). Nuova Eri.

van Fraassen B. (1980). *The Scientific Image.* Oxford: Clarendon Press.

Weyl H. (1963). *Philosophy of Mathematics and Natural Science.* New York: Atheneum.

Chapter 9

MICHEL GHINS ON THE EMPIRICAL VERSUS THE THEORETICAL*

Bas C. van Fraassen
Princeton University.

Abstract Michel Ghins and I are both empiricists, and agree significantly in our critique of "traditional" empiricist epistemology. We differ however in some respects in our interpretation of the scientific enterprise. Ghins argues for a moderate scientific realism which includes the view that acceptance of a scientific theory will bring with it belief in the existence of all those entities, among the entities the theory postulates, that satisfy certain criteria. For Ghins these criteria derive from the criteria for legitimate affirmation of existence for any entities, the directly observable ones not being privileged in that respect. They are roughly that the putatively existing entity should according to the accepted theory manifest itself in our experience, and display a certain permanence and invariance. My disagreement on this topic derives from a larger difference concerning the relation between experience, existence, and theory.

1. Introduction

The empiricist tradition admits of a great variety of views and of diverse links with other traditions. This diversity is evident when I try to compare Prof. Ghins' views to my own, and both of ours to others. We are both empiricists, or rather, trying to be (for neither of us is content with empiricism's past; it is a matter of forging new stages for an old tradition). We are agreed in some of our critiques of "traditional" epistemology, including some items erstwhile beloved of empiricists. I will mention some of our agreements, and then draw attention to some of

*Commentary on M. Ghins, "Empirical versus theoretical existence and truth." I wish to thank Michel Ghins for helpful conversations. This article appeared earlier in *Foundations of Physics, 30* (2000), 1655–1661. It is reprinted with permission.

D. Vanderveken (ed.), Logic, Thought & Action, 175–181.

our differences. I would like to emphasize that for the most part I cannot argue the issues; I will simply display my different way of approaching them, as basis for future dialogue.

One agreement is very clear: while we both wish to give a central role to experience in epistemology, we both reject the myth of the "given" and the phenomenalist views which elaborated on that myth. We are also agreed in our wish to respect the phenomenology of scientific inquiry. We share the hope for a truly non-foundationalist and yet empiricist epistemology.

2. Separating the questions

Concerning existence and truth, I should like to separate questions of epistemology from questions of meaning, reference, and truth. Prof. Ghins writes:

> "These two conditions of presence and invariance constitute jointly the sufficient condition, the criterion, of existence..."

The context makes clear that "presence" refers here to presence in experience, and that the two conditions are being presented as legitimizing affirmations or attributions of existence. That is: if I encounter something in experience, then I may legitimately assert its existence.

The term "invariance," characterizes the (putative) object to which existence is attributed. When Prof. Ghins discusses the reality of the electromagnetic field, he mentions its permanence and the invariance of certain of its properties. Should we think of this as meant to indicate a necessary condition of existence? This would take us outside the area of purely epistemological issues, to issues of ontology. But as I read him, I see this factor mentioned only in adduced grounds for legitimate affirmation of existence. Thus I think that we have here also to do with a partial criterion of legitimacy for assertion of (or belief in) the existence of something.

So if I construe this rightly, we are not discussing a criterion of existence, strictly speaking, but a criterion of warranted or legitimate assertion, affirmation, or belief. The suggested epistemic principle (which may or may not be meant to extend to attributions of properties as well as affirmations of existence) is a sophisticated version of something like the idiom "seeing is believing."

Principles of this sort have indeed been associated with the empiricist tradition. I will discuss this subject further in the next section. In the meanwhile, at the risk of sounding overly pedantic, I would like to emphasize that, in my view, existence has nothing to do with experience, at least not in general. Nor does permanence or invariance. I would not accept any such considerations as bearing on necessary conditions for existence in general. They may bear on the sort of object we are considering: nothing can be a mountain, for example, or a dinosaur, or a horse, without being perceivable and persisting for an appreciable amount of time. This follows from the sorts of things mountains, dinosaurs, and horses are. Therefore it is true about any of them, existent or not. If it implies something about existing mountains (dinosaurs, horses) that is simply because an existing mountain (dinosaur, horse) is a mountain (dinosaur, horse).

The same holds, it seems to me, for the counterfactual assertions implied by certain existence statements, about what we would perceive under different conditions. If you look at a horse from a different angle, you will see a shape predictably related to your first view, by a certain geometric transformation. This is not because it exists, but because it is a horse, and because our vision is thus and so, light is such and such, and so forth. Perhaps there are also entities which are vanishingly transient, one- or two-dimensional, invisible, intangible... I do not know; but the meaning of "existence" will not legislate on that question.

3. Can we reject the epistemic principle?

Even if these remarks did no more than clear the ground for discussion, they serve to raise the question: what of the epistemic principle, apparently suggested here, for legitimate affirmation of existence? Now, as I see this, an affirmation of existence is an assertion, and so the principle must take the form: under the following conditions, one may legitimately assert [believe] that such and such is true. Is it really the case that if we perceive something and it displays a certain permanence in our experience (as well as some invariance in its properties as we or others inspect it from different angles), then we may legitimately assert that it exists?

To me that question remains ambiguous, until we are told what may be substituted in the "it." Are we to read this principle de re or de dicto? The former seems inappropriate, for it would amount to: If something is perceived by a person, etc., then that person may legitimately assert that it exists. To apply that principle in a particular case, the person would have to already believe that there was something s/he was perceiving, and I take it that "there is" means "there exists." But I have even more difficulties with the de dicto reading, on which the "it" can, in effect, be a placeholder for a descriptive phrase.

Suppose that burning is oxidation. Imagine now a person raised on the phlogiston theory, having much evidence for that theory, and as yet very little evidence to support the new rival theory about oxygen. This person encounters some fire [oxidation] and perceives that this phenomenon has some permanence etc. Is this person now entitled to assert or believe that oxidation exists (is taking place)? Is s/he not much more entitled to assert that phlogiston is escaping instead?

We cannot disentangle questions about [belief in] existence from questions about truth. The example I just gave calls into question Prof. Ghins suggestion that a statement about ordinary observable objects can be called true on the basis of simple experiences. The same considerations would seem to me to bear on a criterion of truth in relation to our experience. Statements are true if what they say is indeed so, and false if what they say is not so; isn't that pretty well the end of the matter?

As I said, I am not so much arguing here as displaying a different way of thinking about the same issues, predicated (in my case) on a quite sharp separation between epistemology and semantics. As I see it, the main ambiguity in the philosophical notion of experience is between, on the one hand, what happens to us that we are aware of, and on the other hand, our immediate and spontaneous response to what happens to us. What happens to us, and which of the events that happen to us are noticed by us, those are factual questions whose answers depend on theory-independent factors. But how we respond –and here I include the very first, spontaneous response to those events, prior to any discursive thought– is clearly conditioned by the language in which we live. Any judgement involved in that response (such as "Lo! phlogiston escaping!") always involves some implicit description of the event. This description is historically conditioned –and in general, theory-laden– to a very large extent. An accepted theory may be wrong. If we attend critically to our experience and we have the proper ration of epistemic luck, this falsity will manifest itself in the disappointment of expectations shaped by that theory. Until that happens, however, all those expectations may well be legitimate, warranted, entitled, rational, reasonable, what have you. The grounds adduced for them will be reports on our experience themselves shaped by that very theory, couched in its terms, and implying counterfactuals and predictions via that theory.

4. Moderate realism

I shall leave myself open to suspicions of stone-walling or evasion if I do not reply to Prof Ghins' main challenge, from one sort of empiricist to the other sort of empiricist that I wish to be. Prof. Ghins argues for a moderate scientific realism. If I understand him correctly, and if I may put it to some extent in my own terms, this means that acceptance of a scientific theory will bring with it belief in the existence of all those, among the entities it postulates, which satisfy certain criteria. These criteria derive from the criteria for legitimate affirmation of existence for any entities, the directly observable ones not being privileged in that respect. They are roughly, as we saw, that the putatively existing entity should manifest itself in our experience, and display a certain permanence and invariance in certain respects.

While this is a little abstract, Prof Ghins' example of the electric field, as discussed by Herman Weyl, makes it very clear: An objective reality can thus be conferred on an electric field when actual forces are experienced (condition of presence) and when a systematic variation of some properties is ascertained within the sequence of perceptions (forces). But some of the distinctions I made above may be brought into play here. That is perhaps easier with respect to the metric-gravitational field.[1] Prof. Ghins writes:

> "Moreover, a spacetime geodesic is surprisingly easy to visualize: just let a pen drop on the floor, it will follow an inertial and extremal path in spacetime."

Imagine now three people watching this experiment with the pen: a Stone Age primitive, a Newtonian, and one of our (sufficiently educated) contemporaries who believes in the reality of the metric-gravitational field. The third is indeed entitled to assert that he can, as it were, point to a geodesic line segment. The second is at least entitled to believe that he saw a pen drop; the first is not entitled to believe even that. Perhaps seeing is believing; and undoubtedly all three saw the very same thing (event); but they did not all see that the pen followed a geodesic.

---

[1]The example of the mouse has its intricacies, and is rather different, as I see it, from either the electromagnetic or metric field. Prof. Ghins suggests two hypotheses, each of which fits the findings in my kitchen. One of these implies the presence of one mouse doing many things, and the other implies the presence of many mice doing a little each. It is true that the findings would prima facie support either hypothesis. There are, however, various ways to construe the second hypothesis. If all those mice are real mice, and are as described by current biology, I would not consider the second hypothesis empirically equivalent to the first, for the observable phenomena would not all be the same. (The actually observed phenomena may be the same, if human observers are successfully evaded.) A kitchen with two mice in it is different from one with only one. If the hypothesis is instead of science-fiction mice, who are very transient, springing into and out of existence, there is also an observable difference, though of a different sort. Finally we may entertain a third construal on which the difference is real but not empirical, for instance if empirical mice are construed as sets or series of "time-slices." Such empirically equivalent but "metaphysically" distinct possibilities are of course encountered in certain interpretations of quantum mechanics, e.g., in connection with the "problem of identical particles." I would never wish to forbid theoretical recourse to anything, as long as it is logically consistent, but do not see acceptance of science as requiring a choice to believe in one rather than another of such empirically equivalent hypotheses.

The reality of the event is not in question. Nor is there any question as to how this event should be classified relative to various theories. But that does not logically settle either how much of those theories is true, or how much we need to believe of them when we judge them adequate by all publicly applicable standards.

5. Immoderate empiricism

Empiricism is often suspected of idealist leanings, and Prof. Ghins very clearly lays some of those suspicions to rest. He rejects phenomenalism, the myth of the given and all its ilk, removing the suspicion that he views reality as constituted by or constructed from experience. His moderate realism with respect to science is both genuinely moderate and genuinely realist.

To put it in my own terms, if I may: this moderate realism entails that acceptance of science involves belief in the reality of all those entities among the ones it postulates that bear a certain relationship to what (according to science) can be experienced. In form at least this is very close to the "constructive empiricism" which I advocate. The difference appears when we ask about that "certain relationship." How that question is answered will draw a line in the sand, so to speak, and constructive empiricism is less moderate: it reads "bear a certain relationship to" very strictly as "are among."

But as noted above, this close resemblance between the two positions occurs in a context which may harbor deeper disagreement, and thus qualify the resemblance. I see empirical science as a certain sort of enterprise, oriented to empirical success. To me, this orientation does not have a deeper or privileged basis, either in a connection between reality and experience, or in principles of warrant or legitimacy that confine rational belief to within certain bounds. We engage in an enterprise because we value the results it aims for and because we believe in its adequacy (or superiority to any rivals) as means to that end. There are other sorts of enterprise, distinct from empirical science, distinguished from it by their own aims and not by lesser or greater epistemic warrant.

Am I skeptical of those non-scientific enterprises? I am certainly skeptical of certain forms of metaphysics, both traditional and analytic, and never more than when they purport to be extensions of science in the scientific spirit. On the other hand, I think that much of what we come to understand, know, or find out about ourselves, others, and the human condition is not within the domain of science at all, and the way in which we do so is not at all by means of the sort of "objectifying" inquiry proper to science. Science is a shining example of a cognitive enterprise with a clear and admirable ethic of inquiry. But to follow an example judiciously requires judgement.

Quite possibly Prof. Ghins and I are in agreement on this. But with the emphasis on the distinctive aims of science must come a more nuanced view of experience. The historical character of science is to be taken into account, and the relationship between empirical science and experience –as well as the character of experience in general– is complex. As always, I would insist that what happens to us, what we observe and what we can observe, are all matters which are entirely independent of theory, whether learned, interiorized, or considered hypothetically. But the form of our response is not, and when we try to relate theory to what has shown itself to us in experience, we can only start from the language in which we spontaneously react, with the possibility of self-criticism (including critique of our own prior language and opinion) necessarily posterior to that stage.[2]

[2]For some elaboration of these brief comments, see my "From vicious circle to infinite regress, and back again," D. Hull, M. Forbes, and K. Ohkruhlik, eds., PSA 1992, Vol. 2 (*Proceedings of the Philosophy of Science Association Conference*, Nov. 1992) (Chicago, Northwestern University Press, 1993), pp. 6-29.

III

PROPOSITIONS, THOUGHT AND MEANING

Chapter 10

PROPOSITIONAL IDENTITY, TRUTH ACCORDING TO PREDICATION AND STRONG IMPLICATION*

With a Predicative Formulation of Modal Logic

Daniel Vanderveken
Université du Québec, Trois-Rivières

Abstract

In contemporary philosophy of language, mind and action, propositions are not only *senses of sentences* with truth conditions but also *contents of conceptual thoughts* like illocutionary acts and attitudes that human agents perform and express. It is quite clear that propositions with the same truth conditions are not the senses of the same sentences, just as they are not the contents of the same thoughts. To account for that fact, the logic of propositions according to predication advocates finer criteria of propositional identity than logical equivalence and requires of competent speakers less than perfect rationality. Unlike classical logic it analyzes the structure of constituents of propositions. The logic is *predicative* in the very general sense that it analyzes the type of propositions by mainly taking into consideration the acts of predication that we make in expressing and understanding them. Predicative logic distinguishes strictly equivalent propositions whose expression requires different acts of predication or whose truth conditions are understood in different ways. It also explicates a new relation of strong implication between propositions much finer than strict implication and important for the analysis of psychological and illocutionary commitments. The main purpose of this work is to present and enrich the logic of proposi-

*I am grateful to Elias Alves, Nuel Belnap, Paul Gochet, Yvon Gauthier, Raymond Klibansky, Grzegorz Malinowski, Jorge Rodriguez, Olivier Roy, Ken McQueen, Marek Nowak, Michel Paquette, Philippe de Rouilhan, John Searle and Geoffrey Vitale for their critical remarks. I also thank the Fonds québécois pour la recherche sur la société et la culture and the Social Sciences and Humanities Research Council of Canada for grants that have supported this research. I have developed that logic for the purposes of speech act theory and the formal semantics of natural language in *Meaning and Speech Acts* [1990-91] and other essays.

D. Vanderveken (ed.), Logic, Thought & Action, 185–216.

tions according to predication by analyzing elementary propositions that predicate all kinds of attributes (extensional or not) as well as modal propositions according to which it is necessary, possible or contingent that things are so and so. I will first explain how predicative logic analyzes the structure of constituents and truth conditions of propositions expressible in the modal predicate calculus without quantifiers. The ideal object language of my logic is a natural extension of that of the minimal logic of propositions.[1] Next I will define the structure of a model and I will formulate an axiomatic system. At the end I will enumerate important valid laws. The present work on propositional logic is part of my next book *Propositions, Truth and Thought* which formulates a more general logic of propositions according to predication analyzing also generalization, ramified time, historic modalities as well as action and attitudes.

I will only discuss here modalities such as necessity, contingency and possibility as they are conceived in the broad logical universal sense of S5 modal logic.[2] All the truths of logic and mathematics are necessarily true in this wide sense[3] as are a lot of other analytically true propositions e.g. that husbands are married as well as some synthetically true propositions e.g. that whales are mammals. As Leibnitz pointed out[4], in asserting modal propositions, we consider possible worlds different from the real world in which we are. In the philosophical tradition, the *real world* is just the way things are, while a *possible world* is a way things could be. On one hand, a proposition is necessarily (or possibly) true in the broad logical sense when that proposition is true at all moments in all possible worlds (or at some moment in some possible world). On the other hand, a proposition is contingently true (or false) in the same sense when that proposition is true (or false) at a moment in the real but not in all possible worlds. From this point of view, in thinking that some propositions are logically necessary, possible or contingent we simply proceed to a universal or existential quantification over the set of all *possible circumstances* which are conceived here simply as pairs containing a moment of time and a possible world.

In order to analyze attributes and modalities, I will raise fundamental questions such as these: What is the nature of intensional attributes? What is the structure of constituents of elementary and complex propo-

[1]See my paper "A New Formulation of the Logic of Propositions" in M. Marion & R. Cohen (eds), *Québec Studies in the Philosophy of Science*, Volume 1, [1995]

[2]See C. I. Lewis, *A Survey of Symbolic Logic* [1918].

[3]See A. Plantinga, *The Nature of Necessity* [1974] for a philosophical explanation of the notions of logical necessity and possibility.

[4]See L. Couturat (ed.), *Opuscules et fragments inédits de Leibnitz* [1903].

sitions? In particular, which attributes do we predicate in expressing modal propositions? Moreover, how do we understand truth conditions? How are propositions related by the various kinds of implication (strict, analytic and strong implication) that we can distinguish in logic? We are not omniscient. We do not know the way things are in the real world. So we consider not only how things are but also how they could be. We conceive of many ways actual things could be and we can refer to possible objects which are not actual. We can also try to refer to objects which do not exist. We distinguish between certain necessarily true (or false) propositions and others which are contingently true (or false). Thus we know that it is necessary that $7 + 2 = 9$ and we think that it is contingent that there are nine planets. However we are sometimes inconsistent. We can assert and believe necessarily false propositions in science as well as in ordinary life. We used to believe the paradoxical principle of comprehension in naïve set theory. Some of us still believe that whales are fishes. Are there necessarily true propositions that we *a priori* know and necessarily false propositions that we could not believe? We do not draw all logical inferences. We can believe in the truth of incompatible propositions but these beliefs clearly do not commit us to believing any proposition whatsoever. What kinds of valid theoretical inferences are we able to make by virtue of linguistic competence? We, human agents are *minimally rational*[5] and *paraconsistent* in the use of language and the conduct of thought. Could we explicate rigorously minimal rationality in logic?

1. Principles of the logic of propositions according to predication

As is well known, so called *strictly equivalent* propositions (propositions which are true in the same possible circumstances) are not substitutable *salva felicitate* within the scope of illocutionary forces and psychological modes. We can assert (and believe) that Brasilia is a city without *eo ipso* asserting (and believing) that Brasilia is a city and not an erythrocyte. However, these two assertions (and beliefs) have strictly equivalent propositional contents; they are true under the same conditions. From a philosophical point of view, then, propositional identity requires more than truth in the same possible circumstances. We need a criterion of propositional identity stronger than strict equivalence in

[5]The term of *minimal rationality* comes from C. Cherniak *Minimal Rationality* [1986]

logic. It is a mistake to identify as Carnap advocated[6] each proposition with the set of possible circumstances in which it is true. On the basis of speech act theory, I advocate a finer analysis in terms of predication of the logical type of propositions. As I have pointed out repeatedly, even the simplest elementary propositions whose attribute is extensional and their truth functions have a more complex logical structure than truth conditions. Here are the basic principles of my theory of sense and denotation.[7]

1.1 A finite structure of constituents

Propositions are complex senses provided with a finite structure of constituents. As Frege, Russell, Strawson and many others pointed out, understanding a proposition consists mainly of understanding which *attributes* (properties or relations) objects of reference must possess in order that this proposition be true in a possible circumstance. In expressing and understanding propositions we *predicate* attributes *of* objects in a certain order. Propositional contents are then composed from a finite positive number of *atomic propositions* corresponding to *acts of predication.* Thus the proposition that Paul is wounded and smaller than Mary has two atomic propositions: one predicates of Paul the property of being wounded, the other predicates successively of Paul and Mary the relation of being smaller than.[8]

1.2 No singular propositions

Propositional constituents are senses and not pure denotations. As Frege[9] pointed out, we always *refer to* objects by subsuming them under senses. We cannot have directly in mind *individuals* which are objects of reference of the simplest type.[10] (Persons and material objects of the world which exist in space time are individuals.) We have in mind

[6]Classical logic follows R. Carnap *Meaning and Necessity* [1956]. See R. Barcan Marcus, *Modalities* [1993] and R. Montague, *Formal Philosophy* [1974]

[7]See "Universal Grammar and Speech Act Theory" in D. Vanderveken & S. Kubo (eds.) *Essays in Speech Act Theory* [2001] for propositional universals and *Formal Ontology, Propositional Identity and Truth According To Predication With an Application of the Theory of Types to the Logic of Modal and Temporal Proposition* in *Cahiers d'Épistémologie* [2003] for a more general presentation and axiomatization of my theory.

[8]Predication as it is conceived here is purely propositional and independent on force. To predicate a property of an object is not to judge that it has that property. It is just to *apply* the property to that object in the sense of functional application. We make the same predication when we assert and deny that an object has a property.

[9]See "On Sense and Reference" in P. Geach & M. Black (eds) *Translations from the Philosophical Writings of Gottlob Frege* [1970]

[10]See P.F. Strawson *Individuals* 1959.

concepts of such individuals and we *indirectly* refer to them through these concepts. So expressions used to refer to individuals have a sense called an *individual concept* in addition to sometimes a denotation in each context. When we speak literally we express the proposition that is the sense of the sentence used in the context of utterance. In that case we refer to the objects which fall under the concepts expressed by the referential expressions that we use. It can happen that there are no such objects. This does not prevent us from expressing a proposition. By recognizing the indispensable role of concepts in reference, logic can account for the meaning and referential use of proper names and definite descriptions without a denotation. They contribute to determine propositions which have (according to Russell) or lack (according to Frege and Strawson) a truth value in the context of utterance.

Frege's argument in favor of indirect reference remains conclusive if one accepts that every proposition is the possible content of a thought. From a cognitive point of view, it is clear that the proposition that the morning star is the morning star is very different from the proposition that the morning star is the evening star. We *a priori* know by virtue of linguistic competence the truth of the first proposition while we *a posteriori* learned the truth of the second at a certain period of history. A similar difference of cognitive value exists between the two propositions that Hesperus is Hesperus and that Hesperus is Phosphorus expressed by using the proper names "Hesperus" and "Phosphorus" of the morning star and the evening star respectively.[11] Frege's idea that propositional constituents are the senses and not the denotations of the expressions that we use to refer clearly explains the difference in cognitive value between the two propositions. It also preserves the minimal rationality of speakers. We can make mistakes and believe, as did the Babylonians, that the morning star is not the evening star or that Hesperus is not Phosphorus. But we could not assert or believe the contradictory proposition that the morning star is not the morning star or that Hesperus is not Hesperus. Otherwise we would be totally irrational. So logic has to reject the theory of *direct reference*[12] according to which certain referential expressions, logical proper names (according to the first Wittgenstein and Russell) ordinary proper names (according to Kaplan and Kripke), do not have any sense. There are *no singular propositions* having pure individual objects as constituents in the formal ontology that I advocate contrary to Russell, Quine, Davidson, Kaplan, Kripke

[11]The example was given by David Kaplan in a lecture at McGill University.

[12]The notion of direct reference comes from David Kaplan "On the Logic of Demonstratives", *Journal of Philosophical Logic* [1970].

and others who defend direct reference and externalism. Any *object of reference* is subsumed *under a concept*. Often proper names are introduced into language by an initial declaration.[13] A certain speaker gives the name to an object with which he is acquainted or that he discovers. And the name is adopted by the linguistic community which keeps using it to refer to the same object. Later speakers who do not know much of that object can always refer to it under the concept of being the object called by that name (this is their concept).[14] All propositional constituents are therefore senses: they are concepts or attributes.

1.3 Reference and predication

In my view, as in the logical tradition of Frege, Church, Carnap and Strawson, the two kinds of propositional constituents serve different roles in the determination of truth conditions: *attributes* serve to *predicate* while *concepts* serve to *refer* to objects. Attributes of individuals of degree n are *senses of* n-ary *predicates* while individual concepts are *senses of individual terms* in the formal semantics of the logic of propositions according to predication. So the domain of any possible interpretation of language contains a non empty set *Individuals* of individual objects as well as two non empty sets *Concepts* of individual concepts and *Attributes* of attributes of individuals.

1.4 A relation of correspondence between senses and denotations

There is a fundamental logical *relation of correspondence* between senses and denotations[15] underlying the relation of correspondence between words and things in philosophy of language. To propositional constituents *correspond actual denotations* of certain types in possible circumstances. Thus to each individual concept corresponds in each circumstance the single individual object which falls under that concept in that circumstance whenever there is such an object. Otherwise that concept is deprived of denotation in that circumstance. To each property of individuals corresponds in each circumstance the set of objects under concepts which possess that property in that circumstance. Individual

[13]See S. Kripke *Naming and Necessity* [1980]

[14]The fact that different speakers using a proper name to refer to an object can have very different private mental representations and sensorial impressions of that object as well as very different beliefs about it does not prevent them to have in mind a common concept of that object, for example, the object named by that name in current discourse.

[15]See A. Church "A Formulation of the Logic of Sense and Denotation" in P. Henle & *al* (eds) *Structure, Method and Meaning* [1951]

things change in the possible courses of history of the world. Their properties vary at different moments. So different denotations can correspond to the same concept or attribute in different circumstances. Few senses have a rigid denotation. However individual objects have certain unique *essential properties* (Plantinga 1974) in all circumstances where they exist. For example, each human being has his own genetic code. Speakers who refer to individuals do not know all essential properties.

As is well known, one must take into account the order in which we predicate a relation of several objects of reference. Many relations are not symmetric. Some are even asymmetric. This is why the denotation of a relation of degree n is a *sequence* of n objects under concepts. The order in the sequence shows the order of predication. The first, second, ..., and last element of the sequence are the first, second, ..., and last object of which the relation is successively predicated.

1.5 Intensional attributes

As is well known, many attributes that we predicate of objects of reference are *intensional*; they are satisfied by sequences of objects under certain concepts and unsatisfied by the same sequences of objects under other concepts. One can admire Napoleon under one concept (the winner of the battle of Austerlitz) without admiring him under another concept (the first Emperor of France). For that reason, the actual denotation of a first order attribute of individuals in a circumstance is a set of sequences of individuals under concepts rather than a set of sequences of individual objects. So logic can account for the predication of so-called *intensional attributes* and explain failures of the law of extensionality. *Extensional properties* like the property of being alive have a special feature; they cannot be possessed by an individual object under a concept in a circumstance without being also possessed by the same object under all other concepts of that object in that circumstance. So the truth value of an atomic proposition predicating an extensional property of an object under concept only depends on the denotation of that concept.

1.6 Ignorance of actual denotations

Our knowledge of the world is partial. We do not know by virtue of linguistic competence actual denotations of most propositional constituents in possible circumstances that we consider. We often *refer* to an object under a concept without knowing and being able to identify that object. The police officer who is pursuing the murderer of a certain person called Smith can just refer to whoever in the world is that murderer. Any speaker who refers to an object under a concept *presupposes* that a single object falls under that concept in the context of utterance. The concept gives *identity criteria* for the object of reference (e.g. to be Smith's murderer). But few identity criteria enable us to *identify* the object of reference. Moreover some of our beliefs are false. We can wrongly believe that an object falls under a concept. The presumed murderer is sometimes innocent. In that case the *object of reference* is not the *denotation* of the concept that we have in mind. It can also happen that no object satisfies the identity criteria.[16] (Suppose that Smith's death was accidental.) Or even that several objects satisfy them. (Smith was killed by several men.)

1.7 Many possible denotation assignments to senses

We can ignore who has killed a certain person. But we can at least think of different men who could have committed the crime. Whoever conceives propositional constituents can in principle assign to them possible denotations of appropriate type in circumstances. Our possible denotation assignments to senses are functions that associate with individual concepts one or no individual object at all and with attributes of degree n a set of n-ary sequences of individuals under concepts in possible circumstances. From a cognitive point of view, we often believe that only certain entities could be the denotations of attributes and concepts in circumstances that we consider. Certain possible valuations of propositional constituents are then incompatible with our beliefs. Suppose that the chief of police believes at the beginning of his investigation that Smith's murderer is either Paul or Julius. Then only possible denotation assignments according to which one of these two individuals falls under the concept of being Smith's murderer in relevant circumstances are then compatible with the beliefs of that chief during that period of his investigation.

Among all possible valuations of propositional constituents **there is of course a special one, the *real valuation* (in symbol *val**), that associates with each concept and attribute its actual denotation in any possible circumstance**. Actual circumstances represent a complete state of the actual world at a moment. Possible circum-

[16]Notice that the property of existence is a second order property. The speaker who says that the Golden Mountain does not exist does not refer to that mountain. He could not presuppose the existence of that mountain since he is denying that it exists. So the property of existence does not apply to individual objects under concepts but rather to individual concepts themselves. That property is satisfied by an individual concept in a circumstance when that concept applies to one individual in that circumstance.

stances whether actual or not belong to the logical space of *reality.* We ignore how things are in actual and other possible circumstances of the reality. So we cannot determine which possible valuation is the *real one.* Consider the atomic proposition that attributes to Smith's murderer the property of being wounded. Its concept and property could have many different denotations. According to a first possible denotation assignment a suspect Paul would be Smith's murderer and that suspect would also be wounded in the present circumstance. According to a second, a thief Julius would be the murderer but he would not be wounded now. According to a third no one would have killed Smith. Clearly we need more than linguistic knowledge in order to determine the actual denotation of the concept of being Smith's murderer and the property of being wounded in the actual world. An empiric investigation is required to get that knowledge. However we all know *a priori* according to which possible denotation assignments to its constituents the atomic proposition is true. We know that it is true in actual circumstances according to the first possible denotation assignment considered above and false according to the two others. We also know by virtue of competence that in order to be *true* an atomic proposition must be true according to possible valuations of its constituents which correspond to reality. So we know that the atomic proposition above is true in the present circumstance if and only if a single person really killed Smith and is also wounded now. It does not matter whether or not we know who that person is.

1.8 Meaning postulates

We respect *meaning postulates* in assigning possible denotations to senses and truth conditions to propositions. We assign to propositional constituents denotations of appropriate type. As I said in the last section, possible valuations of propositional constituents associate with each individual concept c_e and possible circumstance c a single individual object or no individual at all. Thus $val(c_e, c) \in Individuals$ or val is undefined for the concept c_e in the circumstance c. In that case, I will for the sake of simplicity like Carnap [1956] identify $val(c_e, c)$ with an arbitrary entity, the empty individual $u_\varnothing$ (rather than the empty set $\varnothing$). The *empty individual* is conceived here as the individual that does not exist at any moment in any possible world.

Possible valuations associate with each attribute R_n of degree n of individuals and possible circumstance a set of n-ary sequences of individuals under concepts. So $val(R_n, c) \in \mathscr{P}(Concepts^n)$. We moreover respect the logical nature of concepts and attributes and internal relations that exist between them because of their logical form. For we apprehend that logical form in conceiving them. Individuals subsumed under two concepts are identical when these two concepts have the same denotation. So the denotation in each possible circumstance of the binary relation of identity between individuals $\| = \|$ is the same according to all possible valuations of senses: it is the set of all pairs of individual concepts applying to the same individual in that circumstance. $<c_e^1, c_e^2> \in val(\| = \|, c)$ when $val(c_e^1, c) = val(c_e^2, c)$. We know *a priori* by virtue of linguistic competence that objects which fall under certain concepts (e.g. the concept of being Smith's murderer) have therefore certain properties (e.g. to be a murderer). And that they could not possess certain properties (to be admirable) without having others (to be admired in a possible circumstance). So possible valuations of logical constants respect traditional meaning postulates.

1.9 Truth according to a possible denotation assignment to constituents

By definition, an atomic proposition of the form $(R_n(c_e^1, \ldots, c_e^n))$ predicating the attribute R_n of n individuals under concepts $c_e^1, \ldots, c_e^n$ in that order is true in a circumstance according to a possible valuation when the sequence of these objects under concepts $c_e^1, \ldots, c_e^n$ belongs to the denotation that that valuation assigns to its attribute in that circumstance. So every possible valuation *val* of propositional constituents associates certain *possible truth conditions* with all atomic propositions containing such constituents. Any atomic proposition of the form $(R_n(c_e^1, \ldots, c_e^n))$ is true in a circumstance c according to a possible valuation *val* of its constituents when $<c_e^1, \ldots, c_e^n> \in val(R_n, c)$. Otherwise it is false in that circumstance according to that valuation.

1.10 Possible truth conditions

Because we ignore actual denotations of most propositional constituents, we also ignore in which possible circumstances most atomic propositions are true. We just know that they could be true in different sets of possible circumstances given the various denotations that their senses could have in the reality. For that reason, in my approach, propositions have *possible truth conditions* in addition to actual Carnapian *truth conditions.* For any atomic proposition one can distinguish as many possible truth conditions as there are distinct sets of possible circumstances where that atomic proposition is true according to a possible denotation assignment to its propositional constituents. Suppose that an atomic proposition is true in a set of possible circumstances according to a certain possible valuation of its constituents. Then clearly it would be true in all and only these circumstances if that valuation of these constituents *were real*, that is to say if it were associating with them their actual denotations. So the corresponding set of these possible circumstances corresponds well to a certain possible truth condition of that atomic proposition. As one can expect, every possible complete valuation of propositional constituents *determines* a unique possible complete valuation of atomic propositions. It assigns to them in accordance with meaning postulates *possible truth conditions* that they could all have together.

In my approach, there are a lot of *subjective* in addition to *objective possibilities* in the reality. When a possible denotation assignment *val* is compatible with the beliefs of an agent in a circumstance, any atomic proposition which is true in that circumstance according to that assignment, is then a proposition that could be true according to him or her in that circumstance. So, for example, according to the chief of police above at the beginning of his investigation Paul could be Smith's murderer.

1.11 Actual truth conditions

Among all possible truth conditions of an atomic proposition there are of course its *actual characteristic* Carnapian *truth conditions* that correspond to the set of possible circumstances where it is true.[17] So among all possible valuations of atomic propositions there is also a special one, let us call it the *real valuation*, that associates with each atomic proposition its actual truth conditions. As one can expect, that *real valuation* of *atomic propositions* is determined by the *real valuation val* of propositional constituents* that we have distinguished above: the one which assigns to each concept and attribute its actual denotation in each possible circumstance. An atomic proposition *is true in a circumstance* when it is true in that circumstance according to all possible valuations of senses that associate with its propositional constituents their actual denotation in the reality. For in that case the sequence of its objects under concepts in the order of predication belongs to the actual denotation of its attribute in each possible circumstance.

1.12 The type of atomic propositions

We can ignore in which circumstances an atomic proposition is true. But we could not *apprehend* one without having in mind its propo-

[17]Carnap did not consider possible truth conditions other than actual truth conditions.

sitional constituents: its single main attribute of degree n and the n individual concepts under which are subsumed the objects of reference. And without knowing under which conditions that atomic proposition is true. From a logical point of view, each atomic proposition of the form $(R_n(c_e^1, \ldots, c_e^n))$ is then a pair whose first element is the set of its n + 1 propositional constituents and whose second element is the set of all possible circumstances where it is true. In symbols, $(R_n(c_e^1, \ldots, c_e^n)) = <\{R_n, c_e^1, \ldots, c_e^n\}, \{c/<c_e^1, \ldots, c_e^n> \in val^*(R_n, c)\}$ where val^* is the real valuation. Notice that the order of predication only matters when it affects truth conditions. The propositions that Hesperus is Phosphorus and that Phosphorus is Hesperus do not differ. For the relation of identity is symmetric. We all know that by virtue of competence.

1.13 A recursive definition of propositions

In my analysis, complete propositions have then *a structure of constituents*: they are composed from a finite positive number of atomic propositions. They also have *possible truth conditions*: they are true in certain sets of possible circumstances according to possible valuations of their constituents. Until now I have mainly analyzed atomic propositions which are the basic units of the structure of constituents of propositions. One can define recursively the set of complete propositions that are expressible in the present modal logic. *Elementary propositions* are the simplest propositions: they are composed from a single atomic proposition and have all its possible truth conditions. Other more complex propositions are obtained by a finite number of applications of truth functional and modal operations to simpler propositions. Complex propositions can be composed from several atomic propositions and, when they are composed from a single atomic proposition, they do not have the same possible truth conditions.

What is the structure of constituents of truth functions and modal propositions? Which attributes do we predicate in expressing them? And how do we determine their possible truth conditions from the possible truth conditions of their constituent atomic propositions?

1.14 Structure of constituents of truth functions

As Wittgenstein pointed out in the *Tractatus,* truth connectives do not serve to make new acts of reference or predication. Truth functions do not change the structure of constituents. Their meaning just contributes to determining truth conditions. Truth functions of various propositions are composed from all and only the atomic propositions of their arguments. Thus the *negation* ¬P of a proposition P is composed from the atomic propositions of P. The *conjunction* (P $\wedge$ Q) and the *disjunction* (P $\vee$ Q) of two propositions P and Q are composed from the atomic propositions of both.

1.15 Structure of constituents of modal propositions

Unlike truth connectives, modal connectives serve to make new predications of so called *modal attributes.* Their meaning contributes to changing both the structure of constituents and the truth conditions of propositions. In thinking the modal proposition that it is impossible that God makes mistakes we do more than predicate of God the property of not making mistakes. We also predicate of Him the modal property of infallibility namely that He does not make a mistake in any possible circumstance. Infallibility is the necessitation of the property of not making mistakes. Modal proposition are then composed from new atomic propositions predicating modal attributes of some of their objects under concept.

Contrary to what Jorge Rodriguez[18] thinks, there is no need to enter into the infinite set of ramified types of propositions in order to analyze in terms of predication the attributes of modal propositions. The new attributes of modal propositions according to which it is necessary that P (in symbols $\Box$P) or that it is possible that P (in symbols $\Diamond$P) remain of the first order. In expressing these modal propositions we do not predicate of their argument, proposition P, the second order modal property of being true in all (or in some) possible circumstances. Rather we predicate corresponding modal attributes of objects under concepts of that argument. In the logic of attributes,[19] modal attributes of individuals are obtained from simpler attributes by quantifying universally or existentially over possible circumstances. The two basic kinds of broad modal operations on attributes associate with any given attribute the *necessitation* and the *possibilization* of that attribute. By definition, an object under concept possesses the necessitation of a property when it possesses that property in all possible circumstances. And it possesses the possibilization of a property when it possesses that property in at least one possible circumstance. (And similarly for

[18] J. Rodriguez Marqueze, "On the Logical Form of Propositions: Some Problems for Vanderveken's New Theory of Propositions" in *Philosophical Issues* [1993].
[19] See G. Bealer *Quality and Concept* [1982].

relations.) Suffixes like "ible" and "able" serve to compose modal predicates in English. Thus the property of being perturbable is the possibilization of the property of being perturbed. Someone is perturbable when he is perturbed in at least one possible circumstance. I will also use the logical constants $\Box$ and $\Diamond$ to express modal attributes. In my symbolism $\Box \mathrm{R}_n$ and $\Diamond \mathrm{R}_n$ are respectively the *necessitation* and the *possibilization* of the attribute R_n. By definition, all possible valuations of propositional constituents respect the following meaning postulates: $<c_e^1, \ldots, c_e^n> \in val(\Box R_n, c)$ when, for every c', $<c_e^1, \ldots, c_e^n> \in val(R_n, c')$. And similarly, $<c_e^1, \ldots, c_e^n> \in val(\Diamond R_n, c)$ when, for at least one c', $<c_e^1, \ldots, c_e^n> \in val(R_n, c')$.

So the formal ontology that I advocate here remains simple. There are only individuals under concepts, attributes of such individuals and first order atomic propositions containing such propositional constituents. There is no ramification of the logical type of propositions. All the modal attributes of the form $\Box R_n$ and $\Diamond R_n$ are of the first order: they are satisfied by (sequences of) individuals under concepts and not by propositions. On the basis of such considerations one can define simply the structure of constituents of modal propositions. A modal proposition of the form $\Box$P or $\Diamond$P contains in addition to any atomic proposition of its argument P predicating an attribute R_n of n individuals under concepts two new atomic propositions predicating in the same order the necessitation $\Box R_n$ and the possibilization $\Diamond R_n$ of that attribute[20] of the same individuals.[21]

1.16 Understanding of truth conditions

How do we understand the truth conditions of propositions? As Wittgenstein pointed out[22], in understanding the conditions under which a proposition is true, we always distinguish between different possible ways in which its objects might be, those which are

[20] There are four modal attributes corresponding to the modal operations of S5 modal logic namely $\Box R_n$, $\Box\neg R_n$, $\Diamond R_n$ and $\Diamond\neg R_n$ where possibility $\Diamond$ is defined as $\neg\Box\neg$. However, the operations of necessitation $\Box$ and possibilization $\Diamond$ are sufficient for my purposes here. For all modal propositions MP where M = $\Box$, $\Box\neg$, $\Diamond$ or $\Diamond\neg$ have the same structure of constituents, no matter how many modal attributes are taken into consideration.

[21] As one can expect, there are four different basic modal functions of a proposition P, namely: $\Box$P, $\Box\neg$P, $\Diamond$P and $\Diamond\neg$P corresponding to the four basic types of modal attributes $\Box R_n$, $\Box\neg R_n$, $\Diamond R_n$ and $\Diamond\neg R_n$ which can be formed from any attribute R_n in the logic of attributes.

[22] See aphorisms 4.3 and 4.4 of the *Tractatus logico-philosophicus*.

compatible with its truth from those which are not. In my approach, we distinguish in understanding a proposition P between two kinds of possible ways in which its propositional constituents might correspond to reality, those according to which P is true from those according to which it is false. In making such a distinction we consider all the atomic propositions of P and draw a large *truth table* more complex than that of Wittgenstein. In the *Tractatus* all propositional constituents are individual objects which are pure denotations. In my logic, they are senses: concepts and attributes to which correspond objects and concepts of objects respectively. Moreover, not all propositions are truth functions. There are modal propositions. So we have to distinguish in drawing a truth table for a proposition P two disjoint sets of possible valuations of its constituents *with respect to* one or more possible *circumstances*: those that assign to atomic propositions possible truth conditions that are compatible with the truth of P in these circumstances from those which do not.

Let me explain this by induction. By definition, an *elementary proposition* is true in a circumstance according to a possible valuation of its constituents when that valuation associates with its attribute in that circumstance a denotation that contains the sequence of its objects under concepts in the order of predication. This is the way objects have to be in order that its single atomic proposition be true according to a valuation in a circumstance. So the possible truth conditions of an elementary proposition are the possible truth conditions of its unique atomic propositions. As one can expect, the negation ¬P is true in a circumstance according to a possible valuation of its constituents when the proposition P is false according to that valuation in that circumstance. In other words, the truth of proposition ¬P in a circumstance is only compatible with possible truth conditions of its atomic propositions that are incompatible with the truth of P in that very circumstance. Furthermore, a conjunction (P ∧ Q) is true in a circumstance *c* according to a possible valuation when both conjuncts P and Q are true in *c* according to that valuation. So the truth of a conjunction in a circumstance is only compatible with possible truth conditions of its atomic propositions that are compatible with the truth of both conjuncts P and Q in that circumstance. Truth functions obey the law of extensionality. Their truth value in a circumstance according to a valuation only depends on the truth value of their arguments in that circumstance according to that valuation. On the contrary, modal operations are intensional. A modal proposition of the form $\Box$P (or $\Diamond$P) is true in a possible circumstance according to a possible valuation of its constituents when its argument P is true in every (or in at least one) possible circumstance c' according to that valuation. So the truth of modal propositions $\Box$P (or $\Diamond$P) in a circumstance is only compatible with possible truth conditions of its atomic propositions that are compatible with the truth of its argument P in every (or at least one) possible circumstance.

1.17 Tautologies and contradictions

There are two borderline cases of truth conditions. In the first case, the truth of a proposition is compatible with all the possible ways in which objects might be. It is a *tautology*. In the second case, its truth is not compatible with any possible way in which objects might be. It is a *contradiction*. In my approach, tautologies are true according to all possible valuations of their constituents while contradictions are true according to none. So the truth of a *tautology* in any possible circumstance is compatible with all the possible truth conditions of its atomic propositions, and the truth of a *contradiction* with none. For that reason, tautologies (and contradictions) are a very special case of necessarily true (and necessarily false) propositions. When we express a tautology and a contradiction we *a priori* know in apprehending their logical form that the first is necessarily true and the second is necessarily false. Tautologies are then unconditionally, *a priori* and analytically true, *contradictions* unconditionally, *a priori* and analytically false.

1.18 The new criterion of propositional identity

Identical propositions have the same structure of constituents and they are true in the same possible circumstances according to the same possible denotation assignments to their propositional constituents. My criterion of propositional identity is much finer than that of modal, temporal, intensional and relevance logics. My logic distinguishes strictly equivalent propositions composed of different atomic propositions. We clearly do not make the same predications in expressing them. So we do not have them in mind in the same possible contexts of utterance. There are a lot of different necessarily true and necessarily false propositions and not only two as classical logic wrongly claims. Tautologies with different constituents are different propositions.

Predicative logic moreover distinguishes strictly equivalent propositions with the same structure of constituents which are not true in the same circumstances according to the same possible valuations of their constituents. When the truth of two propositions is not compatible with the same possible truth conditions of their atomic propositions, we indeed do not understand their truth conditions in the same way. Consider the elementary proposition that the biggest whale is a fish and the conjunction that the biggest whale is and is not a fish. Both are composed from the same atomic proposition predicating of the biggest whale the property of being a fish. And both are necessarily false. In all possible circumstances where they exist, whales are mammals. They all have in common that *essential property*. However the two propositions have a different cognitive value. We recently discovered that whales are mammals. Previously we had believed that the biggest whale was a fish. But we could never have believed that it is and that it is not a fish. Unlike Parry's logic of analytic implication my predicative logic distinguishes such strictly equivalent propositions with the same structure of constituents. Clearly the elementary proposition that a whale is a fish is necessarily false. However it is true according to many possible valuations of its constituents (all those according to which the denotation of the property of being a whale is a subset of the denotation of being a fish). On the contrary, the proposition that a whale is and is not a fish is a pure contradiction: it is not true according to any possible valuation of its constituents. This is why we cannot believe it.

When two propositions are true in the same possible circumstances according to the same possible denotation assignments to their propositional constituents, their truth in each circumstance is by hypothesis compatible with the same possible truth conditions assignments to their atomic propositions. Possible valuations of propositional constituents *determine* by definition all possible valuations of atomic propositions. Thus from a logical point of view one can identify each proposition P with a pair whose first element is the finite non empty set of its atomic propositions and whose second element is the function associating with any possible circumstance the set of possible valuations of its atomic propositions which are compatible with its truth in that very circumstance. Propositions belong to the set $\mathscr{P}\mathrm{U}_a\times$ (*Circumstances* $\Rightarrow \mathscr{P}(\mathrm{U}_a \Rightarrow \mathscr{P}$ *Circumstances*). My theory of sense and propositions is compatible with the current dynamic analysis of meaning according to which the meaning of a sentence in a context of utterance is related to information change potential.[23]

1.19 Truth definition

In the philosophical tradition, from Aristotle to Tarski, *truth* is based on *correspondence* with reality. True propositions represent how objects are in the reality. Objects of reference have properties and stand in relations in possible circumstances. Atomic propositions have therefore a well determined truth value in each circumstance depending on the denotation of their attributes and concepts and the order of predication. However things could have many other properties and stand in many other relations in each circumstance. In addition to the ways in which things are, there are the possible ways in which they could be. Our knowledge is restricted. So we consider a lot of possible truth conditions of atomic propositions different from their actual truth conditions in thinking propositional contents. In our mind, the truth of propositions is compatible with many possible ways in which objects could be. However in order that a proposition be true in a given circumstance, things must be in that circumstance as that proposition represents them. Otherwise, there would be no correspondence. Along these lines, one can say that a proposition *is true in a possible circumstance* when it is true according to any *real* valuation of its propositional constituents assigning to them their actual denotation in each circumstance. In that case its truth in that circumstance is compatible with the actual truth conditions of all its atomic propositions. So a proposition P is true in a circumstance c when it is true according to the real valuation of propositional constituents, that is to say when *val**

[23] Each new sentence in a discourse has to be interpreted in the conversational background of the context in which it is uttered and its interpretation (the illocution that it expresses in that context) updates that background. For the principles of my semantic theory see my paper "Success, Satisfaction and Truth in the logic of Speech Acts and Formal Semantics" in S. Davis & B. Gillan *A Reader in Semantics* [2004]

$\in$ id$_2$P(c). Classical laws of truth theory follow from this concise definition.

1.20 Cognitive aspects in the theory of truth

Each agent *a* has in mind a finite number of propositional constituents in each circumstance c and what he then believes depends on the possible denotations that these constituents have or could have according to him in the reality. So to each agent *a* and circumstance c there corresponds a unique set *Val(a,c)* containing all the possible valuations of senses *compatible with what that agent believes in that circumstance.* Suppose that an agent *a* believes in a circumstance c that no individual could fall under a concept c_e. Then according to all valuations *Val* $\in$ *Val(a,c)* compatible with what he then believes, *Val*(c_e,*c*) $= u_\varnothing$ for any possible circumstance *c*. Any agent having in mind propositional constituents *believes* in the truth of certain propositions containing them. One can now define adequately the notion of *belief* in philosophical logic: *an agent a believes a proposition in a circumstance* c when firstly, that agent has then in mind all its propositional constituents and secondly, that proposition is true in that circumstance according to all possible valuations of constituents f $\in$ *Val(a,c)* that are compatible with his beliefs in that circumstance.[24] As one can expect, tautological propositions are true and contradictory propositions are false according to all agents who have them in mind. But impossible propositions which are not contradictory can be true and necessary propositions which are not tautological can be false according to agents at some moments. These are basic principles of my epistemic logic. So the logic of language imposes different limits on experience and thought. *Objective* and *subjective possibilities* differ. Necessarily false propositions represent impossible facts that could not exist in reality and that we could not experience. In my view there is no need to postulate impossible circumstances where such impossible facts would exist. Impossible facts are objectively impossible. In any possible circumstance where there are whales they

[24]Whenever an agent does not think or act at all (he is in a profound sleep or dead), all possible valuations of propositional constituents are then compatible with his beliefs. But he does not then believe anything by hypothesis. In order to have a conscious belief an agent must have in mind relevant concepts and attributes.

are mammals and not fishes. However there are many more subjective than objective possibilities. Certain objectively impossible facts e.g. that whales are fishes are subjectively possible. Their existence is compatible with certain possible denotation assignments to senses. So we can wrongly believe that exist.

1.21 The notion of strong implication

We, human beings are not perfectly rational. Not only do we make mistakes and have a lot of false beliefs. But we are often inconsistent. Moreover we do not draw all valid inferences. So we assert (and believe) propositions without asserting (and believing) all their logical consequences. Our illocutionary (and psychological) commitments are not as strong as they should be from the logical point of view. We do not even know all logical truths. However we are not completely irrational. On the contrary, we manifest a *minimal rationality* in thinking and speaking that logic can now explain. We know that certain propositions are necessarily false (for example, contradictions): we cannot believe them nor intend to bring about facts that we know to be impossible.[25] Moreover, we always draw certain valid theoretical inferences. When we know *a priori* by virtue of competence that a proposition cannot be true unless another is also true, we cannot believe (or assert) that proposition without believing (or asserting) the other. There is an important *relation of strict implication* between propositions due to C.I. Lewis that has been much used in epistemic logic: a proposition *strictly implies* another whenever that proposition cannot be true in a possible circumstances unless the other is true in that same circumstance. Hintikka[26] and others claim that belief and knowledge are closed under strict implication. However we ignore which propositions are related by strict implication, just as we ignore in which possible circumstances they are true. Moreover we could not know all cases of strict implication. For any proposition strictly implies infinitely many other propositions. We could not think of all of them in a context of utterance.

So we need a relation of propositional implication much finer than strict implication in order to explicate our illocutionary and psycho-

---

[25]See next chapter 15 "Attempt, Success and Action Generation" in this Volume.
[26]See J. Hintikka *Knowledge and Belief* [1962]

logical commitments. Predicative logic can define rigorously that finer propositional implication that I have called *strong implication.* By definition, *a proposition strongly implies another proposition when firstly, it contains all its atomic propositions and secondly, it tautologically implies that other proposition*: whenever it is true in a possible circumstance according to a possible valuation of its propositional constituents the other is also true in that circumstance according to the same valuation. Unlike strict implication, *strong implication is known.* Whenever a proposition P strongly implies another Q, we cannot express that proposition without knowing *a priori* that it strictly implies the other. For in expressing P, we have by hypothesis in mind all atomic propositions of Q. We make all the corresponding acts of reference and predication. Furthermore, in understanding the truth conditions of proposition P, we distinguish all possible valuations of its propositional constituents which are compatible with its truth in any circumstance. These are by hypothesis compatible with the truth of proposition Q in the same circumstance. Thus, in expressing P, we know that Q follows from P. Belief and knowledge are then closed under strong rather than strict implication in my epistemic logic. As I will show later, *strong implication obeys a series of important universal laws.* Unlike strict implication, strong implication is anti-symmetrical. Two propositions which strongly imply each other are identical. Unlike Parry's analytic implication, strong implication is always tautological. Natural deduction rules of elimination and introduction generate strong implication when and only when all atomic propositions of the conclusion belong to the premises. So a proposition P does not strongly imply a disjunction of the form P $\vee$ Q containing new constituents. Moreover strong implication is *paraconsistent.* A contradiction does not strongly imply all propositions. Finally, strong implication is both *finite* and *decidable.*

2. The ideal object-language

The object language $\mathscr{L}$ of my modal predicate calculus is an **extension of that of the minimal logic of propositions.**

2.1 Vocabulary of $\mathscr{L}$

(1) A series of *individual constants* called *individual terms*:
c, c′, c″, ...
(2) for each positive natural number n, a series of *predicate constants of degree* n:
r_n, r'_n, r''_n, r'''_n, ... including the binary identity predicate $=_2$
(3) the *syncategorematic expressions*:
=, >, ∧, ¬, □, ◇, [, (,] and).

2.2 Rules of formation of $\mathscr{L}$

Predicates

Every predicate of degree n of the lexicon is a predicate of degree n of $\mathscr{L}$. If R_n is a predicate of degree n, so are $\Box R_n$ and $\Diamond R_n$. Complex predicates of the forms $\Box R_n$ and $\Diamond R_n$ name respectively the modal attributes of degree n which are the *necessitation* and the *possibilization* of the attribute named by R_n.

The set L_a of predication formulas

If R_n is a predicate of degree n and t_1, ... and t_n are n individual terms, then $(R_n t_1 \ldots t_n)$ is a predication formula which expresses the atomic proposition predicating the attribute expressed by R_n of the n individual concepts expressed by t_1, ... and t_n in that order.

The set L_p of propositional formulas

If $(R_n t_1 \ldots t_n)$ is a predication formula then $[(R_n t_1 \ldots t_n)]$ is a propositional formula. If A_p and B_p are propositional terms, then $\neg A_p$, $\Box A_p$, $(A_p \wedge B_p)$, $(A_p > B_p)$ and $(A_p = B_p)$ are new complex propositional formulas. $[(R_n t_1 \ldots t_n)]$ expresses the *elementary proposition* whose unique atomic proposition is that expressed by predication formula $(R_n t_1 \ldots t_n)$. $\neg A_p$ expresses the *negation* of the proposition expressed by A_p. $\Box A_p$ expresses the *modal proposition* that it is logically necessary that A_p. $(A_p \wedge B_p)$ expresses the *conjunction* of the two propositions expressed by A_p and B_p. $(A_p > B_p)$ expresses the proposition according to which all atomic propositions of B_p are atomic propositions of A_p. Finally, $(A_p = B_p)$ means that propositions A_p and B_p are identical.

2.3 Rules of abbreviation

Parentheses are eliminated according to the usual rules.

Identity: $t_1 = t_2 =_{df} (=_2 t_1 t_2)$
Disjunction: $(A_p \vee B_p) =_{df} \neg(\neg A_p \wedge \neg B_p)$
Material implication: $(A_p \Rightarrow B_p) =_{df} \neg A_p \vee B_p$
Material equivalence: $(A_p \Leftrightarrow B_p) =_{df} (A_p \Rightarrow B_p) \wedge (B_p \Rightarrow A_p)$
Logical possibility: $\Diamond A_p =_{df} \neg\Box\neg A_p$
Strict implication: $A_p \multimap B_p =_{df} \Box(A_p \Rightarrow B_p)$
Tautologyhood: *Tautological*$(A_p) =_{df} A_p = (A_p \Rightarrow A_p)$
Analytic implication: $A_p \rightarrow B_p =_{df} (A_p > B_p) \wedge (A_p \multimap B_p)$
Analytic equivalence: $A_p \leftrightarrow B_p =_{df} (A_p \rightarrow B_p) \wedge (B_p \rightarrow A_p)$
Strong implication:
$A_p \mapsto B_p =_{df} (A_p > B_p) \wedge \textit{Tautological}\,(A_p \Rightarrow B_p)$
Same structure of constitutents:
$A_p \equiv B_p =_{df} (A_p > B_p) \wedge (B_p > A_p))$
Identical individual concepts: $\hat{}t_1 = \hat{}t_2 =_{df} [(r_1\, t_1)] > [(r_1 t_2)]$
Identical attributes: $\hat{}R_n = \hat{}R'_n =_{df} [(R_n\, t_1 \ldots t_n)] > [(R'_n\, t_1 \ldots t_n)]$
for the first n individual constants

3. The formal semantics

A *standard model* $\mathcal{M}$ for $\mathcal{L}$ is a sextuple < *Circumstances*, *Individuals*, *Concepts*, *Attributes*, *Val*, *, $\| \|$ >, where *Circumstances*, *Individuals*, *Concepts* and *Attributes* are four disjoint non empty sets, *Val* is a set of functions and * and $\| \|$ are functions which satisfy the following clauses:

(1) *Circumstances* is the set of *possible circumstances.*

(2) *Individuals* is the set of *individual objects.* For each possible circumstance c, *Individuals*$_c$ is the set of individual objects existing in that circumstance. Let $u_\varnothing$ be the *empty individual* of model $\mathcal{M}$. By definition,

$$\textit{Individuals} = \bigcup_{c \in \textit{Circumstances}} \textit{Individuals}_c \cup \{u_\varnothing\}$$

(3) *Concepts* is the set of *individual concepts* and

(4) *Attributes* is the set of *attributes* of individuals considered in the model $\mathcal{M}$. For each positive natural number n, *Attributes* (n) is a non empty subset of *Attributes* containing all *attributes* of degree n considered in the model $\mathcal{M}$.

(5) $\| \|$ is an interpreting function which associates with each well formed expression *A* of $\mathcal{L}$ its semantic value $\|A\|$ in the model $\mathcal{M}$.

(i) For any individual constant t, $\|t\|$ is a certain individual concept $c_e \in$ *Concepts*.

(ii) For any predicate R_n of degree n, $\|R_n\|$ is a certain attribute of degree n $\in$ *Attributes* (n).
(6) *Val* is the set of all possible *assignments of denotation to propositional constituents* in the model $\mathscr{M}$. It contains a special *real valuation* val $\mathscr{M}$ which assigns to concepts and attributes their *actual denotation* in each possible circumstance according to the model $\mathscr{M}$. The set *Val* is the smallest subset of $(\mathit{Concepts} \cup \mathit{Attributes}) \times \mathit{Circumstances} \rightarrow (\mathit{Individuals} \cup \bigcup_{1 \leq n} \mathscr{P}(\mathit{Concepts}^n))$ which satisfies the following *meaning postulates*:
- For any valuation $val \in Val$ and possible circumstance c, *val* $(\|t\|, c) \in$ *Individuals* for any individual term t and *val* $(\|R_n\|, c) \in \mathscr{P}(\mathit{Concepts}^n)$ for any predicate R_n of degree n.
- $<\|t_1\|,\|t_2\|> \in val(\|=\|, c)$ iff $val(\|t_1\|, c) = val(\|t_2\|, c)$.
- $<\|t_1\|, \ldots,\|t_n\|> \in val(\|\Box R_n\|, c)$ iff, for every $c' \in$ *Circumstances*, $<\|t_1\|, \ldots,\|t_n\|> \in val(\|R_n\|, c')$.
- And similarly $<\|t_1\|, \ldots, \|t_n\|> \in val(\|\Diamond R_n\|, c)$ iff $<\|t_1\|, \ldots,\|t_n\|> \in val(\|R_n\|, c')$ for at least one possible circumstance c'.
(7) For any predication formula $(R_n t_1, \ldots, t_n)$, $\|(R_n t_1, \ldots, t_n)\|$ is the *atomic proposition* predicating the attribute $\|R_n\|$ of the n objects under concepts $\|t_1\|, \ldots,\|t_n\|$ in that order. Formally, $\|(R_n t_1, \ldots, t_n)\|$ is the pair $<\{\|R_n\|,\|t_1\|, \ldots,\|t_n\|\}, \{c \in \mathit{Circumstances} / <\|t_1\|, \ldots,\|t_n\|> \in \mathrm{val}\,\mathscr{M}(\|R_n\|, c)\}>$.
Let $U_a =_{def} \{ \|A_a\| / A_a \in L_a \}$ be the *set of all atomic propositions* considered in the model $\mathscr{M}$.
$\mathscr{P}[U_a]$ is an upper modal semi lattice containing finite sets of atomic propositions which is closed under union $\cup$ and a unary operation * satisfying the following clause: for any $\{\|(R_n t_1, \ldots, t_n)\|\} \in U_a$, $*\{\|(R_n t_1, \ldots, t_n)\|\} = \{\|(R_n t_1, \ldots, t_n)\|\}, \|(\Box R_n t_1, \ldots, t_n)\|, \|(\Diamond R_n t_1, \ldots, t_n)\|\}$ and, for any Γ_1 and $\Gamma_2 \in \mathscr{P}U_a$, $*(\Gamma_1 \cup \Gamma_2) = *\Gamma_1 \cup *\Gamma_2$ and $**\Gamma_1 = *\Gamma_1$. The elements of $\mathscr{P}[U_a]$ represent *structures of constituents* of propositions in the model $\mathscr{M}$.
(8) For any propositional formula A_p, $\|A_p\|$ is the *proposition* expressed by that formula according to the model $\mathscr{M}$. It belongs to the set $(\mathscr{P}U_a) \times (\mathit{Circumstances} \Rightarrow \mathscr{P}\,\mathit{Val})$. As one can expect, the first term, $id_1\|A_p\|$, of proposition $\|A_p\|$ represents the *set of its atomic propositions.* And its second term, $id_2\|A_p\|$, the way in which we understand its *truth conditions*, that is the function which associates with each possible circumstance c the set $id_2P(c)$ of all

possible valuations of propositional constituents according to which that proposition is true in that circumstance c.
The *proposition* $\|A_p\|$ expressed by A_p in the model $\mathscr{M}$ is defined by induction on the length of A_p:
Basis: $id_1\|[(R_n t_1, \ldots, t_n)]\| = \{\|(R_n t_1, \ldots, t_n)\|\}$ and $id_2(\|[(R_n c_1, \ldots, c_n)]\|, c) = \{val \in Val / <\|t_1\|, \ldots, \|t_n\|> \in val(\|R_n\|, c)\}$.
Induction steps:
(i) $id_1\|\neg B_p\| = id_1\|B_p\|)$ and $id_2(\|\neg B_p\|, c) = Val - id_2(\|B_p\|, c)$.
(ii) $id_1\|\Box B_p\| = * \; id_1\|B_p\|)$ and
$$id_2(\|\Box B_p\|, c) = \bigcap_{c' \in Circumstances} id_2(\|B_p\|, c')$$
(iii) $id_1(\|B_p \wedge C_p\|) = id_1(\|B_p\|) \cup id_1(\|C_p\|)$; $id_2(\|B_p \wedge C_p\|, c) = id_2(\|B_p\|, c) \cap id_2(\|C_p\|, c)$.
(iv) $id_1(\|B_p > C_p\|) = id_1(\|B_p\|) \cup id_1(\|C_p\|)$ and $id_2(\|B_p > C_p\|, c) = Val$ when $id_1\|B_p\| \subseteq id_1\|C_p\|$. Otherwise, $id_2(\|B_p > C_p\|, c) = \varnothing$.
(v) $id_1(\|B_p = C_p\|) = id_1(\|B_p\|) \cup id_1(\|C_p\|)$; $id_2(\|B_p = C_p\|, c) = Val$ when $\|B_p\| = \|C_p\|$. Otherwise, $id_2\|B_p = C_p\|(c) = \varnothing$.

Definition of truth and validity

A propositional formula A_p of $\mathscr{L}$ is *true* in a possible circumstance c according to a standard model when it is true in that model according to the real assignment $val\mathscr{M}$ of denotations to senses, that is to say iff $val\mathscr{M} \in id_2\|A_p\|(c)$. A propositional formula A_p of $\mathscr{L}$ is *valid* or *logically true* ($\models A_p$) when it is true in all possible circumstances according to all standard models $\mathscr{M}$ of $\mathscr{L}$.

4. A complete axiomatic system

I conjecture that all and only valid formula of $\mathscr{L}$ are provable in the following axiomatic system **MPC**:[27]
The axioms of MPC are all the instances in $\mathscr{L}$ of the following axiom schemas:

Classical truth functional logic
(t1) $(A_p \Rightarrow (B_p \Rightarrow A_p))$,
(t2) $((A_p \Rightarrow (B_p \Rightarrow C_p)) \Rightarrow ((A_p \Rightarrow B_p) \Rightarrow (A_p \Rightarrow C_p)))$
(t3) $((\neg A_p \Rightarrow \neg B) \Rightarrow (B_p \Rightarrow A_p))$

S5 modal logic
(M1) $(\Box A_p \Rightarrow A_p)$
(M2) $(\Box(A_p \Rightarrow B_p) \Rightarrow (\Box A_p \Rightarrow \Box B_p))$

[27] All these axioms are not independent.

(M3) $(\neg\Box A_p \Rightarrow \Box\neg\Box A_p)$

Axioms for tautologies

(T1) (*Tautological* $A_p) \Rightarrow A_p$

(T2) (*Tautological* $A_p) \Rightarrow$ *Tautological Tautological* A_p

(T3) ($\neg$ *Tautological* $A_p) \Rightarrow$ *Tautological* $\neg$ *Tautological* A_p

(T4) *Tautological*$(A_p) \Rightarrow ($*Tautological*$(A_p \Rightarrow B_p) \Rightarrow$ *Tautological*$(B_p))$

(T5) *Tautological* $(A_p) \Rightarrow$ *Tautological* $(\Box A_p)$

Axioms for propositional identity (I1) $A_p = A_p$

(I2) $(A_p = B_p) \Rightarrow (C \Rightarrow C^*)$ where C^* and C are propositional formulas which differ at most by the fact that an occurrence of B_p replaces an occurrence of A_p

(I3) $(A_p \mapsto B_p \,\&\, (B_p \mapsto A_p)) \Rightarrow (A_p = B_p)$

(I4) $(A_p = B_p) \Rightarrow$ *Tautological* $(A_p = B_p)$

(I5) $\neg(A_p = B_p) \Rightarrow$ *Tautological* $\neg(A_p = B_p)$

Axioms for propositional composition

(C1) $(A_p > B_p) \Rightarrow$ *Tautological* $(A_p > B_p)$

(C2) $\neg(A_p > B_p) \Rightarrow$ *Tautological*$\neg(A_p > B_p)$

(C3) $A_p > A_p$

(C4) $(A_p > B_p) \Rightarrow ((B_p > C_p) \Rightarrow (A_p > C_p))$

(C5) $([(R_n t_1, \ldots, c_n)] > A_p) \Rightarrow (A_p = [(R_n t_1, \ldots, c_n)])$

(C6) $(A_p \wedge B_p) > A_p$

(C7) $(A_p \wedge B_p) > B_p$

(C8) $(C_p > A_p) \Rightarrow ((C_p > B_p) \Rightarrow (C_p > (A_p \wedge B_p)))$

(C9) $A_p \equiv \neg A_p$

(C10) $(\Box[(R_n t_1 \ldots t_n)] > A_p) \Leftrightarrow ((A_p = [(\Box R_n t_1 \ldots t_n)]) \vee (A_p = [(\Diamond R_n t_1 \ldots t_n)]) \vee (A_p = [(R_n t_1 \ldots t_n)]))$

(C11) $\Box\neg A_p \equiv \Box A_p$

((C12) $\Box(A_p \wedge B_p) \equiv (\Box A_p \;\wedge\; \Box B_p)$

(C13) $\Box\Box A_p \;\equiv \Box A_p))$

Axioms for elementary propositions

(E1) $\Box[(R_n t_1 \ldots t_n)] \Leftrightarrow [(\Box R_n t_1 \ldots t_n)])$ And similarly for $\Diamond$.

(E2) $[t = t]$ for any individual term t

(E3) $([t_1 = t_2] \Rightarrow (A_p \Rightarrow A'_p))$ when A'_p differs at most from A_p by the fact that an occurrence of the term t_2 in A'_p replaces an occurrence of the term t_1 which is not under the scope of $\Box$, >, ^ or the sign of propositional identity in A_p.

(E4) $^\wedge t_1 = {}^\wedge t_2 \Rightarrow$ *Tautological* $[(t_1 = t_2)]$

(E5) *Tautological* $[(t_1 = t_2)] \Leftrightarrow [(t_2 = t_1)]$

(E6) $^\wedge R_n = {}^\wedge R'_n \Rightarrow$ (*Tautological*$[(R_n t_1 \ldots t_n)] \Leftrightarrow [(R'_n t_1 \ldots t_n)])$

(E7) $((^\wedge t_1 = {}^\wedge d_1) \wedge \ldots \wedge (^\wedge t_n = {}^\wedge d_n) \wedge (^\wedge R_n = {}^\wedge R'_n)) \Rightarrow ([(R_n t_1 \ldots t_n)] = [(R'_n d_1 \ldots d_n)])$
(E8) $([(R_n t_1 \ldots t_n)] = [(R'_n d_1 \ldots d_n)]) \Rightarrow (^\wedge R_n = {}^\wedge R'_n))$
(E9) $([(R_n t_1 \ldots t_n)] = [(R'_n d_1 \ldots d_n)] \Rightarrow ((^\wedge t_k = {}^\wedge d_1) \vee \ldots \vee (^\wedge t_k = {}^\wedge d_n)))$ where $n \geq k \geq 1$
(E10) $\neg(^\wedge R_n = {}^\wedge R_m)$ when $n \neq m$
(E11) *Tautological* $[(R_2 t_1 t_2)] \Leftrightarrow (^\wedge t_1 = {}^\wedge t_2 \wedge ((^\wedge R_2 = {}^\wedge{=}_2) \vee (^\wedge R_2 = {}^\wedge \Box{=}_2)))$
(E12) $\neg$ *Tautological* $[(R_n t_1 \ldots t_n)]$ when $n \neq 2$
(E13) $\neg$ *Tautological* $\neg$ $[(R_n t_1 \ldots t_n)]$ when $n \neq 2$

The rules of inference of MPC are:

The rule of Modus Ponens:

(MP) From the sentences $(A \Rightarrow B)$ and A infer B.

The tautologization rule:

(RT) From a theorem A infer *Tautological*A.

5. Valid laws

5.1 Laws about the structure of constituents

A proposition is composed from all the atomic propositions of its arguments. $\models A_p > [(R_n c_1, \ldots, c_n)]$ when $[(R_n c_1, \ldots, c_n)]$ occurs in A_p. Modal propositions have all the atomic propositions of their argument. $\models MA_p > A_p$ where $M = \Box, \Box\neg, \Diamond$ or $\Diamond\neg$ Moreover $\models A_p > [(\Box R_n c_1, \ldots, c_n)]$ when $[(R_n c_1, \ldots, c_n)]$ occurs within the scope of $\Box$ in A_p. So $\Box[(R_n t_1 \ldots t_n)]$ is not an elementary proposition.

All the different modal propositions of the form MA_p have the same structure of constituents.

$\models M\Box A_p \equiv M'A_p$ where M and M′ are $\Box$, $\Box\neg$, $\Diamond$ or $\Diamond\neg$. Thus $\models \Box A_p \equiv \Box\neg A_p$ and $\models \Diamond A_p \equiv \Box A_p$. As one can expect, $\models M(A_p \wedge B_p) \equiv (MA_p \wedge MB_p)$; $\models M(A_p \equiv \Box A_p$ and $\models M\Diamond A_p \equiv \Diamond A_p$
Some modal attributes are identical. $\models {}^\wedge\Box R_n = {}^\wedge\Box\Box R_n$ However, $\not\models {}^\wedge\Box{=}_2 = {}^\wedge{=}_2$.

5.2 Laws for tautologyhood

Tautologyhood is stronger than necessary truth and contradiction stronger than necessary falsehood. $\models$ (*Tautological*A_p) $\Rightarrow \Box A_p$. But $\not\models \Box A_p \Rightarrow$ *Tautological*A_p
There are elementary, modal as well as truth functional tautologies and contradictions.

$\models$ *Tautological* [t = t]; $\models$ ^t$_1$ = ^t$_2$ $\Rightarrow$ *Tautological*[t$_1$ = t$_2$] and $\models$ *Tautological* $\Box(A_p \vee \neg A_p)$

5.3 Laws for tautological implication

Tautological implication is much finer than strict implication. $\models$ *Tautological* $(A_p \Rightarrow B_p) \Rightarrow (A_p \multimap B_p)$. But $\not\models (A_p \multimap B_p) \Rightarrow$ *Tautological* $(A_p \Rightarrow B_p)$.

Thus $\models \Box A_p \Rightarrow (B_p \multimap A_p)$. But $\not\models \Box A_p \Rightarrow$ *Tautological*$(B_p \Rightarrow A_p)$. The necessarily true proposition that the biggest whale is a mammal is strictly implied by all propositions. But it is not tautologically implied by any tautology. For it is not tautological. Only tautologies can strongly imply other tautologies.
$\models$((*Tautological* B_p) $\wedge$ *Tautological* $(A_p \Rightarrow B_p)) \Rightarrow$ *Tautological* A_p

Similarly $\models \Box\neg A_p \Rightarrow (A_p \multimap B_p)$. Necessarily false propositions strictly imply all other propositions. But only contradictions can tautologically imply contradictions. $\not\models \Box\neg A_p \Rightarrow$ *Tautological*$(A_p \Rightarrow B_p)$. So only contradictions tautologically imply all other propositions.

All valid laws of material implication of truth functional and S5 modal logic are valid laws of tautological implication. Thus $\models$*Tautological* $(A_p \Rightarrow B_p)$ when $\models(A_p \Rightarrow B_p)$ in S5 modal logic. In particular, $\models$ *Tautological* $(A_p \Rightarrow (A_p \vee B_p))$ and $\models$*Tautological* $(A_p \Rightarrow \Diamond A_p)$. Moreover, $\models$*Tautological* $([\Box R_n c_1 \ldots c_n)] \Leftrightarrow \Box[(R_n c_1 \ldots c_n)])$. And similarly for $\Diamond$. Thus the propositions that John is perturbable and that it is possible that John is perturbed are tautologically equivalent.

Whenever a proposition tautologically implies another, we can have it in mind without having in mind the other. $\not\models$(*Tautological* $(A_p \Rightarrow B_p)) \Rightarrow (A_p > B_p)$) However we could not express both propositions without knowing that the first implies the second. This is why tautological implication generates *weak psychological and illocutionary commitment* in thinking and speaking. Any assertion (or belief) that P *weakly commits* the agent to asserting (or believing) any proposition Q that P tautologically imply.

5.4 Laws for strong implication

Strong implication is the strongest kind of propositional implication. It requires inclusion of content in addition to tautological implication. So there are two reasons why a proposition can fail to imply strongly another. Firstly, the second proposition can require new predications. In that case, one can think the first without thinking the second. $\models \neg(A_p > B_p) \Rightarrow \neg(A_p \mapsto B_p)$. Secondly, the first proposition can fail to imply tautologically the second. In that case, one can ignore even tacitly that it implies the second.

Unlike strict and tautological implications, strong implication is anti-symmetric (Axiom I3). The rule of *Modus Tollens* does not hold for strong implication. $\not\models(A_p \mapsto B_p) \Rightarrow (\neg B_p \mapsto \neg A_p)$

Strong implication is also finer than Parry's analytic implication which is not tautological. $\not\models(A_p \Rightarrow B_p) \Rightarrow (A_p \mapsto B_p)$ For $\not\models(A_p \rightarrow B_p) \Rightarrow \mathit{Tautological}(A_p \rightarrow B_p)$.

5.5 Natural deduction

Valid laws of inference of natural deduction whose premises contain the atomic propositions of their conclusion generate strong implication. Thus when $\models(A_p \Rightarrow B_p)$ in S5 modal logic and $\models (A_p > B_p)$ it follows that $\models (A_p \mapsto B_p)$.

This leads to the following system of *natural deduction*:

The law of elimination of conjunction: $\models(A_p \wedge B_p) \mapsto A_p$ and $\models(A_p \wedge B_p) \mapsto B_p$

The law of elimination of disjunction: $\models((A_p \mapsto C_p) \wedge (B_p \mapsto C_p)) \Rightarrow (A_p \vee B_p) \mapsto C_p$

Failure of the law of introduction of disjunction: $\not\models A_p \mapsto (A_p \vee B_p)$. So strong implication is stronger than *entailment* which obeys the law of introduction of disjunction.

The *law of introduction of negation*: $\models A_p \mapsto O_t \Rightarrow (A_p \mapsto \neg A_p)$ where O_t is any contradiction.

Failure of the law of elimination of negation:
$\not\models(A_p \wedge \neg A_p) \mapsto B_p$
Strong implication is paraconsistent.

The *law of elimination of material implication*:
$\models(A_p \wedge (A_p \Rightarrow B_p) \mapsto B_p$

The *law of elimination of necessity*: $\models \Box A_p \mapsto A_p$

The *law of introduction of necessity*: $\models A_p \mapsto B_p \Rightarrow \Box A_p \mapsto \Box B_p$

The *law of elimination of possibility*: $\models \Diamond A_p \mapsto B_p \Rightarrow A_p \mapsto B_p$

Failure of the law of introduction of possibility: $\not\models A_p \mapsto \Diamond A_p$ because $\not\models A_p > \Diamond A_p$

Strong implication is *decidable.*
For $\models A_p > B_p$ when all predication formulas which occur in B_p also occur in A_p. Moreover, $\models$*Tautological* $(A_p \Rightarrow B_p)$ when all the semantic tableaux of S5 modal logic for $(A_p \Rightarrow B_p)$ close.

There is a *theorem of finiteness for strong implication*: Every proposition only strongly implies a finite number of others. In particular, $\models$*Tautological* $B_p \Rightarrow (A_p \mapsto B_p \Leftrightarrow A_p > B_p)$. A proposition strongly implies all and only the tautologies composed from its atomic propositions.

And $\models$*Tautological* $\neg A_p \Rightarrow (A_p \mapsto B_p \Leftrightarrow A_p > B_p)$. A contradiction strongly implies all and only the propositions composed from its atomic propositions.

The decidability and finiteness of strong implication confirm that it is cognitively realized.

5.6 Laws of propositional identity

Modal propositions are richer than modal predications. In particular, $\not\models \Box[(R_n t_1 \ldots t_n)] = \Box[(R_n t_1 \ldots t_n)]$ For $\not\models \Box[(R_n t_1 \ldots t_n)] > \Box[(R_n t_1 \ldots t_n)]$ The failure of such a law is shown in language. Properties such as being the father of a person are possessed by the same male parent in all possible circumstances. These properties have the same extension as their necessitation. But when we think that someone is the father of someone else, we do not *eo ipso* think that he is necessarily his father.

All the classical *Boolean laws of idempotence, commutativity, associativity* and *distributivity* are valid laws of propositional identity: $\models A_p = A_p \wedge A_p \models (A_p \wedge B_p) = (B_p \wedge A_p) \models (A_p \vee (B_p \vee C_p)) = ((A_p \vee B_p) \vee C_p)) \models \neg(A_p \vee B_p) = (\neg A_p \wedge \neg B_p) \models (A_p \wedge (B_p \vee C_p)) = ((A_p \wedge B_p) \vee (A_p \wedge C_p)) \models \Box(A_p \wedge B_p) = (\Box A_p \wedge \Box B_p)$

So are the laws of *reduction*: $\models \neg\neg\, A_p = A_p \models M\Box A_p = \Box A$ and $\models M\Diamond A_p = \Diamond A_p$ where $M = \Box, \Box\neg, \Diamond$ or $\Diamond\neg$ In particular, $\models \Box A_p = \Box\Box A_p$ and $\models \Box A_p = \Diamond\Box A_p$

Unlike hyperintensional logic, my logic of propositions does not require that identical propositions be *intensionally isomorphic.*[28] Intensional isomorphism is too strong a criterion of propositional identity. However, propositional identity requires more than *co-*

[28] See Max J. Cresswell, "Hyperintensional Logic". *Studia Logica* [1975].

entailment advocated in the logic of relevance. $\not\models A_p \mapsto (A_p \wedge (A_p \vee B_p)$. As M. Dunn pointed out, it is somehow unfortunate that A_p and $(A_p \wedge (A_p \vee B_p)$ co-entail each other.[29] For most formulas of such forms are not synonymous. Co-entailment is not sufficient for synonymy because it allows for the introduction of new sense.

Finally strong equivalence is finer than *analytic equivalence* $\leftrightarrow$. Consider the following law: $\models [(\Box R_1 c)] \Rightarrow ([(\Box R_1 c)] \leftrightarrow ([(\Box R_1 c)] \vee \neg [(\Box R_1 c)]))$. It is not a valid law of propositional identity.

References

Anderson R., Belnap N. & Dunn J.M. (1992). *Entailment The Logic of Relevance and Necessity.* Princeton University.

Bealer G. (1982). *Quality and Concept.* Oxford : Clarendon Press.

Belnap N. (1991). "Backwards and Towards in the Modal Logic of Agency" *Philosophy and Phenomenological Research 51.*

Carnap R. (1956). *Meaning and Necessity.* University of Chicago Press.

Cherniak C. (1986). *Minimal Rationality.* Bradford Books.

Couturat L. (1903). *Opuscules et fragments inédits de Leibnitz*, extraits des manuscrits de la bibliothèque royale de Hanovre. Paris.

Cresswell M. J. (1975). "Hyperintensional Logic", *Studia Logica 34*:25–38.

Fine K. (1986). "Analytic Implication", in *Notre Dame Journal of Formal Logic 27*:2.

Hintikka J. (1962). *Knowledge and Belief.* Cornell University Press.

Kaplan D. (1970). "How to Russell a Frege-Church", *The Journal of Philosophical Logic 8*:1, 716–29.

Kripke S. (1963). "Semantical Considerations on Modal Logic", in *Acta Philosophica Fennica 16.*

— (1975) "Speaker Reference and Semantic Reference" in P.A. French *et al* (eds) *Contemporary Perspectives in the Philosophy of Language.* University of Minnesota Press.

— (1980) *Naming and Necessity.* Harvard Univesity Press.

Lewis C. I. (1918). *A Survey of Symbolic Logic.* University of California Press.

Marcus R. Barcan (1993). *Modalities.* Oxford University Press.

Montague R. (1974). *Formal Philosophy.* Yale University Press.

[29] See his philosophical rumifications in Anderson *et al* [1992].

Nowak M. & Vanderveken D. (1996). "The Minimal Logic of Propositional Contents of Thought: a Completeness Theorem", *Studia Logica 54*, 391–410.

Parry W.T. (1972). "Comparison of Entailment Theories", *The Journal of Symbolic Logic 37.*

Rodriguez Marqueze J. (1993). "On the Logical Form of Propositions: Some Problems for Vanderveken's New Theory of Propositions" in *Philosophical Issues 3.*

Searle J.R. & Vanderveken D. (1985). *Foundations of Illocutionary Logic.* Cambridge Univ. Press.

Strawson P.F. (1959). *Individuals.* Methuen.

— (1974). *Subject and Predicate in Logic and Grammar.* Methuen.

Vanderveken D. (1990-1). *Meaning and Speech Acts*, Volume I: *Principles of Language Use* and Volume II: *Formal Semantics of Success and Satisfaction.* Cambridge University Press.

— (1995). "A New Formulation of the Logic of Propositions", in M. Marion & R. Cohen (eds), *Québec Studies in the Philosophy of Science 1, Logic, Mathematics and Physics.* Boston Studies in the Philosophy of Science, Kluwer, 95–105.

— (2001). "Universal Grammar and Speech Act Theory" in D. Vanderveken & S. Kubo (eds), *Essays in Speech Act Theory.* Benjamins, P&b ns 77, 25–62.

— (2002). "Attempt, Success and Action Generation" in the special issue on Mental Causation of *Manuscrito XXV*:1.

— (2003). "Formal Ontology, Propositional Identity and Truth According to Predication With an Application of the Theory of Types to the Logic of Modal and Temporal Proposition" in *Cahiers d'épistémologie 2003*:03. Université du Québec à Montréal. 29 pages. www.philo.uqam.ca

— (2003). "Attempt and Action Generation Towards the Foundations of the Logic of Action" in *Cahiers d'épistémologie 2003*:02. Université du Québec à Montréal. 43 pages. www.philo.uqam.ca

— (2004). "Attempt, Success and Action Generation A Logical Study of Intentional Action". Chapter 15 of the present Volume. — (2004). "Success, Satisfaction and Truth in the Logic of Speech Acts and Formal Semantics", in S. Davis & B. Gillan (eds) *A Reader in Semantics*, 710–734 Oxford. University Press, in press.

— *Propositions, Truth and Thought New Foundations for Philosophical Logic*, forthcoming.

Whitehead A. & Russell B. (1910). *Principia Mathematica.* Cambridge University Press.

Wittgenstein L. (1961). *Tractatus logico-philosophicus.* Routledge.

Chapter 11

REASONING AND ASPECTUAL-TEMPORAL CALCULUS

Jean-Pierre Desclés
University of Paris-Sorbonne

In the present article, we propose a formal representation of the reasoning expressed in and by natural language sentences like:

(1) *The hunter has killed the deer.*
therefore:

1/ The deer has been killed.
2/ The deer is dead.
3/ The deer is no longer alive.
4/ The deer had been alive.

(2) *Peter has come out of the garage.*
therefore:

1/ Peter was in the garage.
2/ Peter is no longer in the garage.
3/ Peter has come outside the garage from the inside.

(3) *Peter is already back home.*
therefore:

1/ Peter was not at home sometime earlier.
2/ One could expect that Peter was not at home.

(4) *If Peter had been there, Mary would not have left.*

1/ Since Peter was not there, Mary has left.

D. Vanderveken (ed.), Logic, Thought & Action, 217–244.

(5) *One more step and I will shoot.*
therefore:

1/ You have the intention of making one more step.

2/ I don't shoot but I have the capacity of shooting.

How can we infer the sentences (1.1), (1.2), (1.3) and (1.4) from (1)? What are the operations we must execute from the understanding of aspectual and temporal grammatical markers and from the understanding of lexical units? The same questions can be asked about (2), (3), (4) and (5).

1. Theoretical Framework

In order to explain this kind of problem, one has to be able to build metalinguistic representations of the above sentences in a way that such inferences are automatic. The formal model of the metalinguistic representations that we choose is applicative (or functional), which means that it applies operators to different types of operands (Desclés, 1990). These applicative representations take Church's lambda-calculus applicative formalisms with types, as well as Curry's (1958) (see Appendix) Combinatory Logic with types. Since the above reasoning requires aspectual and temporal notions, we will use actualization intervals associated with predicates and sentence relations (that means the intervals of instants between which a predicative relation is considered as actualized or true). Indeed, the analyses of aspects and tenses that we have presented in different publications is based on *topological representations* (Desclés, 1980, 1990b, 1991, 1993; Guentchéva, 1990; Maire-Reppert, Oh, Berri, 1993; Desclés & Guentchéva, 1990, 1995...). Therefore, we attach topological operators interpreted on topological intervals of instants. We associate with an interval of instants two boundaries : a left boundary $\gamma(\mathrm{I})$ and a right boundary $\delta(\mathrm{I})$. A boundary of a topological interval can be "open" (in this case, the boundary does not belong to the interval) or can be "closed" (in that case, the boundary belongs to the interval). An interval is closed when its left and right boundaries are closed; it is open when its left and right boundaries are open; it is semi-open when its left boundary is closed and its right boundary is open.

The knowledge of lexical meanings (verbs in particular) requires knowledge of representation formalisms such as Sowa's conceptual graphs. As far as we are concerned, we use the representations such as the *semantic-cognitive schemes* which we have presented in several previous publications (Desclés, 1990a, 1994, Abraham, 1995). Each of these schemes represents the meaning of a predicate by a typed λ-expression.

We have indicated that the applicative metalinguistic represention constitute a formalism on the basis of combinatory logic and λ-calculus. Combinatory logic with types was used for analysing grammatical problems such as passivization, reflexivization, typology of voices (Shaumyan, 1987; Desclés, Guentchéva, Shaumyan, 1985,1986; Desclés, 1990). We will argue here that this formalism is adequate to analyse the reasoning in natural languages by means of reductions (technically β-reductions). The method of this formalization is divided into several phases:

1/ Observation and analysis of linguistic data;

2/ Conceptualization by means of a concept network (for example, concerning all aspects: process, event, state, perfect, perfective, imperfective...);

3/ Schematization and design of the schemes (for example, the use of semantic-cognitive schemes for the representation of predicate meanings);

4/ Mathematization of concepts, operations and intuitive schemes (for example, the use of topology and basic operations such as application);

5/ Construction of a formal language that must be adequate to formalize the intuitive conceptualizations;

6/ Interpretation of this formal language in a model (Tarski's sense).

Instead of starting from a pre-established formal language (such as, for example, Prior's tense logic), we prefer defining and interpreting a formal metalanguage, this starting from a more or less mathematized model. This approach implies a conceptualization of intuitive notions (for example: progressivity, perfectivity, inchoativity...), which we will later try to formulate in a mathematical way. Many of the logicians (such as, for example, in Montague's approach) start from formal languages (a logic of tenses, a logic of modalities or a logic of indexical terms), and then build corresponding semantic models in order to provide to a certain extent approximations of natural languages. We take the opposite approach: first, we define the model which has already been mathematized (for example, a quasi-topological model of states, events and process or a model of speech-act operations); second, we express the concepts of the model in terms of operators of combinatory logic. This formal language is a metalanguage in the way it describes the semantics of grammatical categories of natural languages.

The theoretical framework in which we develop the following linguistic analyses is that of the Cognitive Applicative Grammar which can be regarded on the one hand as an extension of Shaumyan's Universal Applicative Grammar (1987) with integrations of cognitive representations, and on the other hand as formalizations of speech-act operations from the works of Benveniste (1964), Searle and Vanderveken (1985) and Culioli (1994). In the Cognitive Applicative Grammar (Desclés, 1990a), there are three different representation levels with explicit processes of change of representations from one level into another. These three levels are:

(i) the level of phenotype representations — or morpho-syntactic configurations -, its task is the analysis of the morpho-syntactic data of different languages;

(ii) the level of genotype representations — or logico-grammatical operations -, its task is to exhibit the invariants and the grammatical functions of language;

(iii) the level of semantic-cognitive representations, its purpose is, first, the analysis and the formal representations of the meanings of lexical units, and second, the interaction of language activities with other cognitive activities of human perception and action.

The change of representations from one level into another is similar to the generalized compiling process of high-level programming languages. This compiling process changes units from one representation level to another by means of synthetical "reunitarization" (definition of new units from given units) or by means of analytical "decompositons" of a unit. This device is oriented by an "intelligent" mechanism called "contextual exploration". The purpose of this mechanism is to resolve ambiguities by locating the relevant contextual information during different stages of the process, thus orienting the process towards a decision for solving ambiguities in some grammatical units. The goal of this complex device (compiling directed by contextual exploration) is to establish an explicit relationship between abstract representations and directly observable linguistic configurations.

2. Conceptualizations of Aspect and Tense

We recall some of the theoretical elements of the aspect-tense model that we have developed and presented in several previous publications. A *predicative relation* λ or "propositional content" (which is called a "lexis" by Culioli (1994)) is organized by means of predicate operations.

These predicate operations are very well analyzed and expressed in an applicative formalism. We can consider the following applicative expression with prefixation of the predicate operator:

(*) "to see" "a-deer" "the-hunter"

This applicative expression is obtained by means of two successive applictive operations. First, the predicate "to see" is an operator that applies to a first operand "deer"; the result is a new operator that applies to a second operand "hunter". The result is the predicative relation (*). The building of this predicative relation is represented as follows:

"to see" "a-deer"
————————————>
"to see" "a-deer" "the-hunter"
————————————————————>
"to see" "a-deer" "the-hunter"

Such a predicative relation is tenseless. Before inserting it in a reference space, the speaking subject perceives it in different ways, depending on whether he views it as a progressive process, an event or a resultative state. Thus the subject constructs an *aspectualized predicative relation* (Desclés, 1991) which is considered as true on a *topological interval of instants* I. We note this aspectual predicative relation as follows:

(**) ASP_I ("to see" "a-deer" "the-hunter")

where the aspectual operator *ASP* applies to the predicative relation (*). According to whether the aspect is a state, an event or a process, the interval I will be respectively open, closed or semi-open. We say that the state, the event or the process is true (or actualized) on this topological interval I. Having aspectualized the predicative relation, the subject now has to insert it into his own time reference that is distinct from the external time reference (the clock time, the cosmic time, the calendar time...). The subject will or will not consider the aspectualized predicative relation as concomitant with his own process of speaking, that cannot be reduced to a punctual instant because each process of speaking takes time. In this reference, T^0 denotes "this first non-actualized instant", which means the right boundary (unfinished) of the speaking process in progress (Desclés, 1980, 1990b, 1994; Desclés & Guentchéva, 1990, 1995).

We have therefore several sentences with the same propositional content (*), for example:

(a) *The hunter is looking at the deer at this moment*
(Unfinished) process concomitant with the speaking process in progress

(b) *Finally; the hunter has seen the deer, he is happy*
Resultative state of a finished anterior process

(c) *While the hunter was looking at the deer*
(Unfinished) process non concomitant with the speaking process in progress

(d) *The hunter saw the deer, and then he*
Event inserted into a series of events

With topological representations (which means the diagrams of the actualization of the aspectual predicative relation (**) in accordance with a choice of different aspectual and temporal scope), we will have respectively the differents temporal diagrams given in figure 1.

```
(a')   - - - - - - - - - - - - - [______________________________[T^0
                                 <The hunter sees a deer>
(b')   - - [ - - - - - - - - - - ]______________________________[T^0
                                 <The hunter has seen the deer>
                                          <he is happy>
(c')   - - [______________________________________[T' - - - [T^0
           <The hunter was looking at the deer>
(d')   - - [_________________________________] - - - - - - - - - - [T^0
           <The hunter saw a deer>
```

Figure 1.

In a general way, an aspectualized predicative relation $ASP_I(\Lambda)$ will be actualized (or realized) on an interval I. If this relation is aspectualized as a state, or respectively as an event or as an unfinished process, it will be true respectively on an open interval O, on a closed interval F or on an interval J closed at the left boundary and open at the right boundary (Desclés 1990b). When I designates an interval (open or closed or semi-open), γ(I) and δ(I), the respective bounds are the left or the right of the interval I. If an interval is open, then the two bounds (at left and at right) do not belong to the interval. If the interval is closed, then the two bounds (at left and at right) belong to the interval. We have thus the three diagramatic representations of realization intervals according to the aspects of the predicative relations: a state, an event and an (unfinished) process. We have three more specific aspectual operators *STATE*, *EVENT* and *PROC*. The intervals, with topological boundaries, where the aspectualised predicative relations are realized are shown by temporal diagrams (see figure 2):

Remark: We consider (Desclés, 1980, 1995; Declés & Guentcheve, 1995), the trichotomy state / event / process essential in the analy-

]————[	[————]	[————[
$\text{STATE}_O(\Lambda)$	$EVENT_F(\Lambda)$	$PROC_J(\Lambda)$

Figure 2.

ses of aspects in natural languages (see also: Lyons (1977), Comrie (1979) or Mourelatos (1981)). Note that this trichotomy is different from Vendler's classification of "state", "activity", "accomplisment" and "achievement". Some authors (for instance: Smith (1991), Koceska-Toscewa and Mazurkiewicz (1994), Karolak (1997), Kamp (1981, 1983), Vet (1995)) claim that the dichotomy state / event (punctual) is sufficient. However, several arguments against this theoretical viewpoint have been given (see Desclés and Guentchéva, 1995)).

We formulate different rules about states, events and processes.

1/ **RULE on STATE:** IF a state $STATE_O(\Lambda)$ relating to a predicative relation Λ is true on an open interval O, THEN for each sub-interval O' of O, the state relating to the same predicative relation Λ remains true over O':

$$STATE_O(\Lambda) \ \& \ [O \supset O'] \Rightarrow STATE_{O'}(\Lambda)$$

Remark: One finds a similar rule in Dowty (1979).

2/ **RULE on EVENT:** IF an event $EVEN_F(\Lambda)$ relating to a predicative relation Λ is true, THEN the predicative relation Λ is true at the right boundary δ(F):

$$EVEN_{\mathrm{F}}(\Lambda) \Rightarrow (\Lambda)_{\delta(\mathrm{F})}$$

In general, an event is actualized only at the right boundary δ(F), which is the final boundary of the event actualized on a closed interval F. The same event relating to the same predicative relation Λ is not actualized at the right boundary of each of the sub-interval F' of F, but only at δ(F). An event will only be true if the final instant δ(F) is reached. For example, the sentence *John wrote a letter in one hour* is a predicative relation which is true at its right boundary (final boundary) of a closed interval F, while this predicative relation is not true in a closed sub-interval F' of F. However, in some cases, an aspectualized predicative relation Λ, viewed as an event, can be true both for a closed interval F and for each sub-interval F' of F. An example of this case is: *John ran in the park yesterday afternoon.*

3/ **RULE on PROCESS:** IF an unfinished process $PROC_J$ (Λ) relating to a predicative relation Λ is true on a semi-open interval J, THEN for each semi-open interval J' with the same beginning of J (γ(J) = γ(J')), the predicative process $PROC_{J'}$ (Λ) relating to the same predicative relation remains true for J':

$$PROC_{\mathrm{J}}(\Lambda)\ \&\ [\mathrm{J} \supset \mathrm{J}'] \Rightarrow PROC_{\mathrm{J}'}(\Lambda)$$

Remark: for the justification of this rule, see Desclés and Guentchéva (1995).

4/ **RULE on PROCESS:** IF an unfinished process $PROC_J$ (Λ) relating to a predicative relation Λ is true on an interval J, THEN when the process is finished (in French: "achevé"), it generates:

(i) an event $EVEN_F$ (Λ) (relating to the same predicative relation Λ); this event is realized on a closed interval F which includes the smaller closed interval $\underline{\mathrm{cl}}$(J) including J, called the closure of J;

(ii) a resultative state $RES\text{-}STATE_O$ (Λ) (relating to the same predicative relation Λ) which is true on an open O that is posterior and adjacent to the interval F:

$PROC_J(\Lambda) \Rightarrow$ there is an $EVEN_F(\Lambda)$ and a $RES\text{-}STATE_O(\Lambda)$ such that

$$[\mathrm{F} \supset \underline{\mathrm{cl}}(\mathrm{J})]\ \&\ [\mathrm{O \text{ is adjacent to } F \text{ and posterior to } F}]$$

We represent this rule with the diagram of the figure 3.

```
[ - - - - - - - - J - - - - - - - - - - - >[
[ - - - - - - - - - cl(J) - - - - - - - - ]
[ - - - - - - - - - - - F - - - - - - - - - - - - ]< - - - - - - - - O - - - - - - - - >[
```

Figure 3.

Remark: This rule is designed to capture the meaning of "perfect" in Indo-European languages (see Desclés, 1980; Guentchéva, 1990).

3. Analysis of Tenses and Aspects

Let us start with the following sentence:

(6) The hunter is looking at the deer (at this moment).

The underlying logical form of the sentence is analyzed in the following steps: (i) Formation of an underlying predicative relation expressed by an applicative notation: (("to see" "a-deer") "the-hunter") ; (ii) Aspectualization of this predicative relation as an unfinished process in progress, with the aspectual process operator *PROC* ; (iii) Inclusion of this process, called *predicative process*, into the *speaking process* in progress; (iv) Establishment of a concomitance relation between the unfinished speaking process and the unfinished predicative process.

Generally, a predicative relation Λ that is viewed as an unfinished process is true on an interval J^1 of instants, hence the process operator $PROC_J{}^1$ and the predicative process $PROC_J{}^1(\Lambda)$. This predicative process becomes an argument of the metalinguistic speaking predicate "SAY($\ldots$)S^0", where the symbol S^0 denotes the abstract subject "EGO" of any speaking act: EGO is the origin of the system of persons (I, YOU, HE / SHE) as well as deictic spatial markers as HERE, THERE. The speaking process is true on the interval J^0 of instants, hence the aspectual operator $PROC_J{}^0$ and the following speech-act scheme (see Desclés, 1980):

(***) $PROC^0_J$ (SAY($\ldots$)S^0)

Now we can express explicitly the underlying logical form of sentence (6). To simplify the notations, we use the symbol P_2 to designate the transitive lexical predicate associated to the verb "to see" and respectively the symbols T^1 and T^2 for the terms "the hunter" and "the deer". Formula (7) is an aspectual-tense representation of (6), expressed by a prefixed applicative notation:

(7) $\& \; (PROC^0_J(\mathrm{SAY}(PROC^1_J(P_2T^2T^1))S^0) \; ([\delta(J^1) = \delta(J^0)])$

This formula is a conjunction of the two constituents (7a) and (7b):

(7a) $(PROC^0_J(\mathrm{SAY}(PROC^1_J(P_2T^2T^1))S^0)([\delta(J^1) = \delta(J^0)])$

(7b) $[\delta(J^1) = \delta(J^0)]$

The predicative process $PROC_J{}^1$ $(P_2T^2T^1)$ is in this way considered as being true on the interval J^1. Formula (7a) represents the embedding of the predicative process $PROC_J{}^1(P_2T^2T^1)$ into the speech-act process. Formula (7b) expresses a temporal constraint on the intervals of actualization of the two processes: In an unfinished process concomitant with the speech-act, the right boundaries $\delta(J^1)$ of J^1 and $\delta(J^0)$ of J^0 must be identical. Formula (7) can be read as follows:

(7') "The speaking process "S^0 SAY$\ldots$", which is true on an interval J^0, has as its argument a predicative process "$P_2T^2T^1$" which is true on an interval J^1, with the temporal constraint that the two right boundaries of the two intervals J^0 and J^1 must be identical"

or, in others words:

(7") "the predicative relation "$\mathrm{P}_2\mathrm{T}^2\mathrm{T}^1$" is conceived from an aspectual point of view as an unfinished process which is an argument of the speaking process "S^0 SAY that...", the final sections of the two processes are identical."

Now, we are going to define an *aspectual operator* by integrating the two elementary process operators $PROC_J{}^0$ and $PROC_J{}^1$. into a complex operator. The predicative relation "$\mathrm{P}_2\mathrm{T}^2\mathrm{T}^1$" becomes an argument of this complex operator. With this aim, we use the combinators **B**, $\mathbf{B}^2$ and **C** of the combinatory logic (see Appendix). We present the integration as in Gentzen's "natural deduction" style but with "reunitarization" relations (or definitions of new units from an applicative combination of differents more primitives units). Thus, in the following integration process, step 9. expresses a reunitarization relation: the complex operator SA is defined in terms of the operator SAY and the operand S^0.

Integration of the operator "Speech-act" (SA)

The symbol **<u>SA</u>** designates, in the following integrative process, an operator which means that "S^0 performs a speech-act". We call it an "enunciative operator".

1.	$PROC^0_\mathrm{J}(\mathrm{SAY}(PROC^1_\mathrm{J}(\mathrm{P}_2\mathrm{T}^2\mathrm{T}^1))S^0)$	*hyp*
2.	$\mathbf{B}^2 PROC_\mathrm{J}\mathrm{SAY}(PROC^1_\mathrm{J}(\mathrm{P}_2\mathrm{T}^2\mathrm{T}^1))S^0$	int$\mathbf{B}^2$
3.	$\mathbf{C}(\mathbf{B}^2 PROC^0_\mathrm{J}\mathrm{SAY})S^0(PROC^1_\mathrm{J}(\mathrm{P}_2\mathrm{T}^2\mathrm{T}^1))$	int**C**
4.	$\mathbf{C}(\mathbf{CB}^2\mathrm{SAY}PROC^0_\mathrm{J})S^0(PROC^1_\mathrm{J}(\mathrm{P}_2\mathrm{T}^2\mathrm{T}^1))$	int**C**
5.	$\mathbf{BC}(\mathbf{CB}^2\mathrm{SAY})PROC^0_\mathrm{J}S^0(PROC^1_\mathrm{J}(\mathrm{P}_2\mathrm{T}^2\mathrm{T}^1))$	int**B**
6.	$\mathbf{B}(\mathbf{BC})(\mathbf{CB}^2)\mathrm{SAY}PROC^0_\mathrm{J}S^0(PROC^1_\mathrm{J}(\mathrm{P}_2\mathrm{T}^2\mathrm{T}^1))$	int**B**
7.	$\mathbf{C}(\mathbf{B}(\mathbf{BC})(\mathbf{CB}^2)\mathrm{SAY})S^0 PROC^0_\mathrm{J}(PROC^1_\mathrm{J}(\mathrm{P}_2\mathrm{T}^2\mathrm{T}^1))$	int**C**
8.	$\mathbf{B}(\mathbf{C}(\mathbf{B}(\mathbf{BC})(\mathbf{CB}^2))\mathrm{SAY}S^0 PROC^0_\mathrm{J}(PROC^1_\mathrm{J}(\mathrm{P}_2\mathrm{T}^2\mathrm{T}^1))$	int**B**
9.	$[\underline{\mathbf{SA}} =_{def} \mathbf{B}(\mathbf{C}(\mathbf{B}(\mathbf{BC})(\mathbf{CB}^2)\mathrm{SAY}S^0]$	def of **<u>SA</u>**
10.	$\underline{\mathbf{SA}}PROC^0_\mathrm{J}(PROC^1_\mathrm{J}(\mathrm{P}_2\mathrm{T}^2\mathrm{T}^1))$	repl, 8, 9
11.	$\mathbf{B}(\underline{\mathbf{SA}}PROC^0_\mathrm{J})PROC^1_\mathrm{J}(\mathrm{P}_2\mathrm{T}^2\mathrm{T}^1)$	int**B**
12.	$\mathbf{BB}\underline{\mathbf{SA}}PROC^0_\mathrm{J}PROC^1_\mathrm{J}(\mathrm{P}_2\mathrm{T}^2\mathrm{T}^1)$	int**B**
13.	$(\underline{\mathbf{SA}}\mathbf{0}PROC^0_\mathrm{J}\mathbf{0}PROC^1_\mathrm{J})(\mathrm{P}_2\mathrm{T}^2\mathrm{T}^1)$	def of **0**, 12

Comments: In step 1., it is shown that the predicative process is being embedded into the speech-act process; the first process is true on one interval J^1 and the second is true on another J^0. Steps 2. to 8. introduce the combinators **B**, $\mathbf{B}^2$ and **C** which allow a combination of the speaking operators, hence the definition introduced in step 9. of the operator **<u>SA</u>** which means "the speaker performs a speaking-act" or "the speaker S^0 says that...". We deduce from the definition given at step 9., the expression given at step 10. by remplacement of the *definiens*. By combining the operators **<u>SA</u>**, $PROC_J{}^0$ and $PROC_J{}^1$, we obtain a new complex operator (at step 12. or, equivalently, at step 13.) whose operand is $\mathrm{P}_2\mathrm{T}^2\mathrm{T}^1$. The operators **<u>SA</u>**, $PROC_J{}^0$ and $PROC_J{}^1$

are combined as compositions of functions. From this integrative process it follows that step 13 is an applicative integration of step 1. with the same meaning. The expression at step 13. is deduced from the expression given at step 1. At step 13., it is shown that the complex operator, "($\underline{\mathbf{SA}}$ **0** $PROC_J{}^0$ **0** $PROC_J{}^1$)" is a grammatical operator whose meaning is only aspectual. This applies to the predicative relation "($P_2T^2T^1$)". The two formulas in step 1. and step 13. are considered as being equivalent to the same aspectual meaning.

In the above deduction, we introduce a derived operator $\underline{\mathbf{SA}}$. Its meaning is: "the speaking subject S^0 says that". We call it an "enunciative operator". This enunciative operator is built by means of an application of the abstract operator (a derived abstract combinator) to the elementary operator (saying) and to the operand S^0. Its formal definition is given by the following relation between a *definiendum* (at the left) and a *definiens* (at the right):

(8) $[\underline{\mathbf{SA}} =_{\text{def}} \mathbf{B}(\mathbf{C}(\mathbf{B}(\mathbf{BC})(\mathbf{CB}^2)\text{SAY}S^0]$

At the end of the above reduction, the applicative expression obtained at step 13. represents the result of the application of a complex aspectual operator on the predicative relation ($P_2T^2T^1$), that is to say:

(9) $(\underline{\mathbf{SA}}\ \mathbf{0}\ PROC^0_J\ \mathbf{0}\ PROC^1_J)(P_2T^2T^1)$

Now, we return to the expression (6) again. We substitute in (7) the expression (7a) by (9), since the two expressions (7a) and (9) are equivalent to the same aspectual meaning, hence the expression (10):

(10) $\&\ ((\underline{\mathbf{SA}}\ \mathbf{0}\ PROC^0_J\ \mathbf{0}\ PROC^1_J)(P_2T^2T^1))([\delta(^1_J) = \delta(^0_J)])$

Thus, we can continue the reduction process in such a way as we define an integrated aspectual operator considered as a reunitarized operator. The operand of the first aspectual operator is a predicative relation but the operand of the second aspectual operator is a lexical predicate.

Integrative process of the verbal aspectual operator.

1.	$\&((\underline{\mathbf{SA}}\ \mathbf{0}\ PROC^0_J\ \mathbf{0}\ PROC^1_J)(P_2T^2T^1))\ ([\delta(J^1) = \delta(J^0)])$	hyp
2.	$\mathbf{C}\&([\delta(J^1) = \delta(J^0)])\ ((\underline{\mathbf{SA}}\ \mathbf{0}\ PROC^0_J\ \mathbf{0}\ PROC^1_J)(P_2T^2T^1))$	int**C**
3.	$\mathbf{B}(\mathbf{C}\&([\delta(J^1) = \delta(J^0)])\ (\underline{\mathbf{SA}}\ \mathbf{0}\ PROC^0_J\ \mathbf{0}\ PROC^1_J)\ (P_2T^2T^1)$	int**B**
4.	$(\mathbf{C}\&([\delta(J^1) = \delta(J^0)]))\ \mathbf{0}\ (\underline{\mathbf{SA}}\ \mathbf{0}\ PROC^0_J\ \mathbf{0}\ PROC^1_J)(P_2T^2T^1)$	def **0**
5.	$[\underline{UNF\text{-}PRST} =_{def} (\mathbf{C}\&([\delta(J^1) = \delta(J^0)]))\ \mathbf{0}\ (\underline{\mathbf{SA}}\mathbf{0}PROC^0_J\ \mathbf{0}\ PROC^1_J)]$	def
6.	$\underline{UNF\text{-}PRST}\ (P_2T^2T^1)$	from 5
7.	$\mathbf{B}^2\ \underline{UNF\text{-}PRST}\ P_2T^2T^1$	int $\mathbf{B}^2$
8.	$[prog\text{-}prest =_{def} \mathbf{B}^2\ \underline{UNF\text{-}PRST}]$	def
9.	$(prog\text{-}prest\ P_2)\ T^2T^1$	from 8

Comments: step 1. has exactly the same meaning as in expression (2). The temporal constraints have been added to the proper aspectual conditions. We rearrange the operators so as we can isolate the predicative relation (steps 2 to 4). At step 4. the two operators are combined as two composed functions. Then we introduce the definition of a *grammaticalized aspectual-temporal operator* "Unfinished-Present", designated by *UNF-PRST* at step 5. At step 6., we isolate the two arguments of the predicative relation so that we can define an operator which applies only to the predicate, hence the definition of a *verbal progressive present operator* designated as *prog-prest* at step 8. and in the final expression at step 9. The operator *prog-prest* is the morphological trace of the deep grammatical operator <u>*UNF-PRST*</u> which (i) encodes the temporal constraints (that two intervals of actualization have the same right boundary) and also (ii) combines the two process operators with the enunciative predicate <u>**SA**</u>. Now, the predicate P_2 becomes an argument of the morphological operator *prog-prest*, hence the "aspectualized and tensed" new predicate "*prog-prest* P_2" derived from the lexical predicate P_2.

Finally, we obtain the expressions (11) and (12) which are equivalent to the expression (7); these two expressions have the same meaning as the expression (7):

(11) <u>*UNF-PRST*</u>($P_2T^2T^1$)

(12) (*prog-prest*P_2)T^2T^1

We remark that the *grammatical operator* <u>*UNF-PRST*</u> takes the entire predicative relation ($P_2T^2T^1$) as its operand whereas the *verbal progressive present operator prog-prest* takes only the lexical predicate P_2 as its operand.

Now we look at the sentence (6) *The hunter is looking at a deer (at this moment).* This sentence can be analyzed by means of the prefixed applicative expression (6'):

(6') *is-looking-at a-deer the-hunter*

The verbal operator *is-looking-at* can be analyzed as a complex binary predicate, derived from a lexical predicate "look-at" by means of the morphological operator "progressive present" *prog-prest.* Thus we have the following definition:

(6") [*is-looking-at* = *prog-prest* look-at]

From this definition we can deduce relations between a linguistic configuration *The hunter is looking at a deer* expressed at the phenotype level and the corresponding applicative expressions:

(6''') *The hunter is looking at a deer*

(6''') = *is-looking-at a-deer the-hunter*

(6''') = *prog-prest* (look-at) a-deer the-hunter

By a similar calculus but in a bottom-up way, we get the following reduction of an applicative expression with a verbal aspectual operator applied to a lexical predicate into an underlying applicative expression which describes the grammatical meaning of the verbal aspectual operator:

1. $(\textit{prog-prest}\mathrm{P}_2)\mathrm{T}^2\mathrm{T}^1$
2. $[\textit{prog-prest} = \mathbf{B}^2\,\underline{\textit{UNF-PRST}}]$
3. $[\underline{\textit{UNF-PRST}} =_{\mathrm{def}} (\mathbf{C}\&([\delta(\mathrm{J}^1) = \delta(\mathrm{J}^0)]))\ \mathbf{0}\ (\underline{\mathbf{SA}}\ \mathbf{0}\ \textit{PROC}^0_{\mathrm{J}}\ \mathbf{0}\ \textit{PROC}^1_{\mathrm{J}})]$
4. $\&((\underline{\mathbf{SA}}\ \mathbf{0}\ \textit{PROC}^0_{\mathrm{J}}\ \mathbf{0}\ \textit{PROC}^1_{\mathrm{J}})(\mathrm{P}_2\mathrm{B}^2\mathrm{A}^1))\ ([\delta(\mathrm{J}^1) = \delta(\mathrm{J}^0)])$
5. $\&(\textit{PROC}^0_{\mathrm{J}}(\mathrm{SAY}(\textit{PROC}^1_{\mathrm{J}}(\mathrm{P}_2\mathrm{T}^2\mathrm{T}^1))S^0)\ ([\delta(\mathrm{J}^1) = \delta(\mathrm{J}^0)])$

The expression obtained at step 5. is considered as the *normal form* of the expression given in step 1. These applicative expressions are obtained by successive reductions (in technical terms β-reductions), that is to say, successive eliminations of combinators and replacement by means of definition relations of complex operators. We use the symbol '$_\beta \rightarrow$' to represent the relation of reduction between applicative expressions; we obtain:

(13) $(\textit{pro-prest}\ \mathrm{P}_2)\mathrm{T}^2\mathrm{T}^1\ _\beta\rightarrow \&(\textit{PROC}^0_{\mathrm{J}}(\mathrm{SAY}(\textit{PROC}^1_{\mathrm{J}}(\mathrm{P}_2\mathrm{T}^2\mathrm{T}^1))S^0)([\delta(\mathrm{J}^1) = \delta(\mathrm{J}^0)])$

In replacing P_2, T^2, T^1 by their corresponding lexical units, we obtain:

(14)(*pro-prest* look-at) a-deer the-hunter
$_\beta\rightarrow \&(\text{UNF-PROC}^0_{\mathrm{J}}(\mathrm{SAY}(\text{UNF-PROC}^1_{\mathrm{J}}(\text{look-at a-deer the-hunter}))S^0)([\delta(\mathrm{J}^1) = \delta(\mathrm{J}^0)])$

4. Formal Calculus on Aspectual-temporal Conditions

Let us examine sentence (15):

(15) *The hunter has seen a deer*

(the aspectual value of the present perfect tense is a resultative state)

By an analogous process, we have the following reduction:

(16) $\textit{prest-perf}_{\mathrm{result}}\textit{see})$ a-deer the-hunter
$_\beta\rightarrow \&(\textit{PROC}^0_{\mathrm{J}}(\mathrm{SAY}(\underline{\textit{RESU-PRST}^1_O}\ (\text{to see a-deer the-hunter}))S^0)([\delta(O^1) = \delta(\mathrm{J}^0)])$

In the underlying normal form of the sentence, the tenseless predicative relation "to see a-deer the-hunter" (or in the infixed notation: "the-hunter to see a-deer") is viewed by the speaking subject as a resultative state which is actualized on an interval O^1 concomitant with the speech-act process. We have the temporal constraint: $[\delta(O^1) = \delta(J^0)]$. The present resultative state $\underline{\textit{RESU-PRST}_O}^1$(to see a-deer the-hunter) is adjacent to the occurrence of the event $\textit{EVEN}_F{}^2$(see the-hunter a-deer) and is concomitant with the speaking act. In other words, the event $\textit{EVEN}_F{}^2$(to see the-hunter a-deer) has an occurrence which is located before the speech-act process actualized on the interval J^0. The reulting state is true on the interval O^1; this open interval O^1 is adjacent to the closed interval F^2 and is located after F^2; the two right bounds of O^1 and J^0 are identical.

In order to formalize this resultativity and to relate it to the interval J^0 of actualization of the speech-act process, it is necessary to add further conditions. Let Λ be an arbitrary predicative relation which is considered as resultative state on the interval O; the "present resultative state of Λ is then defined as follows (see Desclés, 1980):

(17) $\underline{\textit{RESU-PRST}_O}(\Lambda) \Leftrightarrow_{\text{def}}$ there exists $\textit{EVEN}_F(\Lambda)$ such as:

(i) $\delta(F)$ is a "continuous cut" (in Dedekind's sense) in the union of the closed interval F and the open interval O, hence:
$\delta(F) = \gamma(O)$;

(ii) $[\delta(O)F < J^0\ ]$ (the interval F is before the instant J^0)

(iii) $\delta(O) = \delta(J^0)$.

We represent this continuous cut $\delta(F)$ by means of the diagram given in figure 4:

```
<- - - - - - - - - - F - - - - - - - - - - >< - - - - - - - - - - - O - - - - - - - - - - - >
[__________________________________________]___________________________________________[
                                        δ(F) =γ(O)
```

Figure 4.

The continuous cut $\delta(F)$ (in Dedekind's sense) means that the two intervals F and O are disjoint: the right boundary $\delta(F)$ of the closed interval F is identical with the left boundary $\gamma(O)$ of the interval O; since the two intervals F and O are topological intervals it follows that the interval F is necessarily closed and O is necessarily open.

In example (15), we have the diagram (figure 5) with intervals corresponding to the realizations of the event (a) "the hunter saw a deer" and the resultative state (b) "the hunter has seen a deer".

<Event $_F$ (to see a-deer the-hunter)><Resulting state$_O$(to see a-deer the-hunter)>

[——————————————]——————————————————————[

δ (F) $=\gamma$(O)

Figure 5.

The definition of the operator $\underline{RESU\text{-}PRST}$ is more complex than the definition of $\underline{UNF\text{-}PRST}$. The reduction of the applicative representation to its normal form is as follows:

$$(18)\quad (\text{prest-perf-}_{result}P_2)T^2T^1 \rightarrow_\beta$$
$$\&\{\&(PROC^0_J(SAY(\underline{\text{RESU-PRST}}^1_O(P_2T^2T^1))S^0)([\delta(O^1)=\delta(J^0)]))\}$$
$$\{\&(\text{EVEN}^2_F(P_2T^2T^1))([F^2<J^0])([\delta(F^2)=\gamma(O^1)])([\delta(O^1)=\delta(\text{J}^0)]))\}$$

Now, look at the sentence (19):

(19) *(Yesterday), the hunter was looking at the deer.*
(with the value of the progressive past tense: past unfinished process)

With an analogous calculus, we get the following reduction:

$$(20)\quad (\textit{pro-past\_}_{\text{proc}}P_2)T^2T^1 \rightarrow_\beta$$
$$\&(PROC^0_\text{J}(\text{SAY}(PROC^1_\text{J}(P_2T^2T^1))S^0)([\delta(^1_\text{J})<\delta(^0_\text{J})])$$

When we replace P_2, T^2 and T^1 by their corresponding lexical units, we obtain:

$$(21)\quad (\textit{pro-past\_}_{\text{proc}}\text{see})\text{ a-deer the-hunter} \rightarrow_\beta$$
$$\&(PROC^0_\text{J}(\text{SAY}(PROC^1_\text{J}(\text{see a-deer the-hunter}))S^0)([\delta(\text{J}^1)<\delta(\text{J}^0)])$$

Finally, we look at the sentence (22):

(22) *The hunter saw a deer*
(with value of past simple tense interpreted as an event)

For this sentence, the reduction is the following:

$$(23)\quad (\textit{simp-past\_}_{\text{event}}\text{ see})\text{ a-deer the-hunter} \rightarrow_\beta$$
$$\&(UNF-PROC^0_\text{J}(\text{SAY}(EVEN^1_F(\text{see a-deer the-hunter}))S^0)([\delta(F^1)<\gamma(^0_\text{J})])$$

In conclusion, there are four definitions of the grammaticalized aspectual operators:

(24) $\underline{\text{UNF-PRST}}$ = unfinished in the present
$\underline{\text{UNF-PAST}}$ = unfinished in the past
$\underline{\text{EVEN-PAST}}$ = past event
$\underline{\text{RESU-PRST}}$ = present resultative state
$\underline{\text{RESU-PAST}}$ = past resultative state

The definitions of these aspectual and temporal operators are:

(25) $[\underline{\text{UNF-PRST}} =_{def} (\mathbf{C}\&([\delta(\mathrm{J}^1) = \delta(^0_{\mathrm{J}})])) \ \mathbf{0} \ (\underline{\mathbf{SA}}\mathbf{0}PROC^0_{\mathrm{J}}\mathbf{0}PROC^1_{\mathrm{J}})]$
$[\underline{\text{UNF-PAST}} =_{def} (\mathbf{C}\&([\delta(\mathrm{J}^1) < \delta(\mathrm{J}^0)])) \ \mathbf{0} \ (\underline{\mathbf{SA}}\mathbf{0}PROC^0_{\mathrm{J}}\mathbf{0}PROC^1_{\mathrm{J}})]$
$[\underline{\text{EVEN-PAST}} =_{def} \mathbf{C}\&([\delta(F^1) < \gamma(\mathrm{J}^0)])) \ \mathbf{0} \ (\underline{\mathbf{SA}}\mathbf{0}PROC^0_{\mathrm{J}}\mathbf{0}EVEN^1_F)]$
$[\underline{\text{RESU-PRST}} =_{def} (\mathbf{C}\&(\&[\delta(O^1) = \delta(\mathrm{J}^0)][])) \ \mathbf{0} \ (\underline{\mathbf{SA}}\mathbf{0}PROC^0_{\mathrm{J}}\mathbf{0}RESUL^1_O)]$
$[\underline{\text{RESU-PRST}} =_{def} (\mathbf{C}\&(\&[\delta(O^1) < \delta(\mathrm{J}^0)][])) \ \mathbf{0} \ (\underline{\mathbf{SA}}\mathbf{0}PROC^0_{\mathrm{J}}\mathbf{0}RESUL^1_O)]$

We can define in the same way the meaning of different aspectual operators as PERF ("perfectivization"), DESC-STATE ("decriptive state"), PERM-STATE ("permanent state"), NEW-STATE ("new state of a referential discourse" in a narrative context), FUTURE-EVENT ("future event"), QUASI-CERTAIN-EVENT ("quasi certain event")... and also the different modalities of action (Aktionsart) as TO BEGIN, TO CONTINUE, and TO FINISH.

5. A temporal and inferential Reasoning

Let us consider the following inference, with lexical variations:

(26) a. *This morning, the hunter killed the deer.*
b. **Therefore** *The deer is dead* (in the speaking act)

In order to explain this "natural inference" encoded by means of linguistic expressions, we have to argue the analysis of the inference carried out by two procedures : on the one hand, this inference relates to a representation of aspectual-temporal conditions as we have demonstrated above, and on the other hand this inference entails a representation of the meaning of the lexical predicate "to kill".

The temporal adverbial phrase *this morning* determines that it concerns an open interval, represented by the interval O^3. This interval is located before the interval J^0 of actualization of the speech-act process, located inside the temporal interval associated to "this day". The interval O^3 includes the interval F^1 associated with the occurrence of the event "the hunter killed a deer". We can deduce the following two coordinated constraints:

$$(27) \quad \&([O^3 \supset F^1])([O^3 < \mathrm{J}^0])$$

These are further constraints to the one already imposed by the speaking coordinates of the event. From the previous section analysis, we can deduce the applicative representation of sentence (26a) as well as its normal form:

(28) (this-morning)((*simple-past*(kill)) a-deer the-hunter) $\rightarrow_\beta$
$\&\{PROC^0_{\mathrm{J}}(\text{SAY}((\text{this morning})_{O^3}(EVEN^1_F(\text{to kill a-deer the-hunter})))S^0)\}$
$\{\&([\delta(F^1) < \gamma(\mathrm{J}^0)])([O^3 \supset F^1])([O^3 < \mathrm{J}^0])\}$

Now we can analyze the lexical predicate "to kill"; the meaning of the word "kill" is represented by a semantic-cognitive scheme (Desclés, 1990) illustrated by the following λ-expression:

(29) $$\text{"to kill"} =_{def} \lambda y.(\lambda x.(\text{TRANS-CHANG}_I(SIT_1[y])(SIT_2[y]))x))$$
with: $SIT_1[y] = (\lambda z.(STATE^1_O(\text{to be-alive z})))y$
$SIT_2[y] = (\lambda z.(STATE^2_O(\text{Neg(to be-alive z)})))y$
and the constraints on topological intervals:
$[O^1 < I < O^2]$; $[\gamma(I) = \delta(O^1)]$ and $[\delta(I) = \gamma(O^2)]$

Comments: The actor x of the event assumes a grammatical role of an agent. Being an agent, he "controls" and "carries out" the change (CHANG) that turns the static situation SIT_1 into a static situation SIT_2. The operator TRANS — in the sense of "semantic transitivity" — expresses the integration of the two operators, one is the "control" (CONTR) and the second one is the "execution" (DOing) which are closely related to an agentive role (see Desclés, 1990) assumed by a term in a predicative relation. The two situations SIT_1 and SIT_2 concern the same patient y which undergoes the change. The interval O^1, on which the situation SIT_1 is realized, precedes the interval O^2, on which the situation SIT_2 is realized. As both of the two intervals are open, they are necessarily separated by a closed interval I which shares the common boundaries: the left boundary γ(I) of I is identical to the right boundary $\delta(O^1)$ of O^1, the right boundary δ(I) of I is identical to the left boundary $\gamma(O^2)$ of O^2. The change that y undergoes is therefore actualized on the interval I, which is nested between the two intervals O^1 and O^2.: the intitial static situation SIT_1 was actualized on the interval O^1 and the final situation SIT_2 will be actualized on the interval O^2.

We represent the intrinsic meaning of "to kill" by a temporal diagram given in figure 6.

]——— O^1——— [<——— I ———>] <——— O^2 ———>[
<- - - the deer is alive - - ->$EVEN_I$ (the hunter kills the deer)<- - the deer is no longer alive - ->

Figure 6.

From the statement:

(30) $$\&\{PROC^0_{\mathrm{J}}(\text{SAY}((\text{this morning})_{O^3}(EVEN^1_F(\text{to kill a-deer the-hunter}))S^0)\}$$
$$\{\&([\delta(F^1) < \gamma(\mathrm{J}^0)])([O^3 \supset F^1])([O^3 < \mathrm{J}^0])\}$$

we deduce, by means of an integrative process, that the event

(31) $(EVEN^1_F(\text{to kill the-deer the-hunter}))$

was actualized at the right boundary $\delta(F^1)$ of the interval F^1 which precedes the interval J^0 of actualization of the utterance. We are also going to deduce that the state:

(32) $((\text{now})^0_J)(STATE^0_J$ (to be-dead the-deer))

is true on the interval J^0. We take a synchronized diagram to represent the temporal intervals of the underlying situation of sentence (26a.

We are now going to present a formal proof of the previous inference (26a, b). To simplify the notations, we use the symbols T^1 for "the hunter" and T^2 for "the deer". The initial statement with the constraints on the temporal intervals is:

(33) $(PROC^0_J(\text{SAY}((\text{this morning})_O3(EVEN^1_F(P_2T^2T^1)))S^0)$
with the coordonates: $[d(F^1)<g(J^0)], [O^3 \supset F^1]$ and $[O^3<J^0]$

So we have a nested event which is actualized on the interval F^1 which occurred before the speech-act and this event is inside the temporal interval O^3 defined by "this morning", hence the conditions on intervals: $[d(F^1) < g(J^0)]$ ("F^1 before the speaking act"); $[O^3 \supset F^1]$ ("F^1 is inside the interval O^3 defined by "this morning"); $[O^3 < J^0]$ ("this morning" is located before the speaking act").

Comments on deduction I: Expression (33) introduces the declarative hypothesis. This declaration allows us to actualize predicative relations on intervals. The underlying predicative relation is true on a closed interval F^1 which precedes the interval J^0 of speech-act. From this hypothesis, one can assert that the event has occurred on an interval F^1 (step 1). Step 2 defines the meaning of "to kill" by a λ-expression. This meaning is used to decribe the lexical meaning of the event. Then follow the substitutions of arguments in the places linked by the abstraction operator λ, which leads to step 6. This result brings about a unification of intervals : the interval F^1 becomes the actualization interval of the event "killing", therefore $[I = F^1]$ (step 8). As the event is completely actualized on the interval F^1, we deduce that the final situation SIT_2 of the event is actualized. This final situation is part of the meaning of the predicate "to kill", so it is actualized on the open interval O^2 which is adjacent to F^1 (step 9). Topological consideration on order of instants concerning the intervals justify the following reasoning (step 10): since F^1 is nested between the two adjacent states O^1 (before F^1) and O^2 (after F^1), and since O^3 includes F^1 (initial hypothesis) and O^3 precedes J^0, we can deduce that O^3 has a non-empty intersection with O^2. By contiguity of intervals in the same referential framework, we deduce that O^2 includes J^0; it follows that the two right boundaries of O^2 and of J^0 must be actualized at the same instant, hence the two right boundaries are identical. Since the state is true on the interval O^2 and the interval O^2 includes the interval J^0, it follows (property of states) that the same state has to be true on the interval J^0 (step 11). Step 12 asserts a lexical equivalence between the one-place predicate "be-dead"

Deduction I:

1.	(this morning)$_{O^3}$)($EVEN^1_F$ (to kill T^2T^1))	hyp.
	$[O^3 \supset F^1]$	hyp.
	$[O^3 < J^0]$	hyp.
	$[F^1 < J^0]$	hyp.
2.	[to kill=$_{def}$ $\Lambda y.(\Lambda x.$(TRANS-CHANG$_I$(SIT$_1$[y])(SIT$_2$[y]))x))]	def. "to kill"
3.	$EVEN^1_F$($\Lambda y.(\Lambda x.$(TRANS-CHANG$_I$(SIT$_1$[y])(SIT$_2$[y]))x)T^2T^1)	repl. 1, 2
4.	$EVEN^1_F$(($\Lambda x.$ (TRANS-CHANG$_I$ (SIT$_1$[T^2])(SIT$_2$[T^2]))x)T^1)	reduction, 3
5.	$EVEN^1_F$ (TRANS-CHANG$_I$ (SIT$_1$[T^2])(SIT$_2$[T^2]))T^1)	reduction, 4
6.	1. SIT$_1$[T^2] = ($\Lambda z.$ ($STATE^1_O$ (to be-alive z))) T^2	def. SIT$_1$
	2. SIT$_1$[T^2] = $STATE^1_O$ (to be-alive T^2)	reduction,6.1
	3. SIT$_2$[T^2] = ($\Lambda z.$ ($STATE^2_O$ (Neg(to be-alive z)))) T^2	def. SIT$_2$
	4. SIT$_2$[T^2] = $STATE^2_O$ (Neg(to be-alive T^2))	reduction,6.3
7.	$EVEN^1_F$ (TRANS-CHANG$_I$($STATE^1_O$ (be-alive T^2)) ($STATE^2_O$ (Neg(be-alive T^2))) T^1)	repl. 5, 6.2, 6.4
8.	$[O^1 < F^1 < O^2]$	unification F^1/I
	$[\delta(O^1) = \gamma(F^1)]$	with [I = F^1]
	$[\gamma(O^2) = \delta(F^1)]$	
9.	$STATE^2_O$ (Neg(to be-alive T^2)) from 7	
10.	1. $[O^1 < F^1 < O^2]$	from 8
	2. $[O^3 \supset F^1]$	initial hyp.
	3. $[O^3 < J^0]$	initial hyp.
	4. $[O^3 \cap O^2 \neq \emptyset]$	from 1, 2
	5. $[O^2 \supset J^0]$	from 1, 2, 3, 4
	6. $[\delta(O^2) = \delta(J^0)]$	from 5
11.	$STATE^0_J$ (Neg(to be-alive T^2))	from 9, 10.5
12.	[to be-dead =$_{def}$ **B** Neg(to be-alive)]	def.
13.	$STATE^0_J$ (to be-dead T^2) repl.,	int. **B**, 11, 12
14.	**B**$STATE^0_J$ (to be-dead) T^2	int. **B**, 13
15.	[*PRST-STATE* =$_{def}$ **B**$STATE^0_J$]	def.
16.	[*prest-state* =$_{def}$ *PRST-STATE*]	def. of *prest*
17.	*prest-state* (to be-dead) T^2	repl., 14, 15, 16
18.	[is-dead =$_{def}$ *prest-state* (be-dead)]	def
19.	[T^2: = the-deer]	instantiation of T^2
20.	is-dead the-deer	repl. 17, 18, 19

and the composition (by the combinator **B**) of the propositional negation, noted "Neg", with the predicate "be-alive". After the substitution, we obtain step 13. Step 14 is the result of the composition of the aspectual operator with the lexical predicate "be-dead". Step 15 defines the grammatical operator *PRST-STATE.* Step 16 defines the morphological predicate *prest-state* ("present state") as being the morphological trace of the grammatical operator *PRST-STATE.* Step 17 is the result of the replacement. Step 18 is the definition of a tensed and aspectualized lexical predicate. Step 19 is an instantiation of the variable T^2, hence the applicative expression at step 20.

Finally, we have the relation of inference:

(34) $(PROC^0_J$ (SAY $(EVEN^1_F$ (to kill the-deer the-hunter))$S^0) \rightarrow_\beta$
$EVEN^1_F$ (to kill the-deer the-hunter)$\rightarrow_\beta$
$STATE^0_J$ (Neg(to be-alive) the-deer)$\rightarrow_\beta$
$STATE^0_J$ (to be-dead the-deer)

this means, at the level of configurations of the phenotype level:

(35) *I say that the hunter killed the deer→*
The hunter has killed the deer→
The deer is not alive→
The deer is dead

6. Other Examples of Inferences

Let us now examine a more complex reasoning:

(36) a. *This morning, the hunter killed the deer*
b. ***therefore*** *the deer was killed* (this morning)
c. ***therefore*** *the deer is* (at the moment)*no longer alive*
d. ***therefore*** *before this morning, the deer was alive*

In the same way, we represent several inferences by formalized deductions.

(37) *This morning, the hunter killed the deer → The deer was killed*

Comments on Deduction II: the hypotheses are given in step 0: the interval O^3 related to *this morning* includes the event realized on F^1 which precedes J^0. This event actualized on F^1 generates a "passive event" actualized on F^2 where F^2 is concomitant with F^1. The "passive predicate" (see the next remark) is derived from the active predicate. So, for the active predicate "to kill", there is the associated passive predicate "to be killed". However, while the agent T^1 is explicit in the active

Deduction II:

0.	$PROC^0_J$ (SAY ($EVEN^1_F$ (to kill T^2T^1))S^0) $[O^3 \supset F^1]\&[F^1 < J^0]\&[O^3 < J^0]$	hyp.
1.	$EVEN^1_F$(to kill T^2T^1)	from 0
2.	$EVEN^2_F$($PASS$ (to kill)) T^2 $[F^2 < J^0]$	passive predicate, 1 unification $[F^2 = F^1]$
3.	$\mathbf{B}EVEN^2_F PASS$ (to kill) T^2	int. **B**,2
4.	[*past-passive*$=_{def}$ $\mathbf{B}EVEN^2_F PASS$]	def. *past-passive*
5.	*past-passive* (to kill)T^2	repl. 3, 4
6.	[was-killed$=_{def}$ *past-passive* (to kill)]	def.
7.	[T^2 := the deer]	instantiation of T^2
8.	was-killed the-deer	repl. 5,6,7

form, it remains implicit in the short passive form. Step 3 introduces the combinator **B** which allows us to compose the aspectual operator with the passivization operator PASS. In step 4, the definition of the grammatical operator "past-passive" is given, which leads to the passive lexical predicate *was-killed* in step 6. After substitution, we obtain the step 8 which represents the passive sentence *the deer was killed*, that is actualized on the interval F^2 concomitant with F^1. This latter event presents an occurrence inside the interval defined by O^3 (*this morning*).

Remark: We have presented an analysis of the passivization in the theoretical framework of Applicative Grammar (see: Desclés (1990), Guentchéva, Shaumyan (1985) and Desclés & Guentchéva (1993)). The basic structure of the passive is a "short" construction, which means "agentless" with an intransitive "passive predicate". The passive predicate is constructed on the basis of the active one. The analysis uses the combinator of conversion **C** and also the existential quantifier noted "Σ": $[P_{pass} =_{def} \Sigma(\mathbf{C}P_{active})]$. From the "short" passive construction $P_{pass}T$, we deduce with the presence of a term denoting a "non-specified agent" x^1 — expressed in French by "on" -, according to a classic elimination rule of the quantifier Σ in natural deduction: P_{active} T x^1. In this way, we have the reduction of the passive construction (for example: *the deer was killed*) to its active equivalent (for example: *one killed the deer*):

In this way, the explicit definition of the grammatical operator of the passivization *PASS* is in order.

(38) *The deer was killed this morning* → *The deer is not alive*

Comments on deduction III: we start with the passive predicate (*the deer was killed, this morning*) and the conditions on the intervals.

1.	P_{pass} T	hyp.
2.	$[P_{pass} = \Sigma(\mathbf{C}P_{active})]$	passive predicate
3.	$(\Sigma(\mathbf{C}P_{active}))T$	repl. 2., 1.
4.	$(\mathbf{C}P_{active})\ x^1 T$	elim. Σ, 3.
5.	$P_{active}T\ x^1$	elim.**C**, 4.

Deduction III:

1.	$EVEN^2_F$ (PASS (to kill)) T^2 $[F^2 < I^0]$ $[O^3 \supset F^2]$ $[O^3 < I^0]$	passive event
2.	[to kill $=_{def} \lambda y.(\lambda x.(\text{TRANS-CHANG}_I(SIT_1[y])(SIT_2[y]))x))$]	def.
3.	$EVEN^2_F(\lambda y.(\lambda x.(\text{TRANS-CHANG}_I(SIT_1[y])(SIT_2[y]))x)T^2x^1)$ $[x^1 =$ non-specified agent]	repl., 1, 2,
4.	$EVEN^2_F(PASS(\text{TRANS-CHANG}_F 2$ $(STATE^1_O((\text{to be-alive})T^2))(STATE^2_O(\text{Neg}((\text{to be-alive})T^2)x^1))$ $[O^1 < F^2 < O^2]$	unification$[I=F^2]$ from 3
5.	$STATE^2_O(\text{Neg}(\text{to be-alive})T^2))$ $[O^2 \supset J^0]$ $[\delta(O^2) = \delta(J^0)]$	from 4
6.	$STATE^0_J$ (Neg((to be-alive) T^2))	from 5
7.	$\mathbf{B}^2 STATE^0_J$ Neg(to be-alive) T^2	int. $\mathbf{B}^2$
8.	[not-be-alive $=_{def}$ $\mathbf{B}^2 STATE^0_J Neg$(to be-alive)]	def.
9.	$[T^2 =$ the-deer]	instantiation of T^2
10.	not-be-alive the-deer	from 8, 9, 10

We introduce the definition of the lexical predicate "to kill" in step 2. After the reductions, the result shows that the event has been actualized on the closed interval F^2 which is nested between the two open intervals O^1 and O^2 (step 4). We obtain the conclusion that the final state of the event is actualized on the interval O^2: the negation of "T^2 is alive" is true on O^2. A reasoning about the intervals shows that the intervals O^2 and J^0 have the same right boundary. According to the property of states which stipulates that the state is actualized on O^2 and that O^2 includes J^0, we conclude that the state is actualized on J^0 (step 6). The introduction of the combinator $\mathbf{B^2}$ allows us to define, in step 8, the complex static predicate "not-be-alive". Finally, after replacement, we get step 10.

(39) *The deer was killed this morning* → *The deer was alive before this morning*

Deduction IV:

1.	$EVEN^2_F(PASS(tokill))T^2$	passive event
	$[\ F^2 < I^0\ ]$	
	$[\ O^3 \supset F^2\ ]$	
	$[\ O^3 < I^0\ ]$	
2.	$EVEN^2_F(PASS(\text{TRANS-CHANG}_F 2$	unification[I=F^2]
	$STATE^1_O((\text{to be-alive})T^2))(STATE^2_O(Neg((\text{to be-alive})\ T^2)))x^1))$	passive reductions
	$[\ x^1 = \text{non-specified agent}\ ]$	
	$[\ O^1 < F^2 < O^2\ ]$	unification
	$[\ F^2 < \text{J}^0\ ]$	
	$[\ O^3 \cap O^2 \neq \emptyset\ ]$	from 1
	$[\ O^2 \supset \text{J}^0\ ]$	
	$[\ \delta(O^2) = \delta(\text{J}^0)\ ]$	
3.	$STATE^1_O((\text{to be-alive})\ T^2)$	from 2
	$[\ O^3 \cap O^1 \neq \emptyset\ ]$	
4.	$[\ O' =_{def} O^1 - O^3\ ]$	def. of O'
	$[\ O' \supset O''\ ]$	
	$[\ O'' < O^3\ ]$	
	$[\ O'' \supset O^1\ ]$	
5.	$STATE_{O''}((\text{to be-alive})\ T^2)$	from 3
	$[\ O'' < \text{J}^0\ ]$	
6.	$\mathbf{B}STATE_{O''}\ (\text{to be-alive})T^2$	int.**B**, 5
7.	$[\ \text{past-state} =_{def} \&(\mathbf{B}STATE_{O''})([O'' < \text{J}^0])$	def. of past state
8.	$[\ \text{was-alive} =_{def} \text{past-state (to be-alive)}$	def.
9.	$[\ T^2 :=\ \text{the-deer}\ ]$	instantiation of T^2
10.	was-alive the-deer	from. 6, 7, 8

Comments on deduction IV: We start from the passive sentence that represents an actualized event on the interval F^2 which is nested in the interval O^3 associated with *this morning*. The meaning of the lexical predicate "to kill" enables us to deduce step 2, with x^1 denoting a non specified agent implied by a passive construction. Since the event is actualized on F^2, it follows that the initial state has been actualized on the interval O^1 which precedes the interval F^2 and is adjacent to it (step 4). We define the interval O' as being before the interval O^3. As the state "T^2 to be alive" was actualized on the interval O^1, it is actualized, according to the property of states, on the interval O', nested in O^1. Hence step 5. The introduction of the combinator **B** in step 6 enables us to define the morphological operator *past − state* in step 7, hence the complex predicate "was-alive" in step 8. After the instantiation of T^2 we finally obtain the applicative expression at step 10.

7. Final Remarks

The above reasonings show how we can relate sentences by means of oriented inferences. Several remarks are now in order:

1) It has been shown how we can formalize the inferential reasonings that are inherent in natural languages by means of aspectual-temporal conditions on the one hand and by means of representations of verbal meanings on the other hand. In this article, we have only illustrated this formalism by a few sentences. However, the other sentences given in the Introduction can be analyzed in the same manner. To formalize those sentences, one would need to introduce other notions, especially the notion of different referential spaces (for speech-act, for narrative texts and for counter-factual events...).

2) The calculus introduced in this article has not been entirely formalized. In order to do this, one would need to better specify the semantics of the topological intervals, the rules on the inter-relations of temporal intervals (Allen's calculus on intervals is clearly not sufficient since it does not take into account the topological bounds) and the relations between states, events and processes. In paragraph 2, we have formulated several rules but more rules need to be added. Furthermore, since the demonstration of Gentzen's natural deduction has been slightly extended in this article, we would need to specify the formal conditions of such an extension.

3) We do not think and have not claimed that the speaker who produces these sentences has to call upon the aspectual-temporal operational processes that have been presented in this article. Yet, the nature of the encountered problem shows the complexity of information representation and processing. There may be a completely different calculus strategy to resolve and control the concrete subject of the analysed sentences. What has been presented in this article is simply a frame of one operational solution.

Appendix: Combinatory Logic

We consider (see Descles, 1990), S.K. Shaumyan (1987) and other linguists that combinatory logic is very useful for analysing the meaning of grammatical operations and the meaning of lexical predicates. The aim of the Combinatory Logic (Curry, 1958) is operational processings by which operators — either elementary or complex — apply to operands. In general, operators and operands are of different types. However, we have not used this notion of type here. The basic constitutive operation of the expressions is called application. An expression that is constructed by the application is called *applicative expression.* An operator X that applies to an operand Y produces a result Z. We note the result of the application of X to Y by a simple concatenation which prefixes the operator to its operand, so: Z = XY. We stipulate: $XYZ =_{def} (XY)Z$. It is evident that $XYZ \neq X(YZ)$.

Some of the abstract operators, called *combinators*, are especially employed to construct complex operators from more elementary ones. The operational action of a combinator (see Fitch, 1974) is given by an introduction rule and an elimination rule, in Gentzen's "natural deduction" style. We will give two examples: the first is the combinator **B** that represents the "functional composition" of operators which would be the expressions of set functions.

Its operational action is defined by the two rules:

1.	X(YZ)		1.	**B**XYZ	
2.	**B**XYZ	int. **B**	2.	X(YZ)	elim. **B**

The operational action of the combinator **C** of "conversion" is defined by the two rules:

1.	XZY		1.	**C**XYZ	
2.	**C**XYZ	int. **C**	2.	XZY	elim. **C**

The third combinator that we introduce here is the combinator of identity **I**. The rules are as follows:

1.	X		1.	**I**X	
2.	**I**X	int. I	2.	X	elim. **I**

Starting from the basic combinators (very small in number), one can generate, by means of the operation of application, an unlimited number of derived combinators. For example, for the combinator $\mathbf{B}^2$, derived from the combinator **B**, one has the definitional relation: [$\mathbf{B}^2 =_{def} \mathbf{BBB}$]. Its operational action is deduced by the successive introductions or eliminations of **B**:

	Introduction of $\mathbf{B}^2$		Elimination of $\mathbf{B}^2$
1.	$X(Z_1Z_2Z_3)$	1.	$\mathbf{B}^2XYZ_1Z_2Z_3$
2.	$\mathbf{B}X(Z_1Z_2)Z_3$	2.	[$\mathbf{B}^2 =_{def} \mathbf{BBB}$]
3.	$\mathbf{B}(\mathbf{B}X)Z_1Z_2Z_3$	3.	$\mathbf{BBB}XYZ_1Z_2Z_3$
4.	$\mathbf{BBB}XYZ_1Z_2Z_3$	4.	$\mathbf{B}(\mathbf{B}X)Z_1Z_2Z_3$
5.	[$\mathbf{B}^2 =_{def} \mathbf{BBB}$]	5.	$\mathbf{B}X(Z_1Z_2)Z_3$
6.	$\mathbf{B}^2XYZ_1Z_2Z_3$	6.	$X(Z_1Z_2Z_3)$

From this we deduce the two relations of β-expansion ($\leftarrow_\beta$) and of β-reduction ($_\beta\rightarrow$):

$$X(Z_1Z_2Z_3) \leftarrow_\beta \mathbf{B}^2XYZ_1Z_2Z_3 \qquad \mathbf{B}^2XYZ_1Z_2Z_3 \;{}_\beta\!\rightarrow X(Z_1Z_2Z_3)$$

A combinator represents an *applicative program of construction* of a complex operator from more elementary operators. Therefore, the combinator $\mathbf{B}^2$ combines the operators X and Y in a way to apply the complex operator ($\mathbf{B}^2$XY) successively to the operands Z_1, Z_2 and Z_3, and this gives the result: $X(Z_1Z_2Z_3)$.

An applicative expression that cannot be further reduced is called a *normal form.* For example, $X(Z_1Z_2Z_3)$ is a normal form associated with the applicative expression $\mathbf{B}^2XYZ_1Z_2Z_3$.

Let X and Y be any two applicative expressions. We stipulate:

X **0** Y = **B**XY

The composition '**0**' between two operators is associative. We have for example:

$\mathbf{B}^2 = \mathbf{B\ O\ B} = \mathbf{BBB}$
$\mathbf{I} = \mathbf{C\ 0\ C}$

References

Abraham M., (1995): *Représentation sémantique des verbes (verbes de mouvement) en vue d'un traitement informatique*, Ph.D. Dissertation, Université de Paris-Sorbonne, March 1995.

Allen J., (1983): "Maintaining Knowledge About Temporal Intervals", *Communications of the ACM, 26*, 832–843.

Benveniste E. (1964). *Problèmes de linguistique générale*. Paris: Gallimard

Bertinetto P.M., Bianch V., Higginbotham J., Squartin M. (eds.), (1995). *Temporal Reference Aspect and Actionality.* Turin: Rosenberg & Sellier

Comrie B. (1976): *Aspect, an Introduction to the Study of Verbal Aspect and Related Problems.* London: Cambridge University Press.

Culioli A. (1990). *Pour une linguistique de l'énoncation, opérations et représentations*. Paris: Ophrys

Curry H., (1958). *Combinatory Logic.* Amsterdam: North Holland.

David J. & Martin R. (1980). *Notion d'aspect.* Paris: Klincksieck.

Desclés J.-P. (1980): "Construction formelle de la catégorie de l'aspect (essai)", in David & Martin *Notion d'aspect*, 198–237. Paris: Klincksieck.

— (1990a). *Langages applicatifs, langues naturelles et cognition.*. Paris: Hermès.

— (1990b): "State, Event, Process, and Topology", in *General Linguistics, 29*, 3:159–200. University Park and London: Pennsylvania State University Press.

— (1991). "Archétypes cognitifs et types de procès", *Travaux de linguistique et de Philologie*, XXIX: 171–195.

— (1993). "Remarques sur la notion de processus inaccompli", *Sémiotique, 5*, December 1993, 31–55. Paris: Didier-Erudition.

— (1994). "Quelques concepts relatifs au temps et à l'aspect pour l'analyse des textes", in Desclés *et alii* (1994), pp. 57–88.

— (1994). "Relations casuelles et schèmes sémantico-cognitifs", *Langages, 113*, 113–125.

— (1997). "Logique combinatoire, topologie et analyse aspecto-temporelle", in Desclés *et alii*, pp. 37–69.

Desclés, J.-P., Guentchéva, Z., Shaumyan S.K. (1985). "Passivization in Applicative Grammar". *Pragmatics & Beyond VI*, 1. John Benjamins.

Desclés J.-P., Guentchéva Z. (1990). "Discourse Analysis of Aoriste and Imperfect in Bulgarian and French". *Verbal Aspects.* Amsterdam : Thelin, Benjamins.

Desclés, J.-P., Guentchéva, Z. (1993). "Le passif dans le système des voix du français". *Langage, 109*, 1993, 73–102. Paris: Larousse.

Desclés J.-P., Guentchéva Z., Karolak S., Koseka-Toszewa V. (1994). *Studia Kognitywne, T.1: Semantyka kategorii aspektu i czasu.* Warszawa: Slawistyczny Osrodek Wydawnczy.

Desclés J.-P., Guentchéva Z. (1995). "Is the Notion of Process Necessary?". *International Conference on Aspect*, Cortona, October 9–12, 1993, in Bertinetto *et alii*, 55–70.

Desclés J.-P., Guentchéva Z., Karolak S., Koseka-Toszewa V. (1997). *Studia Kognitywne, T.2: Semantyka kategorii aspektu i ezasu.* Warszawa. Slawistycny Osrodek Wydawnczy.

Desclés J.-P., Guentchéva Z. (1997). "Aspects et modalités d'action (Représentations topologiques dans une perspective cognitive", in Desclés *et alii*, 145–173.

Desclés J.-P., Jouis C., Oh H.G., Reppert D. "Exploration contextuelle et sémantique : un système expert qui trouve les valeurs sémantiques des temps de l'indicatif dans un texte", in *Knowledge modeling and expertise transfer 1991*, 371–400, D. Herin-Aime, R. Dieng, J.-P. Regourd, J.-P. Angoujard (eds). Amsterdam, Washington DC, Tokyo: IOS Press.

Dowty, D.(1979). *Word Meaning and Montague Grammar.* Dordrecht: Reidel.

Fitch F. B. (1974). *Elements of Combinatory Logic.* New Haven: Yale University Press.

Guentchéva Z. (1990). *Temps et aspect : l'exemple du bulgare contemporain.* Coll. Sciences du langage, Paris: Editions du CNRS.

Kamp H. (1981). "Evénements, représentations discursives et référence temporelle", *Langages 64*, 39–64.

Kamp H. & Rohrer C., (1983). "Tense in texts" in R. Bauerle *et alii* (ed.), *Meaning, use and interpretation of language.* Berlin: de Gruyter, 250–269.

Karolak S. (1997). "Le temps et le modèle de H. Reichenbach", in Desclés J.-P., Guentchéva Z., Karolak S., Koseka-Toszewa V., (eds.) (1997), 95–125.

Koseka-Toszewa V. & A. Mazurkiewicz (1994). "Description de la temporalité et de la modalité au moyen de réseaux" in Desclés *et alii*, 89–112.

Lyons J.(1977). *Semantics, Vol. 2.* London, New York, Melbourne: Cambridge University Press.

Maire-Reppert D. (1990). *Représentation des valeurs sémantiques de l'imparfait français en vue d'un traitement informatique*, Ph.D. Dissertation. Université de Paris-Sorbonne.

Maire-Reppert D., Oh H.G., Berri J. (1991). "Traitement informatique de la catégorie aspecto-temporelle", in *T.A. Informations, 32*, 1:77–90.

Mourelatos A. (1981). "Events, Processes and States", in Tedeschi et Zaenen (eds), 192–212.

Oh-Jeong H.G. (1991). *Représentation des valeurs sémantiques du passé composé français en vue d'un traitement informatique*. Ph.D. Dissertation. Université de Paris-Sorbonne.

Prior A. (1967). *Past, Present and Future*. Oxford: Clarendon Press.

Reichenbach H., (1947). *Elements of Symbolic Logic*. Toronto: The Macmillan Company.

Searle J.R., Vanderveken D., (1985). *Foundations of Illocutionary Logic*. Cambridge University Press

Shaumyan S. K. (1987). *A Semiotic Theory of Language*. Bloomington: Indiana University Press

Smith C.S. (1991). *The Parameter of Aspect*. Dordrecht: Kluwer Academic Publishers.

Thelin N.R. (1990). *Verbal Aspect*. Amsterdam, Philadelphia: John Benjamins.

Vanderveken D. (1991). *Meaning and Speech Acts*, volume II *Formal Semantics of Sucess and Satisfaction*. Cambridge University Press

Vendler Z.(1967). *Linguistics and Philosophy*. Ithaca, New York: Cornell University Press

Vet C. (1995). "The Role of Aktionsart in the Interpretation of Temporal Relations in Discourse" n in Bertinetto *et alii* (1995), 295–306

Chapter 12

PRESUPPOSITION, PROJECTION AND TRANSPARENCY IN ATTITUDE CONTEXTS

Rob van der Sandt
University of Nijmegen

1. Presuppositions and their triggers

'Presupposition' is an ambiguous notion and has been used to describe quite a different number of phenomena. The first and foremost distinction which should be made is the distinction between presuppositions as induced, invoked or triggered by linguistic expressions and presupposition as information taken for granted in a conversation. The first notion is the notion of presuppositions as conventionally associated with linguistic expressions or syntactic constructions. It is found in the literature under a variety of different names as pre-supposition in Gazdar (1979), conventional implicature in Karttunen and Peters (1979) or elementary presuppositions in van der Sandt (1988). The second notion is the notion of presupposition as background information, that is information which is already given or taken for granted in a conversation. It is Stalnaker's context-set or Karttunen's common ground. Though both notions are fundamentally different there is a straightforward connection. If a linguistic element induces presuppositional information, the sentence containing the inducing element will normally only be appropriate in a context which already contains or is suited to accept the information triggered. We should make a further distinction straight away. Presuppositional inferences should be distinguished from suggestions or inferences which are not strictly part of or induced by the linguistic form of an utterance but are invoked by it in view of contextual information, Gricean maxims or principles of discourse coherence.

Presuppositions can be viewed as pieces of information which are induced by lexical items or syntactic constructions. These items or constructions carry descriptive information which, ideally, is taken for

D. Vanderveken (ed.), Logic, Thought & Action, 245–266.

granted by the participants in a conversation. But, regardless of the way this information is induced, it always has to be resolved in context. Sometimes this information is already there, sometimes it is not. If it is there, the information induced may be matched straightaway against the content of the current discourse. If it is not, we have to establish some link and relate the presuppositional information to information which may not be explicitly given in the linguistic context, but which is in some sense assumed as being uncontroversial and which may be added to the discourse without giving rise to infelicity.

The general picture just sketched has some straightforward implications. It means that a theory of presupposition resolution has to consist of two components. Firstly, such a theory has to provide an account of how presuppositional information can be linked to information which is already given in the discourse. Secondly, it has to tell us what happens with the information triggered by some presupposition inducer in case this information is not already present.

With respect to the first component I will rely on the anaphoric account of presupposition,[1] that is I will take for granted that presuppositions are anaphoric expressions, i.e. they are expressions which have to link up to or, put differently, which have to be bound by some previously established antecedent. The object they should link up to should be given as a distinct and identifiable object in the discourse. This feature distinguishes the current account from a whole class of theories which were dominant in the literature of the 70s and the 80s. These theories, originally derive from Karttunen's (1974) and Stalnaker's (1973) work and were taken up later by inter alia by Heim (1983) and Beaver (1993) merely require that the context of utterance contains enough information to satisfy the information triggered by a presuppositional construction. On this account it is not required that there is an entity a presupposition trigger should link up to, but merely that the context entails the information induced by the presuppositional expression.

The second component accounts for the fact that presuppositional information can (under specified conditions) be added to the discourse under construction and thus establish an antecedent in case discourse does not already provide one. I will moreover assume that they are able to do so in view of the informational content which is explicitly or implicitly present in their inducing expressions, that is I will assume an accommodation mechanism in the sense of Lewis (1979). And here again I will, in contradistinction to the theories just mentioned, construe accommo-

[1]Anaphoric accounts of presupposition are found i.a. in Geurts (1995), Kamp and Rossdeutscher (1992), Kripke (ms), and van der Sandt (1992).

dation not just as a mechanism which adds the information required to guarantee the definedness of the inducing sentence. I will take accommodation to do the stronger duty of inserting an identifiable object which after insertion can function as an antecedent for the presuppositional anaphor.

Presuppositional expressions are thus treated as anaphoric expressions. They distinguish themselves from pronouns or other types of semantically unloaded anaphors in two respects. They may (and generally will) be syntactically complex and have internal structure. This means inter alia that they may embed further anaphoric expressions which may give rise to complicated binding structures. It also means that they may carry information attenuated anaphors like pronouns generally lack. The information they carry does a double duty. In the binding process the descriptive content of a presuppositional anaphor has a disambiguating role. It enables the hearer to select an antecedent out of a number of possible candidates. With respect to accommodation their information content has an even more important role. If no antecedent is available, it gives presuppositional expression the capacity to establish an accessible antecedent by means of some default process of filling in information which may be implicitly assumed by the interlocutors in a conversation but is not actually there.

The following examples illustrate both mechanisms:

(1a) Mary has a dog and *her dog* barks
(1b) If Mary has a dog, *her dog* barks

(2a) *Mary's dog* barks.
(2b) It is possible that *Mary's dog* barks.
(2c) If Mary is not at home, *her dog* barks.

(3) Either Mary does not have a dog or *her dog* is in hiding.
(4) ?*Mary's dog* barks and Mary has a dog.

All these sentences contain the description *Mary's dog*. Because it is a presupposition inducer, it triggers a piece of information. The information it is said to induce is that Mary has a dog. It is this information which has to be resolved. Notice that only (2a) through (2c) are presupposing in the intuitive sense of the word. (1a), (1b) and (3) on the other hand are not. In uttering (1a) a speaker does not presuppose that Mary has a dog but asserts it outright. Although this information is entailed by the carrier sentence, it does not have a presuppositional status. The interpretation of (1a) differs from both (1b) and (3). Neither of these sentences entails or presupposes that Mary has a dog. Here any suggestion that the presuppositional information is true is gone. Sentence (4) finally is unacceptable in any context and the question as to its presuppositional status does not even arise.

The explanation I gave in previous work runs roughly as follows.[2] Resolution of a presuppositional expression may ensue either by binding or by accommodation. First consider (1a) and (1b). The presuppositional anaphor *Mary's/her dog* will search for an appropriate antecedent to link up to. In (1a) the first conjunct provides an appropriate antecedent, in (1b) the protasis of the conditional. In both cases the presuppositional expression will be bound to this pre-established antecedent and the information triggered will thus be absorbed by its target. Surely, (1a) will still entail that Mary has a dog. However, this inference is not of a presuppositional nature. This is in conformity with our pre-theoretic intuition according to which we cannot both assert and presuppose the very same proposition by the utterance of the very same sentence. It is also easily checked by applying one of the standard diagnostic tests for presuppositionhood. If we embed the full sentence under a possibility operator or put it in the interrogative mood, any suggestion as to its truth disappears.

(5) Does Mary have a dog and does her dog bark?

On the current account the intuitive notion of presupposition coincides with accommodation of the presuppositional information in the main context. And, since asserted information need not and will not be accommodated, presupposition and assertion come, just as our pretheoretical intuitions require, out as complementary notions.

Resolution proceeds differently in (2a) through (2b). The utterance of any of these sentences will generally be felicitous in a context which contains or is suited to accept the information that Mary has a dog. A co-operative speaker may thus accommodate the presuppositional material. In absence of any indication to the contrary this will indeed happen. Accommodation will establish an antecedent in the main context. Resolution then proceeds as before. The presuppositional anaphor will be bound to the antecedent thus established. Sentence (3) and (4) finally illustrate that accommodation is subject to various constraints. Neither of these sentences provides an accessible antecedent for the anaphoric expression. One might thus try to accommodate the presuppositional information to the main context. However, in both cases this would yield an unacceptable result as (6) and (7) illustrate.

[2]Van der Sandt 1992

(6) ?Mary has a dog. Either Mary does not have a dog or *her dog* is in hiding.
(7) ?Mary has a dog. *Mary's dog* barks and Mary has a dog.

It thus turns out that neither (3) nor (4) can be interpreted in a context which contains the presuppositional information. Consider first (6). Once the first sentence of (6) has been interpreted and the information that Mary has a dog has been established, the second sentence simply conveys that Mary's dog is in hiding. Conveying this information by saying that she either has no dog or that her dog is in hiding would clearly be a rather inefficient, obscure and confusing way to do so. A co-operative speaker should instead simply state that Mary's dog is in hiding. Thus (3) cannot be felicitously uttered in a context which already contains the information that Mary has a dog. So the default strategy of accommodation of this presupposition in the main context is blocked. A similar story can be told with respect to (7). Here the second conjunct is simply superfluous given the information which is already contextually given. Again accommodation of the information triggered by the first conjunct into the main context would result in an unacceptable sequence. In (6) the presuppositional anaphor can nevertheless be resolved. Although accommodation to the main context is blocked, we can accommodate the presuppositional information locally. This yields (9), which has the interpretation desired:

(9) Either Mary does not have a dog or [she has a dog] and her *dog* is in hiding.

The presuppositional expression in (4) on the other hand cannot be resolved. Accommodation of an antecedent for *Mary's dog* would result in infelicity of the resulting discourse. The presuppositional expression thus cannot be bound and the whole sentence will lack an interpretation.

The explanation just given depends on two ingredients. Presuppositional expressions trigger information. This information is of an anaphoric nature and should be resolved in context. One task for presupposition theory thus is to give an explicit account of the information which is often only implicitly contained in presuppositional expressions, that is to give a general format in which this information can be encoded. The second task is to give an account of how this information is resolved, that is to give a resolution algorithm.

2. Binding and accomodation

Accommodation as described in the previous section is an operation on discourse structures. If a piece of presuppositional information is accommodated globally, it modifies the discourse that has been established at a certain point of the conversation. If it is accommodated locally, it modifies some auxiliary context which has been constructed while processing the sentence. In order to account for this I use a representational framework for the actual implementation of the resolution algorithm.

The account is formulated in an extension of discourse representation theory and differs in some crucial respects from the original formulation of Kamp (1981) and Kamp & Reyle (1993). In Kamp (1981) a discourse representation structure or DRS K consists of an ordered pair <U(K), Con(K)>. U(K) is a universe of discourse markers and Con(K) a set of conditions. Sentences are processed in an evolving discourse, the content of which is encoded in the main DRS. Each time a sentence is processed its syntactic tree is broken up top-down and new markers and conditions are added to the main DRS in the course of the parsing process. Indefinite expressions introduce new discourse markers and any marker thus introduced may serve as an antecedent for anaphoric expressions to come. Conditions encode the descriptive content of predicates and assign properties to the members of U(K). Conditions attached to already established discourse markers thus constrain the possibility of anaphoric take up. Pronouns and other anaphoric expressions come with a special instruction. They should be linked up to some pre-established discourse marker. Thus whenever the construction algorithm encounters an anaphoric expression, the latter is resolved straightaway against the universe of the main DRS.

The current account deviates in two respects. It assumes a bottom-up construction procedure and adopts an indirect resolution mechanism. And, most importantly, DRSs are constructed as consisting of a set of markers, and two types of conditions. We extend standard DRT with a class of conditions that encode anaphoric material. The latter conditions are themselves construed out of DRSs. I will refer to them as α-conditions and to the DRS involved as a presuppositional frame. This way of construing anaphoric expressions is motivated by the fact that presuppositions are anaphoric expressions which have syntactic structure and carry information. Since each α-condition is construed as a DRS, it may encode any amount of information. It may moreover embed further anaphoric expressions. It is the content of these conditions which has to be resolved against the content of the incoming DRS.

The general format of a presupposition-inducer is thus an α-condition, involving a marker and a set of conditions that may themselves be α-conditions. A marker in the universe of presuppositional frame has to link up to a previously established antecedent. I refer to it as the anaphoric variable. The conditions encode the descriptive content associated with the trigger. They thus function as constraints on the selection of antecedents in the resolution process. Since each presuppositional frame is itself a DRS, it may contain further α-conditions. We may thus recursively embed anaphoric expressions inside other anaphoric expressions. The construction process proceeds in three stages. First a DRS is constructed for the incoming sentence. Such a DRS is provisional in that it may contain any number of anaphoric expressions which are not resolved at this stage. This DRS is then merged with the DRS which has already been constructed for the discourse up to that point. The result of this procedure yields a DRS which is again incomplete since it will contain the anaphoric expressions of the incoming sentence which are still unresolved. Only in the final stage of the construction process are the anaphoric expressions resolved. Resolution of anaphoric expressions can be achieved by either of two means:

(i) The anaphoric variable may be equated with some previously established accessible marker. The associated conditions will be transferred to the binding site and the antecedent will thus inherit all the descriptive content of the presuppositional expression. Since I take compatibility (of the conditions associated with the presuppositional anaphor and the information associated with its antecedent) to be the minimal condition, a presuppositional anaphor may just like an anaphoric pronoun select one out of a number of suitable antecedents.[3]

(ii) If no suitable antecedent is found the presuppositional expression has to be accommodated. Accommodation consists in transferring the anaphoric variable and the associated descriptive conditions to some accessible position thereby creating an accessible antecedent after all. It effects a modification of the provisional structure which has already been established in the first two stages of the construction process. Accommodation is constrained by various factors, the most important of which are constraints on binding and constraints on acceptability.[4] Accommodation shows a preference for the highest accessible level.

I illustrate the working of the resolution mechanism with a few examples and refer for the technical details to van der Sandt (1992).[5] I will use boldface to represent accommodated material. This is for perspicuity only.

The provisional DRS constructed for (10a) is (10b):

This DRS contains two anaphoric expressions one for the full possessive construction *the farmers dog* and one for the embedded description. This DRS is provisional in that it contains unresolved anaphors and

[5]The encoding of anaphoric expressions in van der Sandt (1992) differs from the encoding given here in that DRSs are construed as triples, the last coordinate of which is a set of DRSs which encode the descriptive material. The idea to encode anaphoric material as a special kind of condition is due to Johan Bos and improves readability when DRSs are represented in linear notation.

(10a) The farmer's dog barks.
(10b) K_0:[:bark(x), αx[x:dog(x), poss(y,x), αy[y:farmer(y)]]]

does not yet allow interpretation. Resolution to a proper DRS, that is a DRS in which all anaphoric expressions have been resolved and which can thus be interpreted according to the standard embedding conditions, proceeds as follows. First, this provisional DRS is merged with the incoming DRS. This operation consists in taking the union of the universes of K_0 and the incoming DRS and putting together the conditions of these DRS s. Assuming that the main DRS is empty the resulting structure will not provide an antecedent for any of the presuppositional expressions, and this forces the resolution mechanism to resort to accommodation. Resolution starts off with the deepest embedded anaphor, that is αy[y:farmer(y)], the presuppositional structure for *the farmer*. And since the DRS thus far constructed does not provide a suitable antecedent the presupposition will be accommodated. Accommodation consists in adding the anaphoric marker to the universe of its target DRS and adding the associated conditions. This yields (11a) (K_0') which is still provisional in that αx[x:dog(x), poss(y, x)], the presuppositional frame for *his dog*, waits for resolution. Note that this condition does not contain any further anaphoric conditions which indicates that it is the deepest embedded anaphoric expression at this stage and which makes it a proper candidate for resolution. Accommodation of this anaphor produces K_0" which does not contain any unbound variables and thus is proper DRS.

(11a) K_0':[**y: farmer(y)**, bark(x), αx[x:dog(x), poss(y,x)]]
(11b) K_0":[**x, y:dog(x), farmer(y), poss(y,x)**, bark(x)]

Suppose on the other hand that (10a) had been processed in a discourse which did already contain the information that there is a farmer who has a dog. Merging the incoming DRS (12a) with K_0 would give K_1. When resolving K_1 we would apply the same processing steps yielding first K_1' and then K_1". There is one difference, though. When resolving K_1 to K_1' we don't accommodate the marker y with the associated condition that it is a farmer. Instead we equate this anaphoric variable with the already established marker v satisfying condition farmer(v), and in the transition of K_1' to K_1" we identify the anaphoric variable x with the marker u which is already established and satisfies the relevant conditions. In both cases the informational content triggered by the presuppositional expression is absorbed by its antecedent. In this and the examples below we won t insert the anaphoric equations but make the relevant substitutions straight away.

(12a) [u, v: farmer(v), dog(u), poss(v,u)]
(12b) K_0:[:bark(x), αx[x:dog(x), poss(y,x), αy[y:farmer(y)]]].
(12c) K_1:[u, v: farmer(v), dog(u), poss(v,u), bark(x), αx[x:dog(x), poss(y,x), αy[y:farmer(y)]]]
(12d) K_1':[u, y: farmer(y), dog(u), poss(y,u), bark(x), αx[x:dog(x), poss(y,x)]]
(12f) K_1":[x, y: farmer(x), dog(y), poss(y,x), bark (x)]

It will be clear that given (12a) as the incoming DRS we obtain the same result regardless of whether we use anaphoric pronouns or of the full descriptions. This means i.a. that the mechanism yields the same output for all of the following discourses.

(13a) The farmer's dog barks.
(13b) There was a farmer. *His dog* barks.
(13c) A farmer had a dog. {*His dog/it*}barks.

In all these cases we end up with K_1" as a properly resolved DRS to which the standard verification conditions apply. (13a) through (13c) will thus be assigned exactly the same truth-conditions that are assigned to (14) the standard representation in first order logic.

(14) $\exists x \exists y(\text{farmer}(x) \wedge \text{dog}(y) \wedge \text{poss}(x,y) \wedge \text{bark}(y))$

The procedure does not, however, yield an interpretation for the second sentences of (13b) and (13c) when these sentences are processed in isolation. The presuppositional structure for an attenuated anaphor like a pronoun would come out as follows αx[x:]. It lacks the descriptive conditions which are generally associated with full descriptions. And given this deficiency pronouns don't have the capacity to accommodate.

3. Projection and scope

In the examples just given accommodation acts as a strategy to adjust the representation structure under construction. If the context of an utterance does not contain an appropriate antecedent for a presuppositional expression, the projection algorithm will try to construct one and will be able to do so in view of the descriptive content associated with the trigger. Viewed this way accommodation thus acts as a repair strategy intended to ensure interpretation even if a presuppositional anaphor cannot be bound. By doing so this mechanism has an important effect: it both generates and constrains the scope of presuppositional anaphors.

The following example illustrates how the algorithm yields the wide and narrow scope readings for definite descriptions.
Consider (15):

(15) It is possible that the king of France is bald.

The provisional DRS constructed for (15) is (16):

(16) [:◇[:bald(x), αx[x:KFx]]

If the incoming DRS provides a suitable antecedent the anaphoric expression will be bound. The α-condition will be absorbed at its binding site, thus giving the presuppositional expression scope over the modal operator. (17) thus resolves to (17'):

(17) [u:KF(u), ◇[:bald(x), αx[x:KFx]]
(17') [x:KF(x), ◇[:bald(x]]

For our present purpose the interesting case is the one in which the incoming DRS is empty. In this case the resulting structure does not provide an antecedent for the presuppositional expression and the latter has to be accommodated. Accommodation may ensue either globally or it will take place locally in the subordinate structure. The first option produces (18a), the second (18b).

(18a) [**x: KF(x)**, [:bald(x)]]
(18b) [:◇[**x: KF(x)**, bald(x)]]

Given the preference for accommodation at the highest level of representation, (18a) is the default option and will *ceteris paribus* be preferred. However, given an incoming DRS which already contains the information that there is no or might not be a king of France, accommodation at top level is blocked and (18b), where the presupposition is accommodated in the subordinate context, is the only possible resolution. In both cases the projection algorithm plugs the anaphoric variable into the universe of its target DRS and adds the descriptive material to its conditions. The *de re* reading is obtained by accommodation of the presuppositional material at the main level of discourse. The *de dicto* reading comes about by local accommodation.

This way of processing definite descriptions clearly has a Russellian flavour. Anyway, it yields the very same readings Russell would predict. There are some crucial differences, however. Firstly, since accommodation at top level is the preferred option and since definites are just one type of presupposition inducers, the mechanism predicts for definites a *ceteris paribus* preference for the wide scope reading. This conforms to their actual behaviour. For, unless there is evidence to the contrary, we will preferably interpret definites as taking scope over all embedding operators. This does not only hold for extensional embeddings, but it applies equally to modals and attitude reports. Since the resolution process is constrained by restrictions on binding as we will see in a moment and since moreover semantic and pragmatic factors interfere during the resolution process, these factors will cut down the number of possible positions where a presupposition can end up. The general preference for accommodations as high as possible will normally single out the preferred interpretation. This prevents overgeneration and obviates the need to take recourse to an independent theory to select the possible or preferred readings out of a much larger set of syntactically generated structures.

Secondly, definites can be projected from any position to any accessible position. Thus, while, on a Russellian account, there is no way to project e. g. a description which is syntactically generated in the consequent of an conditional to its antecedent, the present account can. Any presupposition may end up by accommodation at any position where it could be bound, that is at any position which can be reached by following its accessibility line. (19) is an example. And, finally and perhaps most importantly, the mechanism is applicable to other types of presupposition inducers for which it is at least unclear how they could be handled by a conventional scope mechanism. Presuppositional adverbs (*too, again*), quantifiers, cleft constructions and most significantly lexical presuppositions, like those associated with a noun as *bachelor*, are cases in point. As the literature on presupposition projection shows all follow the same pattern with respect to embedding operators.

I conclude this section with some remarks on binding. As I said, once we assimilate presuppositional expressions with anaphoric expressions, it follows that anaphoric expressions may embed further anaphoric expressions. Presuppositional expressions have internal structure and may thus embed other presuppositional expressions at any level. The resolution mechanism proceeds from left to right. This ordering ensures that the anaphoric variable will not find any unresolved anaphor on its path when processing the anaphoric expression. Furthermore it works bottom up. Thus in case a presuppositional anaphor embeds another presuppositional anaphor, the most deeply embedded one will be processed first. Bottom up processing guarantees that no anaphoric expression will be resolved until all embedded anaphors are resolved. This yields an important constraint on binding. In order to see this we may consider an example where a presuppositional expression contains another expression which is bound by an external quantifier. (19) is a case:

(19) Every German loves his car.

The initial representation is given in (20)

(20) [x:German(x)] <all x>[:love(x,y), αy[y:car(y), poss(z,y), αz[z_{masc}:]]]

The pronoun depends on the quantified expression. The resolution algorithm ensures that the deepest embedded anaphor will be processed first. We thus equate z with the principal variable of the full DRS which yields (21) as an intermediate structure in the resolution process.

(21) [x:German(x)] <all x>[:love(x,y), αy[y:car(y), poss(x,y)]]

Only then we start processing its embedder αy[y:car(y), poss(x,y)]. But note that once the pronoun has been equated with x, the full presuppositional expression cannot be projected any higher than the site where this pronoun is bound since this would leave x free in the presuppositional condition poss(x, y). This excludes the possibility of projecting the α-condition for *his car* to the main context which is what would have happened if the pronoun had been bound to some entity that had already been established at top level. Two possible accommodation sites remain: the restrictor of the quantifier and its scope. The projection algorithm thus yields (22a) and (22a) as the only possible solutions:

(22a) [x: German(x)] <all x>[**y: car(y), poss(x,y)**, love(x,y),]
(22b) [x,y: German(x), **car(y), poss(x, y)**] <all x>[:love(x,y),]

If the presuppositional expression is accommodated in the restrictor, yielding (22a) as its final representation, (19) will be interpreted as meaning that every German who has a car loves it. If, on the other hand, the presuppositional expression is accommodated in its scope as happens in (22b), the sentence will be interpreted as meaning that it holds for every German that he has a car and loves it. In the first interpretation, which is preferred in this case, the sentence is not falsified by Germans who don't own cars, but only by Germans who own a car but don't like it. In the second interpretation it is falsified by any non-car-owning German. The preference for the first interpretation comes about by the fact that there is a conventional association between Germans and cars which enables us to interpret the cars as dependent on the quantified NP and which makes the restrictor a suitable accommodation site. In this respect (19) is on a par with a bridging case like (22). In (22) *the organ* is equally dependent on the churches introduced in the antecedent and this sentence does not seem to be falsified by some non-organ-having Catholic church either.

(22) In all Catholic churches the organ was restored.

Let us take stock. Firstly, in an unresolved DRS presuppositions will always emerge at the site where they are syntactically generated. The resolution algorithm will determine where they end up. This is a direct consequence of the decision to break up the construction algorithm in stages. Only in the final stage of DRS construction are the anaphoric expressions resolved, either by linking them to pre-established discourse markers and thus transferring the associated descriptive material to its binding site or by accommodation of presuppositional material at some level of discourse structure. It is in this stage that the relative scope of presuppositional expressions with respect to syntactically embedding operators is determined. Secondly, resolution of presuppositional anaphors never results in any transformation, weakening or modification of the material triggered neither directly by binding nor indirectly by accommodation. No matter whether resolution is effected by binding or accommodation, the resolution algorithm only searches for a proper place for the presuppositional material to settle down, but never performs any other operation on the presuppositional expression than linking up the descriptive material to the antecedent marker. Nor does the mechanism involve copying routines, thus duplicating material at various levels.[6] On the current account presuppositional material moves through discourse structures and ends up at the site where its antecedent is found or established.

Finally, the truth conditions for the representations derived can either be given along Russellian or Fregean lines. Assuming the standard verification conditions for DRT as found in e. g. Kamp and Reyle (1993) we will, just as on a Russellian analysis assign falsity to (13a) instead of undefinedness as Frege or Strawson would have it. The output structures of (13a) and (13b) would thus be predicted to be truth-conditionally equivalent for the simple reason that both receive the same final representation. Note, however, that the input conditions are different. (13b) is infelicitous in a context which already contains the information that there is a farmer's dog. (13a) is not. (13a) and (13b) thus differ in dynamic meaning. They are moreover derived in different ways. In (13b)

[6]This holds for simple extensional contexts and modal embeddings. Duplication of material may however take place in attitude contexts. For different views on this issue see Zeevat (1992) and Geurts (1995).

we simply insert a discourse marker for the *farmer's dog*. In (13a) it comes about as a result of accommodation. If we distinguish between discourse referents which come about by updating the structure with assertoric material and referents which are created by accommodation, we may revise the verification conditions in such a way that no value will be assigned in cases where an accommodated presupposition has no value in a model. This would restore the Fregean/Strawsonian intuition that presupposition failure gives rise to undefinedness. In the present framework it would also give rise to two different sources of truth value gaps. Undefinedness would come about when a presuppositional anaphor can neither be bound nor accommodated. In that case the construction algorithm would not come to an end and the question as to truth or falsity would not even arise. But undefinedness could also come about in a very different way. A presuppositional anaphor might be resolved by accommodation but the result might simply not fit the world. And in this case we can just assign falsity to (13b) and undefinedness to (13a).

4. A problem about *too*

In the previous section I illustrated the workings of the binding mechanism with respect to definite descriptions. In the present section I will discuss the behaviour of *too*. This presuppositional adverb distinguishes itself from descriptions in two crucial respects. Firstly, the presupposition induced by *too* does not accommodate very easily. Secondly, it is linked in a very different way to its embedding matrix sentence. Both characteristics make it a paradigm example of anaphoric presupposition. The second characteristic has moreover some important consequences with respect to its behaviour in attitude contexts

Descriptions both bind and accommodate quite easily. Their propensity to bind or their reluctance to accommodate is related to the amount of descriptive content they carry, and more importantly to the amount of content they have to carry in a certain context to unambiguously select their antecedent. A descriptively weak description like *the man* is not applicable to non-human beings and suggests an adult male as antecedent. It thus only slightly exceeds the corresponding pronoun in descriptive information. But clearly, if needed, we may expand the descriptive information attached to a description indefinitely in order to enable it to unambiguously select an antecedent. Moreover, the fact that they may encode any amount of information makes them suitable candidates for accommodation, that is they have the capacity to establish an antecedent in case discourse does not provide one.

In this respect presuppositional adverbs distinguish themselves from descriptions. Presuppositional adverbs generally resist accommodation. Traditionally (25b) is given as the presupposition induced by (25a):[7]

(25a) $[\text{Mary}]_F$ lives in London too.
(25b) There is someone other than Mary living in London.

It has been pointed out by Kripke that the informational content of (25b) is so low that it would, under normal circumstances, not add any relevant information to what is already given.[8] Kripke argues that if (25b) were indeed the presupposition contribution of *too*, one should expect that it would easily accommodate in nearly any context. This, however, is not what we find. When uttered out of the blue, (25a) immediately gives rise to some question to specify what particular person or set of persons different from Mary live in London. Thus, if the only requirement on the use of this presuppositional expression were contextual satisfaction of the presupposition triggered, (25a) would be admissible in nearly any context. As Zeevat pointed out this is at odds with the reluctance of this type of presupposition to accommodate. One would moreover expect that the corresponding sentences without this particular particle were admissible in exactly the same contexts which accept their presuppositional variants. This is again at odds with the facts as the difference in acceptability of the following two sequences shows.

(26a) Harry lives in London. Mary lives in London too.
(26b) Harry lives in London. Mary lives in London.

Just the fact that the context satisfies the presupposition (i.e. contains the information that some people different from Mary live in London) will not do. Though most contexts will have this property only few will license the utterance of (25a). But even in these contexts *too* is not simply redundant in the sense that it can safely be skipped without affecting the felicity of its carrier sentence. Kripke's explanation is that *too* contains an anaphoric element which should link up to some parallel information which is foregrounded in the immediate context of utterance.

The presupposition induced by *too* distinguishes itself in another crucial respect from most other types of presupposition inducers. This becomes clear when we look at variable sharing between the anaphoric

[7]The computation of the presuppositions for this type of adverb is generally focus- dependent. Roughly, the rule is to insert a variable for the focused element and to add the condition that some entity (or set of entities) different from the focus value has the relevant property.
[8]As reported in Soames 1989 and Kripke ms.

structure and their matrix. Consider again the classic view on the presuppositional contribution of *too*. According to this view (27) presupposes that someone different from John comes. The representation of this presupposition as commonly formulated (e. g. Karttunen and Peters 1979) will come out in the current format as in (28).

(27) $[\text{John}]_F$ comes too
(28) [x: John = x, come(x), αy[y:come(y), x$\neq$y]

The crucial point is that the variable for the focused NP reoccurs in a condition of the presuppositional condition. We thus observe a similar situation as in (20). The level where x is introduced is the highest position the full presupposition can be projected to. This may seem innocuous when proper names are concerned, for these are projected to top level anyway.[9] It does, however, lead to problems when the referent is introduced at some subordinate level. Consider (29):[10]

(29) My neighbour comes. If a girl comes too,...

The presupposition introduced by *too* should link up to the information introduced at main level. The discourse marker for the girl is, however, introduced in the antecedent of the conditional. If the presuppositional condition requires that some entity different from the marker for the girl has the property of coming (i.e. if the presuppositional condition contains the non-identity requirement) we won t be able to project the presupposition any higher than the place where the girl is introduced. It would thus, in the present case prevent us to bind the presupposition to its intended antecedent.

A simple solution, which, as we will see in a moment, is independently motivated by the behaviour of this type of presuppositions in attitude contexts, is to ensure that the presuppositional condition of *too* does not share any variable with the matrix structure which introduces it. We may account for the non-identity between the focused constituent and the anaphoric variable ensuring that the latter does not have the property that is ascribed to former, that is by encoding this information in the presuppositional predicate. The initial representation for the antecedent of (29) then comes out as (30). Here there is no variable

[9]I assume that proper names are presupposition inducers themselves. In view of their lack of descriptive content they will thus always [or nearly always if we also take names for fictitious objects into consideration] be projected to top level and get maximal scope with respect to embedding operators, though they won't be rigid in the Kripkean sense. See Sommers (1982) for a defense of the view that proper names are anaphoric expressions.
[10]I owe this example to Bart Geurts.

sharing between the presuppositional condition and its inducing matrix. This implies that the presupposition can be processed and interpreted independently of the sentence which triggers it.

(30) [x: girl(x), come(x), αy[y:come(y), ¬ girl (y)]]

The relevant difference between the representation in (30) and the encoding of e. g. descriptions as in (31) is that in the latter the anaphoric marker reoccurs in a condition of its matrix DRS, while in the former the anaphoric marker is fully independent of any marker occurring in its inducing matrix.

(31) [:bald(x), αx[x: KF(x)]]

Note that when the description is projected out, it will still bind a variable in its inducing matrix frame. As we saw in (18a), processing a simple modal embedding of (31) plugs a marker x in the main context, thus binding the modally embedded occurrence in *bald(x)*. Processing the trigger of (29) yields a different result. Projection of the trigger and linking it up to *my neighbour* gives

(32) [x: my__neighbour(x), ¬ girl(x), come(x), [y: girl(y), come(y)]→ [...]

The presupposition has been absorbed by its antecedent without leaving a trace at the position where it was originally generated. It does not have any other semantic effect than adding the condition ¬girl(x) to the main DRS. This correctly predicts that that the full sentence entails that my neighbour is not a girl.

The account just given has some implications with respect to the formulation of Kripke's claims. As I already said, Kripke's central observation is that in sentences like (33) through (35) the usual presupposition does not seem to contribute anything to the content.

(33) If John comes to the party, the boss will come too.
(34) Mary is having dinner in New York too.
(35) Mary is having dinner in New York and the boss is having dinner there too.

Given suitable background knowledge the presupposition of (34) will always be fulfilled. Nevertheless (34) is infelicitous if uttered out of the blue. This is problematic for the standard accounts, since these predict that the presupposition is trivially fulfilled. Kripke concludes that these sentences carry more substantial presuppositions than is usually assumed. His central theses are that

(a) the presupposition of *too* contains an anaphoric element and the presupposition arises from the anaphoric requirement that when someone uses *too* he refers to some parallel information that is either in another clause or in the context;
(b) the presupposition of the consequent of e.g. (33) cannot be determined independently of the antecedent;
(c) (33) does not presuppose that there is someone different from John who will come, but gives rise to the more substantial presupposition that John is not the boss;
(d) this presupposition does not come in addition to, but actually replaces the existential presupposition given by the usual account.

Kripke's first conclusion is certainly right and so are the intuitions which underlie his remaining points. But I disagree with the claim that this forces us to assign a different presupposition to the consequent of e.g. (33) than the standard view does. Firstly, it should be clear that the anaphoric requirement stated under (a) is a more substantial requirement than the standard view demands. It should moreover be clear that, if we view a presupposition as an anaphoric expression, the interpretation of this expression will depend on the way it is resolved. I take this to be the rationale behind Kripke's claim that the presupposition of the consequent of (33) cannot be determined independently of its surrounding context. And if we distinguish between the presupposition triggered and way they are resolved, our actual intuitions about the final interpretation of the inducing sentence will be based on the outcome of the resolution process and thus may involve more substantial inferences than a simple addition of the triggered material to the context would suggest. So I would like to point out that, if we adopt the anaphoric view and we distinguish between the presupposition triggered and the result of the resolution process, (b) through (d) are not needed. We certainly infer from (36) that John is not the boss, but this inference need not be taken to be the presupposition of (36), since it is derivable from a (much weaker) presupposition + Kripkes claim that presuppositions are anaphoric expressions.

(36) John comes to the party and the boss comes too

The presupposition is that there is some x not having the property of being a boss and that this x will come (the presuppositional requirement deriving from the description is independent and will take care of establishing a referent for the boss). The anaphoric requirement is that this variable be bound to some pre-established antecedent. The condition that this variable does not have the property of being a boss will prevent the trivial equation of this variable with the marker established for the description. The condition that x will come will enables us to select John as the antecedent. This antecedent will thereby also inherit the information that John is not the boss. The presupposition postulated thus doesn't replace the existential presupposition assigned by the conventional account, but can be derived from it on the basis of the assumption that a presupposition which is much weaker and which is determined solely on the basis of its inducing sentence, should be resolved to some previously established antecedent.

5. Transparency in attitude contexts

The presuppositional independence between the presupposition of *too* and its inducing matrix has some interesting consequences when we consider projection out of attitude contexts.

It has been observed by Fauconnier (1985) and Heim (1992) that (37) when embedded in an attitude context allows a reading where the presupposition is interpreted as fully independent from the beliefs of the subject of the attitude.

Consider (37). Assume that it is already established that Sue will come. Assume furthermore that there is no indication whatsoever that Sally expects any other person than John to come. All she actually thinks is that John will come.

(37) Sue will come. Sally believes that $[\text{John}]_F$ comes too.

There clearly is an interpretation where the presupposition holds in the main context, but not in Sally's beliefs. Heim imagines two kids taking to each other by phone.[11] This forces the relevant interpretation by background knowledge:

(38) John: I am already in bed.
Mary: My parents think $[\text{I}]_F$ am also in bed.

When hearing this conversation one would normally conclude that Mary does not have the belief that her parents share the information she just got about John's being in bed. Her parents may not even have any belief about John. The presupposition is interpreted fully independent of her parents' beliefs. Heim proposes to interpret such beliefs as a kind of *de re* belief. There is however a problem with this proposal.

The paraphrase Heim puts forward as a tentative *de re* analysis of (39a) is (39b).

[11] Heim (1992) p. 209.

(39a) My parents think I am also in bed.
(39b) Of the property of also being in bed, my parents think that I have it.

With respect to (39b) Heim remarks

> The idea behind this paraphrase is that 'the property of also being in bed'... is just another way of describing the property of being in bed, and that it is a description which fits that property only contingently: it is true just in case John happens to be in bed. And since this latter fact is known to Mary but unknown to her parents, she but not they, can describe it in those words.

This answer would be a legitimate one for a theory which divorces truth conditions from presuppositions as happens in e. g. Karttunen & Peters. According to such a theory *also* does not contribute to the truth conditions. *Being in bed* and *also being in bed* thus denote the same property. This, however, does not hold for Heim's theory which reinterprets Karttunen's heritage conditions as definedness conditions.

Here is an alternative analysis. The central difference between the standard examples of *de re* belief and the external readings of e. g. presuppositional adverbs is the following. In *de re* belief one refers to some individual and uses it to determine a belief attributed to it by the subject of the attitude. On the *de re* analysis of e. g. (40a) the context created by the attitude verb contains a variable which is bound from without.

(40a) Harry believes that the mayor is bald
(40b) There is some x such that x is the mayor and Harry believes that x is bald.

Here it is the speaker of the sentence — not the subject of the attitude — who attributes the descriptive conditions associated with the description to the object the attitude is about. *de re* ascriptions relate a believer to a res and the belief attributed to the subject of the attitude does involve some object. However it is the speaker — but not necessarily the subject of the attitude — who associates the descriptive conditions with this object. The situation is clearly different in the above cases where the information is triggered that there is yet another person different from the object the belief is about who satisfies certain conditions. In the terminology of the theory I presented in this paper: there is no variable linking between the anaphoric variable and the sentential matrix which contains the trigger. While the general format of the relation of a description to its embedding matrix is $[:\varphi(x), \alpha x[x: \psi(x)]$, the relation between the presupposition of *too* and its inducing sentence comes out as $[x:\varphi(x), \alpha y[y: \psi(y)]$. In the former case the presupposition will after exportation to the main context bind the variable in its inducing matrix. In the latter case the presupposition can be exported to the main context (either by binding or by accommodation) without leaving a trace in the attitude context. Nevertheless we still have a self-contained belief as the object of the attitude. In (37) Sally still believes that John comes, though the belief that there is someone different from him who meets the same condition, is interpreted fully externally with respect to her beliefs. In (38) the belief attributed to Mary's parents is that Mary is in bed. This belief can be entertained independently from the information invoked by the presupposition inducer. Since the belief contexts contains no variable which is bound from without, the usual problems involved in the analysis of *de re* belief don't even arise. However, in case a description figures as the subject of the complement, there has to be an object the belief is about and this object has to be linked up to the object the speaker associates the descriptive conditions with. In presuppositional terms, in the case of descriptions the content of the attitude is dependent on the presupposition induced, though the descriptive information associated with its anaphoric variable can be exported and interpreted externally. In the external readings of (37) and (38) on the other hand we may export the presuppositional information that there is some y different from the subject of the belief while retaining a full-fledged belief.

In the representation of the two types of presuppositions this difference is reflected in the different ways the anaphoric variable is linked up to the matrix sentence. With descriptions the anaphoric variable re-occurs in a condition in the matrix sentence, when we consider the presupposition induced by *too* it does not. Resolution of the presuppositional expression in (37) according to the standard mechanism yields a result which is just the same as when we would have processed the corresponding sentence without the anaphoric element.

Just as in the modal case both the anaphoric variable and the associated conditions are projected to the main DRS. The result clearly does not attribute some belief which is *de re* with respect to the object meeting the presuppositional conditions. Sally just believes that John comes. There is no x different from y figuring in Sally's beliefs. Only the speaker commits himself to the existence of the object satisfying the information induced by the presupposition trigger. The proper interpretation does not fall out as a result of quantifying into the attitude context, but as a result of projecting the trigger out.

References

Beaver D. (1993). "What Comes First in Dynamic Semantics", *ILLC*. Amsterdam.

Fauconnier G. (1985). *Mental Spaces*. Cambridge (MA): MIT Press.

Gazdar G. (1979). *Pragmatics. Implicature, Presupposition, and Logical Form*. New York: Academic Press.

Geurts B. (1995). "Presupposing", PhD Thesis. University of Stuttgart.

Heim I. (1983). "On the Projection Problem for Presuppositions", in *Proceedings of the West Coast Conference on Formal Linguistics 2*:114–126. Reprinted in S. Davis (ed.) (1991) *Pragmatics*. Oxford University Press. 397–405.

Heim I. (1992). "Presupposition Projection and the Semantics of Attitude Verbs", *Journal of Semantics 9*:183–221.

Kamp, H. and U. Reyle (1993). *From Discourse to Logic*. Dordrecht: Kluwer.

Kamp H. and A. Rossdeutscher (1994). "DRS Construction and Lexically Driven Inference". *Theoretical Linguistics 20*:165–236.

Karttunen L. (1974). "Presupposition and Linguistic Context". *Theoretical Linguistics 1*:181–194.

Karttunen L. and Peters S.(1979). "Conventional Implicature", in: C.-K Oh and D. Dinneen (eds.) *Syntax and Semantics 11: Presupposition*. New York: Academic Press. 1–56.

Kripke S., "Presupposition and Anaphora. Some Remarks on the Formulation of the Projection Problem", unpublished manuscript.

Lewis D. (1979). "Scorekeeping in a Language Game". *Journal of Philosophical Logic, 8*:339–359.

Soames S. (1989). "Presupposition", in D. Gabbay and F. Guenthner (eds.), *Handbook of Philosophical Logic, Volume IV*. Dordrecht: Reidel. 552–616.

Sommers F. (1982). *The Logic of Natural Language*. Oxford: Clarendon Press.

Stalnaker R. (1973). "Presuppositions". *Journal of Philosophical Logic, 2*: 447–457.

Van der Sandt R. A. (1988). *Context and Presupposition*. London: Routledge.

Van der Sandt R. A. (1992). "Presupposition Projection as Anaphora Resolution", *Journal of Semantics 9*:333–377.

Zeevat H. (1992). "Presupposition and Accommodation in Update Semantics", *Journal of Semantics 9*:379–412.

Chapter 13

THE LIMITS OF A LOGICAL TREATMENT OF ASSERTION

Denis Vernant
Université Pierre Mendès France, Grenoble

It is clear that the concept of assertion has played a crucial role in the construction of contemporary logic. Apart from the central notions of *proposition* and *truth*, assertion raises the issue of *judgment*, which constitutes the very subject of logic. The logical systems put forward by Russell between 1903 and 1925 all attribute the status of a primitive idea to assertion.[1] At the same time, there is considerable variation in Russell's interpretation of assertion. In what follows I will trace these variations. However, I will do so not for purposes of exegesis. Instead, my aim will be to offer an appreciation of the limits of the logical treatment of a complex concept; a concept which, as I will claim, can only be satisfactorily defined through a *pragmatic* formulation of the issues it raises.

The first part of this contribution traces Russell's thematization of assertion from the *Principles of Mathematics* to *Principia Mathematica*, while emphasizing its eminently aporetic character. I concentrate solely on the propositional calculus in order to focus the analysis on what is

[1]The overview of the axiomatics from 1903, 1906, 1908-10, 1925 in Vernant, *Philosophie mathématique de Russell*, p. 460-466 suggests that assertion is the only primitive idea included in all the axiomatics. Jules Vuillemin notes that "in the *Principles of Mathematics*, Russell does not consider assertion to be a primitive notion" (*La Première philosophie de B. Russell*, ch. 1, p. 18) and in *Du discours à l'action* (ch. 2, p. 28, note 1) I claim that assertion is defined in 1903 on the basis of truth. On second thoughts, this may not be the case after all. In the *Principles*, Russell does not explicitly mention assertion in his various lists of primitive ideas. In 1903, he avoids putting an exact number on how many primitive ideas there are. For instance, he points out in § 12, p. 11 that " The number of indefinable logical constants is not great: it appears, in fact, to be eight or nine" and right away follows this with a list of six primitive ideas! But also he declares *de facto* that "the notion of a propositional function [undefinable, DV] and that of assertion are more fundamental than the notion of class", § 44, p. 40; cf. § 99, p 100.

D. Vanderveken (ed.), Logic, Thought & Action, 267–288.

essential.[2] The second part deals with the pragmatic treatment of assertion which began before the second edition of *Principia Mathematica*, with Frege's *Logische Untersuchungen* (1918), and was to continue with Searle's definition of assertive speech acts and their formalization by Daniel Vanderveken. This analysis will show that the contemporary redefinition of the field removes Russell's aporias ; however, it will also show the surviving relevance of certain analyses presented by Russell, such as his treatment of denial.

1. A Logical Account of Assertion

1.1 Assertion in the *Principles of Mathematics*

In 1903, Russell explained assertion in two ways, one logical, the other philosophical.

1.1.1 "A Logical Problem". In the *Principles*, the idea of assertion is introduced along with the initial explanation of the primitive idea of implication, under cover of "a very difficult logical problem, namely, the distinction between a proposition actually asserted, and a proposition considered merely as a complex concept".[3] From a strictly logical viewpoint, such a distinction between *asserting* and *considering* enables one to separate two operations, implication and inference[4]:

– "The proposition "p implies q" *asserts* an implication, though it does not *assert* p or q"[5]. *Implication* only concerns simply considered propositions;

– The inference that bears on the asserted propositions; Russell originally expressed this as follows: "if the hypothesis in an implication is true, it may be dropped, and the consequent asserted".[6]

The passage from one assertion to the other, and the possibility of separating the assertion from the consequent of the implication, resolves Lewis Carroll's paradox of Achilles and the tortoise.[7] "We need, in

[2]For an account of the role of assertion in predicate calculus, and especially of the primitive idea of the assertion of a propositional function, introduced in 1906 and dropped in 1927, see Vernant, *Philosophie mathématique de Russell*, §§ 39-40, p. 261-268. On the latter issue see also P. Hylton, *Russell, Idealism and the Emergence of Analytic Philosophy*, ch. 7, p. 288-298.

[3]Cf. *Principles*, (PoM), § 38, p. 34.

[4]In "Meinong's Theory of Complexes and Assumptions" (MTCA), p. 44, Russell points out that Frege showed that it is "absolutely indispensable" in logic to make room for assumptions next to judgements.

[5]Cf. PoM, § 38, p. 35.

[6]*Ibidem*, § 18, p. 16.

[7]"What the Tortoise said to Achilles" and PoM, § 38, p. 35.

fact, the notion of *therefore*, which is quite different from the notion of *implies*, and holds between different entities".[8] There seems nothing more to add, however, difficulties emerge as soon as one tries to express the operation "therefore" and explain the difference between the entities involved. Let us look at each point separately:

a – Immediately after introducing the principle of inference as described, Russell adds that "This is a principle incapable of formal symbolic statement, and illustrating the essential limitations of formalism".[9] This results from the fact that this principle, presented as the fourth primitive proposition of propositional calculus, constitutes a *rule*, that of "dropping a true premiss".[10] Since this runs the risk of infinite regress, such a rule cannot be part of the process of reasoning: "the rule according to which the inference proceeds is not required as a premiss.... /... In order to apply a rule of inference, it is formally necessary to have a premiss asserting that the present case is an instance of this rule; we shall then need to affirm the rule by which we can go from the rule to an instance, and also to affirm that here we have an instance of this rule, and so on into an endless process".[11] The rule then constitutes a "non-formal principle" that can be expressed in natural language but not formally.[12] This shows us the full extent of the difference between the logical operations expressed by the words "therefore" and "imply".

b – What remains is the difference between the entities involved. If we suppose that an asserted and a merely considered proposition are different in nature, this would lead to the ambiguity sophistry [*fallacia*

[8]PoM, § 38, p. 35.

[9]*Ibidem*, § 18, p. 16.

[10]PoM, § 44, p. 41. See also *Principia mathematica* (PM), Introduction, ch. 1, p. 9: "An inference is the dropping of a true premiss ; it is the dissolution of an implication".

[11]PoM, § 45, p. 70. Here Russell is inspired by one of Bradley's arguments. Note that he uses *to affirm* for *to assert.*

[12]The question here is a metalinguistic one: "Constants such as *truth, assertion*, and *variable* do not occur in the propositions of Russell's logic [...] and the same seems to go for *term* and *relation.* It is tempting, of course, to treat all these as metalinguistic constants, but this is a luxury that Russell does not permit himself. This fact is more important than might be realized, because Russell expects to do his metatheory in the object-language, thus if the logical constants in this second category really are needed in what we would call the metatheory then they are genuinely required for the propositions of logic as well. In other words, Russell's logic is not just a parochial technical exercise which can do without certain terms which might be needed elsewhere. It is a genuinely foundational enterprise, which must include every term needed in any logical investigation; in particular it has to include every term needed in its own metatheoretic treatment", N. Griffin, "Russell on the Nature of Logic", p. 134. See also W. V. Quine, "Whitehead and the Rise of Modern Logic", p. 140-142, who criticises Russell and Whitehead for not having distinguished use and mention sufficiently. See also J. van Heijenoort, "Logic as Calculus and Logic as Language", p. 14, who had already noted that Russell's (and Frege's) logical universalism prevented him from conceiving of a metatheory for assertion.

aequivocationis] because, in inference, the propositions should not be simultaneously considered as members of the implication and asserted as separate premisses. In fact, the propositions remain the same, but depending on the context they may be either considered or asserted: "if assertion in any way changed a proposition, no proposition which can possibly in any context be unasserted could be true, since when asserted it would become a different proposition. But this is plainly false; for in "p implies q", p and q are not asserted, and yet they may be true. Leaving this puzzle to logic, however, we must insist that there is a difference of some kind between an asserted and an unasserted proposition".[13]

One solution to this problem would involve accounting for assertion in terms of Fregean "recognition of truth"[14] or of Meinong's "conviction".[15] Instead of a difference between objects (propositions), we would have a difference in the *subject's attitude* with regard to those objects: "The case of belief and disbelief shows that it is possible to have different attitudes to the same object, and thus allows us to accept the view, which is *prima facie* the correct one, that there is no difference in the object".[16] In 1904 Russell saw this as succumbing to *psychologism.* Appeal to an explanation in terms of states of mind, recognitions or convictions is here forthwith excluded: "In the discussion of inference, it is common to permit the intrusion of a psychological element, and to consider our acquisition of new knowledge by its means. But it is obvious that where we validly infer one proposition from another, we do so by virtue of a relation which holds between the two propositions whether we perceive it or not; the mind, in fact, is as purely receptive in inference as common sense supposes it to be in perception of sensible objects".[17] From the realist perspective inherited from Moore[18] the proposition is not a mental entity, not a sum of meanings, but an existing reality *per se*: "But a proposition, unless it happens to be linguistic, does not itself contain words; it contains the entities indicated by words".[19]

Thus, *contra* Frege and Meinong, the point was to put forward a strictly logical interpretation of assertion and not a psychological one.[20]

[13]PoM, § 38, p. 35. (Here Russell indicates in a footnote that Frege has a special symbol for assertion).

[14]Cf. PoM, App. A, § 478, p. 503.

[15]Cf. MTCA, p. 23.

[16]*Ibidem*, p. 42.

[17]PoM, § 37, p. 33. Cf. also MTCA, p. 21-22 et 44.

[18]Cf. MTCA, p. 23, foot note 2. See also my *Philosophie mathématique de Russell*, § 20, p. 157-166.

[19]PoM, § 51, p. 47.

[20]I make no pronouncement here concerning the relevance of the criticism that Russell levelled at Meinong and Frege.

Assertion has nothing to do with judgement or the subject's recognition of objects but is directly related to the proposition: "when a proposition happens to be true, it has a further quality, over and above that which it shares with false propositions, and it is this further quality which I mean by assertion in a logical as opposed to a psychological sense".[21] It could be objected that this intrusion of truth surreptitiously re-introduces the problem of knowledge. However, for Russell as for Moore, truth is an *intrinsic* property of propositions; something which is nicely expressed by the following metaphor: "there is no problem at all in truth and falsehood; .../... some propositions are true and some false, just as some roses are red and some white".[22]

It would seem, under this theory, that it is impossible to assert the falsehood of a proposition. Russell's response to this was the assertion of the negation of a proposition: "[We shall] regard '*p* is false' as meaning 'not-*p* is true"'[23]. This, however, does not mean that we should believe that the expression – "*p* is true" – correctly translates the operation of assertion, because that expression expresses an *external* relation of the proposition with regards to truth: "If *p* is a proposition, '*p* is true' is a concept which has being even if *p* is false, and thus "*p* is true" is not the same as *p* asserted. Thus no concept can be found which is equivalent to *p* asserted, and therefore assertion is not a constituent in *p* asserted".[24] Truth is a matter neither of knowledge nor of expression, it is an *ultimate datum*, a simple fact: "Thus the analogy with red and white roses seems in the end to express the matter as nearly as possible. We must simply apprehend what is truth and what is falsehood, for both seem impossible to analyse".[25] This time, the explanation runs into the indefinability of the idea of truth. This does not stop Russell from attempting to explain assertion. If the true propositions are the only ones that can be asserted logically, we still need to know what the assertion of a true proposition is.

1.1.2 Contradiction of asserted propositions. In the *Principles,* the philosophical explanation of assertion introduces the "gram-

---

[21] *Ibid.*, § 52, p. 49.

[22] MTCA, p. 75. This response does not resolve anything. The realist conception of propositions adopted by Russell in 1903 renders the question of falsehood aporetic. This is one of the main reasons for the later disqualification of propositions as incomplete symbols (see Vernant, *Philosophie mathématique de Russell*, § 55, p. 375-77).

[23] PoM, App. A, § 478, p. 504.

[24] *Ibidem*, § 478, p. 504. Cf. also PoM, § 1, p. 3: "Mathematics *uses* a notion which is not a constituent of the propositions which it considers, namely the notion of truth".

[25] MTCA, p. 76.

matical" analysis of propositions, notably through the role that explanation assigns to verbs.

First, Russell proposes a distinction between the *subject* and the *assertion* in all propositions. This assertion is defined as follows: "everything that remains of the proposition when the subject is omitted"[26]. This type of analysis is applied to all predicative propositions: in "Socrates is human", one recognizes without difficulty the subject expressed by the proper name "Socrates" and the *assertion* "is human" comprised of the verb and the qualitative adjective.[27]

However, this analytical schema cannot be applied to relational propositions because it allows us to carry out *two* different analyses of a given proposition: "In a relational proposition, say "A is greater than B", we may regard A as the subject, and "is greater than B" as the assertion or B as the subject and "A is greater than" as the assertion. There are thus, in the cases proposed, two ways of analysing the proposition into subject and assertion".[28] The reason for this is that the subject/assertion schema does not account for the *sense* which characterizes relations. This difficulty leads Russell to prefer *functional diagrams*, which have the advantage of being adaptable: they can be applied to predicative propositions – $F(x)$ – as well as to relational propositions – $F(x,y)$ – while preserving the directionality of the relation: "A relational proposition may be symbolised by aRb, where R is the relation and a and b are the terms; and aRb will then always, provided a and b are not identical, denote a different proposition from bRa".[29]

Yet, propositional analysis does not raise merely technical difficulties. The explanation of propositional assertion directly throws Russell's theory of terms into question. In the *Principles*, *terms* are described as self-subsisting entities from an ontological point of view,[30] and as possible subjects from a logical point of view. It is possible then to distinguish between *things*, which can only be used as subjects, from *concepts* (predicates and relations), which have a twofold use: one use as *terms* (in the subject position: "1 is a number"), and one *qua concepts* (in the

[26]PoM, § 81, p. 83.

[27]Cf. PoM, § 48, p. 45.

[28]*Ibidem*, p. 44. If we indicate assertion by means of parentheses, then, from A > B we obtain either A(> B) or (A >)B, which leads back to B(< A). The underlying question is that of the irreducibility of relations to predicates (see Vernant, *Philosophie mathématique de Russell*, §§ 15-17, p. 102-122.

[29]PoM, § 94, p. 95.

[30]This notion of *term* does not designate an expression, but a real thing. In order to avoid ambiguity, M. Sainsbury suggests that we use *meaning-relatum* (*Russell*, ch. 1, p. 20). The linking of terms in the proposition produces a reality, hence the question of truth and the difficulty of accounting for falsehood.

predicate position: "This is one").[31] However, the difference between concept-as-term and concept-as-such is only extrinsic: "The difference lies solely in external relations, and not in the intrinsic nature of the terms. For suppose that *one* as adjective differed from 1 as term. In this statement, *one* as adjective has been made into a term; hence either it has become 1, in which case the supposition is self-contradictory; or there is some other difference between one and 1 in addition to the fact that the first denotes a concept not a term while the second denotes a concept which is a term".[32] In either case, Russell demonstrates that the thesis according to which the concept *qua* concept cannot become the subject of a sentence, and thus a concept-as-term, is self-contradictory.[33] Unlike Frege, for whom the unsaturated concept can never occupy the subject-position, Russell saw the concept as a genuine term that can always become the subject of an assertion through nominalisation. In this way, it would seem that *any term, thing, or concept could become a subject.*

The major philosophical difficulty with assertion is that it leads to a direct contradiction of Russell's thesis. Consider the proposition "Socrates is human". It is an autonomous entity that should be able to become the subject of a new proposition. But its nominalisation by transformation of the verb into a verbal noun *nolens volens* results in transformation of the proposition into a *propositional concept*, which is not equivalent because it results in loss of the assertive force of the initial proposition: "There appears to be an ultimate notion of assertion, given by the verb, which is lost as soon as we substitute a verbal noun, and is lost when the proposition in question is made the subject of some other proposition".[34] The conjugated verb expresses an *effectively relating relation* which is responsible for the the proposition's essential *life-like unity.*[35] When transformed into a verbal noun, the verb translates no more than a *non-relating relation*, reduced to a lifeless element and isolated from the other elements of the proposition: "A proposition, in fact, is essentially a unity, and when analysis has destroyed the unity,

[31] Vernant, *Philosophie mathématique de Russell*, §§ 4-7, p. 30-54.

[32] PoM, § 49, p. 46.

[33] *Ibid.*; see also my analysis of the argument in *La Philosophie mathématique de Russell*, § 6, p. 40-41.

[34] PoM, § 52, p. 48.

[35] "The $\emptyset$ in $\emptyset x$ is not a separate and distinguishable entity: it lives in the propositions of the form $\emptyset x$, and cannot survive analysis", PoM, § 85, p. 88 ; see also my *Philosophie mathématique de Russell*, § 7, p. 41-54. This is the source of the first opposition Wittgenstein makes between saying and showing ; see the *Tractatus* 4.0311, where the proposition is a "*lebendes Bild*".

no enumeration of constituents will restore the proposition".[36] Thus, beyond the logical contradiction produced by the paradox of classes, the specifically *philosophical* contradiction in Russell's writing from 1903 stems from the impossibility of analysing propositions, accounting for propositional unity, and expressing the assertive ingredient that is used to bind the proposition together: "Thus the contradiction which was to have been avoided, of an entity which cannot be made a logical subject, appears to have here become inevitable".[37] Russell, while trying to avoid the Charybdis of unspeakable Fregean concepts, falls to the Scylla of propositions that are entirely unspeakable. From that point onwards, the philosophical contradiction of a proposition that cannot become a subject has to be added to the limit of logical formalism – namely, the impossibility of symbolically translating the principle of inference. This shows the *irremissibly aporetic character* of the account of assertion put forward initially by Russell: in 1903, assertion was not just undefinable, it turned out to be something of a monster.

Was Russell subsequently to raise these logical and philosophical aporias? To answer this question, we need to examine the status of assertion in *Principia Mathematica.*

1.2 Assertion in *Principia Mathematica*

First we will study the logical aspect of the question, the role of the assertion sign. Then we will look at the problem from the philosophical point of view under which a new theory of judgement was to emerge, rejecting the status initially assigned to propositions.

1.2.1 The assertion sign. In the first edition of *Principia Mathematica* (1910), the discussion of assertion begins from the first chapter of the introduction. Russell adopts Frege's assertion-sign: $\vdash$ and maintains his initial thesis according to which only true propositions can correctly be asserted: "For example, if "$\vdash (p \supset p)$" occurs, it is to be taken as a complete assertion convicting the authors of error unless the proposition "$(p \supset p)$" is true (as it is)".[38] From then on, it becomes possible to distinguish asserted propositions from merely considered ones. If we agree to translate "$\vdash p$" by "p is true",[39] we will say

[36] *Ibid.*, § 54, p. 50.
[37] *Ibid.*, § 52, p. 48.
[38] PM, Introduction, ch. 1, p. 8.
[39] In PM, *1, p. 92, Russell takes pains to add (in brackets): "although philosophically this is not exactly what it means". Indeed, the *Principles* objected to the assimilation of assertion to the expression of a proposition's truth (*supra*, § 1.1.1). On the intricate analysis of the

that: "$\vdash: p. \supset .q$" means "It is true that p implies q", whereas "$\vdash .p. \supset \vdash .q$" means "$p$ is true; therefore q is true". The first of these does not necessarily involve the truth either of p or of q, while the second involves the truth of both".[40] Now the *principle of inference* is formulated as follows:

"A proposition "p" is asserted and a proposition "p implies q" is asserted, and then as a sequel the proposition "q" is asserted".[41] Using the new symbol this can be written in the following way:

$$\text{“}\vdash .p. \supset \vdash .q\text{”}$$

which signifies "p, therefore q" and "which is to be considered as a mere abbreviation of the threefold statement:

$$\text{'}\vdash .p\text{' and '}\vdash (p \supset q)\text{' and } \vdash .q\text{”'}.^{42}$$

However, the limitation on formalism, discovered in 1903 still applies because the rule of inference, which allows us to go from the first two assertions to the third, remains recalcitrant to symbolic expression. And although Russell explicitly recognises that "We cannot express the principle symbolically",[43] he retains the ambiguity between primitive propositions and rules, which dates from 1903, by making the "principle" of inference the first primitive proposition of his propositional calculus :

"*1.1 Anything implied by a true elementary proposition is true.Pp (It is the rule that justifies the inference)".[44]

The "principle of assertion" can be formalised as:

$$*3.35 \vdash: p\ .p \supset q. \supset .q$$

which, however, does not concern the truth of p, "but requires merely the *hypothesis* that p is true" and thus does not authorise detachment of the consequent.[45]

---

ambiguities of assertion in *Principia* carried out by Lesniewski, see Vernant, *Du Discours à l'action*, ch. II, § 1, p. 23-24.

[40]PM, *1, p. 92.

[41]PM, Introduction, ch. 1, p. 8-9.

[42]*Ibid.*, p. 9.

[43]Cf. * 1, p. 94, where Russell refers explicitly to PoM § 38. Though Russell speaks of "abbreviation" here, we cannot introduce the new "symbol" by a definition of the type: $\vdash .p. \supset \vdash .q =_{Df} \vdash .p. \vdash (p \supset q). \vdash .q.$

[44]*Ibid.* Wittgenstein criticised the use of ordinary language to express laws of logic (*Tractatus* 5.452). In fact, for Wittgenstein, *Modus ponens* does not need to be expressed because deductions *appear as proof* and result from the formal properties of the propositions involved. B follows from A if and only if the truth of A implies that of B (5.132, 6.1221, 6.1264).

[45]*3, p. 110. The passage shows the difference between the expression of the rule of deduction in the metalanguage, on one hand, and the law that corresponds to it in the object-language, on the other. Note that the "principle" of assertion is in fact a deduced theorem.

Henceforth, apart from definitions that have merely the value of abbreviations and are "the expression of a volition, not a proposition (for this reason, definitions are not preceded by the assertion-sign)",[46] primitive propositions as well as deduced theorems[47] are preceded by the assertion sign.

1.2.2 Judgment. From a purely technical point of view, *Principia Mathematica* does not say anything more on assertion. Let us, then, move on to the philosophical aspects.

Apart from a technical presentation the theory of types, the second chapter of the introduction, "The theory of logical types", puts forward a theory of judgment that radically differs from the 1903 theory of propositions. The analysis of propositions – which, as already mentioned, is connected to the explanation of assertion – is conceptualised in a new way. Sentences that express propositions are now reduced to *incomplete symbols*, and the propositions themselves lose their substantial unity by being dissolved into a chain of "constituents": what "we call a "proposition" (in the sense in which this is distinguished from the phrase expressing it) is not a single entity at all. That is to say, the phrase which expresses a proposition is what we call an "incomplete" symbol; it does not have meaning in itself, but requires some supplementation in order to acquire a complete meaning".[48] Propositional unity is henceforth ensured "from outside", by *judgment.* judgment does not add any verbal elements but unifies propositional elements by virtue of the fact that it is an act: "When I judge "Socrates is human", the meaning is completed by the act of judging, and we no longer have an incomplete symbol".[49] The fundamental *discursivity* of judgment takes the place of propositional unity.[50] The subject's judgment unifies the propositional constituents and when the judgment is true, it corresponds to a per-

[46]PM, Intro. ch. 1, p. 11. Frege was the first to introduce a symbol of the definition: $| \vdash (A \equiv B)$ which means "A and B *have to have* the same content", cf. *Begriffsschrift*, § 24, English translation p. 55. As in Russell's work, the definition given is not a judgement and only has value as an abbreviation. This is the standard conception; the definition does not have the force of assertion but of *declaration*, and belongs to the class of *metadiscursive acts* which concern decisions about the uses of symbols. See Vernant, *Du Discours à l'action*, ch. III, § 3, footnote 2, p. 49. In Lesniewski's systems, the definition is a *rule* that allows for the introduction of a new symbol *as the thesis* of the system; hence its "developmental" capacity (see Denis Miéville, *Un développement des systèmes logiques de Stanislaw Lesniewski,* ch. 2, p. 43-47.

[47]All proofs have the following structure: "$\vdash$. etc. $\supset \vdash$. Prop.", cf. PM *1, p. 103.

[48]PM, Introduction, ch. 2, p. 44.

[49]*Ibid.*

[50]The judgment aRb can be expressed by the *multiple relation* $J(s, a, b, R)$, in which s represents the subject of the judgment. For a closer look at this discursive theory of judgment see my *Philosophie mathématique de Russell,* § 55, p. 370-380.

ceivable fact: "In fact, we may define *truth,* where such judgments are concerned, as consisting in the fact that there is a complex *corresponding* to the discursive thought which is the judgment. That is, when we judge "a in the relation R to b", our judgment is said to be *true* when there is a complex "a-in-the-relation-R-to-b", and is said to be *false* when this is not the case".[51]

From then on, the sentence "Socrates is human" expresses the judgment that carries out the unification of the propositional constituents: Socrates, to be, and human. In 1913, the *Theory of Knowledge* confirms and generalises this analysis by reducing the proposition to a multiple object within what is called a *propositional attitude.*[52] In this way, the entire analytical framework of the 1903 "grammatical" explanation collapses: assertion can no longer concern propositions, but judgments. What, then, happens to the strategy designed to avoid a drift into psychologizing? Are we to say that it is the subject of the judgment that carries out the assertion? Which subject is involved in asserting the theorems and axioms of logical calculus? Should we follow Lesniewski's perfidious suggestion and hold up the *Principia* as a "deductive confession by the authors of the theory"?[53] Such questions make clear the need to re-interpret assertion in the context of the new theory of judgment; but readers of the *Principia* would search in vain for even the beginnings of such an explanation. The logical treatment found in Chapter I of the introduction and the theory of judgment in Chapter II remain entirely distinct. Russell is silent on the matter: it will be remembered that the introduction of *Principia,* due entirely to Russell, consists of re-edited articles including, precisely, a 1906 article, "The theory of implication", for the first chapter[54] and the "La théorie des types logiques" (published in 1910 in the *Revue de métaphysique et de morale*)[55] for the second. Russell simply recycled these articles with some minor changes, and with little concern about whether they converged or not[56].

In the absence of any unification of the themes, the issues clearly and courageously brought forth in 1903 were simply ignored in the *Prin-*

---

[51]PM, Introduction, ch. 2, § III, p. 43.

[52]See N. Griffin, "Russell on the Nature of Logic", p. 178.

[53]On Lesniewski's subtle argument see my *Du Discours à l'action*, ch. II, § 1, p. 23-24.

[54]For the proofs of the propositional calculus see *1 to *5. For a term-for-term comparison of the two versions as differing in insignificant details see D. O'Leary, "The Propositional Logic of *Principia Mathematica* and some of its Forerunners", App. A, p. 108-110.

[55]To the French paper, the *Principia* adds only § 1 on the principle of the vicious circle; § VII is entitled "Raisons pour accepter l'axiome de réductibilité"; and the initial section on the theory of classes is substituted by § VIII on the various contradictions.

[56]This goes to remind us that the cut and paste action is not at all an invention of the world of computers from the researchers at Xerox in Palo-Alto!

cipia. It is true that the main objectives were elsewhere: to carry out an exhaustive logical reduction of all mathematics to the new logic. But where our problem is concerned, unanswered questions remained. These questions can be summed up as the problem of accounting for assertion in the framework of a theory of judgment. The answers which are missing from the *Principia* may be found in Frege's late works.

2. The Pragmatic Description of Assertion

Before examining that solution it should be pointed out that, despite the general title under which Frege's articles appeared ("Logische Untersuchungen"), that solution is essentially not logical but pragmatic. From a strictly logical point of view, the operator of assertion was quickly abandoned by Wittgenstein in the *Tractatus*, as well as by Lesniewski. The "turnstile" assertion symbol today only serves as the symbol for deduction.[57] Frege's thoughts of using the sign no longer apply to the artificial language of the *Begriffsschrift* but to natural language use as it corresponds to basic logical operations. What is at issue is the cognitive usage of language, more precisely and significantly, "propositions with which we communicate or assert things" and "complete interrogative propositions" [*die Fragesätzen*].[58]

2.1 The act of judgment and assertive force

In Frege's writings, the simple idea of "grasping of a thought" [*fassent des Gedankens*] corresponds to Meinong's *Annahme* or Russell's "considering". This simple "act of thought"[59] could concern an affirmative thought (written as $- p$) or a negative thought (written as $\top p$). Affirmation and negation concern thoughts: "For any thought, there exists another one in contradiction with it in such a way that a thought is said to be wrong when the contradictory thought is said to be true. The proposition that expresses the contradictory thought is built from the expression of the initial thought and with the aid of a negation word".[60]

[57]On this area see Vernant, *Du Discours à l'action* (ch. 2, § 3, p. 31–36), where I show that the assertive dimension is sidetracked into the procedures of proof.

[58]"Logische Untersuchungen, I Der Gedanke", p. 34.

[59]*Ibid.*, p. 35.

[60]"Logische Untersuchungen, II Die Verneinung", p. 67. Frege argues at great length against the idea of "negative judgment" corresponding to Russell's denial, saying that negation is not a matter of judgment: "this negation must not be placed on the same rank as judgment, nor must it be interpreted as occupying the pole opposite judgment. Judgment is always about truth" (p. 64). Affirmation and negation therefore pertain solely to the *locutionary* level of propositional content. In his analysis of assertion, Peter Geach follows Frege and refuses Lukasiewicz's operation of rejection as "a futile complication", cf. *Logic Matters*, ch. 8, p.

Thus, *affirmation* (like negation) is a function of truth: "The value of this function will be Truth if Truth is taken as an argument, and Falsehood in all other cases [...]. The value of this function is therefore the argument itself when that argument is a truth value".[61]

Such an act grasping of a thought *can* then result in a *judgment* based on recognition of the truth of the thought in question.[62] *Assertion* is then "die Kundgebung", the making manifest of this judgment.[63] In Frege's mind, since the *Begriffsschrift*[64], assertion does not concern propositions but judgments designed to be the subject's recognition of the truth of a thought. Logically, this is expressed precisely by *assertion sign* (literally the 'bar' of judgment: *Urteilstrich*): $\vdash p$. See the following figure.

For Frege, therefore, judgment is expressed by a declarative sentence [*Behauptungssatz*] endowed with *assertive force* [*behauptende Kraft*]. The key to this assertive *force* is in the subject's *act* of judgment, in its commitment to the truth of the thought in question. Russell spoke of judgments in *Principia* but did not integrate them into his work; moreover, he did not explicitly link judgment to assertion. Frege's great innovation, on the other hand, was that he introduced an *action-based approach* to ordinary language usage under cover of his analysis of assertion. Thus, Frege emphasizes the very *act* of judgment: "We will comply fully with usage if we understand judgment as an act of judging, in the way a leap is an act of leaping".[65] Given this action-based perspective, reference to agent and context can no longer be excluded: "If judgment is an act, then it occurs at a determinate time and subsequently belongs to the

260–261. For our interpretation of rejection in terms of pragmatic denial, cf. "La genèse logique du concept de dénégation de Frege à Slupeski".

[61] "Funktion und Begriff", p. 31-32. Note that the "in all other cases" amalgamates the cases in which the argument is a truthless utterance with that in which *it is not an utterance.* Frege thus achieves total generality, as p could *not* even be a proposition. The Polish logicians introduced an operator for affirmation which Lesniewski inopportunely named '*assertium*'. See Vernant, *Du Discours à l'action,* ch. 2, p. 30.

[62] "I Der Gedanke", p. 35, "II Die Verneinung", p. 59 and "Über Sinn und Bedeutung", p. 49, footnote 2: "Ein Urteil ist mir nicht das blosse Fassen eines Gedankens, sondern die Anerkennung seiner Wahrheit". Of course Frege held back from interpreting this "recognition of the truth of a thought" in terms of a recognition of the truth of a sentence by the speaker ("II, Die Verneinung", p. 63): "The judging individual no more creates a thought when he recognises its truth than the hiker creates the mountain he is climbing". My aim here is not to re-cast Frege's philosophy of logic in a pragmatic light, but only to find the outline of a pragmatic analysis in the writings of 1918.

[63] "I, Der Gedanke", p. 35.

[64] Cf. § 2, Judgment, p. 11-12.

[65] "II Die Verneinung", p. 63, footnote n° 1: "Den Sprachgebrauch des Lebens trifft man wohl am besten, wenn man unter einem Urteile eine Tat des Urteilens versteht, wie ein Sprung eine Tat des Springens ist".

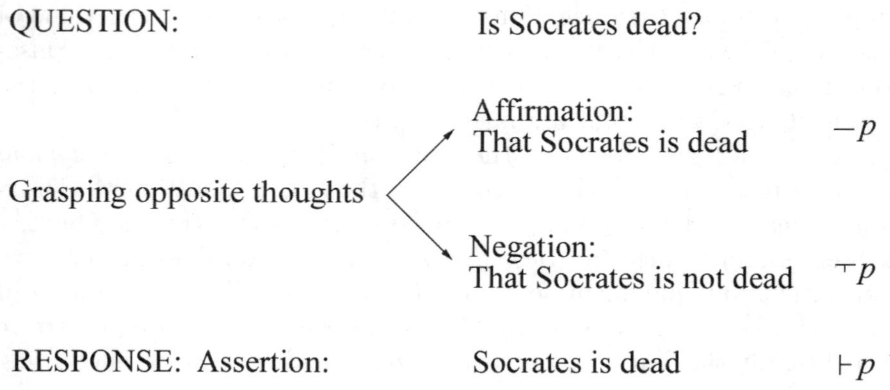

Figure 1. Grasping a thought/assertion of judgment according to Frege

past. An act involves an agent, and the act is not entirely known when the agent is not known".[66]

2.2 Illocutionary logic and force

Thanks to Austin, Frege's English translator, this analysis of judgment and assertive force was soon to constitute the paradigm for an analysis of *illocutionary force* in speech-act theory.[67] Subsequently, Searle defined the act of assertion as one speech act among others on the basis of its illocutionary point, its direction of fit and a specific set of conditions. According to Vanderveken, this definition of assertion in centred on a form of *illocutionary point* characterised by the direction of fit from words to things. Assertion thus "represents a state of affairs as obtaining" and constitutes a propositional content which is governed that by a *preparatory condition*, requiring that "the speaker has reasons to believe in the truth of the propositional content", and a *sincerity condition* under which "the speaker believes the propositional content".[68]

Finally, the differences between the various forms of judgment have to be clarified. Searle, following Frege, distinguishes the assertion of a proposition: $\vdash p$ and the assertion of the negation of a proposition:

[66] *Ibidem.*
[67] *How to do Things with Words*, Eighth Conference.
[68] *Les Actes de discours*, p. 127.

$\vdash\sim p$.[69] To this logical forms of negation, Searle adds a *specifically pragmatic* form of negation which bears on the assertive act itself: *illocutionary negation*, symbolised by: $\neg \vdash p$.[70]

Although such illocutionary negation was not anticipated by Frege,[71] it was anticipated by Russell's use of the term *denial* as early as his critique of Meinong in 1904: "To deny a proposition is not the same as to affirm its denial. The case of assumption can make this clear. Given any proposition *p*, there is an associated proposition not-*p*. Either of these may, as Meinong points out, be merely supposed or assumed. But when we deny *p*, we are not concerned with a mere assumption, and there is nothing to be done with *p* that is logically equivalent to assuming not-*p*. *A direct inspection, I think, will show that the state of mind in which we reject a proposition is not the same as that in which we accept its negation.* Again, the law of excluded middle may be stated in the form: If *p* is denied, not-*p* must be asserted; this form, it is true, is too psychological to be ultimate, but the point is that it is significant and not a mere tautology. Logically, the notion of denying a proposition *p* is irrelevant: it is only the truth of not-*p* that concerns logic. But psychologically, it would seem, there are two states of mind which both have *p* for their object, one affirming and the other denying; and two other states of mind, having not-*p* for their object, one affirming and the other denying".[72] In the "pre-phenomenological" Meinongian context, Russell is naturally led to distinguish between: 1° – truth and falsehood, 2° – an affirmative proposition and its negation, 3° – belief and disbelief, that is to say the assertion of a proposition and its denial. But in those days, such an analysis would inevitably be accused of psychologism and ejected from

[69]The term "deny" persists in 1985 in *Foundations of Illocutionary Logic*, (p. 183). Daniel Vanderveken introduced the correct terminology in 1990 (*Meaning and Speech Acts*, p. 170), distinguishing *assert, negate* and *deny.*

[70]*Les Actes de langage*, 2.4, p. 71. At the time (in 1969), Searle did not distinguish the two types of negation symbolically. Here I adopt the symbolism subsequently proposed by Vanderveken.

[71]See Dummett, *Frege, Philosophy of Language*, ch. 10, p. 316. In 1988, in *Les Actes de discours*, ch. VI, 1, Vanderveken only admitted negation alongside affirmation: "denying a proposition simply means affirming its negation" (p. 168). But this is because he excluded complex illocutionary acts, such as denial, from the outset (see ch. 1, 2, D, p. 30).

[72]MTCA, p. 41. I use italics to emphasise the relevant section. In this passage, Russell often uses *affirming* for *asserting* and *denial* for *negation.* We noted that in Searle and Vanderveken writings there is similar waivering on the expression of these concepts. Peter Hylton, who incidentally points out the opposition between denial and assertion in MTCA, simply adds: "His new concern [with the psychological] represents the beginning of a shift not so much in doctrine as in interest", *Russell, Idealism and the Emergence of Analytic Philosophy*, ch. 6, p. 245. I believe that what we see here is a fruitful intuition on behalf of Russell, one that he was not able to exploit *logically.*

the field of Logic strictly speaking.[73] The pragmatic approach to assertion that accounts for the subject's acts in the framework of discourse allows us to dig up the Russellian distinction that takes *denial* to be the negative form of the act itself of assertion, and not the negative form of the act's propositional content.

But one could go further and, as Russell suggested as early as 1904, attempt to build a logic that does not only account for propositional negation but also for the pragmatic denial of all forms of illocutionary force. In formalizing Searle's Speech Act theory, Daniel Vanderveken's illocutionary logic admits truth-functional operators for affirmation and negation: p and $\sim p$. But over and above this, Vanderveken also authorises *illocutionary denial*[74], written $\neg$A, which generally constitutes the pragmatic negation of a given illocutionary act. Now, although propositional negation complies with all the usual laws of standard logic, the same thing is not true of illocutionary negation. Although illocutionary negation satisfies non-contradiction, it satisfies neither the law of excluded middle nor the law of reduction: $\neg\neg A \neq A$. From this, Vanderveken draws the conclusion that "Illocutionary negation in this respect resembles intuitionistic negation since in intuitionist logic it is also not valid that $P \vee \sim P$ and that $(\sim\sim P \rightarrow P)$".[75] Unfortunately this is not pursued further. But it constitutes a first step for the integration of pragmatic operators of illocutionary force – in this case, those of assertion and denial – into the object-language, and follows in the spirit of Frege's teachings.[76] What is gained is the possibility of formally expressing Russell's proposition concerning the law of excluded middle (If p is denied, non-p must be asserted) as follows: $\neg \vdash p \rightarrow \vdash \sim p$. I propose that this formula, which appears to constitute the first expression of a properly pragmatic law, be baptized "Russell's Law".[77]

However, we still need to understand and formalize the operational value of denial. Here, once again, Russell's intuitions can be of use. Although, for the reasons explained further up, Russell could not have developed a logic of denial, he nevertheless broached its "psychological" aspect by returning to the problem of *disbelief*. Russell's first fundamen-

[73] *Ibid.*, p. 74.

[74] The expression is introduced in 1985 in *Foundations*, ch. 7, p. 152.

[75] *Ibid.*, p. 154.

[76] In *Les Actes de discours*, ch. II, p. 71, Vanderveken points out that he is following in the Fregean tradition by keeping the illocutionary force markers in the object-language. This overcomes the strictures on formalism.

[77] Having said this, it should not be forgotten that Russell was only able to analyse denial in psychological terms of disbelief. This being the case, we can find other areas that show in a more direct manner Russell's relation to the field of pragmatics, the best known being the analysis of egocentric particulars (see *An Inquiry into Meaning and Truth*, ch. VII).

tal idea consists *not* in defining disbelief as the mere negation of belief, but rather in setting it up alongside belief as a new undefinable: "disbelief is a new unanalysable relation, involving merely rejection of the proposition disbelieved, and not consisting in acceptance of its contradictory".[78] Later, Russell was to link the feeling of disbelief to a formal operation of *rejection*: "Rejection of a proposition is, psychologically, inhibition of the impulses which belief in the proposition would generate; it thus always involves some tension, since the impulses connected with belief are not absent, but are counteracted by an opposing force".[79] Finally, in *Human Knowledge*, Russell pointed out that "disbelief [is] a state just as positive as belief".[80]

The idea of the positivity of disbelief, which can appear to be contradictory, is crucial from a pragmatic point of view. Here, we would be well-advised to distrust words – privative prefixes in particular – as well as symbols.[81] *Disbelief is not a simple negation of belief just as denial is not the simple negation of assertion.* In order to account for the curiously positive nature of disbelief and denial, it is appropriate to set up four rubrics for commitment: one can assert (and believe) p, assert (and believe) $\sim p$, deny (reject the belief of) p, deny (reject the belief of) $\sim p$. This would explain why illocutionary negation resembles intuitionistic negation.

We are now in a position to appreciate the degree of complexity of a logic of assertion and denial. I will limit myself to pointing out that if we take the positive nature of denial seriously, combining affirmation/negation and assertion/denial as Russell proposed to do, we then obtain the four forms of judgment as depicted in the following figure. This diagram includes Russell's Law in (1) and another law: $\neg \vdash \sim p \rightarrow \vdash p$, that can be deduced by a simple substitution of $\sim p$ for p in (1).

[78] *Theory of Knowledge*, ch. IV, p. 142. In "On the Nature of Truth and Falsehood", disbelief is admitted as a fundamental "cognitive act", p. 150. In *Analysis of Mind*, ch. XII, Russell mentioned a "feeling of disbelief". More importantly, he once again encountered the difference between considering and believing propositions, and dealt with it simply by appealing to a "feeling of assent". See also our *Bertrand Russell*, §44, p. 167–172.

[79] *An Inquiry into Meaning and Truth*, ch. XVIII, p. 255.

[80] Ch. IX, p. 142.

[81] In this sense, the symbol "$\neg \vdash p$" adopted by Vanderveken can be misleading. It may be better to use the definition $\dashv p =_{Df} \neg \vdash p$ which, apart from having abbreviative value, would avoid confusion, see Vernant, "Pour une logique dialogique de la dénégation"

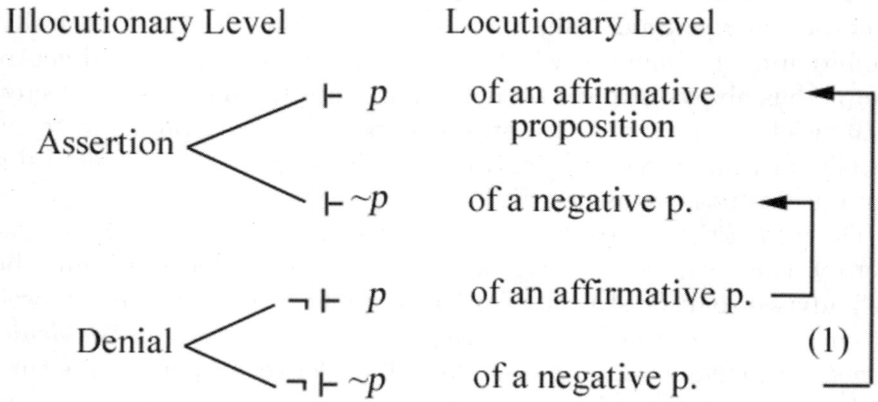

Figure 2. Forms of judgment.

3. Conclusion

It is clear that the adoption of illocutionary logic dissolves the stricture placed on formalization by Russell. By using new possibilities opened up by logics such as those that result from Montague's grammars, illocutionary logic makes possible the formal *expression* of the informal analyses inherited from Austin. Thus, the pragmatic analysis of assertion has the effect of releasing us from the strictures of logical formalism that Russell once encountered and which forced him to relegate his forward-looking analysis of denial to the realm of psychology.

Similarly, the approach in terms of *acts* possessing *assertive force*, inaugurated by Frege, allows us to lift the philosophical aporias that burdened the treatment of assertion from the *Principles of Mathematics* to *Principia Mathematica*, resulting in a profound overhaul of our understanding of assertion.

This, however, does not mean that all problems have been resolved:

1° – illocutionary logic, however promising it may be, is far from complete. In particular, the logic of assertion and denial has barely begun to make the first steps.[82] Research in this area is still too recent for us to possess stable symbolism and standardised axiomatics.

2° – More importantly, the analysis inaugurated by Frege, taken up by Austin, and pursued by Searle and Vanderveken, focuses too exclusively on the acts of single speakers. The congenital fault of this kind of analysis is its *monological* character. The later Wittgenstein showed quite clearly that assertions cannot not be meaningful outside the context of language games in which at least two players took turns in producing them. Even before Wittgenstein, Peirce had drawn up a properly dialogical analysis of assertion.[83] As Frege, too, had clearly understood,[84] assertions are *responses* to foregoing questions. The assertion must therefore be analysed as an *interact* that takes up a distinctive dialogical function.[85] Similarly, denial is not the simple negation of a potential assertion, but a posture of rejection, the outright *refusal* of an implicit proposition, if that proposition has not previously been put forward. Russell takes special note of this in 1940, when he turns denial into a dialogical act pertaining to what he calls "secondary language": "Suppose, for example, you have taken salt by mistake instead of sugar, and you exclaim "this is *not* sugar". This is a denial, and belongs to the secondary language."[86] This dialogical dimension of denial makes possible powerful *rhetorical* effects. Recall the example of Nixon who, when campaigning as candidate for the position of Governor of California, murmured ingratiatingly that he *refused to believe* that his political opponent was

---

[82]I return to this in my article "Pour une logique dialogique de la dénégation". First of all, such a "logic" presupposes a working definition of assertion, a theory of the various combinations of oppositions such as locutionary negation illocutionary denial, and an analysis of the role of illocutionary connectors. This raises an issue about the interpretation of the notation $\vdash p \supset p$, which can be read in Russell's terms of assertion and implication but which, in natural language, also admits an interpretation in terms of the *conditional assertion* of the consequent, see Quine, *Méthodes de logique*, ch. 3, p. 29. Generally, as Dummett points out, betting that "If A, then B" is not the same thing as betting "If A, bet that B", *Frege,* ch. 10, p. 341. For more on the different interpretations of conditionals see Sainsbury, *Logical Forms* ch. 3, p. 103-132. For the symbolisation of conditional speech acts, see Searle & Vanderveken, *Foundations*, ch. 7, § VIII, p. 157 to 160.

[83]J. Brock, "An Introduction to Peirce's Theory of Speech Acts"; C. Chauviré, "Peirce, le langage et l'action, sur la théorie peircéenne de l'assertion"; C. Tiercelin, *La Pensée-signe*, ch. V, p. 296-306.

[84]This point was alluded to earlier. The "Logische Untersuchungen, I & II" begin by situating the problem in question and answer games. The scientific approach is described as a search for answers to questions (p. 184), and a proviso is given: "The very nature of the question demands that we separate our grasp of meaning from that of judgment", p. 55 and "The response to a question is an assertion, based on a judgment, whether the question receives a positive or negative response", p. 54. Hence the format of Figure 1 *supra*.

[85]See Vernant, *Du Discours à l'action,* ch. IV, p. 58-85.

[86]See *An Inquiry into Meaning and Truth,* ch. IV, p. 64. I have pursued Russell's analysis of belief and disbelief, in particular their relation to language and dialogue, in "*From belief to disbelief*: lecture trans- et interactionnelle des phénomènes de croyance chez Russell."

Communist.[87] This denial enabled him to insinuate the proposition denied in an underhand way. This kind of usage shows how much more subtle the logic of assertion and denial is than may at first appear.

Our treatment of assertion, then, stands in need of some improvement. I have contributed to that task elsewhere.[88] In this article, my intention was to show, taking Russell as an example, the crucial role played by the emergence of assertion as a theme for the construction of standard logic, as well as the impossibility of overcoming the philosophical and technical aporias raised by assertion within the original framework of a strictly logical approach.

The concept of assertion raises issues that cannot be answered unless we adopt a resolutely pragmatic and dialogic framework. The complexity that was originally reponsible for assertion's status as something of a monster is today responsible for its richness, a richness that Russell himself had caught a glimpse of.

References

Austin J. (1962). *How to do Things with Words.* Oxford University Press.

Brock J. (1981). "An Introduction to Peirce's Theory of Speech Acts". *Transaction of the C.S. Peirce Society, XVII, 4*:319–326.

Carroll L. (1894). "What the Tortoise Said to Achilles". *Mind, N.S. IV.*

Chauviré C. (1979). "Peirce, le langage et l'action, sur la théorie peircienne de l'assertion". *Les Études philosophiques, 1*: 3–17.

Dummett M. (1973). *Frege, Philosophy of Language.* London: Duckworth. Second edition 1981.

Frege G. (1879). *Begriffsschrift*, L. Nebert, Halle, . English trans. in *From Frege to Gödel*, by Jean van Heijenoort. Harvard University Press, 1967.

— (1980) "Sinn und Bedeutung", *Zeitschrift für Philosophie und philosophische Kritik*, 100, 1892, reed. in *Funktion, Begriff, Bedeutung*, von Günther Patzig. Gottingen: Vandenhoeck & Ruprecht.

— (1993) "Logische Untersuchungen, I; Der Gedanke, II; Die Verneinung", *Beiträge zur Philosophie des deutschen Idealismus*, 1918–1919; reed. in *Logische Untersuchungen,* von Günther Patzig Gottingen: Vandenhoeck & Ruprecht.

Geach P. (1972). *Logic Matters.* Oxford: B. Blackwell.

[87]See Ch. Perelman, *L'Empire rhétorique*, ch. V, p. 60. For a litterary example, see my analysis of Act 3, Scene III of *Othello* in *Du discours à l'action*, ch. IV, p. 83.

[88]See *Du Discours à l'action,* ch. VIII, § 4, p. 166-168.

Griffin N. (1980). "Russell on the Nature of Logic (1903–1913)". *Synthese, 45*:117–188.

Hylton P. (1990). *Russell, Idealism and the Emergence of Analytic Philosophy.* Oxford: Clarendon, Press.

Miéville D. (1984). *Un développement des systèmes logiques de Stanislaw Lesniewski, Protothétique, Ontologie, Méréologie*. Berne: Peter Lang.

O'Leary D. (1989). "The Propositional Logic of *Principia Mathematica* and some of its Forerunners", in *Antinomies & Paradoxes*, Studies in Russell's Early Philosophy, éd. I. Winchester & K. Blackwell. McMaster U.L.P., 92–115.

Perelman C. (1997). *L'Empire rhétorique*, 3e éd. Paris: Vrin.

Quine W.V. (1941)."Whitehead and the Rise of Modern Logic", in *The Philosophy of A.N. Whitehead*, ed. A. Schilpp. Library of Living Philosophers, vol. III.

— (1950). *Methods of Logic.*

Russell B. (1903). *Principles of Mathematics* [PoM]. London: G. Allen & Unwin. 2nd ed. 1983.

— (1904). "Meinong's Theory of Complexes and Assomptions" [MTCA], I, II & III, *Mind*, 1904, N.S. 13, reed in *Essays in Analysis*, ed. D. Lackey. London: G. Allen & Unwin, 1973.

— (1906). "The Theory of Implication", *American Journal of Mathematics,XXVIII*:159–202.

— (1910). "La théorie des types logiques", *Revue de Métaphysique et de Morale, XVIII.* Reed. in *Cahiers pour l'analyse, 10*:53–83. Seuil.

— (1910). "On the Nature of Truth and Falsehood", ch. VII de *Philosophical Essays.* London: G. Allen & Unwin, 1966.

— (1913). *Theory of Knowledge*, ed. E. R. Eames & K. Blackwell, *The Collected Papers of B. Russell*, vol. VII, 1984.

— (1921). *The Analysis of Mind.* London: G. Allen & Unwin.

— (1940). *An Inquiry into Meaning and Truth.* London: G. Allen & Unwin, 1940. London: Routledge, 1992.

— (1948). *Human Knowledge.* London: G. Allen & Unwin.

Russell B. & Whitehead A. *Principia Mathematica* [PM], first ed. vol I, 1910, vol. II, 1912, vol III, 1913, second ed. vol. I, 1925, vol II & III, 1927, Paperback Edition to *56. Cambridge U.P, 1973.

Sainsbury M. (1979). *Russell.* London: Routledge & Kegan Paul.

— (1991). *Logical Forms, An Introduction to Philosophical Logic.* Oxford: Blackwell.

Searle J. & Vanderveken D. (1985). *Foundations of Illocutionary Logic.* New-York:Cambridge University Press.

Searle J. (1969). *Speech Acts.* Cambridge U.P.

Tiercelin C. (1993). *La Pensée-signe.* Nîmes: Ed. J. Chambon.

Vanderveken D. (1988). *Les Actes de discours*. Bruxelles: Mardaga.

— (1990–91). *Meaning and Speech Acts*, vol. 1, *Principles of Language Use*, 1990; vol 2, *Formal Semantics of Success and Satisfaction*. Cambridge U.P.

van Heijenoort J. (1967). "Logic as Calculus and Logic as Language". Reed. in *Selected Essays* Naples: Bibliopolis, 1985.

Vernant D. (1986); *Introduction à la philosophie de la logique* Bruxelles: Mardaga.

— (1993). *La Philosophie mathématique de Russell*. Paris: Vrin.

— (1997). *Du Discours à l'action*. Paris: PUF.

— (2003). "Pour une logique dialogique de la dénégation", in *Du Dialogue au texte, autour de Francis Jacques*, eds. F. Armengaud, M.-D. Popelard, D. Vernant. Paris: Kimé.

— (2003). "*From Belief to Disbelief*: lecture trans- et interactionnelle des phénomènes de croyance chez Russell", in La Croyance en question, *Psychologie de l'interaction*. Nancy.

— (2003). *Bertrand Russell*. Paris: Flammarion, GF n°1192.

— "La genèse logique du concept de dénégation de Frege à Slupeski", in *Philosophie & logique en Pologne (1918–1939)*, Roger Pouivet ed. Paris: Vrin. In press.

Vuillemin J. (1968). *Leçons sur la première philosophie de Russell*. Paris: A. Colin.

Wittgenstein L. (1963). *Tractatus logico-philosophicus*, reed. Suhrkamp Verlag.

— (1953). *Philosophische Untersuchungen*. Oxford: Blackwell.

— (1969). *Über Gewissheit*. Oxford: Blackwell.

IV

AGENCY, DIALOGUE AND GAME-THEORY

Chapter 14

AGENTS AND AGENCY IN BRANCHING SPACE-TIMES*

Nuel Belnap
Pittsburgh University

Abstract *Branching time* puts an indeterminist causal structure on instantaneous but world-wide "super-events" called *moments*. This theory has "action at a distance" as an inevitable presupposition. In contrast, *branching space-times* puts an indeterminist causal order on tiny little *point events*. The benefit of branching space-times theory is that it can represent local indeterminist events with only local outcomes. The "seeing to it that" or "stit" theory of agency developed in Belnap, Perloff and Xu, 2001 employs branching time as a substructure, and thus has the following shortcoming: It is inevitably committed to an account of action-outcomes that makes them instantaneously world-wide. This essay asks how the stit theory of agency can be adapted to branching space-times in such a way that action-outcomes are local.

The aim of this essay is to make some suggestions for the beginnings of a theory of agents and agency in branching space-times. The thought is to combine the ideas of *agency* as developed by Belnap, Perloff and Xu, 2001 against the relatively simple background of branching time with the richer notions of *mere* indeterminism as structured in the theory of branching space-times. My plan is to say a little about agency in branching time and a little about branching space-times, and then ask how the two can be brought together.

1. Stit theory

In this section I offer some brief and general remarks in connection with "stit theory" as described at length in Belnap, Perloff and Xu, 2001

*An earlier version of this essay appeared in the Journal of Sun Yatsen University (Social Science edition), vol. 43, 2003, pp. 147–166.

D. Vanderveken (ed.), Logic, Thought & Action, 291–313.

(henceforth FF).[1] In the course of these remarks, I will occasionally refer to the following designedly boring example.

Peter will drive his car to work one day hence. [1]

What are the chief elements of research strategy for FF? The central aim is to use formal theory to help a little in understanding human agency as it works itself out amid the (we think) indeterministic causal structures of our world. We call it "stit theory." It takes its name from the prominent use of a non-truth-functional connective:

α sees to it that Q,

which we abbreviate with

[α *stit:* Q]

In connection with stit, the analysis moves in several directions, as indicated in the remainder of this section.

1.1 Stit theses

In an effort to tie the formal grammar of stit to natural language, FF offers a series of so-called "theses." The theses, which I paraphrase from the appendix to FF, are these:

- *Agentiveness of stit thesis.* English is ambiguous, and even the English "α sees to it that Q" is ambiguous, but [α *stit:* Q], in contrast, is designed to be unambiguous: It invariably attributes agency to α.

- *Stit complement thesis.* As a matter of grammar, Q can be any sentence whatsoever, including cases in which [α *stit:* Q] turns out trivially false, e.g. [α *stit:* the sun rises every morning].

- *Stit paraphrase thesis.* An English sentence is an agentive if and only if it can be usefully paraphrased as a stit sentence, including the case in which Q is paraphrased as [α *stit:* Q], e.g. [Peter *stit:* Peter drives his car to work].

- *Imperative content thesis.* The content of every English imperative is agentive, hence equivalent to a stit sentence; e.g., the declarative content of "Peter, drive your car to work" is [Peter *stit:* Peter drives his car to work].

[1]See also publications listed there, and in particular, Horty, 2001.

- *Restricted complement thesis.* For many English constructions, e.g. "α promises that__," it is illuminating to restrict their complements to future-tensed agentives, hence to stit sentences; e.g., Peter promises that one day hence [Peter *stit:* Peter will drive his car to work].
- *Stit normal form thesis.* When worried about difficult questions of agency, it is generally useful to paraphrase each agentive as a stit sentence.

Together these theses say that in thinking about the philosophy of action, it is helpful to use [α *stit:* Q] as a normal form for expressing that an agent does something — something that natural languages express in a bewildering variety of fashions. As applied to [1], this gives

One day hence:[Peter *stit:* Peter drives his car to work]. [2]

Note that we have used "*One day hence:*" as a connective in order to express the desired tense structure with complete explicitness.[2] This is important in being clear about indeterminism.

1.2 Stit theory and indeterminism

Stit theory assumes *indeterminism.* Why? Stit theory is equally pre-humanist and pre-scientific. We take it as a fact that our agentive doings together with non-agentive happenings are enmeshed in a common causal order. The sense of "causal order" here should not be picked up from a determinist presupposition. The appropriate causal order is indeterministic. Nor does this involve "laws." Why should it? Bare indeterminism is what is wanted: There are initial events in our world for which more than one future is possible. It is this simple concept of indeterminism that permits progress on the theory of agency. Thousands of pages written by philosophers about agency make (or seem to make) the contrary assumption of strict determinism. Hume and Kant are famous in modern philosophy for what has come to be called "compatibilism", the doctrine that agency is compatible with strict and absolute and perfect determinism. On the other hand there are also hundreds of pages (I'm making up these numbers) assuming some form of indeterministic "free will"; one has to think only of William James and his rejection of the idea of a "block universe." Rarely, however, is an indeterministic view of agency combined with a desire to let the enterprise be guided by a desire for *mathematically rigorous theories.* The aim of our research is

[2]We have dropped the "will" of [1] as logically redundant.

to make an indeterminist account of agency "intelligible" (Kane, 1998). We try to pursue this aim by means of a simple and rigorous theory.

1.3 The metaphysics of agency

The simplest representation of indeterminism is this: A treelike structure that opens toward the future. That is, FF, following Prior and Thomason, represents indeterminism by means of a partial order in which there is no backward branching. We call the elements of the tree *moments.* A moment is an entire slice of history, so to speak, from one edge of the universe to the other. A moment is not a "time"; it is a concrete "super-event" (Thomson, 1977) caught up in the causal order, with its unique past and its future of possibilities. A maximal chain of moments is called a *history*, and represents one particular fully-detailed way in which our world might go or might have gone. Do not in your mind identify a single history with a "world"; instead, it is the entire tree-like assemblage that is to be identified with *Our World.* Philosophical logicians generally use the term *branching time* for such a structure, and we shall do the same, even though "branching histories" would be less misleading. The right side of Figure 1 gives a picture.

In FF we argue at length against the pernicious doctrine that one of the many possible histories through a moment has a special status as "actual" or "real." FF shows that our understanding of assertion, prediction, promising, and the like is much improved if all histories through a moment are treated on a par, with none singled out.

One reaches the "theory of agents and choices in branching time" by adding two elements to the bare theory of branching time. (1) *Agent* is a set whose members are considered to be agents capable of making choices and so acting. (2) *Choice* is a function that is defined on all moments m and agents α, and which delivers a partition of all the histories through m. The partition represents the choices open to agent α at moment m. We may call each member of the partition *a choice that is available to* α at m, or *a possible choice for* α at m. There is one serious theoretical constraint on *Choice*: If two histories h_1 and h_2 in the tree do not split from one another until *after* a certain moment m, then no choice available to α at m can distinguish h_1 and h_2. This we call "no choice between undivided histories." As we often say, you cannot make tomorrow's choices today.

When there are multiple agents under consideration, one naturally (?) assumes that the *simultaneous* choices of two agents α_1 and α_2 are radically independent; technically, we assume that every choice by α_1 is consistent with every choice by α_2 when their choices are simultaneous.

That's it. As you can see, our "metaphysical" theory of the causal structure of agency is about as minimal as can be.

1.4 The semantics of indeterminism

Simple as it is, the branching-time structure does not explain itself, especially not to philosophers who assume determinism as an unquestioned "fact." One has to come to terms with what prediction, and more generally the use of the future tense and other kinds of statements "about" the future, mean in connection with the tree representation of indeterminism. This is a matter not of metaphysics, but of semantics. The problem is how to understand certain linguistic structures under the assumption that they are used in an indeterministic situation. The solution to that problem, including agency as a special case, lies in combining the Prior-Thomason semantics for branching time, in which truth is relativized to both moments and histories, with the Kaplan semantics for indexical expressions.

The Prior-Thomason-Kaplan semantics makes the truth of e.g. [1] relative not only to the moment of utterance, but in addition relative to a moment-of-evaluation parameter needed for understanding tense constructions that take one away from the moment of utterance, and, crucially, relative also to a history-of-evaluation parameter that represents a way in which the future can unfold. The semantics is entirely two-valued, since relative to each moment of utterance, moment of evaluation, and history of evaluation, there is a definite truth value given to the example sentence. If a sentence is true [false] relative to every history through a given moment of evaluation, the sentence is said to be *settled true* [*false*] at that moment. Picture [1] or [2] as asserted by someone on Monday in a situation such that whether or not Peter will drive to work one day hence is still an open question. The semantics, as explained in FF (and in more detail in Belnap, 2002a), makes it possible to say precisely: When asserted on Monday, [1] is neither settled true nor settled false. However, it *is* a settled truth on Monday that, no matter what happens, one of the following holds: Either on Tuesday it will be settled true that [1] was true at the moment of assertion (on Monday), or on Tuesday it will be settled false that [1] was true at the moment of assertion (on Monday). With slightly more brevity (but still with inevitable risk of confusion): The assertion of [1] is not settled one way or the other on Monday; but no matter what happens, on Tuesday it will be definitely settled whether or not the assertion was true at the moment of assertion.

This confusing passage involves "double time references," which are sorted out both in FF and in Belnap, 2002a. It may also be called "bivalence later" (Copley, 2000). The subtle part is that [1] is evaluated with respect to a Monday moment m_0 that is the moment of assertion, and with respect to each history h that passes through not m_0, but instead through some Tuesday moment m_1 at which we are evaluating whether or not the assertion *was* true at the moment of assertion on Monday. In this way, the theory provides a framework for normatively evaluating assertions about the future in terms of what we may call "fidelity": We say that an assertion is *vindicated* or *impugned* at some later moment depending on whether it is settled true or settled false at that later moment that the assertion was true at the moment of assertion.

1.5 The semantics of agency

On the basis of the Prior-Thomason semantics that relativizes truth to moment-histories pairs, we suggest two principal semantic analyses of stit. These analyses depend essentially on an indeterminist view of agency. Both share the idea that action begins in choice, and that there is no choice and therefore no action without the availability of incompatible choices. We work out the semantics of stit in two ways. Both involve a transition from an unsettled situation to settledness due to the choice of an agent.

- The *achievement stit* postulates that the relevant choice occurred at some temporal remove in the past of its settled outcome.

- The *deliberative stit* works through a concept of action based on the choice being in the *immediate* past of its settled outcome.

In either case, the semantics says that [α *stit: Q*] is true at a certain moment m with respect to a certain history h provided (1) a prior (perhaps immediately prior) choice of α guaranteed the truth of Q, and (2) the choice was a real choice among incompatible alternatives. Put technically, the truth conditions for [α *stit: Q*] at a pair m/h when taken as the deliberative stit come to this:

1-1 DEFINITION. *(Stit) Positive condition.* Q is true at m with respect to every history h_1 that belongs to the same possible choice for α at m as does h. (This is the part that says that α's choice at m guarantees the truth of Q.) *Negative condition.* Q is not settled true at m: There is some history h_2 to which m belongs such that Q is false at m with respect to h_2. (This says that α really does have a choice at m that is relevant to the truth of Q.)

The semantic analysis that FF gives to the achievement stit is less satisfying. I nevertheless give a brief version here. First of all, one is required to add a "time" parameter (FF says *Instant*) to the underlying tree structure, with the assumption that all histories are temporally isomorphic. This amounts to the presumption that one can make sense out of comparing the "times" of incompatible moments. The achievement stit relies on this, letting the "positive condition" be, roughly, that [α *stit:* Q] is true at m/h iff Q is settled true at every moment that (1) is co-temporal with m and (2) lies on a history that is in the same choice for α at m as h.

1.6 Strategies

Stit theory allows a sophisticated but simple account of *strategies*, conceived of as a pattern of prescribed choices for future action. In this part of the theory, the principle of "no choice between undivided histories" in modeling agency is critical. The principle, you will recall, denies that our world allows us to choose today between histories that only divide tomorrow. In a nutshell: To choose a strategy is not to choose the later choices that define it.

It follows that we must carefully distinguish the *availability* of a strategy from *having* a strategy. In other words, the stit theory of strategies makes it imperative that one distinguish "acting *in accord with* a strategy" from "acting *on* a strategy." This is a special case of the following: The fundamental notions of stit and strategies may serve as a kind of foundation for intentional notions. If you start that way with the indeterminist causal structure of agency, you *easily* see how difficult it is to sort out the intentional notions in relation to causal notions in any way that keeps them sorted out, and you are more likely to avoid self-deception in appraising your theories.

Almost breathlessly I have highlighted five parts of the stit theory of agency: the stit theses, how stit theory presupposes indeterminism, the metaphysics (causal structure) of agency as we see it, the difficult semantics of indeterminism in general, the truth conditions for stit, and, finally, the stit theory of strategies. It is this apparatus that I suggest should inform our discussion of how agency fits into branching space-times, and I shall presuppose it when I turn to this matter in §3. First, however, I drop agency in order to survey some main ideas of branching space-times.

2. Branching space-times

There is much less written on branching space-times (BST, as I shall say), and it is less accessible, in part because most essays on this topic were written with physical examples such as quantum mechanics in mind.[3] Perhaps the easiest approach to BST starts from a Newtonian picture of the world of events.

Newtonian universe. Non-relativistic and deterministic: World = Line. The Newtonian universe has, as I see it, two features that are so fundamental that they can be described without advanced mathematics, using only the characteristics of the causal-ordering relation and its relata. First, the items that are related by the before-after causal order are momentary (= instantaneous) super-events: Laplace's demon needs total world-wide information concerning what is going on at time t. Second, the causal order is strictly linear: For any two momentary events m_1 and m_2, either m_1 lies in the causal past of m_2, or vice versa: m_2 lies in the causal past of m_1. It is the *linearity* of the causal order that answers to *determinism*, and it is the *global* conception of the items falling into a causal order that answers to *non-relativistic* "action at a distance": An adjustment in positions or momenta here-now can *immediately* call for an adjustment over there in the furthest galaxy. The picture of the causal order in a Newtonian universe is therefore a simple line, with each point representing a world-wide simultaneity slice, all Nature at a certain time t. In such a universe there are not histories (plural), but only a single History, so that we may say that World = History; this makes the Newtonian universe *deterministic.* Furthermore, and independently, the relata of the causal ordering are momentary super-events; this make the Newtonian universe *non-relativistic.* With this in mind, we may say that on the Newtonian view, World = Line as on the left side of Figure 1.

Branching space-times is to be both indeterministic and relativistic. Since the Newtonian universe is both deterministic and non-relativistic, it takes two independent moves to make the transition from the Newtonian universe to branching space-times.

Branching time universe. Non-relativistic but indeterministic: World = Many Lines. In the first move away from Newton

[3] BST theory in the form deriving from Belnap, 1992 is discussed in the following places: Szabo, Belnap, 1996, Rakić, 1997, Belnap, 1999, Placek, 2000a, Placek, 2000b, Belnap, 2002e, Mueller, 2002, Placek, 2002b and Belnap, 2002b. Belnap, 2002d gives an overall view of both stit theory and BST theory.

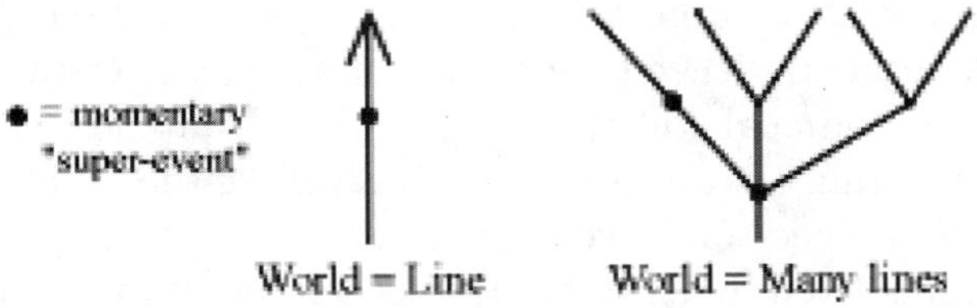

Figure 1. Newtonian universe and Branching-time universe

we *keep* the relata of the causal order as momentary super-events, so that we remain non-relativistic. In order to represent indeterminism, however, we *abandon* linearity in favor of a treelike order. The result of this first transition from the Newtonian universe, when taken alone, is exactly what we have already discussed under the rubric "branching time." In branching time there is indeed a single world, but instead of the equation World = Line, the world of branching time involves many line-like histories, i.e., many possibilities: Branching time is indeterministic. Since, however, we have kept the causal relata as momentary super-events, branching time remains non-relativistic: Splitting between histories in branching time has to be a world-wide matter of "action at a distance" since the consequences of the split are felt instantaneously throughout the farthest reaches of space. We may therefore say that according to branching time, World = Many Lines that split at world-wide momentary super-events as on the right side of Figure 1.

Einstein-Minkowski universe. Relativistic but deterministic: World = Space-time. The other move away from Newton is that made by Einstein in principle, and more explicitly by Minkowski. To obtain the Einstein-Minkowski universe from that of Newton, we *keep* determinism from the Newtonian universe; there is no trace of alternative possible futures. The change is rather that now the terms of the causal relation are no longer simultaneity slices (momentary super-events) that stretch throughout the universe. Instead, the fundamental causal relata are *local* events, events that are limited in both time-like and space-like dimensions. When fully idealized, the causal relata are *point events* in space-time. This, to my mind, is the heart and soul of Einstein-Minkowski relativity. The move to local events is made necessary by Einstein's argument that there is simply no objective meaning for a simultaneity slice running from one edge of the universe to the other. There is no "action at a distance": Adjustments at e_1 influence only events e_2 in "the forward light cone" of e_1, or, as will say, in the causal future of e_1. I wish to urge that not only fancy Einstein physics, but even our ordinary experience (when uncorrupted by uncritical adherence to Newton or mechanical addiction to clocks and watches) shows us that events are not strung out one after the other. Take the event of our being here now at e_1. Indeed some events lie in our causal future, so that there are causal chains from e_1 to them, and others lie in our causal past, so that the causal chains run from them to e_1. But once we take local events as the relata of the causal order, there is a third category, always intuitive, and now scientifically respectable, since we have learned to be suspicious of the idea of (immediate) action at a distance. In this third category are local events e_2 that neither lie ahead of e_1 nor do they lie behind e_1 in the causal order. Letting $<$ be the causal order relation, I am speaking of a pair of point events e_1 and e_2 such that neither $e_2 < e_1$ nor $e_1 < e_2$, Instead, e_1 and e_2 have a space-like relation to each other. Neither later nor earlier (nor frozen into simultaneity by a mythical world-spanning clock), they are "over there" with respect to each other. Einstein makes us painfully aware that space-like relatedness is non-transitive, which is precisely the bar to the objective reality of momentary super-events. Events in their causal relation are not really ordered like a line. Our modern reverence for various parts of Newtonian physics and our related love of clock time delude us.

Since the Einstein-Minkowski relativistic picture is just as deterministic as the Newtonian picture, there are no histories (plural), but only History, so that we have the determinist equation World = History. The difference from the Newtonian picture is with respect to an independent feature: A causally ordered historical course of events can no longer be conceived as a linear order of momentary super-events. Instead, a history is a relativistic space-time that consists in a manifold of point events bound together by a Minkowski-style causal ordering that allows that some pairs of point events are space-like related. Therefore, if we make the single transition from the Newtonian universe to that of Einstein-Minkowski, the result is that World = Space-time as on the left side of Figure 2.

Branching space-times: relativistic and indeterministic: World = Many Space-times. BST now arises by suggesting that the causal structure of our world involves both indeterminism and relativistic space-times; we are therefore to combine two independent transitions from the Newtonian universe. We can already make a certain amount of capital out of that suggestion. For *indeterminism*, we shall expect not World = History but World = Many Histories. For *relativistic* considerations, we shall expect that each history is not a line, but instead

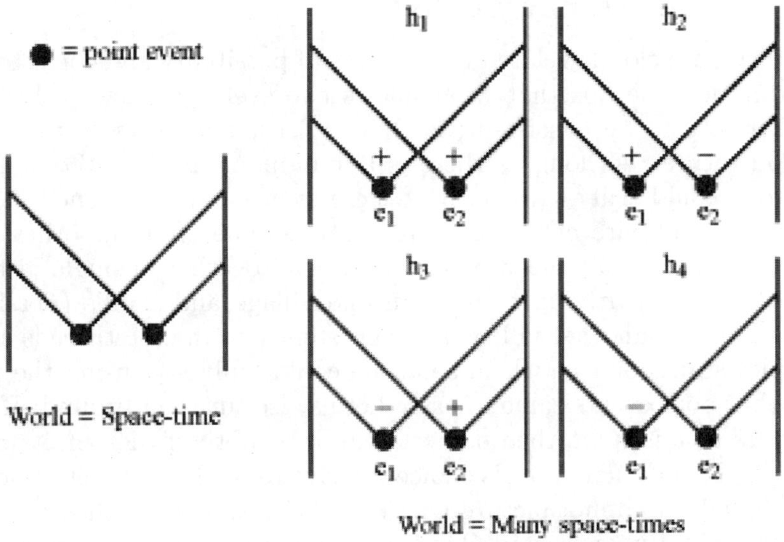

Figure 2. Einstein-Minkowski universe and Branching-space-times universe

a space-time of point events in something like the Einstein-Minkowski sense. Therefore, in BST we should expect that World = Many Space-times as on the right side of Figure 2. Furthermore, just as histories in branching time (each of which is like a line) split at a world-wide momentary super-event, so in branching space-times we should expect that histories (each of which is like a space-time) should split at one or more point events. Technically, we represent our world as *Our World*, which is a set of (possible) point events, and we let e range over *Our World.*

We need, however, more information about how the various histories (= space-times) fit together. What is analogous here to the indeterministic way in which branching-time structured individual Newtonian line-like histories into a tree? A crucial desideratum is this: The theory should preserve our instinct that such indeterminism as there is in our world can be a local matter, a chance event here-now that has no effect on the immediate future of astronomically distant regions of the universe. It needs zero training in mathematics to see that the theory of how BST histories fit together into a single world will be more complicated than the branching-time theory that arranged many lines into a single tree. I mention four key points underlying the theory of "branching space-times."

2.1 Consistency in BST

Histories are closely related to the ideas of possibility and consistency. A guiding idea here is that what allows two events to share a history, and therefore to be consistent, is that at least one event lies in their common future. As long as there is a standpoint in *Our World* from which one could truly say "both of these events have happened," even if the two events are not themselves arranged one after the other, one may be confident that the two events can live together in single history. Peter's (possible) driving to work in one village and Paul's (possible) staying home in another village are consistent just in case there is some (possible) standpoint at which someone could truly say, using the past tense, that both events came to pass. Let us also turn this around: If two events are *inconsistent*, then no event can have both of them in its past. For example, although Peter's choice to drive to work is altogether local, and although we cannot picture *Our World* as a tree, nevertheless, there is no standpoint anywhere in *Our World* that has in its *past* both of the inconsistent events represented by Peter's having driven to work and his having stayed home. These inconsistent possibilities can and must lie ahead of the point at which he makes the choice, but they cannot both lie behind anything whatsoever. A single maximal consistent set of point events will be a space-time; we call it a *history*, we let *Hist* be the set of all histories in *Our World*, and we let h range over *Hist*.

2.2 Choice points local, not global

There have to be choice points, definite local events at which two histories split into radically inconsistent portions. It is presumably not true and we must not assume that when splitting occurs, it occurs in some magical world-wide way. When Peter is given the choice to drive or to stay home on a certain occasion, that occasion is confined in space as well as in time. His little bit of free will is local, not global. And the same might be true when the choice is only metaphorical, a matter of a random outcome of some natural event such as, perhaps, the decay of a radium atom in Paris. It might be that the decay is a strictly local matter, neither influencing nor influenced by contemporary happenings in, say, Manhattan. Whenever there is indeterminism, whether of choice or of chance, a good theory must give meaning to the difficult idea that the indeterminism is local, not worldwide.

Important in BST theory are the same ideas of undividedness and of splitting that are so important for the FF theory of agency. For later reference we indicate some pieces of notation that are useful in speaking of these matters.

2-1 DEFINITION. *(($H_{(e)}$, $h_1 \equiv_e h_2$, $\Pi_e\langle h\rangle$, Π_e, and $h_1 \perp_e h_2$))* $H_{(e)}$ is the set of all histories to which e belongs. $h_1 \equiv_e h_2$ means that histories h_1 and h_2 are undivided at e (they split at some point event that is properly later than e). $\Pi_e\langle h\rangle = \{h_1 : h \equiv_e h_1\}$ is the immediate possibility at e to which h belongs, and Π_e is the set $\{\Pi_e\langle h\rangle : h \in H_{(e)}\}$ of all immediate possibilities at e. $h_1 \perp_e h_2$ means that h_1 and h_2 split exactly at e.

2.3 Prior choice principle

When I put the third point in everyday language, it sounds so obvious that you will yawn. And yet as far as I know a thoroughly controlled statement has never been made apart from the present theory of branching space-times. The BST postulate for locating choice points may be put informally in the following way: Whenever we find ourselves as part of some contingent event instead of in a history that is an alternative to the occurrence of that event, we may always *look to the past* for a choice that serves as a locus of the splitting. This is an axiom of BST theory; I call it the *prior choice principle.*

Example. Suppose that on a certain Tuesday Peter is in his office, having driven to work that morning. Think of this as a particular concrete event, with a definite causal past, and let the event be contingent, i.e., an event that has not been fated from all eternity. Take any history in which that event fails to occur, perhaps a history in which Peter spends the whole of Tuesday high in a tree in the Amazon jungle. The theory guarantees that if you look in the causal past of the given concrete Peter-in-office event (the one that we are supposing occurred), you will find a definite choice point at which things could have gone either in the direction of keeping the Peter-in-office event possible, or in the direction of keeping the Peter-in-tree history possible.[4] You do not need to look in the future, and you also do not need to look far away at events going on "over there." The point is that examination of the causal past of the Peter-in-office event suffices. (The theory does not presume to say if the choice point belonged to Peter or to a lion or to a bit of natural randomness or, perhaps, to some combination.)

[4]The theory will not let you exchange "event" and "history" here; precision of statement is essential.

2.4 Funny business?

The fourth point records a recognition of an important way in which the theory should *not* be strengthened. The following is a principle that an unwary philosopher might easily be inclined to endorse.

Tempting principle (no funny business). If two choice points are related in a space-like way, so that the second is "over there" with respect to the first, then their respective choices are entirely independent of each other. This I call "no funny business." Aristotle gives us a simple example of a causal story without funny business. He tells us how two market-goers meet at the market, as an "accidental" result of their choices that morning, made separately in far-away villages. Their individual choices, we all suppose, are bound to be totally uncorrelated, that is, independent, and that is what the tempting principle of no funny business says *must* be so.

The theory of branching histories, however, resists this temptation. And it does not do so *a priori.* It does so because *Our World* seems, as a matter of fact, to contain violations of the tempting principle. With reference once more to Einstein, quantum mechanics seems to tell us that in fact it is possible for two utterly random choice points to be space-like related, with no hint of a line of causal connection between them, and nevertheless fail to be independent. This is "funny business." In Belnap, 2002e it is shown that this form of no-funny-business is provably equivalent to the following form: *Every* situation that is cause-like with respect to a certain outcome event lies in the causal past of that event.[5]

There is more. Reichenbach has taught us that whenever we find the long arm of coincidence stretching across space, it is in our nature to look for a common cause. Funny business precisely happens when there is objective coincidence — which is to say, a failure of independence — across space, without a common cause. Belnap, 2002c shows that two forms of "no common cause funny business" are equivalent to the two accounts of "funny business" of Belnap, 2002e. Since modern-day physics apparently says that funny business happens, it is good that the theory of branching histories has room for funny business — and indeed has the virtue of permitting us to offer a stable (albeit conjectural) account as to the difference between (1) mere indeterminism but with no funny business and (2) indeterminism with funny business.

[5]If you replace "every" by "some," you have a different statement, one that is serious business instead of funny business, and one that is provably guaranteed by the aforementioned prior choice principle.

2.5 Summary of BST

These necessarily too-brief points allude to a theory of branching histories that gives a satisfying account of how physical indeterminism can be local instead of global. It gives us an account of how choices and outcomes of natural random processes can affect only what lies in their causal future, touching neither their past nor the vast region of space-like related events. But it does so in such a way as to allow plenty of room for individually random space-like-related processes to be, as the physicists say, "entangled." Or, in the phrase I just used, the theory of branching histories helps us to come to terms with funny business.

The result is that the theory of branching histories, in addition to helping us clarify ideas of action and agency, provides low-key suggestions for articulating some of the strangest phenomena uncovered by contemporary physicists. It does this by avoiding careless or fuzzy or sloppy formulations. It does this by insisting on a careful and rigorous account of what it is for indeterminism to be not immodestly global, but modestly local.

3. Agents and agency in branching space-times

Since branching space-times is more realistic and more sophisticated than branching time as a representation of the indeterminist causal structure of our world, it is natural that one should consider agents and agency in branching space-times. The matter is hardly understood at all, which is precisely why it should be investigated. I offer some tentative thoughts on how some of the fundamental ideas might go.

Agents in branching space-times. First off, what shall we do with the concept of an agent? In FF the concept is represented by a set, *Agent*, and to be an agent, α, is merely to be a member of this set. Nothing is said about the "real internal constitution" of an agent beyond membership in the set *Agent*. Instead, the FF postulates describe agents only insofar as agents make choices. FF presumes that every agent has a choice at every moment (for technical convenience counting vacuous choices as choices), where the construction $Choice^{\alpha}_{m}(h)$ represents the choice that agent α makes at moment m on history h. Since the choices of a set of agents Γ at the same moment are to be taken as simultaneous, it is natural that FF postulates such a set of choices to be independent: No combination of individual choices for the members of Γ is impossible. Behind these postulates lie the FF idea that, since moments are taken to be world-wide in extent, many agents can "occupy" a single moment. Recall §2: The chief difference between non-relativistic branching *time* and relativistic branching *space-times* lies in what "the causal order relation" relates. In the former case, *moments* are super-events taken to be "spatially" rich enough to be "occupied" by more than one agent. In the latter case, *point events* would seem to be so small as to admit the "presence" of at most one agent. For this reason it seems reasonable to begin by representing an agent as a set of point events, the set of point events that the agent may be thought of as occupying in the course of his or her life.[6] Continuing to use *Agent* as the set of agents, we may therefore begin with the following.

3-1 TENTATIVE POSTULATE. *(The agent as a set of point events)* Every agent is a set of point events: $\forall\alpha[\alpha \in \mathit{Agent} \rightarrow \alpha \subseteq \mathit{Our\ World}$.[7] When $e \in \alpha$, we may say that the point event *e is part of* the agent α, or that α *is located at* or *occupies* e.[8]

Given that we are going to represent an agent α as a set of point events, what constraints make sense? In the beginning it seems best, since easiest, to think of the life of an agent in a particular history as a portion of a "world line."

3-2 TENTATIVE POSTULATE. *(Agents and world lines)* The portion of the life of an agent in a particular history is a chain of point events: $\forall\alpha\forall h[(\alpha \in \mathit{Agent}\,\&\, h \in \mathit{Hist}) \rightarrow \alpha \cap h$ is a chain in *Our World*.[9]

Since point events in α can belong to many histories, and although the portion of α in each is a chain, it is easy to see (and is provable) that the entire set α will look like a tree, which accurately represents that there are alternative future possibilities for the life of α.[10] I hope it is needless to say that I am claiming for this postulate only that its simplicity makes it a good beginning; it may well turn out to be better to represent an agent in a single history as a cloud of point events rather than as a chain. The thought is that nothing more that Tentative postulate 3-2 is

[6] In contrast, in branching time it would make no sense at all to represent an agent as a set of moments!

[7] In this study I am not after a "fundamental" or "exclusive" ontology of agents. Here a representation counts as useful if its structure leads us to helpful theory.

[8] None of these phrases is entirely happy, but given that in any event we are not trying to say what agents "really are," perhaps it doesn't matter.

[9] A chain in *Our World* is a subset of *Our World* such that each pair of distinct members are comparable by the causal ordering relation. A chain may be empty.

[10] I counsel you to have no patience with those who make fun of this picture by (falsely) describing it as saying that it is a possibility for tomorrow that Peter both be in his office and also stay at home.

desirable for "first purposes." For example, it may be that the track of an agent in any one history is dense or continuous; but at this stage of inquiry I doubt that it matters one way or the other.

Let us think of the joint representation of two or more agents. Since point events are so small, it seems plausible that one should never have more than one agent located at a point event. Perhaps, then, we should enter the following.

3-3 TENTATIVE POSTULATE. *(No agent overlap)* Agents never overlap: $\forall\alpha_1\forall\alpha_2[(\alpha_1, \alpha_2 \in \mathit{Agent}\,\&\,\alpha_1 \neq \alpha_2) \rightarrow \alpha_1 \cap \alpha_2 = \emptyset]$.

That requirement may, however, be too strong; it might be better to say only that no *nonvacuous* point event (no point event with more than one possibility in its immediate future) can be shared by two agents. The weaker restriction would not entirely forbid that the "world lines" of two agents intersect.

Choices and stits in branching space-times. The foregoing tentative postulates suggest (but certainly do not demand) that the choices open to an agent α at a point event e are exactly the same as the immediate possibilities at e in the sense of BST theory. In other words, there may well be no need for imposing a *Choice* function in addition to the possibilities definable from the BST causal ordering alone. Clarity will be heightened, however, if we introduce the *Choice* notation for agents as a separate primitive governing agents in branching space-times.

3-4 NOTATION. *(Choice)* Assuming that agent α occupies a point event e belonging to a history h, $Choice^{\alpha}_{e}(h)$ is a new primitive to be read as "the choice available α at e to which h belongs." Also $Choice^{\alpha}_{e}$, defined when $e \in \alpha$, as $\{Choice^{\alpha}_{e}(h)\colon h \in H_{(e)}\}$, is to be read as "the set of choices available to α at e," and *Choice* is defined as that function defined for every agent/point-event pair α and e such that $e \in \alpha$ that delivers the set of choices available to α at e.

Introducing $Choice^{\alpha}_{e}(h)$ as a primitive leaves open the possibility that some interesting ideas turn out to need *Choice* as a separate concept. Since Π_e (Definition 2-1) is the BST notation for the set of immediate possibilities at e, defined in terms of undividedness at e, the following postulate is to the effect that when a point event e is a part of agent α in history h, then there is no difference between the choice available at e for α that contains h, and the immediate possibility at e to which h belongs.

3-5 TENTATIVE POSTULATE. *($Choice^{\alpha}_{e}(h)$ and $\Pi_e\langle h\rangle$)* Let an agent α occupy a certain point event e. The undividedness-at-e relation between histories of *Our World* determines what is immediately possible at e. We postulate that there is no difference between what is immediately possible at e and what α can choose at e: $\forall e \forall \alpha \forall h[e \in (\alpha \cap h) \rightarrow Choice^{\alpha}_{e}(h) = \Pi_e\langle h\rangle]$.

This postulate is not without content: It says that the choices concerning the immediate future that are available to an agent at a certain point event e are very finely articulated indeed. According to Tentative postulate 3-5, there is no finer articulation that Nature (or other agents) can impose beyond the choosing powers of the agent himself for his immediate future. On the other hand, and with equal realism, choices and happenings in the near vicinity of e can and doubtless will limit the powers of the agent concerning his non-immediate future. Joint agency, for example, will not concern what is immediately possible, as it does in FF; instead, the outcomes of joint agency will come to pass at some spatio-temporal remove from the choices of the various agents involved. Such is of the very essence of the relativistic flavor of agents in branching space-times.

Already we can spell out a notion of agent responsibility for immediate outcomes along the lines of FF. Once we adapt to BST by letting truth be relative to pairs e/h (with $e \in h$) rather than pairs m/h, the definition of the deliberative version of stit seems forced (compare Definition 1-1): [α *stit:* Q] is true at e with respect to h iff $e \in \alpha$ and

3-6 DEFINITION. *(Stit in BST) Positive condition.* Q is true at e with respect to every history $h_1 \in Choice^{\alpha}_{e}(h)$. (This is the part that says that α's choice at e guarantees the truth of Q.) *Negative condition.* Q is not settled true at e: There is some history $h_2 \in H_{(e)}$ such that Q is false at e with respect to h_2. (This says that α really does have a choice at e that is relevant to the truth of Q.)

Agent causation in branching space-times. There should be a rich possibility for notions of agent causation of outcomes adapted from Belnap, 2002b. Let $\mathbf{O}$ be a "scattered outcome event," defined as a consistent set of outcome chains (chains that are nonempty and lower bounded). A "cause-like locus" for $\mathbf{O}$ is a point event e whose occurrence is consistent with the occurrence of $\mathbf{O}$ and which is such that what happens there makes a difference to the occurrence of $\mathbf{O}$: $\exists h[h \in (H_{(e)} \cap H_{\langle \mathbf{O} \rangle}) \,\&\, h \perp_e H_{\mathbf{O}}]$. Funny business happens when some cause-like locus for $\mathbf{O}$ is *not* in the causal past of $\mathbf{O}$, and we know (or, via standard quantum mechanical puzzles going back to Einstein, we think we know) that funny business happens in the natural world. Whether choices of agents can exhibit funny business is presumably not something to be decided by fiat; nevertheless, it seems much the best to begin exploring agency in the absence of funny business, and so we have to be able to say what that means. I am far from sure of the best thing to mean by "no agentive funny business," but the following, though probably inadequate, seems like a reasonable first suggestion to be pondered.

3-7 DEFINITION. *(Agentive funny business)* A pair consisting of an outcome event $\mathbf{O}$ and a point event e counts as *agentive funny business* iff e is part of some agent α and e is a cause-like locus for $\mathbf{O}$ and e is not in the past of any part of $\mathbf{O}$.

3-8 TENTATIVE POSTULATE. *(No agentive-funny-business pairs)* There are no pairs $\mathbf{O}$ and e that count as agentive funny business.

It seems that Tentative postulate 3-8 suffices to rule out at least some cases of superluminal agent causation, and is in the vicinity of saying that the choices of agents make a difference only to the future. It is doubtful, however, that the postulate is strong enough to rule out all forms of agentive funny business; more analysis is needed than we have so far provided. A further strengthening might consider a *joint* choice initial involving an arbitrarily complex (but consistent) set of choice points, each of which belongs to some agent. It would seem that the whole complex should be independent of any joint choice initial, no matter how complex, and no matter if agentive or natural or mixed, as long as the purely agentive joint choice is space-like related to the other complex.

Transitions are important in BST. We can be interested in an "effect" transition $\mathbf{I}\mapsto\mathbf{O}$, which is an ordered pair of an initial event I and a scattered outcome event $\mathbf{O}$, with every member of the initial in the causal past of some member of the outcome. Such a transition is "contingent" if there is a dropping off of histories in the course of the transition, and in such a case we may with profit ask for a causal account of the matter. The answer is to be given in terms the set $cc(\mathbf{I}\mapsto\mathbf{O})$ of all of the *causae causantes* or "originating causes" of $\mathbf{I}\mapsto\mathbf{O}$. For this concept we first define the set $cl(\mathbf{I}\mapsto\mathbf{O})$ of "cause-like loci" for $\mathbf{I}\mapsto\mathbf{O}$ as $\{e\colon \exists h[I\subseteq h \,\&\, h\perp_e H_{\langle\mathbf{O}\rangle}]\}$. These point events are exactly where "the action happens" in keeping $\mathbf{O}$ possible at the expense of ruling out altenative possibilities. Then $cc(\mathbf{I}\mapsto\mathbf{O})=\{e \mapsto (H_{(e)} \cap H_{\langle\mathbf{O}\rangle}) : e \in cl(\mathbf{I} \mapsto \mathbf{O})\}$; these are the transitions that "do the work."

In simple cases, which are the only ones that are so far well understood, we shall find all of $cl(\mathbf{I} \mapsto \mathbf{O})$ in the past of $\mathbf{O}$, with no funny business. Often $cc(\mathbf{I} \mapsto \mathbf{O})$ will involve a mixture of agentive choices and non-agentive happenings; for example, the winning may have been partly caused by how the coin came up (e.g. heads came up), and partly by which bet was chosen (e.g. the bet was on heads). We know that given no funny business, the set of $cc(\mathbf{I} \mapsto \mathbf{O})$ form a set of "inns" conditions for $\mathbf{I} \mapsto \mathbf{O}$: Each member is an insufficient but non-redundant part of a necessary and sufficient condition (whence the acronymn "inns") for the occurrence of the outcome $\mathbf{O}$ given the occurrence of the initial *I*. Non-redundancy means in particular that the complete causal story for $\mathbf{I} \mapsto \mathbf{O}$ cannot leave out any member of $cc(\mathbf{I} \mapsto \mathbf{O})$. Furthermore, we can partition $cl(\mathbf{I} \mapsto \mathbf{O})$ into those cause-like loci that belong to α_1, α_2, etc., and those that do not belong to any agent. In this way we may expect fine control over our causal descriptions. For example, if $cl(\mathbf{I} \mapsto \mathbf{O}) \subseteq \alpha_1 \cup \alpha_2$, we can say that the two agents α_1 and α_2 were between them entirely responsible for the transition $\mathbf{I} \mapsto \mathbf{O}$.

Message-sending in branching space-times. Such joint responsibility is not of course the same as joint action. Our primitives, since being purely causal, do not permit discussion of factors such as "joint intention" that may be thought to underlie joint action. We can, however, say a little bit about the causal aspect of the communication that seems to be required for joint action. Look at Figure 3.

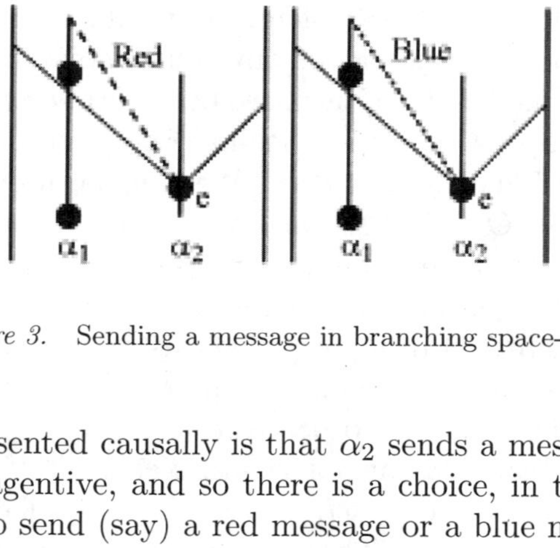

Figure 3. Sending a message in branching space-times

What is represented causally is that α_2 sends a message to α_1. Message-sending is agentive, and so there is a choice, in this case a choice by α_2 whether to send (say) a red message or a blue message. There is indeterminism in the life of *each* agent, but only the sender is agentive in the matter; α_1 is entirely passive. If the two are to "collaborate," one would expect that a message sent by α_2 to α_1 would (not dictate but) limit the choices open to α_1 at a subsequent choice point, depending on red vs. blue; and of course further collaboration would involve agency on the part of α_1 in choosing a message to α_2. If an "intentional" story involving the minds of α_1 and α_2 cannot be told as an enrichment of the bare causal tale told by Figure 3, then it is hard to see how it could be useful.

It is striking that Tentative postulate 3-8 says nothing about non-agentive funny business such as occurs, or seems to occur, in EPR-like funny business. I should think that it is essential to think through the combination of "no agentive funny business" with "some EPR-like non-agentive funny business." Even physicists and philosophers of physics seem to distinguish between the independence constraints put on the chance outcomes of measurements on the one hand and on the deliberate settings-up of measurements. There is, however, too little talk about why there should be a difference, and its exact nature. It is plausible that the structure of agency in BST can help in thinking about this.

Choice and state in branching space-times. One last thought. In developing an evaluative theory of choices by an agent α in branching time, Horty, 2001 introduced the idea of a "state," which Horty defined in effect by freezing the *simultaneous* choices of all agents except for the target agent α. Since in branching time the simultaneous choices of distinct agents are always independent (every combination is possible), he was in a position to consider the familiar two-dimensional diagram in which rows represent the choices available to α, columns represent the various possible "states" (independent of the choices of the agent), and the intersections are labeled with evaluative information. In transporting this idea to BST, one must consider from the beginning that space-like relatedness in BST, unlike simultaneity in branching time, is definitely *not* a transitive relation, any more than is the same relation in Minkowski space-time. It follows that a choice by an agent α can well be space-like related to each of two choices by a second agent, those two choices being causally ordered (successive), so that although each is independent (by space-like relatedness) of the choices of α, they are not independent of each other. Worse (if that is the right word), branching space-times allows that choice points for other agents, even though space-like related to the given choices of α, are outright inconsistent with each other. One is therefore going to have to put in extra work — and work that should be carried out — in order to locate a satisfying account of "state" that is eligible to do the same work as Horty's account in branching time. It seems to me that in working through this nest of considerations, one would naturally be led to a useful theory of "games in branching space times" that would take with utmost seriousness the causal structure of the players and the plays in a fashion that sharply separates (as von Neumann's theory does not) causal and epistemic considerations. It is one thing for two choices to be independent of each other, and a different thing for there to be certain relations of (say) ignorance. So much is clear. It is not of course self-evident that the difference makes a difference. If it does not, however, that should be the result of serious reflection, not an unconsidered presupposition.

4. Summary

I have tried to sketch some of the main ideas of stit theory, drawing on FF, and I have given a kind of overview of branching-times theory as discussed in the publications listed in note 3. Then I have raised but definitely *not* answered some questions concerning how to use these materials in order to fashion a theory of agents and their choices in branching space-times. I have suggested with all too much brevity some possible lines of investigation.

References

Belnap N. (1992). "Branching Space-Time". *Synthese* 385–434.

— (1999). "Concrete Transitions." In *Actions, norms, values: Discussions with Georg Henrik von Wright.* G. Meggle, Ed. Berlin: Walter de Gruyter. 227–236. A "postprint" (2002) may be obtained from `http://philsci-archive.pitt.edu`

— (2002a). Double Time References: Speech-Act Reports as Modalities in an Indeterminist Setting. In Wolter *et al*, 37–58. A preprint of this essay may be obtained from `http://www.pitt.edu/~belnap`

— (2002b). A Theory of Causation: *Causae* causantes (Originating Causes) as inus Conditions in Branching Space-Times. A preprint of this essay may be obtained from `http://philsci.archive.pitt.edu`.

— (2002c). No–Common-Cause EPR-Like Funny Business in Branching Space-Times. Forthcoming in *Philosophical studies.* A preprint of this essay may be obtained from `http://philsci-archive.pitt.edu`

— (2002d). Branching Histories Approach to Indeterminism and Free Will. This essay may be obtained from `http://philsci-archive.pitt.edu`

— (2002e). EPR-like 'Funny Business' in the Theory of Branching Space-Times. In Placek and Butterfield 2002, 293–315. A preprint of this essay may be obtained from `http://philsci-archive.pitt.edu`

Belnap N., Perloff M. and Xu M. *Facing the Future: Agents and Choices in our Indeterminist World.* Oxford: Oxford University Press. 2000.

Copley B. (2000). "Conceptualizing the Futurate and the Future". In *MIT working papers in philosophy and linguistics: The linguistics / philosophy interface*, Bhatt R., Hawley P., Kackl M. and Maitra I. eds. Cambridge MA.: MIT Press. 45–72.

Horty J.F. (2001). *Agency and Deontic Logic.* Oxford: Oxford University Press.

Kane R. (1998). *The Significance of Free Will.* Oxford: Oxford University Press.

Müller T. (2002). Branching Space-Time, Modal Logic and the Counterfactual Conditional. In Placek and Butterfield, 273–291.

Placek T. (2000a) "Stochastic Outcomes in Branching Space-Time: Analysis of Bell's Theorem". *British journal for the philosophy of science, 51*:445–475.

— (2000). *Is Nature Deterministic?.* Kraków: Jagiellonian University Press.

— (2002b). Partial Indeterminism is Enough: a Branching Analysis of Bell-Type Inequalities. In Placek and Butterfield 2002, 317–342.

Placek T. and Butterfield J. (20020). *Non-Locality and Modality.* Dordrecht: Kluwer Academic Publishers.

Rakić N. (1997). Common-Sense Time and Special Relativity. Ph. D. thesis. University of Amsterdam.

Szabo L. and Belnap N. (1996). Branching Space-Time Analysis of the GHZ Theorem. *Foundations of physics, 26*, 8:989-1002.

Thomson J. (1977). *Acts and Other Events.* Ithaca: Cornell University Press.

Wolter F., Wansing H., de Rijke M. and Zakharyaschev M. "Advances in Modal Logic". *World Scientific Co. Pte. Ltd, 2002:3.* Singapore.

Chapter 15

ATTEMPT, SUCCESS AND ACTION GENERATION: A LOGICAL STUDY OF INTENTIONAL ACTION*

Daniel Vanderveken
Université du Québec, Trois-Rivières

Abstract Contemporary philosophers have broadly studied intentional actions that agents attempt to perform in the world. However, logicians of action have tended to neglect the intentionality proper to human action. I will present here the basic principles and laws of a logic of individual action where intentional actions are primary as in contemporary philosophy of action. In my view, any action that an agent performs unintentionally could in principle have been attempted. Moreover any unintentional action of an agent is an effect of intentional actions of that agent. So my logic of action contains a theory of attempt and of action generation. As Belnap pointed out, action, branching time and historic modalities are logically related. There is the liberty of voluntary action. I will then work out a logic of action that is compatible with indeterminism.

Propositions with the same truth conditions are not the contents of the same attitudes of human agents. For that reason I will exploit the resources of a non classical modal and temporal predicative propositional logic capable of distinguishing the contents of intentional actions which are different. My primary purpose is enrich the logic of agency so as to adequately characterize attempts, intentional actions and the different kinds of action generation.

I will only consider here *individual actions* that a single agent performs at one moment. Examples of such actions are intended body movements

*A first draft of this paper has been published in the special issue on *Mental Causation* of *Manuscrito Vol XXV*, 2002 pp 323-356). I thank *Manuscrito* for granting permission to republish the paper here. I am also grateful to Elias Alves, Nuel Belnap, Jean Caelen, Paul Gochet, Hans Kamp, J-Nicolas Kaufmann, André Leclerc, Ken MacQueen, Raymond Klibansky, Michel Paquette, Giovanni Queiroz, John Searle, Philippe de Rouilhan, Candida Jaci de Sousa Melo and Denis Vernant for their critical remarks.

D. Vanderveken (ed.), Logic, Thought & Action, 315–342.

like voluntarily raising the arm, some effects of these movements like touching something, mental actions like judgements and elementary illocutionary acts such as assertions and questions which are performed at one moment of utterance. Individual actions performed at a single moment are part of all other kinds of action: they are part of longer actions like deliberations which last during several moments of time and of collective actions like debates performed by several agents.

In my ideal language, formulas representing actions are of the canonical form: individual agent a *does that* A (or *acts so as to bring about that* A), where A represents what the agent does (the content of his or her action). In order to contribute to the foundations of the logic of action, I will attempt to answer general philosophical questions: What is the logical form of proper intentional actions? What are their success conditions? And what are the logical relations that exist between our intentional and unintentional actions? Some types of action contain other types of action. An agent cannot perform an action of the first type without performing an action of the second type. Thus it is not possible to shout without producing sounds. Moreover certain action tokens *generate* others in certain particular circumstances. An agent who expresses at a moment an attitude that he or she does not have lies. He or she could be sincere at another moment. What are the basic laws governing agentive commitment and action generation? In particular, how can an agent perform certain actions by way of performing other actions? Are all actions performed by an agent at a moment generated by a single *basic intentional action* of that agent at that moment? If yes, what is the nature of that basic action? What are the different kinds of agentive generation and how can we explicate them?

Furthermore, what kind of theory of truth do we need in the logic of action? By way of performing actions agents bring about facts in the world. They make true propositions representing these facts. How are success and truth related? Which predications do we make in attributing actions to agents? What is the nature of propositions representing actions? How do we determine in thought their truth conditions?

The structure of this paper is the following. I will first make philosophical remarks regarding the nature of propositions and actions. I will state basic criteria of adequacy for the theory of action and I will try to explicate the intrinsic intentionality of action. In contemporary philosophy of action[1], philosophers are mainly concerned with intentional actions. By definition, *intentional actions* are actions that agents *at-*

[1]See Goldman [1970], Davidson [1980], Searle [1983] and Bratman [1987].

tempt to perform in the world. However, our intentional actions have unintended effects in the world. Thus in walking intentionally on ice an agent might unintentionally slip and fall on the ground. **I will formulate a logic of action where intentional actions are primary** as in contemporary philosophy of action. In my view, any action that an agent performs unintentionally could in principle be intentional. Moreover any unintentional action of an agent is generated by intentional actions of that agent. However, not all unintended effects of intentional actions are the contents of unintentional actions. But only those that are historically contingent and that the agent could attempt to perform. So many events which happen to us in our life are not really actions.

In order to analyze adequately the contents of intentional actions I will use a non classical predicative modal and temporal propositional logic containing that the logic presented in chapter 10. That propositional logic takes into consideration the acts of predication that we make in expressing propositions. It analyzes both their structure of constituents and the effective way in which we understand their truth conditions. So my logic of agency is able to distinguish strictly equivalent propositions which do not have the same cognitive values.

As Belnap [1988,1991] pointed out, action, branching time and historic modalities are logically related. Our intentional actions are not fully determined. Whenever we do something, we could have done something else. Moreover, our present actions can have many different incompatible future effects. So I will use the logic of ramified time that is compatible with indeterminism. According to indeterminism, several incompatible moments of time might follow the same moment in the future of this world. Any moment of time can then belong to several histories representing possible courses of the world with the same past and present but different historic continuations of that moment.

On the basis of my philosophical considerations about truth and action I will further develop Chellas [1992]' and Belnap [1991-2]'s classical logics of agency. I will use a richer ideographic object language containing an additional logical constant of attempt. I will also state important valid laws governing purposes, actions and action generation.

1. Philosophical considerations on proposition and truth

In classical philosophical logic (whether modal[2], temporal[3], intensional[4], agentive[5] or epistemic), propositions are reduced following Carnap [1956] to their truth conditions. So strictly equivalent propositions (which are true in the same possible circumstances[6]) are identified. However it is clear that such propositions are not substitutable *salva veritate* within the scope of verbs of action and attitudes. Whenever we act so as to put a stone on the table, we do not *eo ipso* act so as to bring about that the stone is on the table and a material object in space. In order to act intentionally an agent must know what he or she is trying to do and under which conditions he or she would succeed. We cannot do what we could not intend to do. So the propositional content conditions of intentions and attempts are success conditions of our actions. Any content of a successful action must satisfy these propositional content conditions. Human agents are minimally rational. They never intend to perform actions of bringing about a fact that they know to be unpreventable. So we could not act so as to bring about that an existing stone is a material object in space. For we know that this is necessarily the case no matter what we would do. Similarly we cannot act so as to bring about something in the past. For our intentions are essentially directed towards the present and the future.

From a philosophical point of view, then, we need a criterion of propositional identity stronger than strict equivalence in the logic of action. We cannot identify, as it is commonly done in classical logics of action, each proposition with the set of circumstances in which it is true. We need to consider the structure of constituents of propositions in order to analyze adequately intentional actions. Jocasta is Oedipus' mother. So by way of marrying Jocasta Oedipus *eo ipso* married his mother. However Oedipus did not know then that Jocasta was his mother. So he did not intentionally married his mother when he married Jocasta. In order to account for such facts, I will proceed here to a finer analysis in terms of predication of the logical type of propositions.

---

[2] See R. Barcan Marcus [1993] and S. Kripke [1963].

[3] See Prior [1967], Thomason [1984], Belnap [1992].

[4] See R. Montague [1974].

[5] See the special issue 51 on action of *Studia Logica* in 1992.

[6] In the logic of branching time, possible circumstances are pairs containing a moment of time and a history to which that moment belongs.

I have already presented my logic of propositions according to predication in chapter 10. I will now rapidly repeat its basic principles. Readers who already know them can skip the rest of this section.

We make acts of reference and of predication in expressing propositions. So propositions have a more complex logical structure than truth conditions. First, they have *propositional constituents*: *concepts* which serve to refer and *attributes* (properties or relations) which are predicated. They are composed from *atomic propositions* which attribute properties or relations to objects of reference under concepts[7]. Propositions composed from different atomic propositions are by nature different. We have to make different acts of predication in order to have them in mind. This is why the proposition that a stone is on the table is different from the proposition that it is on the table and in space.

Moreover, in understanding the truth conditions of propositions we do not determine their truth value in all different possible circumstances, as logicians influenced by Carnap wrongly believe. Rather, we only determine that their truth in each circumstance is compatible with certain possible denotation assignments to their constituents and incompatible with others. Thus in understanding an elementary proposition we know that it is true in a circumstance when its unique atomic proposition is true in that circumstance. But we do not *eo ipso* know whether it is true or false in that very circumstance. Simplest atomic propositions are true in a circumstance when the objects which fall under their concepts satisfy their attribute in that very circumstance. However we often refer to an object under a concept without knowing which object falls under that concept. We moreover often do not know which objects of reference possess the properties or entertain the relations that we predicate. So we can assign to expressed concepts and attributes other denotations that they actually have in reality. From a cognitive point of view, atomic propositions have therefore many *possible truth conditions* according to agents. They could be true in a lot of sets of possible circumstances given the different possible denotations that could correspond to their senses in reality. Suppose that in a given circumstance Smith's wife is a suspect (she could have killed Smith) according to the chief of police. Then the atomic proposition that attributes to her the property of being Smith's murderer could be true in that circumstance according to the chief of police. This is an epistemic possibility. From a logical point of view, **each possible truth condition of an atomic proposition**

[7]In my propositional logic, two atomic propositions are identical when they have the same propositional constituents (the same attribute and objects under concepts) and the same truth conditions (they are true in the same circumstances).

corresponds to (and can be identified with) a unique particular set of possible circumstances where that proposition could be true given at least one possible denotation assignment to its attribute and concepts. So any interpretation taking into consideration a number n of possible circumstances has to consider 2^n different possible truth conditions for atomic propositions.

Among all possible truth conditions of an atomic proposition there are of course its *actual truth conditions*, which correspond to the set of all possible circumstances where the objects which fall under its concepts satisfy its attribute. Objects of reference have properties and stand in relations in each circumstance. Atomic propositions have therefore a well determined truth value in any circumstance given the extension of their attribute and concepts and the order of their predication. But we are not omniscient. Our objects of reference could have according to us many other properties and stand in many other relations. **So in our use and comprehension of language we consider a lot of possible truth conditions of expressed atomic propositions** and not only their proper actual truth conditions, as Carnap advocated.

We *a priori* know the truth (or falsehood) of few elementary propositions. For few contain a tautological or contradictory atomic proposition. *Tautological* atomic propositions attribute to an object of reference an property that we *a priori* know that it possesses e.g. that an existing stone is a material object in space. Their only possible truth condition is the set of all possible circumstances. On the contrary, *contradictory* atomic propositions attribute to an object a property that we *a priori* know that it does not possess. Their only possible truth condition is the empty set of all possible circumstances. **Moreover, the truth of most complex propositions is compatible with various possible ways in which objects could be**. Think of disjunctions, past and future propositions, historic possibilities, etc.[8]

As Wittgenstein pointed out in the *Tractatus*, they are however two limit cases of propositions: tautologies that we *a priori* know to be necessarily true and contradictions that we *a priori* know to be necessarily false by virtue of linguistic competence. In my conception of truth, ***tautologies*** are propositions whose truth in any circumstance is compatible with all possible denotation assignments to their propositional

[8]Consider the past proposition that the actual pope was attacked. In order that it be true in a given circumstance, it is sufficient that the actual pope be attacked in at least one previous circumstance. So the truth of that past proposition in any circumstance c is compatible with a lot of possible truth conditions of the atomic proposition attributing to the pope the property of being attacked (namely all those which contains at least one circumstance anterior to c).

constituents. And *contradictions* are propositions whose truth in any circumstance is not compatible with any. Tautologies (and contradictions) are important kinds of necessarily true (and false) propositions for the purposes of the logic of action. For they represent facts that we *a priori* know to be respectively inevitable and impossible.

When the truth of two propositions is compatible with different possible denotation assignments to their constituents, these propositions do not have the same cognitive values. We do not understand in the same way their truth conditions even when they are strictly equivalent and have the same atomic propositions. In other words they represent according to us different facts. So we need in philosophical logic a finer explication of truth conditions than that of Carnap. In particular, **we have to distinguish universally true (and false) propositions** – which are true (and false) in all circumstances – **from tautologies (and contradictions)** composed of the same atomic propositions. Consider the elementary proposition (1) that Oedipus is the son of Jocasta and the tautological proposition (2) that Oedipus is or is not the son of Jocasta. Both are composed from the same atomic proposition which attributes to Oedipus the property of being the son of Jocasta. And they are strictly equivalent. Both are necessarily true. For it is an essential property of any living person to have at any moment a unique mother according to all possible histories. However it is clear that the two propositions in question have different cognitive values. We all *a priori* know by virtue of competence that the tautological proposition (2) is true but we might believe like Oedipus did that Oedipus is not Jocasta's son. The elementary proposition (1) could be false; it is not tautological.

Unlike traditional logic, my logic explains easily such a cognitive difference in terms of predication. The truth of these propositions is not compatible with the same possible truth conditions of their single atomic proposition. In my approach, propositions have then two distinct (but logically related) features. First, they are composed of a finite positive number of atomic propositions. Second, their truth in each circumstance is compatible with a unique set of possible denotation assignments to their propositional constituents.

In the philosophical tradition from Aristotle to Tarski, the truth of a proposition is based on its *correspondence* with reality. In order that a proposition be true in a circumstance, the things which fall under its concepts in that circumstance must be as that proposition represents them in that very circumstance. Otherwise, there would be no correspondence. Along these lines, **a proposition is by definition *true in a circumstance* when its truth in that circumstance is compat-**

ible with the special *denotation assignments* that associate with its propositional constituents their actual denotation in each possible circumstance. As one can expect such denotation assignments determine the actual truth conditions of all its atomic propositions. One can derive from my concise truth definition all the classical laws of the theory of truth.

Speakers often rightly or wrongly believe at a moment that certain objects could fall under concepts and could satisfy attributes in possible circumstances. According to them particular atomic propositions could then be true in certain sets of possible circumstances. Suppose a particular set *Val(a,m)* of possible denotation assignments to propositional constituents is compatible with what the speaker *a* believes at the moment *m*. We can determine which propositions that speaker then *believes to be true*. For we can define exactly the notion of truth according to a speaker in my approach: a proposition *is true in a circumstance according to a speaker a at a moment m* when the truth of that proposition in that circumstance is compatible with all possible denotation assignments *Val(a,m)* that the agent *a* at that moment considers for its propositional constituents. As one can expect, tautological propositions are true and contradictory propositions are false according to all agents who have them in mind. But impossible propositions which are not contradictory can be true and necessary propositions which are not tautological can be false according to some agents at some moments. Moreover whenever the modal proposition that it is then necessary that A is true in a circumstance according to a speaker at a moment that proposition is also true according to that speaker at that moment in all coinstantaneous circumstances. These are basic principles of my epistemic logic.

2. Action, time and modalities in philosophical logic

In order to analyze adequately the logical form of temporal, modal and agentive propositions, we must pay attention to the following facts:

2.1 As regards their structure of constituents

Unlike truth functions, modal, temporal and agentive operations on propositions introduce more atomic propositions. We make new predications in expressing them. Thus in asserting that someone is making the hostages free we attribute to an agent the agentive property of freeing hostages. Prefixes like "en" serve to compose agentive predicates in English. To enable is to make able and to enrich is to make rich. Similarly in asserting that someone is making an attempt to be elected we attribute to him or her the agentive property of being a candidate for an election.

2.2 As regards truth conditions

The truth values of many propositions depend on both moments of time and histories. In the logic of branching time, a *moment* is a possible complete state of the world at a certain instant and the *temporal relation* of *anteriority / posteriority* between moments is partial rather than linear because of indeterminism. On the one hand, the past is unique: each moment m is immediately preceded by at most one past moment m'. Moreover all moments are historically connected: any two distinct moments are preceded by a common past moment. On the other hand, there are multiple future routes: several incompatible moments might be immediately posterior to a given moment. Consequently, the set of moments of time has the formal structure of a *tree-like frame*:

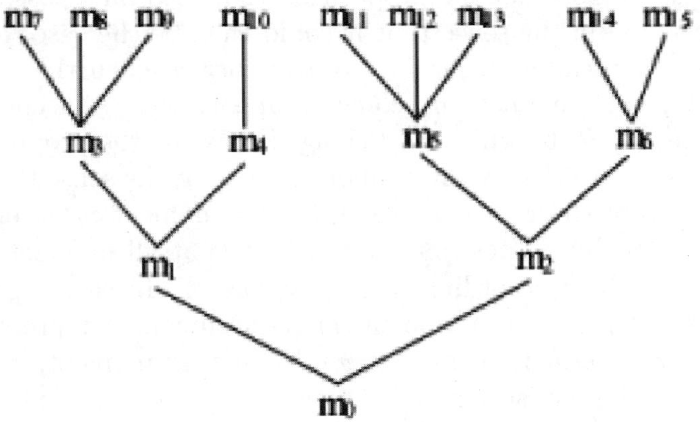

A maximal chain h of moments of time is called a *history*. It represents a *possible course of history of our world*. The truth of certain propositions is *settled at each moment* no matter how that moment continues. So are past propositions because the past is unique. The past proposition that it was the case that A (in symbols: *WasA*) is true at a moment m when A is true at a moment m' anterior to m. Its truth value does not depend on histories. For all histories passing through a moment have the same past at that very moment. The proposition that it is settled that A (in symbols *Settled*A) is by definition true at a moment m according to a history h when the proposition that A is true at that moment m according to all histories to which it belongs. Unlike what is the case for past propositions, the truth of future propositions is not settled at each moment; it depends on which historical continuation h of that moment is under consideration. Like Belnap [1994] let us say that the future proposition that it will be the case that A (in symbols *Will*A) is true at a moment m according to a history h when the proposition that A is true at a moment m' posterior to m according to that very history h.[9]

Two moments of time are said to be *alternative* when they belong to histories which have the same past before these moments. For example, moments m_7, m_8 and m_9 are alternative in the last figure. They represent how the world could be immediately after the moment m_3. The set of all *instants* is a partition *Instant* of the set *Time* of all moments containing exactly one moment of each history and respecting the temporal order of histories. For example, moments m_3, m_4, m_5 and m_6 of the last figure are *coinstantaneous.* They belong to the same instant.

Thanks to instants, the logic of agency can analyze the modal notions of *historic possibility* and *historic necessity* (in the sense now of *inevitability*)[10]. Consider the proposition that *it is then possible that* A (in symbols $\Diamond$A) in the sense that it could then be the case that A. $\Diamond$A is true at a moment m according to a history h when the proposition that A is true at a moment m' coinstantaneous with m according to at least one history h' to which m' belong. Similarly, the proposition that *it is then necessary that* A (in symbols $\Box$A) — in the sense that it could not have been otherwise than A- is true at a moment m according to a history h when the proposition that A is true at all moments m' coinstantaneous with m according to all histories h'. In case a proposition of the form $\Box$A is true at a moment m, its argument A represents a fact that is not only *settled* but also *inevitable* at that moment.

Agents can repeat actions of the same type at different successive moments in a possible course of the world. They can drink and eat again. Agents also perform actions of the same type at alternative moments. Suppose that a player is in a winning position at a moment in a chess game: that player wins the game if he or she plays. In that case the player is a winner at all alternative moments where he or she makes any move in playing that game. As one can expect, **moments of time are related by virtue of the actions of agents at these moments.** According to the logic of action, to each agent a and moment m there always corresponds the set $Action^a_m$ of alternative moments m' which

[9]In the logic of branching time and action, circumstances are pairs of a moment of time m and history h where $m \in h$. So when I say that a proposition is true at a moment m according to a history h, I always assume that m belongs to h.

[10]As Prior [1967] says, now unpreventable propositions are "those outside our power to make true or false"

are *compatible with all the actions* that agent a performs at moment m. They are all, as Chellas [1992] would say, "under the control of - or responsive to the actions of" agent a at the moment m. Suppose that an agent a does not do anything at a moment m then all alternative moments to that moment are compatible with moment m. Suppose that he does something A. Then the proposition that A is true at all alternative moments $m' \in Action_m^a$.

In my view, in order that a moment be compatible with all the actions of an agent at another moment, **that agent must perform exactly the same actions at these moments**. So by definition, the *relation of compatibility with actions* that I consider is reflexive, symmetric and transitive. Of course the same actions of an agent can have different physical effects (that are not actions) in the world at different moments which are compatible with what that agent does at that moment. **Every agent persists in the world**. What an agent does at each moment depends on how the world has been up to that moment. The possible causes and effects so to speak of the actions of an agent at a moment are limited to those which are possible outcomes of the way the world has been up to that moment. This is why, the relation of compatibility with actions has to satisfy the so called *historical relevance condition.* As Belnap and Perloff [1990,1992] pointed out, in order that a moment *m'* be compatible with all the actions that agent a performs at another moment m, both must belong to histories with the same past.

Thanks to the new compatibility relation, the logic of action can start to analyze individual action. The proposition that *A is true given what agent a does* (in symbols ΔaA) is true at a moment m according to a history h when the proposition that A is true at all moments m' compatible with the actions of agent a at m according to all histories *h'*. By hypothesis, all histories h to which a moment m belongs, are responsive to all actions of each agent at that moment. Whenever an agent does something at a moment, he or she does it at that moment, no matter how that moment continues. So the truth of the proposition ΔaA is settled at each moment in my logic of agency.[11] Chellas [1992] tends to identify the very notion of action with the normal modal operation corresponding to Δ. However any proposition of the form ΔaA is true whenever A is historically necessary. But it is quite clear that no agent could act so as to bring about an inevitable fact. Inevitable facts exist no matter what we do. So in order that the proposition that an agent a do something, we have to require furthermore that the thing in question

[11]My conception of action at a moment is then incompatible with that of the deliberative *sees to it* of von Kutchera [1986], Horty [1989] and Belnap, Perloff and Ming Xu [2001].

be not then necessary. As Belnap pointed out, the proposition that an agent *sees to it that* A (in symbols [a *stit* A]) implies that it is false that $\Box A$.

In their logic of agency Belnap and Perloff use the logic of branching time and von Neumann [1944]'s theory of games. Agents make free choices in time. The notion of acting or choosing at a moment m is thought of as constraining the course of events to lie within some particular subset of the possible histories available at that moment. Belnap and Perloff [1992] first studied actions that are guaranteed by a past choice of the agent. (They made a theory of the so-called *achievement stit.*) However most often agents succeed to do things that they had no prior intention to do. They spontaneously attempt to do them. I never planned to use the words that I am typing right now. Many human actions are only due to a present choice of the agent at the moment of the action. So Belnap, Perloff and Xu [2001] came to study later actions directed at the future that are guaranteed by a present choice of the agent. (This is their theory of the *deliberative stit.*) In my logic of agency, I will study individual actions which are made at the very moment of the agent's choice. It does not matter to me whether they are oriented towards the present or the future. By definition *attempts* correspond to a present rather than to a prior choice. Every intentional action contains an attempt, few execute a prior intention. In my conception of time, most successful attempts by an agent to move one's body cause the movement at the very moment of the attempt. So speakers utter words at the very moment where they try to utter them in contexts of utterance.

Belnap's logical analysis of action in terms of ramified time and historic modalities has the merits of taking very seriously into consideration the temporal and causative order of the world. His logic is compatible with science. I will follow his approach under many aspects. Unfortunately Belnap tends to neglect the *intentionality* proper to action. For that reason agents carry out too many actions in his logic of agency. Suppose a proposition strictly implies another proposition which is not then necessary. According to Belnap an agent cannot make the first true without *eo ipso* making the second true even when the second proposition has nothing to do with what that agent could do or try to do at that moment. For example, an agent who repeats an action sees to it that he or she does and has done it in Belnap's logic.

I will try to work out a logic of action that takes into account the intrinsic intentionality of action so as to explicate adequately agentive commitment. On my account, there is no **action without attempt. So the logic of action must incorporate a *logic of attempt*.** We need a new logical constant of attempt in the lexicon of the logic of action. Let formulas of the form $aTriesA$ express the proposition that *agent a attempts to bring about that A*. Before stating truth conditions, let us make a philosophical analysis of the nature of attempts. Clearly attempts and actions are logically related in the philosophy of mind: **every attempt contains an intention**. However, unlike *prior intentions* which are *mental states* that agents *have*, attempts are *mental actions* that agents *make*. An attempt to do something contains an *intention in action*. For to make an attempt is to do something with the intention of achieving a purpose. By raising the arm an agent can make an attempt to greet someone and start a conversation.

No attempt is determined. There is the freedom of the will. So agents could attempt to do something else or make no attempt at all. Moreover **each attempt is personal and subjective**. Only an agent a can attempt that he or she does something. Someone else cannot. So when two different agents succeed to do the same thing (e.g. to drink), they do it by making different personal attempts. From a philosophical point of view, **attempts are a very special kind of action** that philosophers and logicians have tended to neglect until now. On the one hand, **all attempts are *intentional actions***. An agent cannot make an attempt without intending to make that attempt. On the other hand, all individual **attempts are also *successful* actions** in the sense that no agent can fail to make the attempt that he or she is trying to make at a moment. For in trying to make an attempt the agent *eo ipso* makes that very attempt. This is tautological. An attempt is essentially a mental act. An agent who tries to raise the arm could fail. (He or she could be prevented by an external force or his or her arm could already be up.) But that agent has at least mentally tried to make that movement. He or she had in mind the corresponding *intention in action*. So direct attempts by an agent to move parts of one's body are *basic actions* in the sense of Goldman [1970]. If an agent really wanted to make a direct attempt to move such an attempt would result from his or her want no matter whether he or she is in *standard conditions* or not.[12] Notice that we often have an *experience of the attempt* when that attempt fails.[13] Such an experience presents or represents the success conditions of the attempted action.[14]

[12]See Goldman [1970] page 65.

[13]The notions of direction of fit, intention in action and experience of an action are explained in Searle [1983]. Searle like other philosophers of mind has not sufficiently taken into account the fact that attempts (or intentions in action) are themselves actions.

[14]Attempts of moving the body contain a *presentation* and attempts of making an act of conceptual thought a *representation* of their success conditions.

From a philosophical point of view, both intentions and attempts have the same world-to-mind direction of fit and related conditions of satisfaction. An intention is *satisfied* when it is *carried out*, an attempt when it is *achieved*. Each attempt is directed at an *objective* or *aim* and serves a certain *purpose*. It *succeeds* when that agent achieves his or her purpose. Otherwise it is a *failure*.[15] An agent can **have various types of purposes. So there are various kinds of attempt**. A first and basic kind of attempt is to do something at the moment of the attempt. (So are direct basic attempts to move one's body at a moment.) By hypothesis, any agent who has such a basic purpose either succeeds or fails to achieve it at the very moment of the attempt. As we have seen, the very performance of an individual action at a moment is settled at that moment no matter what happens later. An agent who attempts to make a movement at a moment could fail. So there are alternative moments where that agent does not make the attempted movement. However if he or she makes the movement then his or her present choice at the moment of the attempt corresponds to the set of whole histories passing through that moment. Human agents persist in the world and they live in society. For that reason they often also have future and collective objectives in addition to present ones. They do something at a moment in order to bring about future things. I am now typing on my computer with the intention of revising this chapter. Human agents also often act in order to do their part in a collective action that they want to carry out with others. I am now working in order to edit a collective book in a certain collection of Springer. In such cases, it is not settled at the moment of the attempt whether agents will or not reach their objectives. They can succeed according to one possible historic continuation of the world and fail according to another. All depends on how future things will be and how other agents will act. So **the achievement of many attempts depends on both the moment of the attempt and the historic continuation of that moment**. This always happens when the purpose is future or collective.

Given the fact that attempts have purposes and conditions of achievement, moments of time and histories are also logically related by virtue of the attempts of agents at these moments. To each agent a and moment m there also corresponds a (possibly empty) set of alternative moments m' where that agent succeeds to achieve according to at least one possible course of history h' the attempts that he or she makes at the moment m. Such alternative mo-

[15]The notions of success and failure of an attempt are relative to satisfaction (and not success) conditions; they concern the achievement (and not the making). of attempts.

ments m' are said to be *compatible with the achievement of all attempts* of agent a at moment m.

Attempts have the characteristic world-to-mind direction of fit. In order that an attempt be achieved in a possible course of history that attempt must be made in that course of history. This is part of the satisfaction conditions of attempts. So all moments m' which are compatible with the achievement of the attempts of an agent a at a moment m are alternative and coinstantaneous with that moment. For that agent a has to make then all the attempts that he or she makes at the moment m. As one can expect, achievable attempts of agent a directed at the present are achieved at all such compatible moment m', while achievable attempts directed towards the future are achieved at a later moment m" posterior to m'.

Most of the time we try to do possible things. In that case, there are a lot of alternative moments which are compatible with the achievement of our attempts. However we can wrongly believe that an objective can be reached. So we can try to do impossible things. In that last case, there does not exist by hypothesis any such compatible moment. In the logic of ramified time, moments represent *possible states* of the world (objective possibilities). Impossible objectives cannot be achieved just as necessarily false proposition cannot be true at any moment according to any history in the logic of ramified time. **How can we deal formally with unachievable attempts**? In philosophy of mind, human agents are minimally rational. They know that successful actions have to bring about facts in the world. And that impossible facts cannot happen. So whenever agents try to do something they at least believe that they could do it. This is part of the sincerity conditions of any attempt and intention. The logic of attempts and intentions is then much stronger than that of desires. Agents can have desires that they believe unsatisfiable, for example to be at Paris and Rome at the same moment for different reasons. But they could never have similar intentions or make similar attempts.[16] Any rational agent who makes an attempt believes that he or she could succeed. So to each agent a and moment m there always corresponds a non empty **set $Attempt^a_m$ of alternative moments m' coinstantaneous with m which are compatible *according to the agent a* with the achievement of his or her attempts at that very moment m.** Suppose that an agent a tries something A at m. Then at all moments $m' \in Attempt^a_m$ he or she also tries A and A is true

[16]See Searle's chapter "Desire, Deliberation and Action" in the book.

according to him or her at these moments $m' \in Attempt^a_m$ according to at least one history.

As we saw earlier, each attempt is intentional. Attempts are always attempted. So the relation of compatibility with the achievement of attempts that we consider is transitive. In each model of the logic of action: when $m' \in Attempt^a_m$ and $m'' \in Attempt^a_{m'}$ it follows that $m'' \in Attempt^a_m$. Moreover, as attempts are actions, an agent makes the same attempts at all moments which are compatible with what he or she does at a moment. So whenever $m' \in Action^a_m$, both $m' \in Attempt^a_m$ and $Attempt^a_{m'} = Attempt^a_m$.

By nature **attempts** are intentions in action. So they **have** like intentions **strong propositional content conditions** that the logic of action must determine. As one can expect, the set $Goals^a_m$ of propositions representing *possible goals* of an agent a at a moment m is for that reason provided with the following logical structure in each standard model. **In order to make an attempt an agent must exist**. So the set $Goals^a_m$ is empty when agent a does not exist at moment m. Because individual actions at a moment (in particular direct basic personal body movements) are constitutive of all other kinds of action, **agents always attempt *to do something in the world at the moment of an attempt***. Consequently, propositions of the form ΔaA representing actions of the agent a at moment m always belong to the set $Goals^a_m$ of a model when that set is not empty. As **attempts are *personal***, $[b\ tries\ A] \notin Goals^a_m$ when $b \neq a$. Furthermore, **as agents are minimally rational, they never attempt to bring about something that they know *a priori* to be necessary or impossible**. So $A \notin Goals^a_m$ when the proposition that A is tautological or contradictory. Finally, **attempts are directed at a present or future purpose**. So propositions of the form $WasA \notin Goals^a_m$. We never attempt to do something in the past. On the basis of previous considerations on the nature of attempts, I will say that a proposition $[aTriesA]$ is *true at a moment m* according to a history when, firstly, the proposition that A represents a possible goal of agent a at moment m (i.e. that $A \in Goals^a_m$) and, secondly, the proposition that A is true according to agent a at all moments $m' \in Attempt^a_m$ according to at least one history h'.

As Searle pointed out in the third chapter, the logic of desire and intention is very different from that of belief. Agents can both intend to do something and believe that their intended action will have a certain effect without *eo ipso* desiring and intending to produce that effect. Someone who rejects an offer can believe the he or she will irritate the hearer without desiring and intending to provoke such an attitude. In that case there is a conflict between the intentions and beliefs of an agent at a moment. Certain moments compatible with the execution of the agent's intentions at a moment are not compatible with his or her beliefs at that very moment. For the unwanted effect of the intended action does not occur at these moments according to at least one history. Agents know that some of their beliefs could be false. This can even occur when the agent believes that it is settled or even inevitable that his or her action will have a certain unwanted consequence. Bratman and Searle have given a lot of convincing examples. A prior intention to do something (that A) and a belief that it is then necessary that if A then B do not commit the agent to a prior intention to do that B. We know that we can wrongly believe that certain facts are inevitable. We would then be happier if such facts would not occur. So there is something wrong with Kant's principle: "whoever intends to achieve an end thereby will the necessary means or effects that he or she knows to be part of the achievement of that end" This principle does not work for *prior intentions.*

However because agents are rational they have to minimally coordinate their cognitive and volitive states in trying to act in the world. So a restricted form of Kant's principle "Any agent who wills the end is committed to willing the necessary means" works for attempts which are intentions in action.[17] Suppose that an agent trying to do something knows that in order to succeed he or she has to do intentionally something else. Then that agent is going to try to do that other thing. In other words, an attempt to do something and a knowledge that one could not do it without intentionally doing something else *commit the agent to* an attempt to do that other thing.

Such a restricted Kantian principle is valid in my logic of action. For as I said earlier, whenever a modal proposition of the form $\Box A$ is true according to an agent a at a moment m in a circumstance the same proposition $\Box A$ is also true according to that agent at that moment m at all coinstantaneous circumstances containing all moments $m' \in Attempt^a_m$, which are compatible according to that agent with the achievement of his or her attempts at moment m. Let me give two examples. Any agent knows that in order to supplicate someone at a moment he or she has to make a request. So whoever tries to make a supplication *eo ipso* tries to make a request. His or her attempted request then *constitutes* his or her attempted supplication. Every agent also knows that in order to supplicate a person one has to tell him or her what one desires. So whoever tries to supplicate someone also tries to send him or her a message. His

[17] See Searle [2001] page 266.

or her attempted emission of signs (or utterance act) then *generates* his or her attempted supplication.

Let us now come to the explication of *success*. As philosophers of action pointed out, the successful performance of an intentional action requires more than the existence of an attempt and the truth of its content. In order that an agent succeed to bring about a fact, it is not enough that he or she try and that the fact occur. It is also necessary that the fact occurs *because of* his or her attempt. The agent does not succeed in doing something in the case in which someone elsedid it. The attempt of the agent must be the *cause* of what is done. Along these lines, one can **define simply as follows the logical form of intentional actions**. An agent *a succeeds to do that* A (in symbols: $\delta_i a$A) when firstly, the agent a attempts to do that A, secondly, A is true given what he or she does and thirdly, it is not then necessary that A. Notice that $\delta_i a$A entails $\delta a([aTriesA] \Leftrightarrow \text{A})$. This is a step towards the explication of *intentional causation.* In case someone else does what an agent a attempts to do, that agent a does not do it. For there is then a moment compatible with what that agent does at the moment of his or her attempt where it is not the case. I have given this first explication of success in [2003]. However, for a full account of intentional causation, we need, I think, the *counterfactual conditional.*[18] That conditional enables us to state an important additional necessary condition of success: If agent a had not tried to do that A then it would not be true that A given what that agent does. For that purpose, I will add the counterfactual conditional $\Box\!\!\rightarrow$ to the lexicon of my logic of action. Formulas of the form A $\Box\!\!\rightarrow$ B mean that if it were the case that A then it would be the case that B. So $\delta_i a\text{A} =_{def} ([aTriesA]) \wedge (\Delta a\text{A}) \wedge (\neg\Box\text{A}) \wedge (\neg[aTriesA] \;\Box\!\!\rightarrow \neg\Delta a\text{A})$ in the present logic of action.

How could we now explicate the general notion of an individual action (whether intentional or not)? I propose the following definition: an agent *a acts so as to bring about that* A (in symbols δaA) when firstly, A is true given what he or she does, secondly, it is evitable that A, thirdly, that agent a could attempt or have attempted to bring about that A, and fourthly, he or she brings about that A because of a present attempt. Thanks to the counterfactual conditional one can state precisely the condition of mental causation by saying that if the agent a had not made such a present attempt then he or she would not have done that A. For short, in symbols: $\delta a\text{A} =_{def} (\Delta a\text{A}) \wedge (\neg\Box\text{A}) \wedge (\Diamond([a\textit{Tries}\text{A}] \vee$

[18]One can incorporate a logic of counterfactuals within the logic of ramified time by introducing a relation of comparative similarity between moments or histories in the sense of Lewis [1973]. See Thomason & Gupta [1980].

$Was[aTries\mathrm{A}]) \wedge (\exists \mathrm{p}(\delta_i a\mathrm{p} \wedge (\neg\delta_i a\mathrm{p} \; \Box\!\!\rightarrow \neg\Delta a\mathrm{A})))$. In my conception of action, there is no action without a simultaneous attempt of the agent. So dead agents do not act any more. What agents do at each moment has to be the effect of their intentional actions at that very moment.

By definition, **the notions of success and failure are relative to intentional actions**. No agent can succeed or fail to do something unless he or she makes an attempt. So we do not properly succeed to perform our unintentional actions. It just happens that we perform them. As philosophers of action pointed out, some of our actions, called *basic actions*, are by nature intentional. So are attempts, voluntary body movements, meaningful utterances and illocutionary acts. In order to perform a basic action an agent must make an attempt to perform it. Basic actions are then always successful when they are performed. Of course some intentional actions are *more basic than* others. For example, utterance acts are made by way of voluntarily emitting sounds or producing marks. Attempts of performance of illocutionary acts are made by way of making meaningful utterances. Such attempts cause the successful performance of illocutionary acts when they are made in appropriate contexts. Acts of communication occur when hearers understand illocutionary acts. They can provoke intended perlocutionary effects on such hearers. And so on. Following Goldman I will say that an agent *basically does* something at a moment m when he or she performs at that moment all his or her intentional actions by way of doing that thing. In my view, **all intentional actions that an agent performs at one moment are consequences of a unique action** that he or she basically performs at that moment. **That basic action is always an irreducibly personal attempt of moving parts of his or her body**. In particular, all public speech acts of an agent at a moment are generated by one's attempt to emit tokens of signs at that moment.

3. The ideal object-language

The **ideal object language L** of the present logic of action contains in its lexicon:
(1) A series of **individual constants** naming *agents*
(3) a series of **propositional variables and constants** and
(4) the **syncategorematic expressions**:
Tautological, >, $\wedge$, $\Box$, *Tries*, Δ, *Will*, *Was*, *Settled*, $\neg$, $\rightarrow$, $\exists$, [,], (and).

Rules of formation of formulas of L

Propositional variables and constants are formulas. If A_p and B_p are formulas, x and y are individual constants and p is a propositional variable, then *Tautological*(A_p), ($A_p > B_p$), $\neg A_p$, $\Box A_p$, *Will*A_p, *Was*A_p, *Settled*A_p, [*xTries*A_p], $\Delta x A_p$, $\exists p\ A_p$, ($A_p \wedge B_p$) and (A $\Box\!\!\rightarrow$ B) are new complex formulas. Closed formulas have the following meaning:

Propositional constants express and propositional variables indicate propositions. *Tautological*(A_p) expresses the proposition that A_p is tautological. ($A_p > B_p$) expresses the proposition that all atomic propositions of B_p are atomic propositions of A_p $\neg A_p$ expresses the negation of the proposition expressed by A_p. $\Box A_p$ expresses the modal proposition that A_p is then necessary (i.e. that it could not have been otherwise than A_p). *Will*A_p expresses the future proposition that it will be the case that A_p. *Was*A_p expresses the past proposition that it has been the case that A_p. *Settled*A_p expresses the proposition that the truth of A_p is settled. [*xTries*A_p] expresses the proposition that agent x attempts to do A_p. $\Delta x A_p$ expresses the proposition that A_p is true given what agent x does.[19] ($A_p \wedge B_p$) expresses the conjunction of the two propositions that A_p and that B_p. $\exists p\ A_p$ means that at least one proposition p satisfies A_p. Finally, (A $\Box\!\!\rightarrow$ B) means that if it were the case that A then it would be the case that B.

Rules of abbreviation

I will sometimes eliminate the subscript p. So A is short for A_p. I will eliminate exterior parentheses and introduce truth, modal and temporal connectives and the universal and unique existential quantifiers according to usual rules of abbreviation.

So ($A_p \Rightarrow B_p$) $=_{df}$ $\neg(A_p \wedge \neg B_p)$ and similarly for *material equivalence* $\Leftrightarrow$;

Was-alwaysA_p $=_{df}$ $\neg$ *Was*$\neg A_p$ and *Will-always*A_p $=_{df}$ $\neg$ *Will*$\neg A_p$;

AlwaysA_p $=_{df}$ *Was-always*A_p $\wedge$ A_p $\wedge$ *Will-always*A_p;

LaterA_p $=_{df}$ *SettledWill*A_p and *Before*A_p $=_{df}$ *SettledWas*A_p ;

Historical possibility: $\Diamond A =_{df} \neg\Box\neg A$;

Universal necessity: $\blacksquare A =_{df}$ *Always*$\Box$A ;

Universal possibility: $\blacklozenge A =_{df} \neg\blacksquare\neg A$;

Strict implication: A $-\!\!\in$ B $=_{df}$ $\Box$ (A $\Rightarrow$ B);

Strong implication: $A_p \mapsto B_p =_{df} (A_p > B_p) \wedge$ *Tautological* ($A_p \Rightarrow B_p$);

Propositional identity: $A_p = B_p =_{df} (A_p \mapsto B_p) \wedge (B_p \mapsto A_p)$

Intentional action:

$\delta_i x A_p$ $=_{df}$ [*xTries*A_p] $\wedge$ ($\Delta x A_p$) $\wedge$ ($\neg\Box A_p$) $\wedge$ ($\neg$[*xTriesA*] $\Box\!\!\rightarrow$ $\neg\Delta x$A))

x fails to do A_p $=_{df}$[*xTries*A_p] $\wedge$ (($\neg\Delta x A_p$) $\vee$ $\Box A_p$) $\vee$ $\neg$($\neg$[*xTriesA*] $\Box\!\!\rightarrow$ $\neg\Delta x$A))

Action (intentional or not): $\delta x A_p$ $=_{df}$ ($\Delta x A_p$) $\wedge$ $\neg\Box A_p$ $\wedge$ ($\Diamond$([*xTries*A_p]

[19] δ is the logical constant of Chellas' [1992] logic of agency.

$\vee$ *Was*[*xTries* A_p]) $\wedge$ $\exists$p(*xTries* p $\wedge$ ($\neg$[*xTries p*] $\Box\!\!\rightarrow$ $\neg\Delta x$A))
x basically does $A_P =_{df}$ *xTries* A_p $\wedge$ $\forall$p ($\delta_i x$p $\Leftrightarrow$ ($\neg x$ *Tries*A_p $\Box\!\!\rightarrow$ $\neg\delta_i x$p))
Identity of agents $(x = y) =_{df} \Delta x A_p = \Delta y A_p$

In my ideal object language, propositions representing an action of an agent are of the canonical form $\delta x A_p$. Any proposition of the form $\delta x A_p$ is *agentive for the agent* a [20] in the sense that it represents an action of that agent, no matter whether A_p is itself agentive for x or not. So the sentence "Oedipus killed Laïus" represents an action of Oedipus. For it can be paraphrased as "Oedipus acted so as to bring about that Laïus is dead". What agent x does is represented by A_p in $\delta x A_p$. From an ontological point of view, the content of an action can be a state of affairs, an event or even an action.

4. Basic laws of the logic of action

Fundamental laws governing elementary propositions, truth functions, universal modalities, tautologies and propositional identity have been stated in chapter 10. Laws governing historic modalities and ramified time are stated in my work *Attempt and Action Generation Towards the Foundation of the Logic of Action* (2003). Here are basic proper laws of my logic of action.[21]

As usual, $\models$A means that A is logically true or valid in my logic.

First, there is a **normal logic for the Chellas connective Δ**

(C1) $\models (\Delta x A_p \Rightarrow A_p)$
(C2) $\models (\Delta x(A_p \wedge B_p) \Rightarrow (\Delta x A_p \wedge \Delta x B_p))$
(C3) $\models ((\Delta x A_p \wedge \Delta x B_p\ ) \Rightarrow \Delta x(A_p \wedge B_p))$
(C4) $\models (\Box A_p \Rightarrow \Delta a A_p)$
(C5) $\models (\Delta x A_p \Rightarrow \Delta x \Delta x A_p\ )$
(C6) $\models (\neg\Delta \mathrm{x} \neg A_p \Rightarrow \Delta x \neg\Delta \mathrm{x} \neg A_p\ )$[22].
(C7) $\models (\Delta x A_p \Rightarrow \mathit{Settled}\Delta x A_p)$

The basic laws for attempts are the following

(A1) Any attempt of an agent contains an attempt to perform an individual action. $\models ([x\ \mathit{Tries}\ A_p] \Rightarrow \exists p[x\mathit{Tries}\Delta x\ p])$

(A2) Any attempt is an intentional action of the agent.
$\models$ *Tautological* $([x\ \mathit{Tries}\ A_p] \Rightarrow \delta_i x[x\mathit{Tries}\ A_p])$ So $\models ([x\ \mathit{Tries}\ A_p] \Rightarrow \Delta x[x\mathit{Tries}\ A_p])$,

[20] The terminology is due to Belnap & Perloff [1990].
[21] For the model-theoretical semantics and a full axiomatization of my logic of action see *Attempt and Action Generation Towards the Foundations of the Logic of Action* in *Cahiers d'Épistémologie* Université du Québec à Montréal n°293, 2003.
[22] Axiom schema (C6) is not valid in Chellas logic for Δ

$\models ([xTriesA_p] \Leftrightarrow [xTries[xTriesA_p]])$ and $\models [x\ Tries\ A_p] \Rightarrow \neg\Box A_p$
(A3) Each attempt is personal. $\models [xTries\delta y[yTriesA_p]] \Rightarrow x = y$
Agents are minimally rational.
(A4) They do not attempt to do something tautological or contradictory. $\models (\mathit{Tautological}\ \mathrm{A}_p \vee \mathit{Tautological}\ \neg\mathrm{A}_p\ ) \Rightarrow \neg\Diamond[xTries\mathrm{A}_p]$
(A5) They do not attempt to change the past. $\models \neg\Diamond\ [xTriesWas\mathrm{A}_p]$
(A6) Whenever they attempt to do one thing and they attempt to do another thing they attempt to do both. $\models ([xTries\mathrm{A}_p] \wedge [xTriesB_p]) \Rightarrow [xTries(\mathrm{A}_p \wedge \mathrm{B}_p)])$
(A7) The converse is true when propositional content conditions are preserved.[23]
$\models ([xTries(\mathrm{A}_p \wedge \mathrm{B}_p)] \Rightarrow (([xTries\mathrm{A}_p\ ] \wedge \Diamond\ [xTriesB_p]) \Rightarrow [xTriesB_p]))$
(A8) Any attempt is generated by the basic action of the agent.
$\models [xTries\mathrm{A}_p] \Rightarrow \exists p\ (x\ \mathit{basically\ does}\ p)$

One can derive the following laws in my logic of attempts. Firstly, agents really *make attempts.* So the making of an attempt is always settled at each moment. Secondly, *attempts can fail.* In order to achieve a purpose an agent must make the right attempt in the right circumstance. Suppose you want to threaten someone at a moment. You must speak to the right person and utter appropriate words. Otherwise your utterance is a wrong attempt. Moreover the context must be appropriate. If it is mutually known that you are unable to do what you say, your attempt is made in a wrong circumstance. So $\not\models[aTries\delta aA] \Rightarrow \delta aA$. Failure can happen even when the agent believes the contrary. An agent might have wrong beliefs about objects at which his or her action is directed. Furthermore, when an attempt is directed towards the future, the agent can succeed according to a possible continuation of the moment of the attempt and fail according to another. $\not\models[aTriesWillA] \Rightarrow SettledWillA$. All depends on what will happen later.

Agents can believe in the truth of impossible propositions for example that whales are fishes. So their objectives are sometimes impossible. In the past fishermen were trying to catch a big fish while trying to fish a whale. However they remain minimally rational. What they try to do is not contradictory. The truth of propositions representing their objective is compatible with some of their valuations at the moment of their attempt.

Notice finally that **the set of our purposes is not partially closed under strict but under strong implication**. In philosophical logic, a

[23] Any proposition is identical with a conjunction of that proposition with a tautology. Thus $\models \mathrm{A}_p = \mathrm{A}_p \wedge (\mathrm{A}_p \vee \neg\mathrm{A}_p)$. However an attempt to make it true could not contain an attempt to make true a tautology.

proposition *strictly implies* another proposition when it cannot be true unless that other proposition is also true. Agents ignore actual truth conditions of most propositions. So they also ignore how propositions are related by strict implication. For that reason, they can attempt to make true a proposition without *eo ipso* attempting to make true another proposition that the first strictly implies. Moreover attempts like intentions have strong propositional content conditions. Each proposition strongly implies many others which could not be the content of an attempt. So $\nvDash\Box(A \Rightarrow B) \Rightarrow ([aTries\delta aA] \Rightarrow [aTries\delta aB])$. However, as I said earlier, because agents are minimally rational, whoever attempts to achieve an end attempts to use means that he or she knows to be necessary. In short, if we introduce in the logic of action the epistemic connective K of knowledge ([KaA] means that *a knows A*) we could assert the following law in my logic of action: $\vDash[\mathrm{K}a\Box(\Delta aA \Rightarrow \Delta aB)] \Rightarrow ([aTries\Delta aA] \Rightarrow [aTries\Delta aB]$.

Given my analysis of propositions, one can explain which propositions agents *a priori* know to be necessarily true and necessarily false. Any agent who has in mind a tautology or a contradiction *a priori* knows that the first represents an inevitable fact and the second an impossible fact. So agents *a priori* know that certain facts could not exist without others. As I have explained in Chapter 10, **there is a much finer logical propositional implication than strict implication called *strong implication* that agents *a priori* know by virtue of competence.** By definition, a proposition *strongly implies* another when firstly, it has all its atomic propositions and secondly, all possible denotation assignments to propositional constituents which are compatible with its truth in a circumstance are also compatible with the truth of that other proposition in that circumstance. In symbols: $\mathrm{A}_p \mapsto \mathrm{B}_p =_{df} (\mathrm{A}_p > \mathrm{B}_p) \wedge$ *Tautological* $(\mathrm{A}_p \Rightarrow \mathrm{B}_p)$. **Strong implication is important for the purpose of the logic of knowledge and action.** Whenever a proposition strongly implies another proposition, an agent cannot have it in mind without having the other in mind and knowing that the first implies the second. For in case the proposition A strongly implies that B, it is identical with the conjunction (A $\wedge$ B). So according to any agent the fact represented by A contains the fact represented by B: the first could not exist without the second. Consequently any attempt by an agent to bring about the first fact A is also an attempt to bring about the second B in any circumstance where B is also a possible goal of the agent.

So $\vDash (\mathrm{A}_p \mapsto \mathrm{B}_p)] \Rightarrow (([xTries\mathrm{A}_p\ ] \Rightarrow (\Diamond\ [xTriesB_p] \Rightarrow [xTriesB_p])))$ For $\vDash(\mathrm{A}_p \mapsto \mathrm{B}_p) \Rightarrow (([\mathrm{K}x\Box(\mathrm{A}_p \Rightarrow \mathrm{B}_p)]$.

5. Other important valid laws

First of all, there is a normal logic of action. As one can expect, some basic laws governing Δ are also valid for the action connective δ. In particular:
(D1) By acting agents bring about facts in the world.$\models (\delta x \mathrm{A}_p \Rightarrow \mathrm{A}_p)$
(D2) To do something and to do something else is to do both.
$\models ((\delta x \mathrm{A}_p \wedge \delta x \mathrm{B}_p) \Rightarrow \delta x(\mathrm{A}_p \wedge \mathrm{B}_p))$
(D3)Actions are also facts that agents bring about in the world.
$\models (\delta x \mathrm{A}_p \Rightarrow \delta x \delta x \mathrm{A}_p)$
(D4) Any action that is done at a moment changes the world at that very moment.
$\models (\delta x \mathrm{A}_p \Rightarrow \mathit{Settled} \delta x \mathrm{A}_p)$ Consequently, $\models \delta x \mathrm{A}_p \Rightarrow \mathit{Settled} \mathrm{A}_p$.

So our individual **actions really change the world**. Whenever we do something at a moment, it is settled that it is done. $\models \delta x A_p \Rightarrow \delta x \mathit{Settled} A_p$. We can even bring about now how things will be in the future. In that case, we act in such a way that they will be so later no matter how the world continues. By making a move in a chess game a player sometimes puts his or her adversary in an inevitable losing position. In that case it is settled that if the game is pursued the adversary will lose. So $\models \delta x \mathit{Will} \mathrm{A}_p \Rightarrow \mathit{Later} \mathrm{A}_p$ where $\mathit{Later} A_p =_{df} \mathit{Settled} \mathit{Will} A_p$. Otherwise, the fact that A_p will be the case is not the result of a present action. It depends on something else. As one can expect, an agent cannot change the past of the world. $\models \neg \delta x \mathit{Was} A_p$. Secondly, **agents are free.** $\models (\delta x A_p \Rightarrow \Diamond \neg \delta x A_p)$ and $\models \neg \Box \delta x A_p$ We are not determined to do what we do.

Thirdly, **any action can be attempted** in at least one circumstance. $\models (\delta x \mathrm{A}_p \Rightarrow \Diamond([x \mathit{Tries} \mathrm{A}_p] \vee \mathit{Was}[x \mathit{Tries} \mathrm{A}_p]))$. So $\models \neg \Diamond([x \mathit{Tries} \mathrm{A}_p]) \Rightarrow \neg \Diamond([\delta x \mathrm{A}_p])$. Consequently, **our *mistakes* and *failures* are not really actions that we perform but rather events which happen to us**. For we cannot really attempt to make a mistake or fail.[24] Paradoxical sentences of the form "I am trying not to try anything", "I am doing nothing", "I am trying to fail" and "This very action is a failure" are logically false. $\models \neg [x \mathit{Tries}\ \forall p\ \neg [x\ \mathit{Tries}\ p]]$. $\models \neg \delta x\ \forall p\ \neg\ \delta x\ p$. $\models \neg \exists p\ [x \mathit{Tries}(p \wedge \neg [x \mathit{Tries}\ p])]$ and $\models \neg \exists p\ \delta x\ (p \wedge (\neg\ \delta x\ p))$.

Fourthly, **any action is generated by a simultaneous intentional action.** $\models (\delta x \mathrm{A}_p \Rightarrow (\exists \mathrm{p}[a \mathit{Tries}\ \mathrm{p}] \wedge \Delta a(a \mathit{Tries}\ \mathrm{p} \Rightarrow \mathrm{A}_p)))$ Consequently, in order to act an agent must exist and be conscious at the

[24] According to Goldman [1970], there seems to be act properties like misspeaking, miscalculating, miscounting that preclude intentionality. In my view, such properties are not really act properties. We "suffer" mistakes and failures. We do not make them.

moment of his or her action. No agent can act after death. However past actions of a dead agent can still have effects now. Beautiful works of art of the past continue to provoke admiration.

There is a *law of foundation for intentional action.* An agent can only make a finite number of intentional actions at a moment. For he or she can only refer to a finite number of things and have in mind a finite number of attributes. In my logic, **an agent performs all his or her intentional actions at each moment by way of performing a unique basic intentional action**. $\models \delta_i \mathrm{xA}_p \Rightarrow \exists! p$ *x basically does p.* Two agents can perform individual actions of the same type at the same moment. But their individual intentional actions are always different. They contain different personal attempts. The very basic action of an agent at a moment is an individual attempt of moving parts of one's own body. If he or she had not made that attempt he or she would not have done anything.

By acting so as to bring about that A an agent does not act so as to bring about any effect B of A. $\not\models \Box(\mathrm{A} \Rightarrow \mathrm{B})) \Rightarrow (\delta x\mathrm{A} \Rightarrow \delta x\mathrm{B})$ for various reasons. First, as I have said repeatedly, an agent can only do things that he or she could attempt to do. Moreover, as medieval philosophers already pointed out, **no agent can do something which is inevitable**. By way of moving his or her body any agent inevitably moves invisible subatomic particles in the air. However, that event is not in and of itself an action. For he or she could not have done otherwise. $\models \neg\Diamond\delta x\Box\mathrm{A}$. So agents cannot do necessary or impossible things. $\models (\Box\mathrm{A} \vee \Box\neg\mathrm{A}) \Rightarrow \neg\delta x\mathrm{A}$. They only can attempt without success to do such things when they believe them to be possible. So $\not\models(\Box\mathrm{A} \vee \Box\neg\mathrm{A}) \Rightarrow \neg[x\mathit{Tries}\mathrm{A}]$.

Laws of action generation

By carrying out some actions in certain situations agents carry out other actions. My logic of action explains why certain action tokens *generate* others (causally, conventionally, simply and by extension) in the sense of Goldman [1970]. In order that an action by an agent at a moment generate another action of that agent at that moment, we have to require that the agent would not have made the second action if it were not the case that he had made the first. So I propose to explicate *generation* as follows:

$(\delta x\mathrm{A}\ \mathit{generates}\ \delta x\mathrm{B}) =_{def} \delta x\mathrm{A} \wedge \exists \mathrm{p}\ \delta x((\mathrm{A} \wedge \mathrm{p}) \Rightarrow \mathrm{B}) \wedge (\Delta x\mathrm{p}) \wedge (\neg\Box\mathrm{p}) \wedge (\neg\blacksquare(\mathrm{A} \Rightarrow \mathrm{p})) \wedge (\neg\Box\mathrm{B}) \wedge (\Diamond[x\mathit{Tries}\mathrm{B}] \vee \mathit{Was}[x\mathit{Tries}\mathrm{B}]) \wedge (\neg\delta x\mathrm{A} \Box\!\!\rightarrow \neg\Delta x\mathrm{B})$

Physical causal generation: Sometimes by doing something an agent also does something else for what he or she brings about physically causes that effect. For example, by flipping the switch an agent can turn on the light. By making a fire he or she can get burned. In such cases, the first action *causally generates* the second. We can explain such instances of physical causal generation. The agent x acts then at a moment in a situation C (that philosophers of action call a *circumstance*) where the premise ΔxC is true. For example, the electricity is on when the agent flips the switch. In making the fire he or she touches something very hot. These are circumstances C. In case what the agent brings about A is a cause of B, the other premises: $\Box((A \wedge C) \Rightarrow B)$, δxA, $\neg\Box$B and $(\neg\delta x A \Box\!\!\rightarrow \neg\delta x B)$ are also true. The first represents *a law of nature.* So when agent x could attempt B, (in symbols: $\Diamond x$ *TriesB*), one can conclude that agent x also does B.

Conventional generation: Sometimes by doing something at a moment in a certain situation an agent does something else at that moment because there is a convention according to which the first action in that situation counts as constituting the second. For example, by checkmating his opponent a player wins the game of chess. In such cases, the first action *conventionally generates* the second. In the case of conventional generation, what the agent x does (A) *counts as* doing something else B in the situation C where he or she acts because of a *collective acceptance of a previous declaration.* We need illocutionary logic (that contains a logic of declarations and acceptances) in order to fully explicate conventional generation.

Simple generation: Sometimes by doing something at a moment in a certain situation an agent also does something else because any performance of the first action in such a situation would be the performance of the second action. For example, by expressing a mental state that he or she does not have, an agent *lies.* By asserting a proposition that is future with respect to the moment of utterance a speaker makes a prediction. In such cases, the first action *simply generates* the second in Goldman's terminology. In case an action token δxA simply generates another δxB, this is due to the law: $\models(\Delta x C \wedge (\neg\Box C) \wedge (\blacksquare((A \wedge C) \Rightarrow B)) \wedge (\neg\Box B) \wedge ((\Diamond x\ \mathit{TriesB}) \vee \mathit{Was}[x\mathit{Tries}B]) \wedge (\neg\delta x A \Box\!\!\rightarrow \neg\Delta x\ B)) \Rightarrow (\delta x A \Rightarrow \delta x B)$

In my view indirect performances of speech acts are simply generated by the performance of literal speech acts in certain contexts of utterance where additional non literal success conditions are obviously fulfilled in the conversational background.[25]

Generation by augmentation: A special case of simple generation occurs when the generated action *strongly commits the agent to* the gen-

[25]See Vanderveken [1997].

erating action. For example, by putting a part of one's own body on something one touches that thing. By begging very humbly from someone in power one makes a supplication. In such cases, the generating action is augmented by a certain way or means or fact which is part of the circumstance C in which the agent acts. And any token of the generated action is also a token of the generating action. So in the case of generation by augmentation the following law holds: $\models (\Delta x \mathrm{C} \wedge (\neg\Box \mathrm{C}) \wedge (\blacksquare((\mathrm{A} \wedge \mathrm{C}) \Leftrightarrow \mathrm{B})) \wedge (\neg\Box \mathrm{B}) \wedge ((\Diamond[x\mathit{Tries}\mathrm{B}]) \vee \mathit{Was}[x\mathit{Tries}\mathrm{B}]) \wedge (\neg\delta x\ \mathrm{A} \;\Box\!\!\rightarrow \neg\Delta x\ \mathrm{B})) \Rightarrow (\delta x \mathrm{A} \Leftrightarrow \delta x \mathrm{B})$ All performances of elementary illocutionary acts whose force F is stronger than a primitive force F* are generated by augmentation from the very performance of an illocutionary act with that primitive force F* and the same propositional content. In such cases of generation by augmentation, the generated illocutionary act of the form F(P) *strongly commits the speaker to* the generating illocutionary act F*(P). Any successful performance of the first act is also a successful performance of the second.[26]

Few intentional action are generated by our basic actions. By succeeding to do something an agent also succeeds in doing something else only if he or she attempts to do it, that thing is evitable, he or she knows that the first action generates the second action. Intentional generation requires much. So the number of intentional actions is finite and very limited. **However our basic actions generate many more (an indefinite open number of) unintentional actions**. Many of them are unexpected. We are not aware of most contingent effects which follow from what we do. Whenever we do something to a woman we do it to the mother of her children whenever she has children no matter whether we know that or not. By way of marrying Jocaste, the queen of Thebes, Oedipus also married unintentionally his mother. So by doing something intentionally we can also do other things unintentionally that we might not at all want to do.

References

Belnap N. & Perloff M.(1990). "Seeing to it: a Canonical Form for Agentives", in Kyburg H.E. et al, *Knowledge Representation and Defeasible Reasoning.* Dordrecht: Kluwer Academic Press.

Belnap N. & Perloff M. (1992). "The Way of the Agent", *Studia Logica, 51.*

[26] See Vanderveken [1990-91] for the notions of a stronger force and of strong illocutionary commitment.

Belnap N., Perloff M. & Ming Xu (2001). *Facing the Future Agents and Choices in Our Indeterminist World.* Oxford: Oxford University Press.

Bratman M. (1987). *Intentions, Plans and Practical Reason.* Harvard University Press.

Carnap R. (1956). *Meaning and Necessity.* University of Chicago Press.

Chellas B.F. (1992). "Time and Modality in the Logic of Agency". *Studia Logica, 51.*

Davidson D. (1980). *Essays on Action & Events.* Oxford: Oxford University Press.

Goldman A. (1970). *A Theory of Human Action.* Princeton University Press.

Horty J.F. "An Alternative stit Operator" Unpublished seminar notes.

Marcus R. Barcan (1983). *Modalities.* Oxford University Press.

Montague R. (1974). *Formal Philosophy.* Yale University Press.

Prior A.N. (1967). *Past, Present, Future.* Oxford: Clarendon Press.

Searle J.R. (1983). *Intentionality.* Cambridge University Press.

— (2001). *Rationality in Action.* A Bradford Book, MIT press.

Searle J.R. & Vanderveken D. (1985). *Foundations of Illocutionary Logic.* Cambridge University Press.

Thomason R. (1984). "Combinations of Tense and Modality", in D. Gabbay & F. Guenthner (eds.), *Handbook of Philosophical Logic, Vol. 2.*

Thomason R. & A. Gupta (1980). "A Theory of Conditionals in the Context of Branching Time", *Philosophical Review, 89.*

Vanderveken D. (1990–91). *Meaning and Speech Acts*, Volume I: *Principles of Language Use* and Volume II: *Formal Semantics of Success and Satisfaction.* Cambridge University Press.

— (1997). "Formal Pragmatics of Non Literal Meaning", in the special issue Pragmatik of *Linguistische Berichte* edited by E. Rolf, 135–148.

— "Propositional Identity, Truth According to Predication and Strong Implication", Chapter 10 of the present volume.

— (2001). "Universal Grammar and Speech Act Theory" in D. Vanderveken & S. Kubo, *Essays in Speech Act Theory*, Benjamins, P&b ns 77, 25–62.

— (2003). *Attempt and Action Generation Towards the Foundations of the Logic of Action* in *Cahiers d'espistémologie 2003*:02. Montréal: UQAM. 43 pages. `www.philo.uqam.ca`

— (2004). "Success, Satisfaction and Truth in the Logic of Speech Acts and Formal Semantics", in S. Davis & B. Gillan (eds), *A Reader in Semantics*, forthcoming at Oxford University Press.

— *Propositions, Truth and Thought.* Forthcoming.

Von Neumann J. & Morgenstern O. (1944). *Theory of Games and Economic Behavior.* Princeton University Press.

Chapter 16

PRAGMATIC AND SEMIOTIC PREREQUISITES FOR PREDICATION

A Dialogue Model

Kuno Lorenz
Saarbrücken University

Rather than starting my presentation of how to construe predication with a discussion about propositions, and independent of the ongoing debate whether propositions should be understood as propositional kernels of full sentences (having a *force*, expressing a *thought*, and denoting a *truth-value*) or should themselves be considered as a particular force of a sentence, i.e., its constative force as an assertion, I would like to invite you to a journey into the prepropositional state where the task to provide for propositions or sentences is still to be accomplished.

The primary means for changing one's state, or for realizing that such a change has occurred, I consider to be a dialogue, better still: a dialogue-situation. This is a situation where, using a Peircean term, a >habit-change< occurs which should be construed as the acquisition of an action-competence.[1] It turned out that such an acquisition procedure will most profitably be modelled using the conceptual frame of a *two-person-game*, and such a game may be considered as a generalized Wittgensteinean language-game or, rather, its pragmatic basis, not yet with an explicit linguistic activity. In the beginning the game is not an *object of study* but a *means of study*.

[1] Cf. K. Lorenz, Pragmatics and Semiotic: The Peircean Version of Ontology and Epistemology, in: G. Debrock/M. Hulswit (eds.), *Living Doubt.* Essays Concerning the Epistemology of Charles Sanders Peirce, Dordrecht: Kluwer. 1994, 103-108.

D. Vanderveken (ed.), Logic, Thought & Action, 343–357.

Before giving a sketch of the dialogical constructions,[2] which lead from modelling simple activity to modelling the growth of more complex activities up to elementary verbal utterances, at first a few general remarks are in order to put my suggestions into a broader perspective.

I

In accordance with C. S. Peirce, I consider pragmatics to have become the modern heir of ontology with semiotics being its counterpart as the modern heir of epistemology. Yet, in this context both disciplines should not be understood as two newly established empirical sciences, but as ways of investigation where empirical procedures are combined with reflexive procedures. Using such a broader perspective both actions and sign-actions are not only treated as *objects* of research and representation, as, e.g., in Ch. Morris' and U. Eco's approach, but also as a *means* or tool of research and representation. You not only observe and describe these entities according to certain standards, but you also produce them in a perspicuous fashion in order to arrive at some kind of approximating reconstruction of what you take to be available, already. Wittgenstein has used the term 'language-game' for this kind of activity which aims at disclosure of what is going on by providing tools of comparison, though in his description of language-games pragmatic and linguistic activity is not accounted for by separate steps.

Hence, the constructions serve cognitive purposes in the sense of delineating the very areas of (particular) objects one proceeds afterwards to investigate in the more usual way. Language-games as well as the generalized ones of acquiring simple action competences have to count as paradigm cases of perceptual knowledge, because they exhibit a significative function if understood as icons in the sense of Peirce. An area of internally structured objects is found by inventing a prototype.
It should be obvious, therefore, that even the distinction of action and sign-action – a special case of the basic and embarrassing distinction between world and language – which still is prevalent in Wittgensteinian language-games where simple action competence is presupposed, has to be relativized in view of a purely functional account of both what it

[2]For further details, cf. K. Lorenz, Artikulation und Prädikation, in: M. Dascal/D. Gerhardus/K. Lorenz/G. Meggle (eds.), *Sprachphilosophie.Philosophy of Language.La philosophie du langage.* Ein internationales Handbuch zeitgenössischer Forschung II, Berlin-New York: de Gruyter 1995, 1098-1122; K. Lorenz, Rede zwischen Aktion und Kognition, in: A. Burri (ed.), *Sprache und Denken. Language and Thought*, Berlin-New York: de Gruyter 1997, 139-156; K. Lorenz, Sinnbestimmung und Geltungssicherung. Ein Beitrag zur Sprachlogik, in: G.-L. Lueken (ed.), *Formen der Argumentation*, Leipzig: Leipziger Universistätsverlag 2000, 87–106.

means to be an object and what it means to be a sign (of an object). In fact, it belongs to one of the basic tenets of, e.g., Nelson Goodman's approach that the seemingly clear-cut division of world and language – non-verbal language included – as a division between the given and the constructed, between that which is found and that which is made, between the fact and the artefact, is outdated, and that it has even been challenged once and again since the time of the pre-socratics. But, only rarely is history looked at in this way. Any matter we are concerned with, Goodman tells us, is dependent on some manner as the means by which we deal with it. So worlds are but versions and worldmaking begins with one version and ends with another. The message we should learn runs thus: "never mind mind, essence is not essential, and matter doesn't matter".[3] Goodman goes on in claiming that we choose the facts as much as the frameworks, though this statement should better be split into two complementary statements: We produce the facts as much as the frameworks *and* we experience the frameworks as much as the facts. Constructions, when serving cognitive purposes, are always *re*constructions.

The last two Aristotelian categories, ποιεῖν and πάσχειν, which seemed forgotten throughout most of modern philosophy in the tradition of Descartes, though they play an important role both in Spinoza and Leibniz, will enjoy a lively comeback as the two sides we are concerned with when doing something: you do it yourself (active) and you recognize others (including yourself!) doing the same (passive [with respect to the content of recognition]). In fact, the two sides reoccur in the model of an elementary dialogue-situation with two agents being engaged in the process of acquiring an action-competence. At each given instant just one of the agents is active – a >real< agent – and the other agent – the >potential< agent or >patient< – is passive. The agent in active role is performing an action, i.e., he is able to produce different tokens of the same type, while the agent in passive role is recognizing an action, i.e., he sees different tokens as belonging to the same type. One has learned an action, if one is able to play both roles: while acting you know what you are doing, or, conversely, if you don't know what you are doing, you don't act. Another way of saying this would be: Each action appears in two perspectives, in the I-perspective by performing the action (= producing an action token) – it should be called the *pragmatic* side of an action, or its >natural< side – and in the You-perspective when recognizing the action (= witnessing an action type) which should be called

[3]N. Goodman, *Ways of Worldmaking*, Hassocks (Sussex): The Harvester Press 1978, p 96.

its *semiotic* or >symbolic< side. We have come across the first step to execute the program of >naturalizing language< and other symbol systems, and, at the same time, of >symbolizing world<, in order to bridge the gap between the two.

For further guidance we may turn to Peirce again. He sketches a way of deriving signs out of objects in more or less the same manner as I just did, the difference being that he proceeds upside down.

> "If a Sign is other than its Object, there must exist, either in thought or in expression, some explication or argument or other context, showing how – upon what system or for what reason the Sign represents the Object or set of Objects that it does. Now the Sign and the Explanation together make up another Sign, and since the Explanation will be a Sign, it will probably require an additional explanation, which taken together with the already enlarged Sign will make up a still larger Sign; and proceeding in the same way, we shall, or should, ultimately reach a Sign of itself, containing its own explanation and those of all its significant parts; and according to this explanation each such part has some other part as its Object. According to this every Sign has, actually or virtually, what we may call a *precept* of Explanation according to which it is to be understood as a sort of emanation, so to speak, of its Object."[4]

The argument of Peirce calls for something which is a sign of itself, that is, which combines object status and sign status, or better: which functions in both ways. The basic point of his pragmatic foundation of semiotics was to give an account of the process of separation between sign and its object within the framework of his Pragmatic Maxim.[5] And the arguments used for this purpose are themselves to be understood as sections of an open sign-process on the level of reconstruction. And it is these sections that may be looked at as conceptualizations of generalized Wittgensteinian language-games. Now, the descending sequence of interpretants ends with an ultimate logical interpretant[6] which is identified as a habit-change. As stated in the beginning, already, such a habit-change, in contemporary terminology, is nothing else but the acquisition of an action-competence such that all the ways of dealing with the object in respect of what is signified by the initial sign are included. And within the process of acquisition, if it is modelled as an elementary dialogue-situation, the two perspectives of agent and patient may count, respectively, as action on the object-level in performing the action, and action on the sign-level in recognizing the action through the

[4]C. S. Peirce, Meaning [1910], in: *Collected Papers of Charles Sanders Peirce* I-VI, C. Hartshorne/P. Weiss (eds.), Cambridge Mass.: Harvard University Press 1931-1935, 2.230.
[5]Cf. B. M. Scherer, *Prolegomena zu einer einheitlichen Zeichentheorie*: Ch. S. Peirces Einbettung der Semiotik in die Pragmatik, Tübingen: Stauffenburg Verlag 1984.
[6]Cf. C. S. Peirce, *Collected Papers* 5.476.

performance functioning as a representative of any other performance. Thus, habit-changes are, indeed, entities which are signs of themselves. We may finally conclude that a *verbal* sign of an object *signifies* a range of possibilities of dealing with that object. Even more generally, by deleting the dummy term 'object', it might be said that having the competence for such a sign-action – being a verbal sign-action, it functions symbolically – is tantamount to >knowing<, by that very action, of a whole range of further actions which may be said to be signified by the (symbolic) sign-action.
We arrive at the following equivalence: Knowing an action, in the sense of being acquainted with it, is knowing ways of dealing with that action. And this implies that knowing an arbitrary object is equivalent to treating this object as a sign of its distinctions, i.e., of its internal structure which is exhibited step by step in an open sign-process.

II

Now, within the model of acquisition of an action-competence by an elementary dialogue-situation it is important to make some further distinctions. They are based on the observation that producing an action-token and witnessing an action-type, i.e., I-perspective and You-perspective of an action, are inseparably bound together and cannot be treated in isolation from each other. The model of acquisition of action-competence is a model of actions as a means and not yet of actions as objects which, in order to be accessible, will in turn be dependent on other actions as a means of dealing with objects. Dialogical construction as a means of study asks for self-application such that the interdependence of the status of being-a-means and the status of being-an-object, hence of >epistemology< and >praxeology< on the one hand, and of >ontology< on the other hand, is laid bare. Actions as a means are characterized by their two sides as they arise from the two perspectives, from *singular* performance in I-perspective and *universal* recognition in You-perspective. Yet, when performing is understood to be a case of producing (an action-token) and, analogously, recognizing to be a case of witnessing (an action-type), the action in question is treated as an object, in fact, sometimes even two objects, the token as an external or >corporeal< particular and the type as an internal or >mental< particular. But, even if action particulars, i.e., individual acts, are treated uniformly without being split into external and internal entities, particularity is to be kept strictly distinct from singularity and universality. Usually, in the terminology of type and token, where types are treated logically as generated >by abstraction< out of tokens, and where tokens originate >by concretion< from types (rather than looking at the relation of tokens to types in a psychological fashion as a relation of external to internal particulars), both types and tokens are (individual) objects, yet of different logical order, which are related in standard notation as sets to their elements. At the lowest level, if there is one, the final universe of discourse is located, i.e., a world of elementary individual objects, the *particulars*, to which everything else will have to be reduced. Such an account, by neglecting the distinction between particularity and singularity as well as universality, violates the inseparability of (producing a) token and (witnessing a) type in the context of actions as a means, or, rather, it exhibits an equivocation in the use of 'type' and 'token'. It is necessary to relinquish both the equivalence of 'performing an action' with 'producing an action token' and the equivalence of 'recognizing an action' with 'witnessing an action type'.

Instead, performation is performation of something singular and recognition is recognition of something universal, whereas producing (a token) together with its twin activity of witnessing (a type) occur with respect to something particular. Now, if types and tokens are not in this way construed as particulars that are produced or witnessed, respectively, they should be identified, in tune with action as a means, with universal features and singular ingredients of particulars that are exhibited by actions which deal with them. Particulars together with the situations (of acting) of which they occupy the foreground are *appropriated* by performing an action which deals with them, and they are *objectified* by recognizing such an action. It should be noted that neither universal features nor singular ingredients have object status by themselves; they remain means with respect to (particular) objects. Universals cannot be appropriated and singulars cannot be objectified. Hence, in performances of an action that is dealing with a particular you (pragmatically) *present* one of the (singular) token ingredients of this particular, whereas in recognitions of an action that is dealing with a particular you (semiotically) *represent* one of its (universal) type features. Switching from the language of means – pragmatic means are singular, semiotic means are universal – to the language of objects (= particulars) you may say that it is individual acts that provide both services, of presentation with respect to its performance perspective and of representation with respect to its recognition perspective. With recourse to a traditional terminology it may be said, in a >spiritualistic< version, that an individual act has been >aimed at< in a singular performance and will >originate< from a universal recognition, but it could as well be said, in a >naturalistic< version, that an individual act was >caused< by a singular performance and is >conceived< by a universal recognition. In appropriation as well as in objectivation of particulars of arbitrary category, like individual acts, individual things or events, groups of individuals or other non-individual particulars, etc., the actions of dealing with particulars are used as a means, of presentation (of singular tokens – the way a particular is present) in the case of appropriation and of representation (of universal types – the way a particular is identified) in the case of objectivation.

Particulars may be said to act as appearances of >substances<, i.e., some part of the whole out of like singular tokens is a *part* of the particular, and as carriers of >properties<, i.e., the particular is an *instance* of a universal type.[7] Therefore, in order to avoid misunderstandings, instead of 'perform' we will, henceforth, say 'actualize', and we say 'schematize' instead of 'recognize'. Within the model of an elementary dialogue-situation where two agents are engaged in the process of acquiring an action-competence, the activities of actualizing and schematizing should not be understood as performances of two separate actions; it is one action the competence of which is acquired by learning to play both the active and the passive role. Active actualization makes the action appear in I-perspective, passive schematization lets it appear in You-perspective. Any action as a means is characterized by its pragmatic and its semiotic side, and it doesn't make sense as yet to speak of the action as an >independent< object(-type) split into particulars, i.e., some set of individual acts. In order to achieve the switch from action as a means to action as object, it is essential to iterate the process of acquiring an action-competence by turning the two sides of an action into proper actions by themselves, i.e., into actions of dealing with the original action under its two perspectives such that the (secondary) action-competences additionally required will have to be modelled in turn by means of (now non-elementary) dialogue-situations. Such a further step may be looked at as an application of the *principle of self-similarity*.

What has to be done is to schematize and to actualize the elementary dialogue-situation, i.e., to create a He/She-perspective towards the I/You-situation such that, on the one hand, He/She becomes a (secondary) You-perspective with respect to I/You as I, and, on the other hand, He/She becomes a (secondary) I-perspective with respect to I/You as You. In the first case you gain an >exterior view< of the original action by acquiring a second level action (with respect to the original

[7] A particular wooden chair, for example, acts as a carrier of all the properties conceptualized by 'wooden', and as an appearance of the substance >wood<, inasmuch as a part of >the whole wood< may be considered to be a part of the particular wooden chair; cf. the entry 'Teil und Ganzes' in: *Enzyklopädie Philosophie und Wissenschaftstheorie* IV, J. Mittelstraß (ed.), Stuttgart-Weimar: Metzler 1996, 225-228.

action) which functions as one of the indefinitely many *aspects* of the original action: The You-perspective is turned into the schema of a second level action out of an indefinite series of second level actions. In the second case you gain an >interior view< of the original action by acquiring a second order action (with respect to the original action) which functions as one of the indefinitely many *phases* of the original action: The I-perspective is turned into an actualization of a second order action out of an indefinite series of second order actions. The semiotic side of an action is split into a multiplicity of aspects or (secondary) You-perspectives, and the pragmatic side of an action likewise into a multiplicity of phases or (secondary) I-perspectives.

By (dialogical) construction, it is in its active role that an aspect-action is I-You-invariant and, in this sense, >objective<, whereas a phase-action is I-You-invariant in passive role, only. Hence, by applying the principle of self-similarity once again to both aspects and phases, the pragmatic side of an aspect-action is split into a multiplicity of objective *articulations* or *sign-actions*, and the semiotic side of a phase-action into a multiplicity of objective *mediations* or *partial actions*. Any one of the sign-actions is a means to designate the original action, and any one of the partial actions is a means to partake of the original action, where designating and partaking function with the proviso that the original action itself is turned from a means into an object. In fact, an action as object – things, events, and other categories of entities are included among actions by identifying an entity[-type] with the action[-type] of dealing with the entity – is constituted, on the one hand >formally<, by *identification* of the schemata of the aspect-actions, i.e., of their >subjective< semiotic side (when turned into a multiplicity of full-fledged actions, one would get perceptual actions), and, on the other hand >materially<, by *summation* of the actualizations of the phase-actions, i.e., of their >subjective< pragmatic side (when turned into a multiplicity of full-fledged actions, one would get poietic actions). On the one side, through identification, an action as object is a semiotic (abstract) invariant of which one partakes by means of a partial action, and on the other side, through summation, it is a pragmatic (concrete) whole which one designates by means of a sign-action. With respect to the additional dialogue-situations modelling the acquisition of second-order-action-competences as well as second-level-action-competences the original action as object occurs within a situation which, in fact, is responsible for individuating the original action as object. The move of objectivation from action as a means to action as object is accompanied by a split of the action into (action-)particulars such that the respective invariants may be treated as *kernels* of the schemata of aspects (= universalia), and the respective wholes correspondingly as *closures* of the actualizations of phases (= singularia). An objectival foreground together with a situational background will semiotically be a constant foreground against a variable background (= the same particular in different surroundings, i.e., its varying external structure), and it will pragmatically be a variable foreground against a constant background (= different particulars [of the same kind] by their varying internal structure, in the same surrounding). Both together, kernel and closure – >form< and >matter< in philosophical tradition[8] – make up a *particular within a situation*. Working backwards again, i.e., making the countermove of appropriation, the schemata of the kernel and the actualizations of the closure, are realized in representations and presentations, respectively, by schematizing and actualizing a particular (in a situation) as explained above.

Dialogical construction of particulars being dependent on identification of schemata of aspects and on summation of actualizations of phases, implies the establishment of mutual independence between objectival foreground and situational background. In order to achieve this, a specially chosen articulation has to act as a substitute for arbitrary aspects with respect to some partial action – such a function of *substitution* may be articulated by *rules of translation* among aspects – and will be called *symbolic articulation*. Constant foreground and variable background will thus become independent of each other. Analogously, any mediation will have to acquire the function of having the phase to which it belongs extended by arbitrary other phases with respect to some sign-action – such a function of *extension* may be articulated by *rules of construction* for phases – and will be called *comprehensive mediation*. In this case, constant background and variable foreground are made independent of each other. Both constructions together guarantee that particulars contrast with their surroundings.[9] By symbolic articulation, a symbolic sign-action, you arrive at a semiotically determined particular in actualized situations, i.e., the particular is *symbolically represented*, whereas by comprehensive mediation, a comprehensive partial action, you arrive at pragmatically determined particulars in a schematized situation, i.e., the particulars are *symptomatically present*.

The semiotic side of partial actions (>what you do<) and the pragmatic side of sign-actions (>how you speak<), together they make up the *ways of life* (of the agents). Correspondingly, the pragmatic side of par-

[8]The treatment of particulars as *mixta composita* (σύντετα) out of εἶδος or *forma*, and ὕλη or *materia*, is due to Aristotle as explained, e.g., in the commentary of Alexander of Aphrodisias on Aristotle's *Metaphysica*, cf. *Comm. in Arist. Graeca* I, p 545, line 30ff; 497, line 4ff.

[9]For an explicit dialogical construction of both identification and summation, cf. my 'Rede zwischen Aktion und Kognition', op.cit. [note 2], p 145ff.

tial actions (>how you act<) and the semiotic side of sign-actions (>what you say<), together they make up the *world views* (of the agents).

III

Articulation is signified canonically by the result of a sign-action, an *articulator* (= >signifiant< in the sense of F. de Saussure); articulation has a pragmatic side, i.e., it is a sign-action in its being an action, and a semiotic side, i.e., it is a sign-action in its being a sign. Semiotically, articulation is effected by uttering an articulator that has to be taken as a (verbal) type, in a speech situation; and if it is treated as functionally equivalent with any other way of articulation, including non-verbal ones, it acts as a *symbolic articulator*. Again semiotically, i.e., as a *sign*(-action), it shows its two sides, a pragmatic one and a semiotic one. The pragmatic one is to be called *communication*, or the side with respect to persons or subjects, and the semiotic one is to be called *signification*, or the side with respect to particulars or objects (these two sides in their function being reminiscent of Plato's λέγειν and ὀνομάζειν). By iteration, communication splits into (content of) *predication* on the semiotic side, and *mood* (of predication) on the pragmatic side, whereas signification splits into (intent of) *ostension* on the pragmatic side, and *mode of being given* on the semiotic side. Any predication can take place only by using a mood, and any ostension is effected only by using a mode of being given. We have strictly to distinguish: content and mood of predication, intent and mode of ostension. The moods of predication are, of course, speech acts, and only with respect to a mood a predication contains a claim, e.g., a truth claim.

Articulation of a mood of predication yields *performators* on the semiotic side of the articulation, whereas articulation of a mode of ostension yields *perceptions* (= Wahrnehmungsurteile) on the pragmatic side of the articulation. Without such a second order articulation of mood and mode, we have arrived at one-word sentences 'P' (in a mood and using a mode of being given) by uttering the articulator 'P'.

With the next step we introduce the separation of significative and communicative function, two functions that coincide with showing and saying in the terminology of Wittgenstein's *Tractatus*. Separation with respect to predication, i.e., the semiotic side of communicative function, yields: $\delta \mathrm{P} \varepsilon \mathrm{P}$ (this P [= something done] is P[-schematized]), or, alternatively, $\sigma \mathrm{P} \pi \mathrm{P}$ (the universal P [= something imagined] is P-actualized), whereas separation with respect to ostension, i.e., the pragmatic side of significative function, yields: $\delta \mathrm{P} \zeta \mathrm{P}$ (this P belonging to P), or $\kappa \mathrm{P} \xi \mathrm{P}$ (the whole P [= something intuited] being P-exemplified).

The operators: demonstrator 'δ' and attributor 'ε' (= copula), respectively, neutralize the communicative function and the significative one; 'δ' keeps the significative function and 'ε' the communicative one, with the result that 'δP' plays a singular role and 'εP' a universal one. In the terminology of logic or semiotics, 'δP', which is used to >ostend< P, is an *index* of an actualization of the action articulated by 'P', whereas 'εP', which is used to >predicate< P, is a predicator serving as a *symbol* of the schema of action P. *Predication* εP and *ostension* δP with its respective associates: form of a proposition '$__\varepsilon$P' and form of an indication 'δP$__$', are the modern equivalents of the traditional >forms of thinking< and >forms of intuition<. Proceeding dually with respect to the predication oriented and, hence, semiotic distinction 'singular-universal', it is also possible to use another pair of operators, universalisator 'σ' and presentator 'π', where 'σP' has only significative function with universal role and 'πP' only communicative function with singular role. In the second case which works with respect to the ostension oriented and, hence, pragmatic distinction 'active-passive', either demonstrator 'δ' and partitor 'ζ', or, dually, totalisator 'κ' and exemplificator 'ξ', serve the same purpose: 'δ' and 'κ' keep the significative function in active and passive role, respectively, whereas 'ζ' and 'ξ' keep the communicative function, here in passive and active role, respectively.

What is not yet available up to now and what would not even make sense, are >propositions< of kind δPεQ and >indications< of kind δQζP. The reason why these expressions don't make sense, is simply the following: 'δP' is not the kind of expression to occupy the empty place in a propositional form '$__\varepsilon$Q' with Q$\neq$P, and 'ζP' is not the kind of expression to occupy the empty space in an indicational form 'δQ$__$' with Q$\neq$P. Instead, we have to introduce *individuators* 'ιP' in order to refer to particulars, i.e., the situation-dependent units of the action articulated by 'P'; >things< as well as objects of other categories, any one (type) of them being identified with the action(-type) of arbitrary dealings with an object(-type), hence, any of the so-called >natural kinds<, are, of course, included among the P. Particulars, be they individual things or events, individual acts or processes, are composed out of kernels of schemata of aspects: $\sigma(\iota$P) (= invariants), together with closures of actualizations of phases: $\kappa(\iota$P) (= wholes). Hence, particulars may be considered to be half thought and half action. Using individuators we, now, may write down *eigen-propositions* ιPεP as well as *eigen-indications* δPιP (short for: δP$\zeta\iota$P), and it is possible to render these versions of saying and showing with the help of the four operators: demonstrator, attributor, universalisator, and totalisator, in the following traditional way:

(i) In the case of saying ($\iota P \varepsilon P$): the universal σP is *predicated of* a P-particular by means of 'εP' (or: within the proposition $\iota P \varepsilon P$, the individuator is a sign of an indication, and, hence, functions as a nominator of a P-particular, i.e., within the proposition $\iota P \varepsilon P$, *nomination by 'ιP'* is shown), and

(ii) In the case of showing ($\delta P \iota P$): *ostending* the whole κP *at* a P-particular by means of 'δP' (or: within the indication $\delta P \iota P$, the individuator is a sign of a proposition, and, hence, functions to say that participation in a P-particular holds, i.e., within the indication $\delta P \iota P$, *participation in ιP* is said).

Hence, *reference* to particulars ιP includes both nomination of $\kappa(\iota P)$, i.e., of the *matter* of ιP, and participation in $\sigma(\iota P)$, i.e., in the *form* of ιP. As a remark, it may be added that nominating is the articulation of designating by symbolic articulation, and, analogously, participating is the articulation of partaking by comprehensive mediation.

The composition of P, e.g., wood, and Q, e.g., chair, is a result of separating speech-situation and situation-talked-about. It can be realized by analyzing and reconstructing what happens when, e.g., in a Q-situation you are uttering 'P'. In the foreground of the situation-talked-about which is articulated by 'P', there are two particulars to be welded. It may come about in either of two possible ways:

(i) An aspect (with its schema being) out of $\sigma(\iota P)$ coincides with a phase (actualizations of which being) out of $\kappa(\iota Q)$, e.g., sitting on a wooden chair as a phase-action with respect to chair is simultaneously an aspect-action >sitting on the wood of the chair< with respect to wood;

(ii) A phase out of $\kappa(\iota P)$ coincides with an aspect out of $\sigma(\iota Q)$.

In the first case you may articulate the coincidence *predicatively* by εP_Q (= is a wood of [a] chair), in the second case *ostensively* by $\delta(QP)$ (= this wood with the form of [a] chair).
Instead of $\delta P_Q \ \varepsilon \ P_Q$ we may write $\iota Q \varepsilon P$(= ιQ *is* P, or: this [particular] chair is wooden), and likewise, instead of $\delta(QP) \ \zeta \ (QP)$, it is possible to write $\delta P \iota Q$ (short for: $\delta P \zeta \iota Q$) (= δP *at* ιQ, or: this dealing with wood belonging to this [particular] chair). Hence, 'εP' acts as a *symbol* for the result of schematizing ιQ, whereas 'δP' acts as an *index* for the result of actualizing ιQ.

Actualizations ostending $\kappa(\iota Q)$ are simultaneously actualizing the universal σP [= $\delta Q \varepsilon P$; equivalent with: $\sigma P \pi Q$]; schemata predicating a universal out of $\sigma(\iota Q)$ exemplify simultaneously the whole κP [= $\delta P \zeta Q$;

equivalent with: κPξQ] by being the form of an element of a partition of κP into a class ϵP. Hence, $\sigma(\iota$Q) is an *appearance* of the whole or *substance* κP, and $\kappa(\iota$Q) is a *carrier* of the universal or *property* σP. An *indication* δPιQ shows that the substance κP is ostended at ιQ by means of 'δP'; a *proposition* ιQεP says that the property σP is predicated of ιQ by means of 'εP'.

Involution as transformation of phase-structures (= internal structure) into aspect-structures (= external structure), and vice versa, can now be proved to exist in a one-to-one way.[10] So, it makes indeed sense to say: ιQ consists both of phases such that the closure of their actualisations is $\kappa(\iota$Q), and of aspects such that the kernel of their schemata is $\sigma(\iota$Q).
And, taking our example, a phase-action with respect to chair which is the pragmatic side of a dealing with chair, being turned into an independent action, can be mapped one-one onto (>seen as<) an aspect-action with respect to wood which is the semiotic side of a dealing with wood, being turned into an independent action.

As a historical remark, it may be added that the two sides of a particular ιQ, the concrete whole $\kappa(\iota$Q) and the abstract invariant $\sigma(\iota$Q), correspond neatly to >body< or >phenomenon< and >soul< or >fundament< of a monad as it is conceived in the *Monadologie* of Leibniz.[11] It may also be useful to observe that the identification of $\delta P_Q \; \varepsilon \; P_Q$ with ιQεP, i.e., the introduction of (one-place) elementary propositions, is closely related to Reichenbach's transition from a thing-language to an event-language articulated with the help of an asterisk-operator which moves the predicative ingredients of a subject term of an (one-place) elementary proposition into its predicate term, e.g., from 'this man is smoking' you arrive at 'smoking of [this particular] man', or: $(\iota Q\varepsilon P)^{*} = P_Q$.[12]

Now, we have reached the usual account of (one-place) predication where >general terms< 'P' – they should more appropriately and in line with the Fregean analysis of general terms as propositional functions or predicators be rendered as 'εP' – serve to attribute properties to particulars of an independently given domain of Q-objects, in the simplest case referred to by deictic descriptions 'ιQ' that are special cases of >singular terms< [another use of 'singular'!] or nominators.

[10] Cf. K. Lorenz, On the Relation between the Partition of a Whole into Parts and the Attribution of Properties to an Object, in: *Studia Logica* 36 (1977), 351-362.
[11] Cf. for further corroboration various essays in: *Leibniz and Adam*, M. Dascal/E. Yakira (eds.), Tel Aviv: University Publishing Projects 1993.
[12] H. Reichenbach, *Elements of Symbolic Logic*, Toronto: Macmillan 1947, § 48.

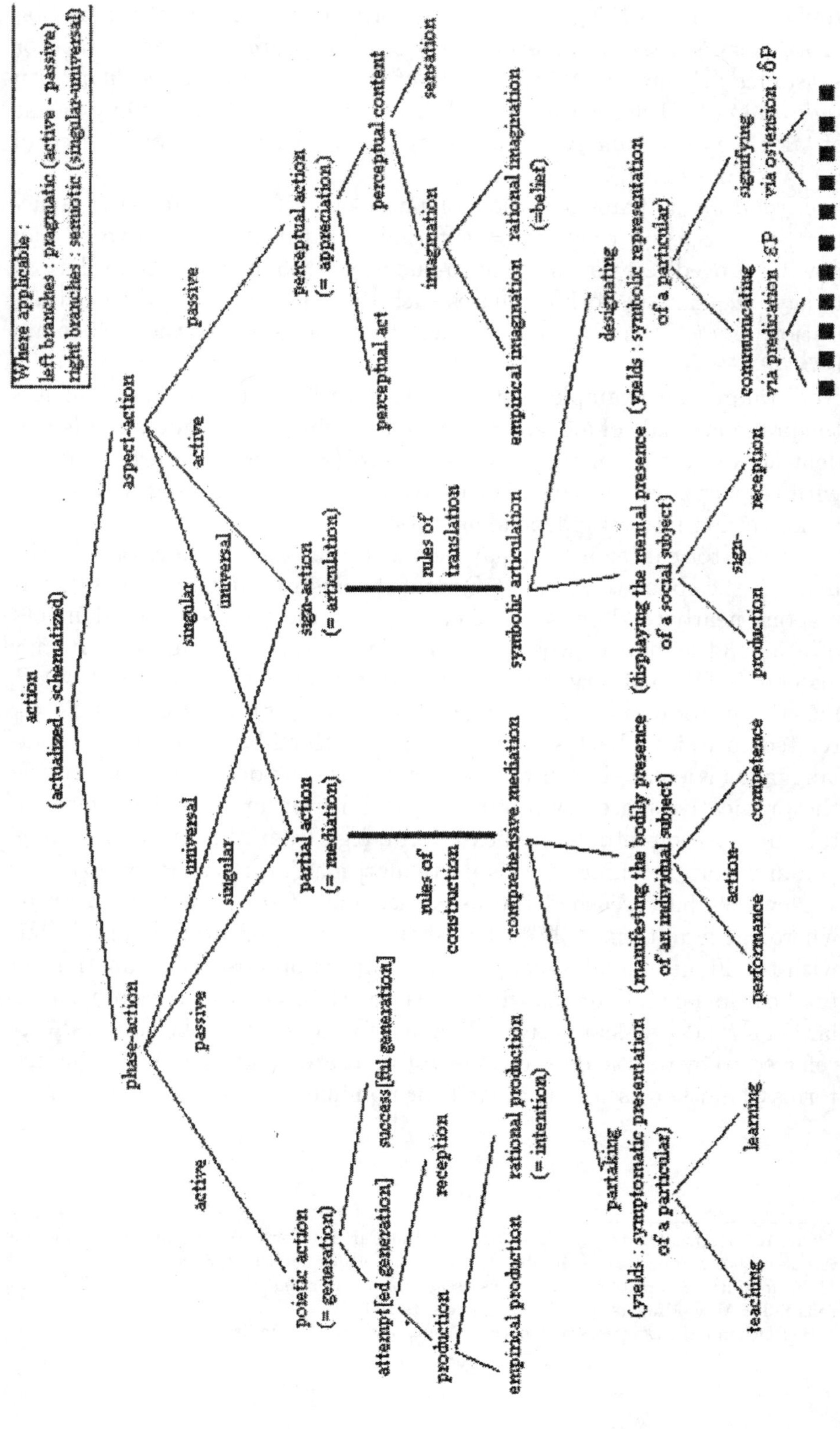
Where applicable :
left branches : pragmatic (active - passive)
right branches : semiotic (singular-universal)
action
(actualized - schematized)
phase-action
aspect-action
active
passive
universal
singular
singular
universal
passive
active
poietic action
(= generation)
partial action
(= mediation)
sign-action
(= articulation)
perceptual action
(= appreciation)
attempt[ed generation]
success[ful generation]
production
reception
empirical production
rational production
(= intention)
rules of
construction
rules of
translation
comprehensive mediation
symbolic articulation
perceptual act
perceptual content
imagination
sensation
empirical imagination
rational imagination
(=belief)
partaking
(yields : symptomatic presentation
of a particular)
(manifesting the bodily presence
of an individual subject)
(displaying the mental presence
of a social subject)
designating
(yields : symbolic representation
of a particular)
teaching
learning
performance
action-
competence
production
sign-
reception
communicating
via predication : εP
signifying
via ostension : δP

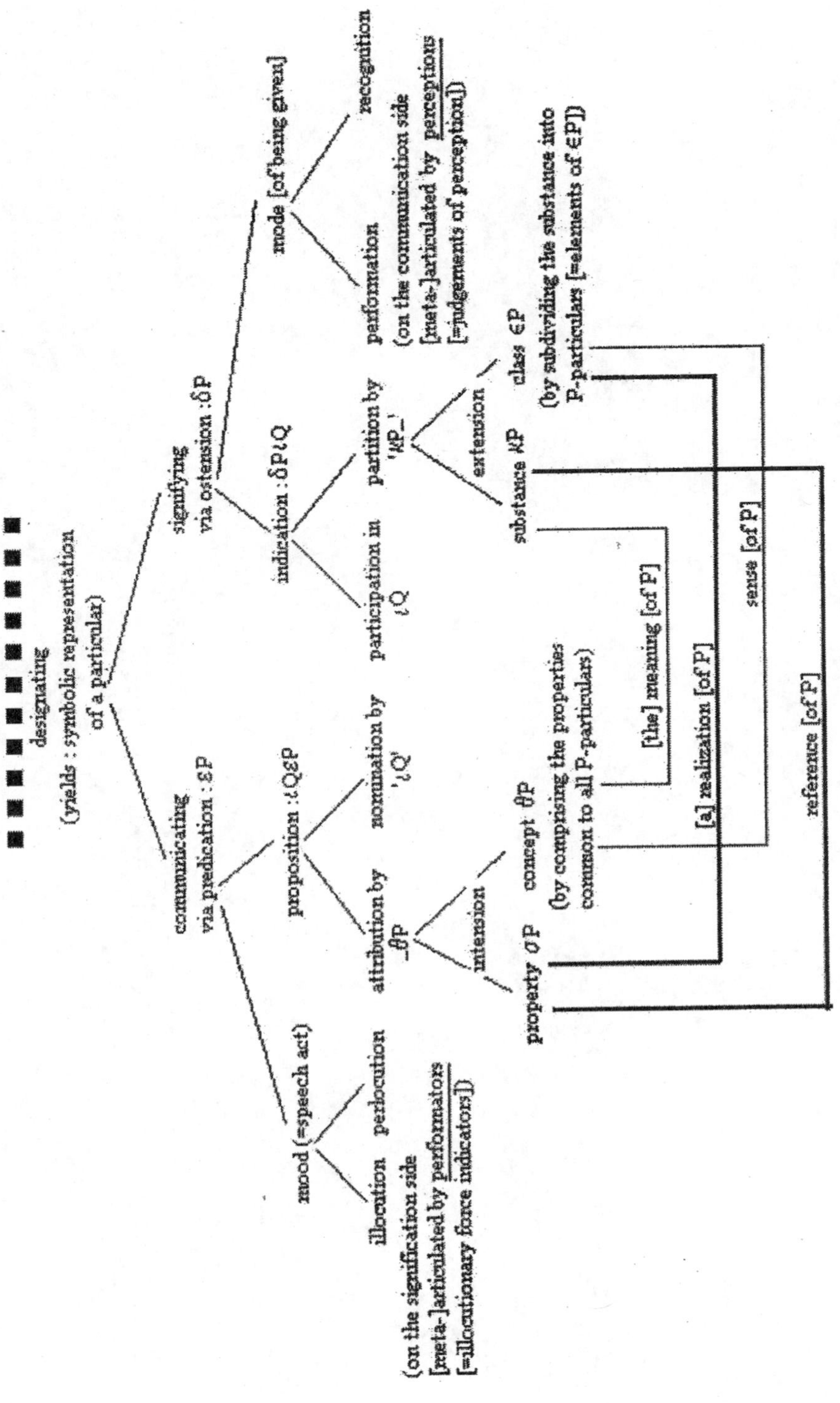
designating
(yields : symbolic representation of a particular)
communicating via predication : εP
signifying via ostension : δP
mood (=speech act)
proposition : ιQεP
indication : δPιQ
mode [of being given]
illocution
perlocution
attribution by _θP
nomination by 'ιQ'
participation in ιQ
partition by 'κP_'
performation
recognition
(on the signification side [meta-]articulated by performators [=illocutionary force indicators])
(on the communication side [meta-]articulated by perceptions [=judgements of perception])
intension
extension
property σP
concept θP
substance κP
class ∈P
(by comprising the properties common to all P-particulars)
(by subdividing the substance into P-particulars [=elements of ∈P])
[the] meaning [of P]
[a] realization [of P]
sense [of P]
reference [of P]

Chapter 17

ON HOW TO BE A DIALOGICIAN

A Short Overview on Recent Developments on Dialogues and Games

Shahid Rahman and Laurent Keiff
Université Lille 3

Abstract We will take as one of the main issues of this paper the challenge which the dialogical approaches offer to the relation of semantics and pragmatics concerning the concept of proof (strategy) and proposition (game). While our aim here will be to present the main technical and philosophical features of what can be seen as the dialogical approach to logic, illustrated through both very well known and new dialogics, we would also like to delineate the common *pragmatic attitude* which constitutes the cohesive force within the dialogical universe.

The paper aims to answer a question asked by Daniel Vanderveken at the Workshop on *Dialogue and Logic* Grenoble 2002. In that occasion, he asked whether there is a general framework for the study of the various interactions between Dialogue and Logics, which undoubtedly have today a vigorous research momentum. The question is interesting, because while a lot of ink has been displayed concerning the differences between these approaches, much less has been said about their common conceptual roots — with the notorious exception of Johan Van Benthem's program, who proposes to use the very concept of game as the keystone of a general framework. To start, let us present here very briefly a way to systematise the recent history of the origins of these various developments, conceived as different outcomes from a general dialogical approach, without pretending to be impartial. Indeed, taking as ours Graham Priest's words, we concede that "philosophy is contentious".

D. Vanderveken (ed.), Logic, Thought & Action, 359–408.

1. A brief survey of dialogic.

The first paper on dialogic by Paul Lorenzen was published more than forty years ago. Since the publication of *Logik und Agon* [1958] different dialogical systems and related research programmes have been developed. Among them two main approaches can be distinguished which followed two main targets:

The first group could be characterised as seeking the dialogical (or argumentative) structure of logic. Namely:

The constructivist approach of Paul Lorenzen and Kuno Lorenz, who sought to overcome the limitations of *Operative Logic* by providing dialogical foundations to it. The method of semantic tableaux for classical and intuitionistic logic as introduced by Evert W. Beth [1955] could thus be identified as a method for the notation of winning strategies of particular dialogue games (cf. Lorenzen and Lorenz [1978], Lorenz [1981], Felscher [1986]).

The game-theoretical approach of Jaakko Hintikka – known as GTS – who recognised at a very early stage that a two-player semantics offers a new dynamic device for studying logical relations in the framework of games in which loss of information could happen. This approach is better known and opened many new lines of research developed by Hintikka and co-authors, specially the *semantic games* which offer a deep and thorough insight into the notion of scope implemented by the *Independence-Friendly Logic*, the *interrogative games* which are essentially epistemic games and the *formal games* of theorem-proving which deal with the logical truth of propositions and not with their material truth (as in the semantic games) or with one's knowledge of their truth (as in the epistemic games) (cf. Hintikka and Sandu [1996], Hintikka [1996], [1996–1998]).

The linear logic approach of Yves Girard's *Linear Logic*, in the dialogical version suggested in 1992 by Andreas Blass. In this approach the dialogical concept of propositions as proof-theoretical games has been combined with the games of imperfect information of the game theoretical approach. (Cf. Girard [1993], [1998a], [1998b], Blass [1992], [1997-2002], Hyland [1997]).

The second group could be conversely characterised as seeking the logic –including informal logic– and mathematics of dialogues and argumentation. Namely:

The argumentation theory approach of Else Barth and Erik Krabbe [1982], cf. also Gethmann [1979], who sought to link dialogic with the informal logic or *Critical Reasoning* and rhetoric originated by the seminal work of Chaim Perelman (cf. Perelman and Olbrechts-Tyteca [1958]),

Stephen Toulmin [1958], Arne Naess [1966] and Charles Hamblin [1970] and developed further by Ralph Johnson [1999], Douglas Walton [1984], John Woods [1988] and associates. As will be detailed at the end of this section, in this approach the structural rules of dialogic implement kinds of games at the border between logics and rhetoric.

These groups, however, followed in their origins separate paths and, with some occasional exceptions, did not actually pool their results in a common project. More recently two very important lines of research, coming from computer sciences and linguistics, aimed to combine the lines of these different approaches:

Logic in Games, the approach of Johan van Benthem and his group who aim to study the many interesting interfaces between logic and games (including those of mathematical game theory) including models for multi-agent activities. Logic *in* Games should be understood too as the games of logic and the logic of games (cf. Van Benthem [2001-04]).

Argumentation models for Non-Monotonic reasoning: Henry Prakken, Gerard Vreeswijk and Arno Lodder stressed the argumentative and pragmatic structure of non-monotonic reasoning which seems to have a purely dynamic character (cf. Prakken/Vreeswijk [2001], Lodder [1999]).

Quite recently a new demand for *a diversity of logical systems* which could serve various applications has arisen from artificial intelligence, computer science and linguistics, as well as from legal reasoning, philosophy and psychology. This demand has caused extensive research in different new and old logical systems and rendered a gradually increasing interest in the study of dialogic from different research areas.[1] Several factors are responsible for this new impulse which has a sharpened focus in computer science and artificial intelligence. Let us delineate some of these factors:

- The question of how a *common general frame* for studying most of these logics could be formulated was now of particular importance. One relevant step in this direction was taken by the formulation of a condition which is sometimes known as *Dosen's Principle* and which proposes that alternative logical systems could be obtained by modifying structural rules against the background of a fixed set of rules for the logical particles (cf. Dosen [1988]).[2] Actually, an analogous principle represents a characteristic feature of dialogic.

[1] Cf. Walton [1985], 259-274.

[2] Cf. Wansing [1994], 128. 'Dosen's Principle' plays a central role for example in *Display Logic and Substructural Logics* (cf Belnap [1982] and Wansing [1998]).

The set of rules in dialogic is divided into particle rules and structural rules. The particle rules determine how the corresponding formulæ can be attacked and defended for each logical particle in a not yet determined kind of game, whereas the structural rules determine the general course of a dialogue fixing the kind of game. Now, it is easy to see that one can obtain different logical systems by changing only the set of structural rules while retaining the same set of particle rules. For example, classical and intuitionistic logic only differ dialogically in a single structural rule. The converse is also possible: one can generate different logics by introducing new particles. A prominent example of producing logics with the help of new particles is Jean-Yves Girard's *Linear Logic*. We can thus call this way of producing logics *Girard's Principle.*[3] Hence, the differentiation of rules into particle and structural rules of dialogic makes it simple to generate new logics by a systematic variation and combination of the structural and particle rules (cf. Rückert [1999]).[4]

Another way to see this is that the structural rules determine how to label formulæ – number of the move, player (Proponent or Opponent), formula, name of move (attack or defence) – and how to operate with these labelled formulæ. Dov Gabbay has proposed a general theory of labels called *Labelled Deductive Systems*, where object-language features reside side by side with metalogical ones (cf. Gabbay [1996]).[5]

- In addition to these procedures another aspect should be considered, namely interactions. As observed by Henry Prakken and Gerard Vreeswijk, the 'players' of the argument games of computer science models are often not real actors but stand for the alternate search for arguments and counter-arguments that is required by the proof theory at stake. An embedding of argumentation systems in dialogic would yield an account of how their input theories are constructed dynamically during disputes, instead of being fixed and given in advance (cf. Prakken and Vreeswijk [1999], 88–89, 2001, 219-318).[6]

---

[3]Cf. Girard ([1993], [1998a], [1998b]), cf. also Blass ([1992], [1997], [2002]) and Hyland [1997].

[4]The application of these principles motivated the publication of papers on various dialogics (cf. Rahman [1993], Rahman, Rückert and Fischmann [1997], Rahman [2001], Rahman and Carnielli [2000], Rahman and Van Bendegem [2002]).

[5]Cf. Fitting ([1983], [1993]), Rahman and Rückert [1999], Rahman [2002].

[6]This has been stressed before by Gabbay ([1996], 20–34).

The dynamic aspect of the semantics of dialogic builds a bridge to the field of informal logic, particularly in the case when structural rules "complete" the local semantics of the particle rules into a kind of game where e.g. the aim is persuasion rather than logical validity. Indeed structural rules have the role of implementing the meaning of a logical constant determined by the particle rules relative to a given context to which this logical constant will apply. On this view, structural rules complete the meaning schematised by the particle rules analogously to the case of the use of pronouns in natural language, which, though they have a meaning, only obtain a full-fledged meaning when an appropriate context has been described.[7]

2. Logic as a game.

2.1 The language.

Our aim here is to build the conceptual kernel of dialogic[8] in the context of the dialogical reconstruction[9] of first order propositional calculus, in its classical and intuitionist versions.

Let our language **L** be composed of the standard components of first order logic (with four connectives $\wedge$, $\vee$, $\Rightarrow$, $\neg$, and two quantifiers $\forall$, $\exists$), with small letters (a, b, c,...) for prime formulæ[10], capital italic letters (A, B, C, ...) for formulæ that might be complex, capital italic bold letters ($\boldsymbol{A}$, $\boldsymbol{B}$, $\boldsymbol{C}$, ...) for predicators, let our constants be noted τ_i, where i $\in$ N, and our variables the usual (x, y, z, ...). We will also need some special force symbols: $?_{...}$ and $!_{...}$, where the dots stand for indices, filled with some adequate information that will be specified by appropriate rules. An *expression* of **L** is either a term, a formula or a special force symbol. **P** and **O** are two other special symbols of **L**, standing for the players of the games. Every expression e of our language can be augmented with labels **P** or **O** (written **P**-e or **O**-e, called (*dialogically*) *signed expressions*), meaning in a game that the expression has been played by **P** or **O** (respectively). We use X and Y as variables for **P**, **O**, always assuming X $\neq$Y. Other more specific labels will be introduced where needed.

[7]Cf. Rahman/Rückert [2001].

[8]We use the term *dialogic* when we consider the general frame of concepts for any logic.

[9]I.e. sound and complete with respect to these systems. Cf. Felscher [1986] and Rahman [1993].

[10]By *formula*, we understand the traditional wff.

2.2 Particle rules.

An *argumentation form*[11] or *particle rule* is an abstract description of the way a formula, according to its principal logical constant,[12] can be criticised, and how to answer the critics. It is abstract in the sense that this description can be carried out without reference to a determined context. In dialogic, we say that these rules state the *local semantics*, for they show how the game runs locally, in the sense that what is at stake is only the critic and the answer to a given formula with one logical constant rather than the whole (logical) context where this formula is embedded.[13] Hence, the particle rules fix the dialogical semantics of the logical constants of **L** as depicted in Table 2.2.

	$\wedge$	$\vee$	$\rightarrow$
assertion	X - $A \wedge B$	X- $A \vee B$	X-$A \rightarrow B$
attack	Y-$?_L$, or Y-$?_R$	Y- $?_\vee$	Y-A
defence	(respectively) X-A or X-b	X-A, or X-b	X-b

	$\forall$	$\exists$	$\neg$
assertion	X-$\forall xA$	X-$\exists xA$	X-$\neg A$
attack	for any τ Y may choose, Y- $?_{\forall/\tau}$	Y-$?_\exists$	Y-A
defence	for any τ chosen by Y, X-A(x/τ)	for any τ X may choose, X-A(x/τ)	– (i.e. no defence)

(Where A and B are formulæ, and $A(x/\ \tau)$ is the result of the substitution of τ for every occurrence of the variable x in A.)

One more formal way to stress the locality of the semantics fixed by the particle rules is to see these rules as defining a state of a (structurally not yet determined) game. Namely:

Definition (*state of the game*): A *state of the game* is an ordered triple <ρ, σ, A> where:

- ρ stands for a role assignment either R, from players X, Y to only one element of the set {?(attack), !(defence)} determining which player happens to occupy the challenger and which the defender role, or R', inverting the role assignment R of both players (e.g. if R(X)=? and R(Y)=!, then R'(X)=! and R'(Y)=?). The players

[11] Felscher [1985], 218-221.

[12] Determined as usual.

[13] Obviously there is no particle rule for prime formulæ. But particle rules could be augmented with an assignment of atomic (predicate) games to prime formulæ. See the note on GTS at the end of 2.2.

perform their assigned role as challengers (defenders) by stating an attack (or performing a defence) fixed by the corresponding rule.[14]

- σ stands for an assignment function, substituting as usual individuals by variables.
- A stands for a dialogically labelled subformula A with respect to which the game will proceed.

Particle rules are seen here as determining which state of the game S' follows from a given state S without yet laying down the (structural) rules which describe the passage from S to S'.[15] What state follows of S=<R, σ, F>for the X-labelled formula F?

- *Negation particle rule*: If F is of the form $\neg A$ then S'=<R', σ, A>, i.e. Y will have the role of defending A and X the role of (counter)attacking A.
- *Conjunction particle rule*: If F is of the form $A \wedge B$ then S'=<R, σ, A> or S" =<R, σ, B>, according to the choice of challenger R(Y)=? between the attacks $?_L$ and $?_R$.
- *Disjunction particle rule*: If F is of the form $A \vee B$ then S'=<R, σ, A> or S'=<R, σ, B>, according to the choice of defender R(X)=!, reacting to the attack $?_\vee$ of the challenger R(Y)=?.
- *Subjunction particle rule*: If F is of the form $A \Rightarrow B$, then S'=<R', σ, A> and the game might proceed to the state S"=<R", σ, B>, or even the other way round according to the choice of the defender and reacting to the attack A of the challenger R(X)=?.
- *Universal quantifier particle rule*: If F is of the form $\forall x A x$ then S'=<R, $\sigma(x/\ \tau)$, A> for any constant τ chosen by the challenger R(Y)=? while stating the attack $?_{\forall/\tau}$.

[14]We note R" an assignation inverting R', and so on.

[15]The concept of state of a game suggests a connection to GTS. Indeed to establish the connection between these particle rules and the GTS corresponding notions, it suffices to extend the notion of state of the game. A first step would be to add an assignment of so-called atomic games to some prime formulæ relevant for the principal formula of the game, yielding *material games*. The next step would be to replace this assignment by another one, a valuation function giving these prime formulæ a truth value (introducing hereby in dialogic the notion of truth as a primitive), yielding what we call *alethic games*. This extension process makes of the games something increasingly determined by initial conditions (which is precisely the point of GTS, i.e. reasoning under conditions). We can then, once more, see the local semantics of dialogic as the result of an abstraction process, transforming the initial conditions into general rules to free the games from the constraints of a model and displaying the schematic meaning of logical constants for any other logics. In fact the structural formal rule accomplishes this transformation: every prime formulæ could be seen as encoding an atomic (predicate) game which can be retrieved and even "copied" when demanded. This strategy is known in the literature as "copy-cat strategy": X: how do you justify your prime formula a? Y: In the same way (i.e., with the same atomic game) you justify the move **n**, where you conceded that a.

- *Existential quantifier particle rule*: If F is of the form $\exists xAx$ then S'=<R, $\sigma(x/\ \tau)$, A> for any constant τ chosen by the defender R(X)=! reacting to the attack $?_{\exists}$ of the challenger R(Y)=?.

Philosophical remarks, games as propositions.

Particle rules determine dynamically how to extend a set of expressions from an initial assertion. In the game perspective, one of the more important features of these rules is that they determine, whenever there is a choice to be made, who will choose. This is what can be called the pragmatic dimension of the dialogical semantics for the logical constants. Indeed, the particle rules can be seen as a proto-semantics, i.e. a game scheme for a not yet determined game which when completed with the appropriate structural rules will render the game semantics, which in turn will build the notion of validity.

Actually by means of the particle rules games have been assigned to sentences (that is, to formulæ). But sentences are not games, so what is the nature of that assignment? The games associated to sentences are meant to be *propositions* (i.e. the constructions grasped by the (logical) language speakers). What is connected by logical connectives are not sentences but propositions. Moreover, in the dialogic, logical operators do not form sentences from simpler sentences, but games from simpler games. To explain a complex game, given the explanation of the simpler games (out) of which it is formed, is to add a rule which tells how to form new games from games already known: if we have the games A and B, the conjunction rule shows how we can form the game $A \wedge B$ in order to assert this conjunction.

Now, particle rules have another important function: they not only set the basis of the semantics, and signalise how it could be related to the world of games – which is an outdoor world if the games are assigned to prime formulæ, but they also show how to perform the relation between sentences and propositions. Sentences are related to propositions by performing illocutionary acts such as demands and assertions, the content of the later are propositions. Assertions are thus linguistic acts which have a propositional content and endowed with a theory of force. The forces performing this connection between sentences and propositions are precisely the attack (?) and the defences (!). An attack is a demand for an assertion to be uttered. A defence is a response (to an attack) by acting so that you may performed the assertion (e.g. that A). Actually the assertion force is also assumed: perform the (conditional) assertion that A only if you know how to win the game A.[16]

[16]Cf. Ranta [1988]. Thanks to Daniel Vanderveken for fruitful remarks on this point. For further discussion, see Vanderveken [1991]

Certainly the "know" introduces an epistemic moment, typical of assertions made by means of judgements. But it does not presuppose in principle the quality of knowledge required. The constructivist moment is only required if the epistemic notion is connected to a tight conception of what means that the player X *knows that there exists* a winning game or strategy for A. So the next two sections will be dedicated to a presentation of (i) the structural rules, which allow the connection of those argumentation forms to produce well defined games, and (ii) the game theoretical[17] notion of validity as the existence of a winning strategy for one of the players.

2.3 Structural rules.

A dialogue can be seen as a sequence of labelled expressions, the labels carrying information on the game significance of these expressions. In other words, the set of expressions which is a complete dialogue can be dynamically determined by the rules of a game, specifying how the set can be extended from the original thesis formula. Particle rules are part of the definition of such a game, but we need to set the general organisation of the game, and this is the task of the *structural rules.*[18]

We first present here one version of the usual dialogical structural rules for the standard predicate dialogic, which we will refer to as SD. In the next paragraph (2.5) we will change the rules, motivated by some technical difficulties which arise in the formulation of the structural rules as applied to logics without losing – that is the claim – the pragmatic features of dialogic.

Each expression of a dialogue sequence (with the exception of the first member of the sequence called the *thesis*) is said to be a *move*, a move is either an attack or a defence expression – note that moves such as $?_L$ and $?_\exists$ do not have a propositional content which could be asserted. This leads us to the first structural rule:

> **(SR-0)** (***starting rule***): Expressions are numbered and alternately uttered by **P** and **O**. The thesis is uttered by **P**. All even numbered expressions including the thesis are **P**-labelled, all odd numbered ex-

[17] It seems that Lorenz was the first to apply this use of the concept of winning strategy (as opposed to Hintikka's use, where winning strategy means truth in a model).

[18] Jean Yves Girard developed thoroughly one point, which is one of the possible ways to perform dialogical semantics. Most of the stipulations by the structural rules can also be expressed at the propositional level, by extending the set of particles. Indeed, when the conditions of logical inference are the *object* which one is reasoning about, they should be expressed in the object language, but when they are the *modalities of the act* of inference, they have to stay at the metalinguistic level, otherwise the rational process would become viciously circular. Structural rules determine the way one can infer, while particle rules are inference patterns.

pressions are **O** moves. Every move below the thesis is a reaction to an earlier move with another player label and performed according to the particle and the other structural rules.

(SR-1) ***(winning rule)***: X wins the dialogue if the last move (the move carrying the highest numerical label) is an X-expression and there are no more Y-expressions allowed according to the particle and the other structural rules.

Actually, one of the functions of (SR-0) is to situate the abstract definition of game situations in a concrete game: it establishes who is X and who is Y, assuming that in the first situation **P** has the role of the defender and asserts a formula which fixes the topic of the dialogue. Now, every other situation developed from the thesis by means of determined rules is called a *round* in the game. Rounds are *opened* by an attack and *closed* by a defence. We now need some rules in order to be able to determine how to pass from one round to another. Here dialogic shows one of its salient features: the ability to determine by different means the differences between logics, in order to be compared and sometimes even combined. In the present reconstruction classical (SDC) and intuitionist (SDI) standard dialogic[19] differ only by a structural rule concerning the closure of rounds:

(SR-2I) ***(intuitionist ROUND closing rule)***: In any move, each player may attack a (complex) formula asserted by his partner or he may defend himself against *the last not already defended* attack. Defences may be postponed as long as attacks can be performed. Only the latest open attack may be answered: if it is **X**' s turn at position n and there are two open attacks m, l such that $m < l < n$, then **X** may not at position n defend himself against m.[20]

These rules define an intuitionistic logic. To obtain the classical version simply replace (SR-2I) by the following rule:

(SR-2C) ***(classical ROUND closing rule)***: In any move, each player may attack a (complex) formula asserted by his partner or he may defend himself against *any attack* (including those which have already been defended).

(SR-3) Neither player has to defend himself against an attack unless this attack has been defended of a counterattack.

Dialogical games can be seen, among other things, as a way to test the validity of a formula. In this case, they must be considered independently

---

[19]More precisely, SDC rules are (SR-0, 1, 3, 4, 5) with (SR-2C), and SDI rules are (SR-0, 1, 3, 4, 5) with (SR-2I). When irrelevant, we will omit the distinction, to keep the exposition as simple as possible.

[20]Notice that this does not mean that the last open attack was the last move.

of any model. This means that in this type of dialogues there is no set of prime formulæ conceded from the departure by **O** – or in the terminology of alethic semantics there is no assignment of truth values to prime formulæ as in GTS. Thus, we need a rule for prime formulæ:[21]

> **(SR-4)** (***formal use of prime formulæ***): **P** cannot introduce prime formulæ: any prime formula must be stated by **O** first. Prime formulæ can not be attacked.

Up to now, the rules allow the same expression to be challenged or defended an infinite number of times. This would prevent any dialogue from ending. Thus, we need to rule out the redundant repetitions. The following definitions and rule constitute one way of avoiding this kind of looping – in Lorenz' versions of dialogic this was solved with a convention concerning the number of repetitions allowed.[22] Unfortunately, when producing dialogical reconstructions of logic(s), while Lorenz's conventional device must be given up, the devices introduced herewith may seem awkward and arbitrary. Another simpler way will be justified in 2.4.

Definition (*strict repetition*): We speak of the *strict repetition of an attack* iff:

- A move is being attacked although the same move has already been attacked with the same attack before (notice that though choosing the same constant is a strict repetition, the choice of $?_L$ and $?_R$ are in this context different attacks).

In the case of moves where a universal quantifier has been attacked with a new constant, the following type of move has to be added to the list of strict repetitions:

- A universal-quantifier move is being attacked using a new constant, although the same move has already been attacked before with a constant which was new at the time of that attack.

Definition (*strict repetition*): We speak of the *strict repetition of a defence* iff:

- A challenging move (=attack) λ which has already been defended with the defensive move μ (=defence) before, is being defended against the challenge at λ once more with the same defensive formula (notice that the left part and the right part of a disjunction are in this context two different defences).

[21] For more commentaries on this issue see 2.5 and footnote 15.
[22] See Lorenz [2001], 260.

In the case of moves where an existential quantifier has been defended with a new constant, the following type of move has to be added to the list of strict repetitions:

- An attack on an existential quantifier is being defended using a new constant although the same quantifier has already been defended before with a constant which was new at the time of that defence.

(Notice that according to these definitions neither the new defence of an existential quantifier nor a new attack on a universal quantifier using a constant, not new but different from the one used in the first defence (or in the first attack), represents a strict repetition).[23]

(SR-5) (***no delaying tactics rule***):

- While playing with the *classical* structural rule (SR-2C) **P** may perform a *strict* repetition of a defence stating a prime formula a twice (or more) if and only if **O** has conceded a twice (or more). No other strict repetitions are allowed.
- While playing with the *intuitionistic* structural rule **P** may perform a *strict* repetition of an attack (SR-2I) if and only if **O** has introduced a new prime formula (see R1 below) which can now be used by **P.**

Validity is defined in dialogic via winning strategies of **P**:

Definition (*validity*): In a given dialogical system the proposition expressed by the formula stating the thesis is said to be valid iff **P** has a (formal) winning strategy for it, i.e. **P** can in accordance with the appropriate rules succeed in defending the thesis against all possible allowed criticism by **O.**

Examples of games: In Table 1 the outer columns indicate the numerical label of the move, the inner columns state the number of a move targeted by an attack. Expressions are not listed following the order of the moves, but writing the defence on the same line as the corresponding attack, thus showing when a round is closed. Recall, from the particle rules, that the sign "—" signalises that there is no defence against the attack on a negation. In this example, **P** wins because, after the **O**' s last attack in move 3, **P**, according to the (classical) rule SR-C, is allowed to defend (once more) himself from the attack in move 1. **P** states his defence in move 4 though, actually O did not repeat his attack – we signalise this fact by inscribing the not repeated attack between square brackets.

In the game of Table 2, **O** wins because, after the challenger's last attack in move 3, **P**, according to the intuitionistic rule SR-I, is not allowed to defend himself (once more) from the attack in move 1.

[23] Take the well-known case of $\exists x(Ax \Rightarrow \forall yAy)$.

Table 1. SDC rules. **P** wins.

O			**P**	
			$a \vee \neg a$	0
1	$?_{\vee}$	0	$\neg a$	2
3	a	2	—	
[1]	[$?_{\vee}$]	[0]	a	4

Table 2. SDI rules. **O** wins.

O			**P**	
			$a \vee \neg a$	0
1	$?_{\vee}$	0	$\neg a$	2
3	a	2	—	

2.4 Strategy games (SG) and the repetition rule.

2.4.1 An asymmetric formulation of SG. As already mentioned in the preface structural rules can, while implementing the local semantics of the logical particles, determine a kind of game where e.g. the aim is persuasion rather than logical validity.[24] But *when the issue at stake is indeed testing validity*, i.e. when **P** can succeed with the use of the appropriate rules in defending the thesis against all possible allowed criticism by **O**, games should be thought of as furnishing the branches of a tree which displays the games relevant for testing the validity of the thesis.[25] As a consequence of this definition of validity, each split of such a tree into two branches (dialogue games) should be considered as the outcome of a propositional choice of **O**. In other words when **O** defends a disjunction, he reacts to the attack against a conditional,[26] and when he attacks a conjunction, he chooses to generate a new branch (dialogue). Dually **P** will not choose to change the dialogue (branch). In fact, from the point of view of games as actual (subjective) procedures (acts), it could happen that the subject playing as **O** (**P**) is not clever enough to see that his best strategy is to open (not to open) a

[24]See Prakken H. and Vreeswijk G. [2000].

[25]The strategical games perspective suggests an interesting dynamical notion of logical form. It seems indeed reasonable to compare two formulæ by considering only those parts of them which are relevant for the games in testing their validity: following this line, $(a \wedge b) \Rightarrow a$ is the same as $(a \wedge C) \Rightarrow a$, for b and C are redundant. So our point here is that the consideration of redundancies relative to a given game determines the concept of logical form. Changing the way one arguments a formula amounts indeed to having another concept of its form.

[26]Which means either defending X-$a \Rightarrow b$ with X-b, or counterattacking the attack Y-a.

new dialogue game (branch) anytime he can, but in this context where the issue is an inter-subjective concept of validity, which should lead to a straightforward construction of a system of tableaux, we simply assume that **O** makes the best possible move. As we will see in 2.5.2 another type of SG, called *symmetric* for reasons which will be clear below, can be formulated too, which are more congenial with the dialogical general approach to semantics. Let us first describe the asymmetric structural rules for SG:

> **(SR-ST0)** (***starting rule***): Expressions are numbered and alternately uttered by **P** and **O**. The thesis is uttered by **P**. All even numbered expressions including the thesis are **P**-labelled, all odd numbered expressions are **O** moves. Every move below the thesis is a reaction to an earlier move with another player label and performed according to the particle and the other structural rules.

> **(SR-ST1)** (**winning** ***rule***): A dialogue is closed iff it contains two copies of the same prime formula, one stated by **X** and the other one by **Y**, and neither of these copies occur within the brackets "<" and ">" (where any expression which has been bracketed between these signs in a dialogue either cannot be counterattacked in this dialogue, or it has been chosen in this dialogue not to be counterattacked). Otherwise it is open. The player who stated the thesis wins the dialogue iff the dialogue is closed. A dialogue is finished if it is closed or if no other move is allowed by the (other) structural and particle rules of the game. The player who started the dialogue as a challenger wins if the dialogue is finished and open.

> **(SR-ST2I)** (***intuitionist ROUND closing rule***): In any move, each player may attack a (complex) formula asserted by his partner or he may defend himself against *the last not already defended* attack. Defences may be postponed as long as attacks can be performed. Only the latest open attack may be answered: if it is **X**'s turn at position n and there are two open attacks m, l such that $m < l < n$, then **X** may not at position n defend himself against m.

> **(SR-ST2C)** (***classical ROUND closing rule***): In any move, each player may attack a (complex) formula asserted by his partner or he may defend himself against *any attack* (including those which have already been defended).

> **(SR-ST3)** (***strategy branching rule***): At every propositional choice (i.e., when **O** defends a disjunction, reacts to the attack against a conditional or attacks a conjunction), **O** will motivate the generation of two dialogues differentiated only by the expressions produced by this choice. **O** will move into a second dialogue iff he loses the first chosen one. No other move will generate new dialogues.

(SR-ST4) ***(formal use of prime formulæ)***: **P** cannot introduce prime formulæ: any prime formula must be stated by **O** first. Prime formulæ can not be attacked.

(SR-ST5) ***(no delaying tactics rule)***:

(i) While playing with the classical structural rule **P** may perform once a new defence (attack) of an existential (universal) quantifier using a different constant (but not new) iff the first defence (attack) compelled **P** to introduce a new constant. No other repetitions are allowed.

(ii) While playing with the intuitionistic structural rule **P** may perform a repetition of an attack if and only if **O** has introduced a new prime formula which can now be used by **P**.

Definition (*Validity*): A tableau for $(\mathbf{P})A$ (i.e. starting with $(\mathbf{P})A$) proves the validity of A iff the corresponding tableau is closed. That is, iff every dialogue generated by $(\mathbf{P})A)$ is closed.

Remark: Notice that these re-formulations solve most of the criticisms concerning the repetition rule. The very point of these dialogues is that the notion of finishing coincides with the usual notion of closing a branch of the usual tableau systems. Hence, repetitions will not be of any use if **P** does not manage to close the dialogue with such a repetitive move. Certainly, there are some procedural aspects concerning "repetition" which are still a problem. But these are no other problems than the very well known problems related to finding a procedure for completing tableaux with quantifiers. Figure 3 shows an example concerning Peirce's Law.

Table 3. **P** wins.

	O			**P**	
				$((a \to b) \to a) \to a$	0
1	$(a \to b) \to a$	0		a	4
5	a		1	$a \to b$	2
3	a	2			
[1]	$[(a \to b) \to a]$	[0]		a	6

In standard dialogues, this kind of formula motivated the formulation of the first part of the no delaying rule SR-5. The move 3 is the counterattack a against $a \Rightarrow b$ stated by **P** in move 2, but **O** can also state a in move 5, defending 1 $(a \Rightarrow b) \Rightarrow a$, compelling **P**, in a classical dialogue, to defend the thesis twice *with the same move.*

In the version of strategy dialogues what actually happens is that **O** generates two dialogues one defending and the other counterattacking. Both dialogues will be closed and thus won by **P**.

Table 4. **P** wins.

	O			**P**	
				$((a \to b) \to a) \to a$	0
1	$(a \to b) \to a$	0		a	4
			1	$a \to b$	2
2	a	2			

Table 5. **P** wins.

	O			**P**	
				$((a \to b) \to a) \to a$	0
1	$(a \to b) \to a$	0		a	4
3	a		1	$< a \to b >$	2

This type of dialogues also facilitates the comparison with the extensive form usual in GTS. In the extensive form every choice (propositional or not) of each of the players causes a split. Thus our point here is that strategy trees build up from strategy dialogue games can be seen as a filter on the extensive trees.[27]

Helge Rückert who read a first version of these rules remarked that these rules seem to be conceptually related to what Paul Lorenzen called, in the context of intuitionistic logic, asymmetrical rules. Indeed in these games the rights of **P** and **O** are asymmetrical (**O**'s and **P**'s choices are different), but this asymmetry is a import of the level of validity as argued in Rahman/Rückert [1999]. Moreover, from a dialogical point of view, validity should not determine the semantics but rather the other way round should be the case: in analogy to the game of chess, to play a game the players need to understand the rules fixing the aims of the game and the way to play (i.e. the local and structural rules) but they do not need to be able to always find a winning strategy in order to show that they understood those rules. Indeed, one can see the dependency of the notion of validity on the theory of meaning as being one of the major virtues of dialogic, seen as a pragmatic method for generating tableaux validity – we will come to this point in 2.5, giving results even when the

[27]Lorenzen [1989], pp. 43-69.

model-theoretical semantics is not yet available. The asymmetric rules of SG seem to contravene this cherished piece of dialogical philosophy. For the moment let us remark again that the motivation of strategy dialogues is indeed that of validity and this determine in some way the structural rules. A purely semantic formulation of the structural rules in the pragmatic sense of dialogic actually amounts to Lorenz` conventional device: The number of repetitions being part of a pre-agreement before the dialogue really starts. In this sense any other repetition rule as the conventional seems to be motivated by validity considerations. However symmetric rules come nearer, one could argue, to the dialogical ideal than the asymmetric ones. In fact, a formulation for symmetric SG is easy to fix.

2.4.2 Symmetric rules for SG. The change concerns the branching rule:

> **(SR-ST3/SY) (strategy branching rule)**: At every propositional choice (i.e., when X defends a disjunction, reacts to the attack against a conditional or attacks a conjunction), X may motivate the generation of two dialogues differentiated only by the expressions produced by this choice. X might move into a second dialogue iff he loses the first chosen one. No other move will generate new dialogues.

These SG retain all the advantages of the others concerning the repetition rules. Let us take once more the case of Peirce's law in classical dialogic. Here **O**, knowing that no repetition of a propositional formula is allowed, would prefer to stay at the same dialogue. But, he lost this dialogue. Moreover, the dialogue is finished and he has to move to a second one. Assume now, that in our example, the a's are not prime but complex formulæ. In this case, **O** might well state A twice in the same dialogue. But this, does not lead to any success since **P** might finish the dialogue and win by simply sticking to one of the occurrences of A stated by **O**.

2.5 Tableaux for validity.

As already mentioned, the strategy dialogical games introduced above, furnish the elements of building a tableau notion of validity. Following the seminal idea at the foundation of dialogic, this notion is attained via the game-theoretical notion of *winning strategy.* X is said to have a winning strategy if there is a function which, for any possible Y-move, gives the correct X-move ensuring the wining of the game.[28]

[28]For a precise formulation of GTS see the paper of Pietarinnen on this volume.

Indeed, it is a well known fact that the usual semantic tableaux for intuitionistic and classical logic, as reformulated 1968 in a tree-shaped structure by Raymond Smullyan and 1969 by Melvin Fitting, are directly connected with the tableaux (and the correspondent sequent calculus) for strategies generated by dialogue games, played to test validity in the sense defined by these logics. E.g. table below.

(**O**)-cases	(**P**)-cases
Σ,(**O**)$A \to B$	Σ,(**P**)$A \to B$
Σ,(**P**)A,... $\vert\Sigma$,$<$(**P**)$A>$ (**O**)B	Σ,(**O**)A, Σ,(**P**)B

The vertical bar "|" indicates alternative choices for **O**, **P**'s strategy must have a defence for both possibilities (dialogues). σ is a set of dialogically signed expressions.

Intuitionistic tableaux are generated with an extra notational device. Some of the expressions are labelled with $_{(\mathbf{O})}$, for instance $(\mathbf{P})_{(\mathbf{O})}A$. the intuitionistic deduction rule includes this: the totality of the previous **P**-formulæ on the same branch of the tree are eliminated. The $_{(\mathbf{O})}$ label marks every assertion of **O**. However the resulting tableaux are not quite the same. A special feature of dialogue games is the notorious formal rule (SR-ST4), which is responsible for many of the difficulties of the proof of the equivalence between the dialogical notion and the truth-functional notion of validity. The role of the formal rule, in this context, is to induce dialogue games which will generate a tree displaying the (possibly) winning strategy of **P**, the branches of which do not contain redundancies. Thus the formal rule actually works as a filter for redundancies, producing a tableau system with some flavour of natural deduction. This role can be generalised for all types of tableau generated by the various dialogics. Once this has been made explicit, the connection between the dialogical and the truth-functional notion of validity becomes transparent. Here we will only discuss the simpler propositional case. Let us thus go into the task of making the effects of the role of the formal rule explicit. We start with some definitions.

Definition (*truth determinant*): We call a set of signed (occurrences of) prime formulæ, occurring on a branch ψ of a (truth-functional) tableau for a formula A ψ ***truth-determinant set for the formula A*** (ψ-TD(A)) iff the elements of this set are sufficient to determine whether the branch is closed or not.[29]

[29]See Rahman [1999].

Definition (*TD redundancy*): We call occurrences of prime formulæ *redundant concerning TD* iff those occurrences are not elements of the TD-set(s).

Here are some examples, were the different occurrences of the prime formulæ are noted as indices.

E.g. 1:
$F(a_1 \wedge b) \rightarrow a_2$
ψ-TD: $\{Ta_1, Fa_2\}$
(i.e. The branch $\{F(a_1 \wedge b) \rightarrow a_2, Ta_1 \wedge b, Ta_1, Tb, Fa_2\}$ closes using only the TD elements)
ψ-Redundancy: b

E.g. 2:
$F(a_1 \vee (b \wedge c)) \rightarrow a_2$
ψ-TD: $\{Tb, TC, Fa_2\}$
ψ-Redundancy: a_1
The branch $\{F(a_1 \vee b) \rightarrow a_2, Ta_1 \vee (b \wedge c), Tb, TC, Fa_2\}$ remains open.

E.g. 3:
What about $F(a_1 \wedge (b_1 \vee c_1)) \rightarrow ((a_2 \wedge b_2) \vee (a_3 \wedge c_2))$?
ψ-TD: $\{Tb_1, Fb_2\}$
φ-TD: $\{TC_1, FC_2\}$
ψ- Redundancy: a_1, a_2
φ- Redundancy: a_1, a_3
The branches are closed

Theorem: A tableau generated by branches without redundant prime formulæ (let us call them non-redundant tableaux, for short NR-tableaux validates (or refutes the validation of) the same formulæ as a tableau which contains branches with redundant formulæ. Proof: this follows simply form the definition given above.

Theorem: Dialogical strategy tableaux induced by the formal rule (SR-ST4) are the dialogical equivalent to the NR-tableaux. The proof follows from the following:

(i) Strategy games build the branches of the dialogical strategy tree.

(ii) **T** (**F**)-assignments in a NR-tableau correspond to **O** (**P**)-assignments in the dialogical strategy tree.

(iii) Every different occurrence of a prime formula in a branch of a NR-tableau corresponds to a different move in the corresponding game of the dialogical strategy tree.

(iv) The formal rule induces **P** to demand prime formulæ following the principle of propositional-variable-sharing (i.e., he will try to show that the prime formulæ he states are exactly those which **O** has already conceded and the justification of which can thus be considered to be "shared"). **O** will follow a dual strategy trying to avoid any use of variable sharing.

(v) Every game won by **P** ends with a variable-sharing move (every such game will close). Dually every game won by **O** ends with the demand for a prime formula which cannot be assured by variable sharing (every such game will remain open).

(vi) **SR-ST1** corresponds to closing a branch on the generated NR-tableau.

Let us run a dialogical strategy tree for $(a \wedge (b \vee c)) \Rightarrow ((a \wedge b) \vee (a \wedge c))$ – we will not use the dialogical table-notation here but the tree-shaped one instead.

I	**P**-$((a \wedge (b \vee c)) \rightarrow ((a \wedge b) \vee (a \wedge c)))$	Thesis
II	**O** -$(a \wedge (b \vee c))$	attack on I
III	<**P**-$?_L$>	attack on II
IV	**O** -a	defensive answer to III
V	<**P**-$?_R$>	attack on II
VI	**O** -$(b \vee c)$	defensive answer to V
VII	<**P**-$?_\vee$> attack on VI	
VIII	**O** -b	**O** CHOOSES*
IX	**P**-$((a \wedge b) \vee (a \wedge c))$	defence against the attack on II
X	<**O** -$?_\vee$>	attack on IX
XI	**P**-$(a \wedge b)$	defence against the attack on X

* (between b and c, opening two different branches — here we only display the b-option)

Notice here the idea behind the strategy of **P**. He waited to answer to the attack stated on II until **O** chose to state the prime formula at VIII which at this stage of the dialogue will be strategically determinant and which cannot be determined by **P**. The continuation is clear: **O** will attack with, say, left, and afterwards right. **P** will be able in both cases to answer because of **O** moves IV and VIII. More to our point, **P** can in the context of these choices of **O**, consider C to be redundant. Of course another choice of **O** at VIII will yield another TD, but this will yield also another winning strategy.

Remark: If the assignments usual to GTS games are taken from the corresponding TD, then we will immediately have values of the choice function defining the winning strategies which will represent the branches of the NR-tableau. Moreover, in a game under conditions, the assignations define the TD.

One important last point (just) to be mentioned here is that tableaux for intuitionistic logic as presented above have an awkward feature: one states first some formulæ and afterwards they will be eliminated because they are not allowed to be considered when closing a branch. The combination of the formal rule and the intuitionistic structural rule for strategy dialogues induces trees where this type of redundancy does not appear. By applying a similar method as for TD the connection between intuitionistic NR and corresponding strategy trees can be expressed more precisely and in general this applies too for all other dialogical reconstructions of various logics.

Philosophical remark: Hintikka's point re-visited

It has been stressed that structural rules can, while implementing the local semantics of the logical particles, determine a kind of game with other aims than testing (logical) validity. Now the point here is that the structural rules should, from a dialogical point of view, extend the local meaning of the particle rules in a conservative way – in the sense that the structural rules should be formulated in accordance with the particle rules and validity in accordance with the structural rules. This point becomes central while (re)constructing logics. Indeed, the distinctive feature of the pragmatical semantics of dialogic is its connection to proof, both in its elementary and its complex level. In its elementary level it is associated to prime formulæ which express a proposition understood as encoding an informal proof, conceived as the predicate atomic language games. In this context, the particle rules show how to play a game fixing the semantics for complex formulæ based on the atomic (game) meaning assigned to prime formulæ – but (possibly) independently of any determinate assignation). Now, one very important point, and a quite often misunderstood tenet of the dialogic, is that this theory of meaning should set the basis for the formal concept of proof leading to the notion of validity. Once more, the meaning in use of the notion of informal proof underlying the local semantics should furnish the way to formal proof, and not the other way round. Hintikka argued quite often against dialogic because of their supposedly "in-door" – or purely formal approach to meaning as use. He argues that the notion of formal proof is certainly connected with "out-door" games but actually formal-proof games are not of very much help in accomplishing the task of connecting the linguistic rules of meaning with the real world. The point we would make here is that he is indeed right if we understand his critics as claiming the need for a tight connection between the informal notion of proof (or even between the alethic conception of semantics) and the formal notion of proof. From the dialogical point of view the formal concept of proof cannot deliver a concept of meaning.

Structural rules have the role of implementing the meaning of a logical constant relative to a given context to which this logical constant will apply[30]. This could be conceived in an analogous way as the case of determining the meaning of "I" in "I am a French cook",[31] though it

[30] A similar point has been argued in Dubucs [2002].

[31] The example is due to Recanati [2001].

has a meaning it gets a full-fledged meaning or (less drastically) gets it's meaning in a broader context than the local one (i.e. in "I am a French cook") once the context has been explicitly described – in our case we give it an identity: François Recanati. Moreover, structural rules should establish the connection between meaning and validity. In the case of the aforementioned additive constants, the meaning of such linear constants may be completed or applied with the help of structural rules or new particles expressing structural properties.

3. The notion of formal use.

As already discussed in 2.5 the notion of formal use induces a kind of dialogical relevance at a procedural level. In a model-theoretical language we would say that, in this exchange, when validity is the issue, **O**'s task is to make sure that the model in which the game takes place is relevant to test the validity (non validity) of the formula. **P**'s task is to choose, from every statement that **O** concedes while constructing the (counter)model, only that which is relevant for proving the validity.[32] In order to differentiate between the respective tasks of **P** and **O**, it is necessary to introduce an asymmetry in the rules of the game, limiting the moves of **P** in order to make it possible for him to choose – following a "variable sharing principle" (recall the remarks in footnote 15) – only those resources needed. This is what we call the *formal use* of resources.

There are several notions of formal use in dialogic, according to the resource considered. Namely: formal use of constants (yielding free dialogic), of atomic *falsities* for negative literals (yielding a kind of paraconsistent dialogic) and of contexts of argumentation (yielding modal dialogic). The formal restriction can be also used to express meta-linguistic properties in the object language (allowing the formulation of connexive particles). In all these systems, redundant information is restricted by introducing asymmetry of information access in the externalised agonistic tension between the players of the game. At the end of the paper we will suggest how the dialogical version of those logics offers a bridge to IF-logic, opening new ways of exploration. Let us first have a very brief look at the dialogical version of these logics.

3.1 Free dialogic (FD).[33]

Free logic is the result of the serious consideration of the classical problem of the ontological commitment of quantifiers. So, following Ben-

[32] Van Benthem [2001-2004] calls the opponent the *Builder*, underlining this very point.
[33] For details, see Rahman/Rückert/Fischmann [1997] and Rahman [2001].

civenga,[34] we will define free logic as a "formal system of quantification theory [...] which allows for some singular terms in some circumstances to be thought of as denoting no existing object, and in which a given set of quantifiers is thought of as having an existential import."

Free dialogic stems from the regulative restriction of the use of constants.

Definition (*constant introduction*): A constant τ is said to be *introduced* in the dialogue when and only when it is used to defend an existential quantifier, or to attack a universal quantifier, and has not been used in the same way before.

The idea is to give **O** the sole right to introduce constants. This is the object of a new structural rule, vernacular to free dialogical systems:

> **(SR-ST6F)** (***formal use of constants***): only **O** may introduce constants.

The non-existent objects one could want to be part of the theory of quantification are symbolised here by the constants *used* in the dialogue, but not *introduced* in the sense of the definition given above. The tableaux for FD are straightforward. New constants are labelled with a star (e.g. τ^*), and **P** cannot use a constant which has not been labelled with such a star.

FD can be easily extended in order to cope with a more complex domain of non-existent objects. The first step (yielding FD^4) is to distinguish between two pairs of quantifiers, one with existential import ($\forall$ and $\exists$), and one without ($\forall^0$ and $\exists^0$), with the following particle rules (the rest of the rules remain unchanged):

> $\forall$*particle rule*: from $\forall xA$ follows a state <R, $\sigma(x/\ \tau)$, A> responding to the attack-$?_{\forall/\tau}$ of the challenger who chooses τ), but τ has to be chosen under the restriction SR-ST6F.
>
> $\exists$*particle rule*: from $\exists xA$ follows a state <R, $\sigma(x/\ \tau)$, A> responding to the attack $?_{\exists}$ of the challenger and the defender chooses τ under the restriction SR-ST6F.

It is then possible to generalise to n pairs of quantifiers, distinguishing n levels of reality (or fiction). This system has been called FD^n. An interesting fact is that in FD^n the dialogical differentiation between quantifiers that are ontologically charged or not is made only at the structural level. We could even go a step further and say that in such sets of quantifiers it is not the local meaning that is at stake but rather the question where (in which domain) the given quantification applies. As we will see at the end of the paper, one could even see the difference

[34] Bencivenga [1983].

between these quantifiers as being due to a phenomenon of an imperfect exchange of information.

3.2 Paraconsistent dialogic (PD).[35]

A formal system is said to be *inconsistent* if it contains as a theorem a formula and its negation, in other words when it contains a contradiction. A formal system is said to be *trivial* if every wff of it is a theorem. The conceptual motivation of paraconsistent logics is to differentiate between the notions of inconsistency and triviality: a system is said to be *paraconsistent* if it is inconsistent without being trivial.[36]

Paraconsistency is achieved by splitting the set of contradictions into two different sets, namely the set of *explosive* and the set of *non explosive* contradictions. The latter set contains contradictions the truth of which are considered not to lead to the triviality of the system which contains this set. Dually, a contradiction is *explosive* in a given logical system when its truth trivialises this system. In the "Brazilian interpretation" of Newton da Costa and associates the distinction between these two sets is implemented by means of a modification of the standard semantics of negation.

Until now dialogical paraconsistency has been studied within the approach of da Costa. This dialogic is based on the idea that it might make sense, in some context, to assume that a and $\neg a$ can both be asserted, but that it does not mean that *everything* could be asserted (by the use of *ex contradictione sequitur quodlibet*). As medieval Scholastics already prefigured, the assertion of a contradiction may have been stated *for the sake of the argument*, and in this case it should not provide an argument to prove anything.

More precisely, while reconstructing dialogically Sette's logic P1[37] one cannot *assume* that any prime formula a and its negation $\neg a$ – considered in this logic not to be explosive – are in conflict with each other. They are if their assertion reveals an *internal contradiction*, i.e. if a player who has stated $\neg a$ then attacks the same formula $\neg a$ stated by the other player. Again, the idea is to use **O**'s antagonism to reduce, according to a pre-agreed definition of the sets of explosive and non explosive contradictions, to a minimal the inference possibilities from a and $\neg a$. This is achieved through a new structural rule extending the

[35] For details, see Rahman [2001], Rahman/Carnielli [2000], Rahman [1999] and Rahman/Van Bendegem [2002].

[36] cf. da Costa [1977], and Priest & alia [1989] for a general survey.

[37] See A. M. Sette [1973].

formal rule to negative literals – and which make it possible to formulate a new tableau system for P1:

> **(SR-ST7L)** (***negative literal formal rule***): **P** has the right to attack a negative **O**-elementary statement[38] (the so-called *negative literal*) iff **O** has already attacked the *same* negative literal before.

From this idea stems another more general way to produce paraconsistent dialogic, inspired by a Gentzen-style formulation of the meaning of negation. As is very well known, the introduction rule for negation in natural deduction systems amounts to defining $\neg A$ as $(A \Rightarrow \bot)$, where $\bot$ is the *falsum*, i.e. a formula referring to an arbitrary absurdity. Now, in dialogic the prime formula $\bot$, because of the formal restriction rule, cannot be asserted by **P** unless **O** asserted it first, but this is too weak a restriction for controlling the explosive power of absurdity. If we want the dialogues to behave paraconsistently beyond minimal logic, we must conceive some negated formulæ, which might build non explosive contradictions, as generating its own particular absurdity. In order to implement this in the dialogic for P1 we will index the $\bot$ symbol with the corresponding formula: e.g. $(a \Rightarrow \bot_a)$. The structural rule for paraconsistent positive dialogues is as follows:

> **(SR-ST7P)** (***elementary absurdities formal rule***): for any formula a, **P** may only use the elementary absurdity $\bot_a$ iff $\bot_a$ has been stated before by **O**.

Let us run an example in Table 6.

Table 6. Fig. 3. PEP rules. **O** wins.

	O			**P**	
				$a \rightarrow ((a \rightarrow \bot_a) \rightarrow (b \rightarrow \bot_b))$	0
1	a	0		$(a \rightarrow \bot_a) \rightarrow (b \rightarrow \bot_b)$	2
3	$a \rightarrow \bot_a$	2		$b \rightarrow \bot_b$	4
5	b	4			
7	$\bot_a$		3	$< a >$	6

O wins because his move 7, where he concedes $\bot_a$, does not allow **P** to state $\bot_b$.

Changing the subscripts by indices, i.e., replacing $\bot_a$ by $\bot_i$, which could be quantified, also opens the way to IF- applications.

The corresponding tableaux for PDL (paraconsistent dialogic for P1) are a very straightforward restriction of the classical (PDLC) or intu-

[38] I.e. prime formulæ or predicative statements of the form $a\tau$, with τ a constant.

itionist (PDLI) rules. There is, in each case, a restriction to apply to the closing branching rules:

PDL closing restriction: check after finishing the tableau and before closing branches that for every elementary **P**-statement which follows from the application of an **O**-rule to the corresponding negative **O**-literal (i.e. for every attack on a negative **O**-literal), there is an **O**-attack on the *same* negative literal, stated by **P**. If there is not, the branch remains open.

3.3 Connexive dialogic (XD).[39]

The structural relevance yielded by the notion of formal use, up to now, has been strictly applied to **P**. Moreover, the validity of a formula is expressed in a dialogue via the notion of formal restriction: validity is a *winning strategy under formal restriction.* Now, we might want to express some metalogical features in the object langage. But to perform this amounts to introducing a new logical constant, i.e., a particle, which can be played by both players and which allows both players to change sides concerning the formal restriction during the game. Let us be more precise.

The first step in connexive dialogic is to add to the game two new logical constants, expressing the metalogical (and unpalatable) properties of *attackability* (**V**) and *defensibility* (**F**) of a formula. By playing **V**A, X asserts that A can be defended under certain conditions. Attacking **V**A, Y asserts that A cannot be won, whatever the conditions are. Hence Y is committed to winning a dialogue (actually a *subdialogue* of the initial dialogue) with X-A as a thesis, *where he has to play formally*, i.e. where the formal restriction applies to him. The fact that a structural rule is part of the particle rule for the **V** operator is a consequence of the meta-theoretical nature of what **V** expresses. Dually, by playing **F**A, X asserts that A can be successfully attacked under certain conditions. Attacking **F**A, Y asserts that A cannot be lost in a game, thus committing himself to defending A in a subdialogue, playing under the formal restriction – notice that **F** is closely related to the GTS definition of negation.

Subdialogues are labelled parts of a dialogue, determined by the player who opened them, and labelled to order them. The labelling introduces a tree-based order in the following way: (i) the subdialogue where the thesis is asserted is called the *initial dialogue*, and is labelled 1. (ii) The mth subdialogue opened in the nth subdialogue is labelled $n.m$. (iii)

[39] For details, see Rahman/Rückert [2001].

The subdialogue m where a subdialogue n has been opened is called the *upper section* of $m.n$.

To formulate the particle rules one must extend the notion of state of the game in order to reflect the distribution of the formal restriction on the players. A *state of the connexive game* is then $<\mathrm{R}, \sigma, A, \lambda>$ where R, σ, A are as before, and λ is an assignation of subdialogues to formulæ. We write $\lambda_{a/m}$ to express that the formula A is played in the subdialogue m. The particle rules are:

V*particle rule*: from a formula of the form $\mathbf{V}A$ follows a state of the connexive game $<\mathrm{R}, \sigma, A, \lambda_{a/m}>$, responding to the attack $?_{\mathbf{V}}$, of the challenger who plays under the formal restriction and where m is a subdialogue opened by the player with the role of the defender.

F*particle rule*: from a formula of the form $\mathbf{F}A$ follows a state of the connexive game $<\mathrm{R}, \sigma, \neg A, \lambda_{a/m}>$, responding to the attack $?_{\mathbf{F}}$ of the challenger who plays under the formal restriction and where m is a subdialogue opened by the player with the role of the defender.

The motivation of connexive logic is that classical entailment cannot discriminate between the trivially true conditionals[40] and those which express some determinate kind of meaning connection linking the if-part to the then-part. So connexive dialogic uses another logical constant to express the non-trivial entailment: the *connexive If-Then* ($\Rightarrow$). The corresponding particle rule goes thus:

$\Rightarrow$*particle rule*: from $A \Rightarrow B$ follows at least one of the following states $<\mathrm{R}, \sigma, \mathbf{V}A>$, $<\mathrm{R}, \sigma, \mathbf{F}B>$ or $<\mathrm{R}, \sigma, B>$, responding respectively to each of the attacks $?_{if}$, $?_{then}$, A, chosen by the challenger.

Some structural rules need a modification:

> **(SR-ST4X)** (***connexive formal rule***): At the start of the dialogue for A it is **P** who plays under the formal restriction. If X plays under the formal restriction in a given section (subdialogue) he cannot use a prime formula if Y did not utter this formula first in the same section. Dually, Y can introduce a new prime formula in the given section anytime he wants, according to the other rules. The only changes in this distribution of the formal rule are those regulated by **V** and **F**.

> **(SR-ST8X)** (***connexive subdialogues relations***): in a given subdialogue, X may choose as a target of his attacks any (complex) Y-formula of this section (in so far as the other rules allow it). X may also choose conditionals (Y-$A \Rightarrow B$) of the corresponding upper section, provided he attacks with the classical X-A. No other type of attacks is allowed.

The formal restriction is used here to discard contradictions and tautologies. More precisely, the player X, when asserting X-$A \Rightarrow b$, must

[40] I.e. conditionals with a contradictory if-part and/or contradictory then-part.

have the possibility of choosing the correct resources to show that A and b are not trivially connected. Moreover, this logic is able justify the formulæ$\neg(\neg a \Rightarrow a)$ and $\neg(a \Rightarrow \neg a)$ which were considered valid by Aristotle, Boethius, Kant, Strawson and many others but were subsequently rejected by most of the logicians. These formulæ and have been related to the well-known problem of subalternation in traditional syllogistic via the dependency of $\neg((a \Rightarrow b) \Rightarrow \neg(a \Rightarrow \neg b))$ on $\neg(\neg a \Rightarrow a)$. A further feature of this dialogic is that its object language makes it possible to express via **V** that the formula a is (contingently) true.

The idea behind the tableau method for connexive dialogic is to keep track of the formal restriction by labelling the expressions with f and *nf*, standing for formal and non-formal. The tableau rules are no longer determined by **O** and **P**, but by the more general labels $\mathbf{X}_f$ and $\mathbf{Y}_{nf}$. A tableau for a starts with $\mathbf{P}_f - a$, and a closed tableau proves that $\mathbf{P}_f$ has a winning strategy for a.

(Y_{nf})-cases	(X_f)-cases
$\Sigma, (\mathrm{Y}_{nf})A \to B$	$\Sigma, (\mathrm{X}_f)A \to B$
$\Sigma, (\mathrm{Y}_{nf})A \to B\Sigma,$ $<(\mathrm{X}_f)?_{if}>(\mathrm{Y}_{nf})\mathbf{V}A\Sigma,$ $<(\mathrm{X}_f)?_{then}>(\mathrm{Y}_{nf})\mathbf{F}B$ $\Sigma, (\mathrm{Y}_{nf})\mathbf{V}A$	$\Sigma, (\mathrm{X}_f)A \to B \vert \Sigma, (\mathrm{X}_f)A \to B\Sigma,$ $<(\mathrm{Y}_{nf})?_{if}>(\mathrm{X}_f)\mathbf{V}A \vert \Sigma,$ $<(\mathrm{Y}_{nf})?_{then}>(\mathrm{X}_f)\mathbf{F}B$ $\Sigma, (\mathrm{X}_f)\mathbf{V}A$
$\Sigma_{[\to]}, <(\mathrm{X}_f)?_{\mathbf{V}}>(\mathrm{Y}_{nf})A$ $\Sigma, (\mathrm{Y}_{nf})\mathbf{F}A$	$\Sigma_{[\to]}, <(\mathrm{Y}_f)?_{\mathbf{V}}>(\mathrm{X}_{nf})A$ $\Sigma, (\mathrm{X}_f)\mathbf{F}A$
$\Sigma_{[\to]}, <(\mathrm{X}_f)?\mathbf{V}>(\mathrm{Y}_{nf})\neg A$	$\Sigma_{[\to]}, <(\mathrm{Y}_f)?\mathbf{V}>(\mathrm{X}_{nf})\neg A$

The line "▬" signalises the opening of a new subdialogue (a new one will be opened with every application of a **V**- or **F**-rule). The point here is that the formal restriction changes from X to Y in the $(\mathrm{X}_f)\mathbf{V}A$ and $(\mathrm{X}_f)\mathbf{F}A$ rules.

The restrictions for attacks which cross subdialogues, described at the game level have to be expressed at the tableau level too. For this, we need a device which, in a subdialogue, erases from the σ set all the non relevant expressions of the upper section of the dialogue. This is what we write as $\Sigma_{[\Rightarrow]}$. The corresponding rule for $\Sigma_{[\Rightarrow]}$ is the following:

In the set σ of the subdialogue containing $\sigma_{[\Rightarrow]}$, replace those formulæ of the upper section in which the connexive $\Rightarrow$ occurs as the principal logical constant with corresponding formulæ with the standard $\Rightarrow$ as principal logical constant, invert the f and *nf* labels when necessary, according to the change of the formal restriction which has taken place in the subdialogue, and delete all the other formulæ.[41]

3.4 Modal dialogic (MD).[42]

As already discussed in the different dialogics, statements uttered in a game are always thought contextually. Modal dialogic is a systematic account of an explicit notion of context, in the sense that the latter is introduced at the propositional level of the object language.[43] Modal moves are hence dialogical expressions with a supplementary label, indicating the context in which the move has been made. This means that the state of the game induced by a modal formula is $<R, \sigma, a, \lambda>$, where R, σ, a are as before, and λ is an assignation of contexts to formulæ. The usual modal operators for *necessity* "N" and *possibility* "P" are then defined in the following way:

> **N**-*particle rule*: From **N**A follows $<R, \sigma, A, \lambda_{a/m}>$, responding to the attack $?_{N/m}$ of the challenger where $\lambda_{a/m}$ is the assignation of context m to the formula A, and m is a context chosen by challenger.

> **P**-*particle rule*: From **P**A follows a state of the game $<R, \sigma, A, \lambda_{a/m}>$, responding to the attack $?_P$ of the challenger, where $\lambda_{a/m}$ is the assignation of context m to the formula A, and m is a context chosen by the defender.

Now, as well known, each modal logic can be distinguished by the properties of the accessibility relation existing between the contexts defined by this logic. In dialogic these relations are defined by the structural rules, specifying what contexts are accessible from a given context. Before we give these structural rules, let us precise the notion of (dialogical) context.

Dialogical contexts always constitute a set of moves. These contexts may have a finite number, or a countable infinity of elements, semi-ordered by a relation of succession, obeying the very well known rules which define a tree. The thesis is assumed to have been stated at a dialogical context which constitutes the origin of the tree. The initial dialogical context is numbered 1. Its n immediate successors are numbered $1.i$ (for i=1 to n) and so on. An immediate successor of a context $m.n$ is said to be *of rank +1*, the immediate predecessor m of $m.n$ is said to be *of rank -1*. And so on for arbitrarily higher (lower) degree ranks.

---

[41]In connexive logics, the conditional $\Rightarrow$ is part of the local meaning of $\Rightarrow$.
[42]See Rahman / Rückert [1999] and [2001].
[43]Our modal propositional language $\mathbf{L}_m$ is syntactically defined in the standard way.

Modal dialogics share the formal approach of all other dialogics in a double sense: this rule determines the use of prime formulæ in a given dialogical context and the generation of new dialogical contexts. The following rule concerns the former:

> **(SR-ST4M)** (***modal formal use of prime formulæ***): only **O** may introduce prime formulæ. **P** cannot use a prime formula **O** did not utter first *in the same context.* **O** can introduce a new prime formula anytime he wants, according to the other rules.

In fact, on the contrary to the model theoretical approach, neither the relevant prime formulæ nor the dialogical contexts are in principle given before the dialogue begins – however it is not difficult to formulate a conditioned version in GTS-style: one simply adds the prime formulæ with the corresponding context-labels as hypotheses to the thesis. With this method one could add to the set of hypotheses even the relevant accessibility conditions, provided the language is extended with help of an hybrid notation (see 4.2). A simpler method would be to add structural rules describing the accessibilities required by the model.

Since, as already remarked, the available dialogical contexts for the argumentation, exactly like prime formulæ, are not statically pre-arranged conditions, but dynamically determined resources, which appear in the course of the dialogue there must be a formal rule which restricts the context opening possibilities in the course of a modal dialogue and establishes that **O**'s task is to consider the sole contexts which are relevant for the validity of the formula at stake:

Definition (*choice of dialogical contexts*): a context m is said to be *chosen* by X when X chooses the dialogical context label m when performing an attack against a modal formula of the form $\mathrm{N}A$, or when defending a formula of the form $\mathbf{P}A$ (for any formula A). The dialogical context label m is said to be *new* if it has never been chosen before. A dialogical context with label m is said to have been *introduced* iff the dialogical context label m is new. The initial context is considered to be given while stating the thesis and, though it might not have been chosen before, it is not new.

> **(SR-ST9.1)** (***formal rule for contexts***): **O** may introduce a context anytime the other rules let him do so. **P** cannot introduce a context, and his choices when opening a context are restricted by the adequate (SR-ST9.2) rule which reconstructs the properties of the accessibility relation particular to the modal dialogic in question.

Restrictions on the possibilities for **P** to choose dialogical contexts are the dialogical reconstruction of what the standard possible worlds semantics expresses with help of accessibility relation properties. Suppose **P** is at $1.n$. The dialogical equivalent to reflexivity amounts to allowing **P** to choose $1.n$ (i.e. to stay in the context he is playing in). Symmetry is reconstructed as allowing **P** to choose a context of rank -1 relative to $1.n$. Transitivity is reconstructed as allowing **P** to choose a context of rank >+1 relative to $1.n$. It is then straightforward to define the correct SR-ST9.2 rule for the main modal systems, namely K, T, B, S4 and S5.[44]

> **(SR-ST9.2K)** (***K***): **P** may choose a (given) dialogical context of rank +1relative to the context he is playing in.
>
> **(SR-ST9.2T)** (***T***): **P** may choose either the same dialogical context where he is playing in or he may choose a (given) dialogical context of rank +1 relative to the context he is playing in.
>
> **(SR-ST9.2B)** (***B***): **P** may choose a (given) dialogical context of rank -1 (+1) relative to the context he is playing in, or stay in the same context.
>
> **(SR-ST9.2S4)** (***S4***): **P** may choose a (given) dialogical context of rank >+1 relative to the context he is playing in, or stay in the same context.
>
> **(SR-ST9.2S5)** (***S5***): **P** may choose any (given) dialogical context.

Seriality cannot be reconstructed in a similar way, for this property transgresses the formal rule for context SR-ST9.1, allowing **P** to introduce a new context. What we need here is serial variant for SR-ST9.1:

> **(SR-ST9.1D)**: **O** may introduce a dialogical context anytime the other rules let him do so. **P** may introduce a dialogical context of rank +1 relative to the dialogical context he is playing in, and his choices are restricted otherwise by SR-ST9.2K.

Interesting is the fact that there is a real affinity between modal dialogic and the GTS version of it. Indeed, while fixing rules for modal logic Hintikka distinguishes the rules which generates new worlds from the rules in which a player fills actually with formulæ an already generated world. This distinction is nowadays quite often implemented in tableau systems for modal logic. It should be clear that, dialogically, the moves in a modal game feature an application of the same notions: introducing a dialogical context corresponds to generating one, and the defence of a necessity or a possibility by **P** amounts to filling an already generated dialogical context with a formula.

---

[44]The corresponding dialogic has been obtained using the SG rules (either in the classical or intuitionist version), replacing SR-ST4 by SR-ST4M, together with SR-ST9.1, the adequate SR-ST9.2 rule, and the particle rules for the modal operators. It is straightforward to use the structural rules of strategy games instead.

4. Intrinsic pluralism and it's dialogical language

Up to now, we have shown how some logical systems, coming from very different traditions, find a natural dialogical reconstruction allowing to exhibit in the same framework some of the connections between them.

In this section, we will turn to the problem of the unity and the diversity of the notion of logical consequence within the frame of dialogic. This leads indeed to a *nexus* of related difficulties, which we want first to delineate, before we present what dialogic can do about it.

There are ongoing discussions about three apparently different topics, namely the tension between: (i) explicit / implicit modalities, (ii) metalogical / object language level rules, (iii) propositional and non propositional knowledge. Our point here is that all these tensions, which are closely related to the question concerning the unity of logical consequence, can be seen as different expressions of a same phenomenon, and that (in some cases) the tension could be dissolved in the way pragmatics philosophy conceives the interaction between propositional and non propositional levels as developed in the work of Gilbert Ryle for example. Let us present these discussions very briefly:

(1) Göran Sundholm, as many other intuitionistic philosophers, thinks that the standard formulation of modal logics do not do justice to the epistemic motivation of modalities. His point is that the propositional translation of the epistemic character of logical inference by means of the necessity operator of standard modal logics brings into the object level the epistemic necessity of logical inference, transforming this relation between a subject and (a) proposition(s) into the pure relation between propositions called logical consequence. In his view this move induces either a rejection of the epistemic nature at the metalogical level or a kind of *regressus ad infinitum*, for one needs either some kind of non-epistemic notion of logical consequence at the metalogical level or some other epistemic notion of inference of higher degree.

(2) Another important point, discussed by substructuralists and linearists concerns the change of logics. When defining the components of a logic, substructuralists differentiate between: (i) fixed particles (the usual introduction and elimination rules for logical constants) and (ii) structural rules (which apply to these logical constants). Structural rules belong to another level (Hilbertians love to say that there are meta-propositions) expressing, from the point of view of logical consequence, general properties of the logical constants in question. One can see the choice of the structural rules as a way to say how to apply the logical connectives to a given context – e.g. in a given argumentation context (say temporal) one might not wish to have commutativity: we simply

say that the correspondent structural rule does not apply here. In fact, what many substructuralists do is to concede that different structural logics do define different notions of logical consequence.

On the other hand pro-linearists (followers of the French computer-logician Jean Yves Girard) claim that every structural rule can be expressed at the object level via particles, or in Gentzen terms, with the help of appropriate introduction and elimination rules. Changing logics is, in Girard's view, a mistaken way to express that for some (mathematical) purposes a new and more adequate combination of particles rules has been formulated.

If we generalise point (1) extending the scope of the notion of implicit knowledge as including structural conditions and put this together with (2) we might say that while the substructuralists use of particle rules is propositional and explicit, their use of substructural rules is either implicit and non propositional or explicit but meta-logical.

In Ryle's philosophy, there is a well-studied distinction between propositional and non-propositional knowledge. One of his points is that propositional level arises when something at the non propositional level does not work: you do not usually need to read a book to ride a bike but sometimes, perhaps when the shape of the bike has changed drastically or when you want to teach (in a dialogical situation) someone, or even teach to anyone: you need a level where the knowing-how (implicit knowledge) becomes propositional (explicit knowledge). Narahari Rao wrote an excellent book (Rao [1994]) about the relation between these two types of knowledge. The very point of the distinction is, as developed by Rao, that there should be an interface between these levels: a subtle feedback, producing a special kind of looping: indeed, if the non propositional level is the object of the propositional one, the later finds its *ratio essendi* only when it has, as a consequence, the improvement of the non propositional.

We would like to suggest that Rao's interpretation of Ryle could be used to solve the tension of (1) and (2). To put it very briefly: there are some occasions (argument contexts) where we would like to reflect and discuss the conditions under which an argument schema is considered to be valid. In these (critical) moments, the implicit structures assumed with the validity of the argument in question will be brought to the propositional level examined as a new kind of logical constant which induces a determinate notion of logical consequence– this obviously presupposes other implicit structural rules with the help of which the (by now explicit) conditions of validity at stake are being examined, but which might be too the start of another cycle of implicit/explicit exchange.[45] And those circumstances which motivates a reflexive moment seem quite often to motivate a *change in the way we reason.* Pluralism, as the reckoning that the words "logical inference (consequence)" have more than one meaning, is then a condition of intelligibility of what happens when this type of implicit/explicit exchange occur.

Much of the research in the field of non-classical logic has inherited the classical idea that there is *one and only one* real logic (i.e. definition of what is a correct inference), a position which we will refer to as *logical monism.* Our position, on the other hand, is pluralist in the sense that we accept that there is an arbitrary number of distinct correct notions of inference, for which there seems to be no reason to always assume that they are reducible. The dialogical frame has proved itself a valuable means of confronting these different notions, via the confrontation of different sets of structural rules, notions that could be studied too at the particle level in the way suggested just above. In fact, from the dialogical point of view proving is a special kind of action in context, the rules of which might change when the context changes.

The ability to produce (m)any logic(s) within the same conceptual frame, conserving the same general notion of validity and the same way of fixing the required semantics, allows the dialogician to compare very precisely the different logics. That is why dialogic *is not* a logic, but a frame in which formal languages can be expressed (among many other things) and studied. If admittedly the first dialogical system was an attempt by Lorenzen, who was a pure bred monist, to delineate the foundations of intuitionist mathematics and logics, nevertheless, since the philosophical work of Lorenz, the dialogician is not that much concerned with the defence of a specific formalism as *the* way to solve all the (metamathematical) problems of mankind. (S)he is rather interested in what dialogic does best: studying the relations between logics or even more generally between argumentation systems, which implies accepting the multiplicity of logics and argumentation systems.

In the following section (4.1), we will show that there is an adequate logic for the reasoning about different logics, a kind of in-built logic of pluralism, and how to express it as a dialogic.

In the next section (4.2), we will present a special kind of dialogues, namely: structure seeking dialogues (SSD), showing with some detail how one can conceive a game where some of the rules of the game are at stake. This should shed some light on the interplay between implicit and explicit forms of logical structures.

[45]This might shed some light on the passages in Aristotle concerning the interpretation of his use of modalities, which in some occasions seems to be implicit and in others explicit.

4.1 Non-normal contexts and logical pluralism.

4.1.1 The maxim of pluralism. When Kripke[46] introduced the non-normal worlds to the formal apparatus of the possible worlds semantics, it was to gain a discrimination between the notions of validity and of necessity. In some modal logics it could be desirable to get rid of the *necessitation* rule, which establishes that if a proposition a is valid, then it is necessary that a. The aim of this section is to relate logics without necessitation rule with logical pluralism.

In this perspective, a really pluralist view of logic is based on the claim that, with respect to some formal system λ_m there is always another formal system λ_n which both: (i) does not count an arbitrary proposition a, which is a theorem of λ_m, to the number of its (λ_n) theorems, and (ii) makes sense with respect to some context of rational thinking. The idea is that it sometimes seems sensible to make it explicit that another kind of logic is always conceivable, applying somewhere else in the universe of the arguments. Indeed, though we might assume that many of the arguments in our world of everyday are logical, we are not thereby assuming in general that the logical necessity of these arguments extends to all type of arguments. Moreover it very much looks as if in such argumentation contexts, we are prepared to concede that no logical necessity of these arguments is necessarily necessary – which was one of the major points stressed by Descartes while discussing the issue of eternal truths. Our claim is that the explicit propositional formulation of this pluralistic tenet can be expressed by the so-called *non-normal modal logic.*

4.1.2 Non-normal modal dialogic. Let us call *non-normal* such argumentation contexts or "worlds", where a different logic holds relative to the logic holding in a given world defined as *normal.* Logicians have invented several logics capable of handling logically arguments which are aware of such a situation which seems to threaten the very idea of logical necessity. The main idea of their strategy is simple: logical validity is about normal contexts and not about logically weird ones: we only have to restrict our arguments to the normal ones. Now, the point of this strategy is not to ignore the non-normal contexts, but to take into consideration no other context than the normal ones while deciding about the validity of a given argument. Actually there is a less

[46]First attempts to formulate non-normal modal logics were made by Hugh MacColl [1906], and axiomatised by C.I. Lewis. A possible-worlds semantics is due to Kripke [1965] and a GTS one to Hintikka [1975]. Non-normal modal logics are used to solve the omniscience problem in epistemic logic, see for instance the classical paper of Veikko Rantala [1975]. For a new interpretation of non-normal modalities, see Rahman [2004].

conservative strategy: namely, one in which a formula is said to be valid if it is true in all worlds whether normal or non-normal. The result is notoriously pluralistic: no logical argument could be proven in such systems to be unconditionally necessary.

Anyway if we have a set of contexts, how are we to recognise the normal ones? The answer is clear in the dialogical context: those contexts in which a logical necessity of a particular kind has been *explicitly assumed* (conceded) during an argument to hold, cannot be considered as non-normal in the same argument and relative to the same kind of necessity. Or, to express it the other way around, if there is some loss of information concerning previous choices or concessions of a necessary formula of a particular kind, one cannot assume that the context is a normal one relative to this kind of necessity.

The point seems to be now, while testing the validity of a given argument, to find a device to check whether the argument loses its logical force when non-normal contexts are taken into account. For instance, as already pointed out, the rule of necessitation fails in every non-normal logic.

As we have seen, in dialogic and in game-theoretical approaches to modal logic, the introduction or generation of new contexts (subdialogues, subgames, indexed sets of formulæ, possible worlds, or whatever type of contexts your ontology can stomach) are seen as complementary to rules which allows to profit from these generated contexts by filling them with formulæ, or simply by proving in them. From the dialogical point of view they are seen as concessions of **O** which can be used by **P**: as already mentioned, the introduction of a dialogical context works in an analogous way to the formal rule for prime formulæ. Let us detail now what happens to modal dialogic when non-normal contexts are considered.

Validity concerning normality. The major issue here is to determine dynamically – i.e., during the process of a dialogue – which of the contexts are the normal ones. This must be a part of the dialogue's structural rules (in the case that we are not dealing with dialogues where the dialogical contexts are supposed to have been given and classified from the start). Let us formulate the following general rule implementing the required dynamics:

(SR-ST10.05) (S05-*rule*): **P** may attack a formula of the form $\mathbf{P}A$ or defend a necessity-formula of the form $\mathbf{N}A$ stated in the context m if and only if m is a normal context.

Two further assumptions will complete this rule: (i) the dialogue's initial context is normal, and (ii) no other context than the initial is normal.

The dialogic resulting from these rules is a dialogical reconstruction of a logic known in the literature as S.05. In this logic validity is defined relative to normality and has the constraint that no other newly introduced world is normal. Certainly $\mathrm{N}(a \Rightarrow a)$ will be valid (the newly generated context, which has been introduced by the challenger while attacking the thesis has been generated from the normal starting context). $\mathrm{NN}(a \Rightarrow a)$, on the contrary, will not be valid: the attack of O in the second context cannot be responded by **P** since this context is not normal (and thus we cannot assume that the external and the internal necessity operators are of the same kind – i.e. correspond to the same logic).

Let us produce a dialogical reconstruction of another logic, known as S2, where we assume not only that the first context is normal and the rule introduced above, but also:

> **(SR-ST10.2)** (***context normality rule***): If **O** has stated in a context n a formula of the form $\mathbf{N}A$ (or if **P** has stated in n a formula of the form $\mathbf{P}A$), then the context n can be assumed to be normal$_{Li}$.[47]

This is, for our purposes, a more appealing logic than S.05 because it makes of the status of the contexts at stake a question to be answered within the dynamics of the dialogue. One can even obtain certain iterations such as $\mathbf{N}(\mathbf{N}(a \Rightarrow b) \Rightarrow (\mathbf{N}a \Rightarrow \mathbf{N}b))$ which is not valid in S.05, but it is in S2 (the first context is normal by definition, the second context will be normal because **O** will concede $\mathbf{N}a$ there. Now, because the second context is normal, **P** can answer in the third one which has been generated from the second normal one). Adding transitivity to S2 renders S3.

Validity concerning normality and non-normality. The point of the logics presented in the chapter before was not to ignore the non-normal contexts, but only to take into consideration the normal ones while deciding about the validity of a given argument. We will motivate here a less conservative concept, namely, one in which a formula is said to be valid if it is true in all worlds whether normal or non-normal. These logics are known as E. In no E system will $\mathbf{N}a$ be valid for any formula a. Isn't this challenging?

[47]Notice that this rule introduces normality when **P** has the choice while attacking or defending a modal formula. See details in Rahman [2004].

Suppose one modifies S.05 in such a way that no context is assumed from the start to be normal. This logic, called E.05, is unfortunately not of great interest: a formula will be valid in E iff it is valid in non-modal logics (think of $\mathbf{N}(a \Rightarrow b) \Rightarrow(\mathbf{N}a \Rightarrow\mathbf{N}b)$, which in this logic cannot be proven to be valid). Modality seems not be of interest there, and this logic can be thought as a kind of a modal lower limit.

Now the elimination of the assumption of the first context to be normal in S2 yields an interesting dialogic for our purposes. $\mathbf{N}(a \Rightarrow b) \Rightarrow(\mathbf{N}a \Rightarrow\mathbf{N}b)$ is valid there, signalising a more minimal condition for the validity of this formula as K (for it does even not assume, as K does, that validity concerns only normal worlds), which in turn will prove of importance for the structure-seeking dialogues. Similarly one could produce D versions, etc. Indeed E2 seems to be the appropriate language where the logical pluralist might make explicit his arguments against the *eternity of logical truths*.

4.2 Structure-seeking dialogues. (SSD).[48]

SSD are meant to let the player discuss, within the object language of the dialogue, some structural rules specific to modal propositional dialogic or more generally, in a more standard language, the aim here is to study how the assumed validity of some propositional modal formula uttered in a given field of argumentation can be said to impose some determinate frame conditions. An interesting fact is that the formal devices applied by Patrick Blackburn [2001] while displaying an anti-pragmatist interpretation of modal dialogic can be used to study what we see as the passage from dialogical pragmatics to dialogical semantics. Let us deploy some few lines about this issue in order to explain what we see as the philosophical background of SSD: Charles William Morris' distinction between pragmatism and semantics is still very influential today. Blackburn (2001), for instance, seems to think within this conceptual framework while discussing the pragmatical content of dialogues. Morris' view amounts to defining semantics as a relation between signs and objects, and pragmatics as a relation between signs and utterers using these signs. When applied to dialogic, this distinction misses the point: from a dialogical point of view, both, semantics and pragmatics, are conceived as a relation between signs and utterers using these signs But this does not mean that in dialogic we deny the pragmatics/semantics distinction, it just applies somewhere else. Dialogical semantics define

[48]This section is the result of fruitful exchanges during the elaboration of E. Genot's predoctoral thesis (DEA).

meaning by the playing of a game *where the rules (in use) are fixed.* Dialogical pragmatics, on the other hand, deals with the games with *not yet fixed or not yet determined rules* – that does not mean that degrees between these moments are not possible: the difference between structural and local meaning witnesses such a graduation.

Besides this general background concerning the dialogical theory of meaning, SSD are also related to the passage from explicit to implicit modalities in logics. In effect, as already mentioned, from the dialogical point of view, making explicit a given rule of the game within a game, i.e. expressing it *within the same propositional object language*, amounts to make this rule (which up to that point has been implicitly agreed on) the object of the argument. Thus some part of the rules which was fixed in an implicit agreement and which fixed a part of the global (structural) meaning of the formula in question is now at stake, hence (pragmatically) open to some (global) meaning changes. However, once the game is finished, a global semantics, perhaps even new, will be fixed.

We will discuss here a particularly simple case of such a movement, where the rule at stake is the structural rules restricting the accessibility relation in modal propositional dialogic. Now these implicit rules can in turn be made explicit, under the motivation of a very frequent situation: given a (modal) formula a, one wonders under which conditions one could win a game for A. *It is exactly this case we would like to explore.*

4.2.1 Nominals. We first need to enrich our language in order to be able to refer to the contexts and their accessibility at the level of the games, i.e. at the object-language level. We will use an adapted version of Blackburn's *satisfaction operator* @.[49]

We need a device to refer to a given context, or a class of contexts, within the formulæ of the games: let NAME be a function assigning a prime formula ν_n to every context n introduced during the dialogical process. The prime formulæν_i assigned by NAME are called *nominals*, and the context n is said to be the *denotation* of the prime formula ν_n, since any such formula uniquely determine the context it is bound to. This calls for a structural rule for the use of nominals:

> **(SR-ST1SSD)(*nominals use rule*)**: X-ν_n can only be played in the dialogical context n. The player under formal restriction can (as usual) use a nominal if, and only if, this nominal has been introduced before by the player who is not formally restricted.

Now this rule calls for a precise definition of the notion of *introduction* of a nominal:

[49] Blackburn [2001] attributes the idea to Prior [1967].

Definition (*introduction of a nominal*): a nominal is *introduced* by a player when: (i) it is asserted by the player in the same way the other prime formulæ are asserted; (ii) the label of the dialogical context it denotes has been used by the player to attack a modal operator of necessity; or (iii) when the player defends against an attack aimed at a modal operator of possibility in the context denoted by the nominal.

> **(SR-STSSD)** (***Formal formal rule of accessibility***): Once a context n has been introduced by **O** as a reaction to a modal formula in m, n could be used by **P** to attack or defend another modal formula stated in m.

Let us now define the syntax for the @ operator, extending our modal language $\mathbf{L}_m$ into a *hybrid* language $\mathbf{L}_H$:

Let m be a dialogical context label, i.e. a finite sequence of positive integers of the form $n.o.p\ldots$ and i a free variable ranging over the set of dialogical contexts. $@m$ or $@i$ can be added to a wff of $\mathbf{L}_m$ to form a new wff: if A is a wff of $\mathbf{L}_m$, possibly complex, $@mA$ and $@iA$ are wff. For any A and B two wff of $\mathbf{L}_m$ and * any dyadic connector (i.e. $* \in \{\wedge, \vee, \Rightarrow\}$), $@mA*@mB$ can be written $@m(A*B)$, and $@iA*@iB$ can be written $@i(A*B)$.

The idea behind the @ operator is to distinguish the assertion that a given formula a can be defended in the dialogical context m (or in *any* dialogical context) (X-$@mA$ (or X-$@iA$)) from the dialogical context n where the assertion has been uttered – which could be different from m.

Thus, X-$@mA$ (or X-$@iA$) can be asserted in any dialogical context n (possibly different from m), and its dialogical local meaning amounts to the assertion that the player X is ready to defend a in the dialogical context m (or in any dialogical context m chosen by Y). So there are two particle rules for the @ operator, according whether its index is a constant or a variable:

> @ *particle rule* (*constant*): from a formula of the form $@mA$, where m is a constant, follows a state of the modal game <R, σ, A, $\lambda_{a/m}$ >, responding to the attack $?_@$ of the challenger, where m is the label of the dialogical context stipulated by $@m$.
>
> @ *particle rule* (*variable*): from a formula of the form $@iA$, where i is a free variable ranging over the set of labels of dialogical contexts, follows a state of the modal game <R, σ, A, $\lambda_{a/m}$ >, responding to the attack $?_{i/m}$, of the challenger where m is the label of a dialogical context chosen by the challenger.

Now we need a structural rule to define what dialogical contexts are eligible while performing the attack $?_{i/m}$ against $@iA$:

> **(SR-ST2SSD)** (***@ attack rule***): while performing an attack against @i, X can choose the label of any context which the other structural rules of the modal dialogue let him choose.

4.2.2 Structure seeking dialogues. To allow the players to discuss the rules of the game in the sense mentioned above let us introduce some formal devices, namely the Δ sequence, and the operators $\{\Delta\}$ and **Min**.[50]

The first point to precise is to determine which of the rules of the game can be challenged. In our case, what is at stake is the structural rule restricting the context choice of the proponent (SR-ST9.2).[51] The properties of this rule can be expressed within the propositional modal language by means of Blackburn's *hybrid language* [2001] in the following way (e.g.):

> reflexivity: $@i\mathbf{P}\nu_i$, symmetry: $@i\mathrm{NP}\nu_i$, transitivity: $\mathbf{PP}\nu_i \Rightarrow \mathbf{P}\nu_i$, density: $\mathbf{P}\nu_i \Rightarrow \mathrm{P}\mathbf{P}\nu_i$, euclidianity: $(\mathbf{P}\nu_i \wedge \mathbf{P}\nu_j) \Rightarrow (@i\mathbf{P}\nu_j \vee @j\mathbf{P}\ \nu_i)$, seriality: $\mathbf{P}\nu_n$

Any of these formulæ will generate a dialogue where the corresponding property will be said to have been made an *explicit structural condition* for the validity of a given formula So the SSD will run within the limits of a sequence Δ of structural conditions. The sequence Δ is introduced in the language as the follows:

Given an arbitrary positive integer k, let Δ be a finite sequence[52] of k structural conditions, which we assume not to be empty. Let Δ_i (for i=0 to k-1) be the elements of Δ. We also assume an order between the subsets of Δ.

Let us introduce now $A_{\{\Delta\}}$, expressing that a formula a will become the thesis of a SSD with respect to some given sequence Δ. The operator$_{\{\Delta\}}$ in $A_{\{\Delta\}}$ can be understood as a kind of existential quantifier restricting the structural conditions under which a is assumed to be valid (see note 54 for an explicit and formal use of such a quantifier). The player who states $A_{\{\Delta\}}$ asserts hereby that in the sequence Δ, there is at least one (non-commutative conjunction of) structural condition(s) which is minimal and sufficient to win any dialogue for a. Let us precise the particle rule for this operator:

> $\{\Delta\}$ *particle rule*: from $A_{\{\Delta\}}$ follows the SSD game <R, σ, $\mathbf{Min}(\delta_n A)$, λ>, reacting to the attack $?_{\{\Delta\}}$ of the challenger, where $\delta_n \in \Delta$ and has been chosen by the defender.

The role of the operator **Min** is to allow to build formulæ such as **Min**(Δ n,a), which we call *structural statements* (STS). By uttering an STS formula a player makes explicit which of the structural conditions he

[50] We assume that the $A_{\{\Delta\}}$ and $\mathbf{Min}(\alpha,a)$ are wff extending our basic language.
[51] We use the rule number without a suffix to designate any of the variants of the (SR-ST9.2) rule.
[52] We express it as a sequence for reasons which will rapidly become clear.

assumed while stating $A_{\{\Delta\}}$. In other words, the player claims herewith that he assumes that a determined element (or conjunction of elements) δ_i of Δ is the minimal structural condition for the validity of a – the point behind the minimality claim involved by STS is that **P** should be obvious: if no minimality is claimed **P** would always choose S5 and win. Informally, the idea is that structural statements can be attacked by the challenger in two distinct ways. *First*, by conceding the condition δ_i, claimed by the player X to be minimal, and asking X to prove the thesis. *Second*, by (counter)claiming that the thesis could be won with a (subset of) condition(s) of lesser rank in Δ. In that case, the game proceeds in a subdialogue, started by the challenger who now will claim that the formula in question can be won under the hypothesis δ_j, where δ_j is different from δ_i and has a lesser rank as δ_i. Since the challenger (Y) starts the subdialogue *he now has to play formally.* Obviously, the player that now plays under the formal restriction in the subdialogue cannot state extra STS: he has fixed the rules with his attack and hereby completed the relevant semantics for the formula in question: the subdialogue is not played as a SSD but simply as dialogue under hypotheses. [53]

Notice that one could see the SSD as a way to extend the local semantics of certain modal formulæ. Before stating the local semantics of structural statements let us extend the notion of *state of a modal game*:

Such a state following a formula A is a tuple $<R, \sigma, A, \lambda, \psi>$, where R, σ, A and λ are defined as in modal dialogic, and ψ is a assignment of SSD-subdialogues to formulæ (we write ψd to express that ψ assigns the SSD-subdialogue d to the formula A).

Here the local semantics for structural statements:

> **Min** *particle rule*: from $\mathbf{Min}(\Delta_m, A)$, where A is a formula and $\Delta_m \in \Delta_m$, follow either a state of the SSD game $S=<R, \sigma, \Delta_m, \lambda, \psi>$, consecutive to an attack $?\Delta_m$ of the challenger, or a state $S'=<R', \sigma, (\Delta_n \Rightarrow A), \lambda, \psi/d>$, where d is a new SSD-subdialogue. Whether the game proceeds to S or S' is the choice of the challenger.

Remark: Actually the idea is to find procedurally during the game, by means of a sort of explicit abduction, which are the adequate conditions for winning. As already mentioned **P** starts by choosing which of

[53]We introduced this notation in order to keep matters simpler but actually one could express all this at once and more formally using restricted quantification explicitly. Indeed, instead of writing $A_{\{\Delta\}}$ we could, e.g., write: $\exists(\delta_i \subseteq \Delta)((\delta_i \Rightarrow a) \wedge \forall(\delta_{j\neq i} \subseteq \Delta)(\mathbf{F}((\delta_j < \delta_i) \Rightarrow a)))$, where δ_i is a non empty subset of structural conditions in Δ: When δ_i contains more than one element, it constitutes a non commutative conjunction. The operator **F** is the attackabiliy operator introduced in the chapter on connexive logic. The whole formula reads: There is at least one subset Δ_i of structural conditions sufficient to win any dialogue for A; and no dialogue for A could be won with a subset $\Delta_{j\neq i}$ of conditions of (pre-agreed) lesser rank in Δ.

the conditions of Δ he assumes to be the minimal one. But it may well happen that though this condition is indeed minimal it is still insufficient for winning, i.e.. it might happen that the subset of conjuncts should contain more elements. In that case, **P** will have to defend again against the $?_{\{\Delta\}}$ attack, with a stronger condition (i.e. of higher rank).[54] This means that the part of the SSD where conditions are at stake is always played with the classical rules. This is a consequence of the fact that Δ is finite: there is always a decision procedure which grants that classical rules are safe. Let us fix this with a structural rule:

> **(SR-ST3SSD)** ($_{\{\Delta\}}$ ***attack rule***): even when the dialogue is played with intuitionist rules, **P** is allowed to defend himself several times – but with different choices – against the same attack $?_{\{\Delta\}}$ on his structural statement.

Since during a SSD changes of players concerning the formal restriction are possible, we will follow here the device introduced in connexive logics using the expression X_f (or Y_{nf}) to identify who of the players is the one playing under the formal restriction (or not under formal restriction). Let us start redefining the formal rule for SSD games.

> **(SR-ST4SSD)** (***SSD formal rule***): At the start of the dialogue for A it is **P** who plays under the formal restriction. If X plays under the formal restriction in given SSD-subdialogue he cannot use a prime formula Y did not introduced before in the same subdialogue. Dually, for Y who does not play under the formal restriction. There is no other change of formal restriction than the one induced by the rule SR-ST6SSD (see below).

What we need now is to establish those structural rules which connect the attack against structural statements and which regulate the changes of the formal restriction:

> **(SR-ST5SSD)** (***Min attack rule***): only Y_{nf} is allowed to attack a formula of the form $\mathbf{Min}(\delta_m, A)$ with δ_m, or with $\delta_n \Rightarrow A$, where $\delta_n \in \Delta$ and n<m.

> **(SR-ST6SSD)** (***formal restriction changes rule***): when a formula of the form $\mathbf{Min}(\delta_m, A)$ is attacked by $\delta_n \Rightarrow A$, the game proceeds in a new subdialogue where Y now plays under the formal restriction (i.e., Y now assumes the role Y_f), and X now plays as Xn_f.

NOTE: Notice that the notion of Δ given here is abstract. We kept it abstract, because there is no universal definition of structural minimality: each context and each interpretation of the modalities carries its own notion of minimality. That is why the nature and the order

[54]We express this by numbering the STS in the attack/defence column, using S1, S2, ...

of the various structural conditions are pragmatically determined by a convention before each SSD. A good example of the context-dependence of the order of the rules is that of the question of the rank of seriality. On the one hand, many deontic interpretations of the modalities would have D as a theorem but without reflexivity. Thus in a deontic SSD, one would say seriality is minimal with respect to reflexivity. On the other hand, the serial rule is not formal, in the sense that the player under formal restriction is nonetheless authorised to open a new context. In some other context, formality may be a very important property, and reflexivity would hence be seen as minimal. One can easily imagine that things get even worse when trying to order combinations of structural conditions, or structural conditions from foreign domains (think for instance of including in SSD the choice between normal and non-normal structural conditions).

Table 7. SSD for D, where $\Delta=\{\nu_1 < \mathrm{P}\nu_n < @i\mathrm{P}\nu_i\}$ and *ctx* reads *dialogical context*. **O** wins.

ctx		**O**			**P**		*ctx*
					$(\mathbf{N}_a \rightarrow \mathbf{P}_a)_{\{\Delta\}}$	0	1
1	1	$<?_{\{\Delta\}}>$	0	S1	$\mathbf{Min}(\nu_1, (\mathbf{N}a \rightarrow \mathbf{P}a))$	2	1
1	3	$<\nu_1>$	2		$\mathbf{N}a \rightarrow \mathbf{P}a$	4	1
1	5	$\mathbf{N}a$	4		$\mathbf{P}a$	6	1
1	7	$<?_{\mathbf{P}}>$	6				
[1]	[1]	$[<?_{\{\Delta\}}>]$	[0]	S2	$\mathbf{Min}(@i\mathbf{P}\nu_i, (\mathbf{N}a \rightarrow \mathbf{P}a))$	8	1
1	9	$(\mathbf{P}\nu_n) \rightarrow (\mathbf{N}a \rightarrow \mathbf{P}a)$	8				
1	11	$\mathbf{N}p \rightarrow \mathbf{P}a$		9	$\mathbf{P}\nu_n$	10	1
1	13	$\mathbf{P}a$		9	$\mathbf{N}a$	12	1
1.1	19	a		13	$<?_{\mathbf{P}}>$	14	1
1	15	$<?_{\mathbf{P}}>$	10		$<\nu_{1.1}>$	16	1.1
1.1	17	$<?_{\mathbf{N}/1.1}>$	12		a	18	1.1

The example depicted in Figure 7 shows how an SSD would run for D. **P** makes the mistake of choosing reflexivity in move 8 where seriality is sufficient to win relative to the sequence $\Delta=\{\nu_1 < \mathbf{P}\nu_n < @i\mathbf{P}\nu_i\}$. As in connexive dialogic, the shaded areas of the dialogue indicate the subdialogues *d.i* where the player who stated the thesis of *d.i* plays under the formal restriction.

The epistemic features of these dialogues are expressed by the change of the formal restriction. This dialogical device has been elaborated to yield the connexive dialogic, where the meaning relevance connecting the terms of the dyadic operators can be expressed by using contingency assertions, granting that no winning strategy stems from triviality. Now a contingency assertion cannot be uttered independently of the knowledge of a situation (a context), since it amounts to *describing* a situation where the formula at stake is true (or false). It is then reasonable to understand the use of the contingency operators **V** and **F** by a player as the player's claim to *know* a situation where the formula is true (or false). A lost subdialogue about this claim is then evidence that the player *believed* he knew, and was wrong. In the very same way, SSD are epistemic. By using the **Min** operator, **P** claims he knows the minimal structural conditions under which the thesis can be won. But this claim can be contested, and become the subject of an argument. Now when contesting such a claim, the challenger has to allow the defender to use his knowledge of the situation and free him of the formal restriction.

It is important to notice that there is some implicit knowledge supporting **P**'s claim which should not be propositionalised in the same SSD. It is an implicit epistemic notion acting at the level of judgement. That said, is should be clear that it could be the object of another structure seeking game, but then a higher degree of non-propositional knowledge will be needed.

5. Conclusion and the way ahead.

One of the major points of this paper was to stress the convergences and correspondences existing between different approaches to logics. Our hope here is that it is possible to transfer from one way of doing logic to another, so as to be able to profit from the best that each has to offer. Pluralism is nothing but the acknowledgement that, as Frege insisted all his life (although in a very monist way), formalism is not neutral to the thought, and (as Frege would never have accepted) that we need several Begriffsschriften, in order to say all that has to be said about our objects. Now, plurality of languages become a wealth if one possesses the Rosette's stones, allowing one to understand how the dialogues are to be built. Let us suggest some steps in that direction.

5.1 The way to IF (1): Normal and non-normal contexts.

One could see the logics of section 4.1.2 as a problem of quantification over restricted domains. If one accepts the interpretation of the modal operators as quantifiers ranging over contexts, then it is straightforward to divide between operators ranging over normal contexts and others ranging over non-normal contexts. But restricted domains emulate in a static way a dynamic process, as described by IF-logic. Actually, restricted domains have been used to dispense with IF, but one could go the other way round too, especially if interested in studying how these restricted domains arise as the product of a dynamical process.

Let us distinguish two types of normality concessions: generated normalities (**GNor**) and fixed normalities (**FNor**):

> **Definition** (*generated normality*): if **O** states a formula **N**A in context $d.n$ it is said that **O** has generated an **N** -normal context - i.e. he states herewith that $d.n$ is an element of **Gnor**.

> **Definition** (*fixed normality*): the context $d.n$ is said to have been fixed as normal iff it has been established at the start of the dialogue that $d.n$ belongs to the set of normal contexts **FNor**.

*If **O** generates a context d.m* from d, then **P**, who before was in d may place moves in $d.m$ if and only if the new generated context has been generated from either **GNor** or **FNor**.

Now let us suppose that we are in the context of IF-logic where independence of information is possible. Assume that the player X has to play under some loss of information concerning previous choices or concessions of a necessary formula of a particular kind. In this case the player X cannot assume that the context is a normal one relative to this kind of necessity – note that the player might, e.g., not lose information about the fixed normality but about the generated ones. This opens the field to applying all the subtleties of IF in the exploration of this type of logics. It suggests too another possible way to understand the classification between normal and non-normal in a non-ontological way. Non-normality becomes the sign of an epistemic failure concerning the borders where the concept of necessity involved applies.

5.2 The way to IF (2): Free and paraconsistent logics

The main point of free logics is that from $a\mathrm{k}_1$ one might not deduce $\exists xAx$, because this quantifier might have an existential import, which is missing in the individual constant. In fact we could also express free logic in the framework of restricted quantifiers in, e.g. the formulation $\exists_{\{DE\}}xAx$ – i.e. $\exists x$ ranges over DE (the domain of existents, as opposed to DF, the domain of fictions). Now, free logic can also be understood, in the dialogical frame, as the result of an information independence problem. While **O** has conceded A_{k1}, **P** still does not know in which domain (DE or DF – or even in any other) **O** chose it. So **P** cannot use A_{k1} to attack or defend a quantifier until **O** explicitly concedes some information about the domain A_{k1} comes from.

The situation is similar with paraconsistency. In one main interpretation paraconsistency seeks to differentiate between two domains of negations: in one domain anything follows from a contradiction, while in the other it does not. Once again one can easily prefigure a game situation where one player does not know if the other's use of a given negation in a contradiction allows *quodlibet* to be derived from that contradiction or not.

These brief lines are intended to suggest how to develop a dialogue between paradigms. The implicit claim throughout the paper is that the pragmatic philosophy inherent to dialogic is a sufficient condition for achieving unity in diversity. Nevertheless, and more modestly, dialogic could also be seen as a heuristical stage on the way to a semantics of the logic of a given argumentation context before the (dynamic) features of the standard model semantics for this logic have been formulated. With this in mind, the dialogician should set a basis for co-operation between philosophers and logicians in the same dialogical way as philosophers have always done.

Acknowledgements. We would like to thank for enriching discussions Jean Caelen (Grenoble), Alain Lecomte (Grenoble), Narahari Rao (Saarbrücken), Helge Rückert (Saarbrücken), Gabriel Sandu (Helsinki), Daniel Vanderveken (Montreal) and Denis Vernant (Grenoble). We would also like to thank André Laks, director of the Maison des Sciences de l'Homme du Nord-Pas de Calais (MSH). In fact this work is part of the research projects *Preuve* and *La science et ses contextes*, attached to the MSH-Nord-Pas de Calais. Many thanks to Florian Ferrand, for a most needed help in editing the chapter. Also thanks to E. Genot and H. Tahiri, students of the Lille 3 team of Logic.

References

Barth E. M. and Krabbe E. C. W. *From Axiom to Dialogue. A Philosophical Study of Logic and Argumentation.* Berlin: de Gruyter. 1982.

Belnap N. (1982). "Display Logic", *Journal of Philosophical Logic 4*, 375–418.

Blackburn P. (2001b). "Modal Logic as Dialogical Logic". In S. Rahman and H. Rückert 57–93.

Bencivenga E. (1983) "Free Logics". In D. M. Gabbay and F. Guenthner (eds.), *Handbook of Philosophical Logic*, Vol. III, 373–426. Dordrecht: D. Reidel.

Beth E. W. (1955) "Semantic Entailment and Formal Derivability", *Mededelingen van de Koninklijke Nederlandse Akademie van Wetenschappen, Afdeling Letterkunde 18*:309–342.

van Benthem J. *Logic in Games*, 2001–2004. (Typescript available online at `http://turing.wins.Uva-nl/~johan/Phil.298.html`).

Blass A. (1992). "A Game Semantics for Linear Logic", *Annals of Pure and Applied Logic 56*:183–220.

Blass A. (1997). "Some Semantical Aspects of Linear Logic", *Journal of the Interest Group in Pure and Applied Logic 5*:487–503.

da Costa N. C. A. and Alves E. II. (1977) "A semantical Analysis of the Calculi C_n", *Notre Dame Journal of Formal Logic XVI, 4*:621–630.

Dosen K. (1988). "Sequent Systems and Groupoid Models I", *Studia Logica 47*:353–389.

Dubucs J. (2002). "Feasibility in Logic". *Synthese 132*: 213–237.

Felscher W. (1986). "Dialogues as a Foundation for Intuitionistic Logic". In D. M. Gabbay and F. Guenthner (eds.), *Handbook of Philosophical Logic, Vol. III*:341–372. Dordrech: D. Reidel.

Fitting M. (1983). *Proof Methods for Modal and Intuitionistic Logic.* Dordrecht: D. Reidel.

Fitting M. (1993). "Basic Modal Logic". In D. M. Gabbay, C. J. Hogger and J. A. Robinson (eds), *Handbook of Logic in Artificial Intelligence and Logic Programming.* Oxford: Clarendon Press. 365–448.

Gabbay D. M. *Labelled Deductive Systems.* Oxford: Oxford University Press. 1996.

Gethmann C. F. *Protologik. Untersuchungen zur formalen Semantik von Be-gründungsdiskursen.* Frankfurt a. M.: Suhrkamp. 1979.

Girard J.-Y. (1993). "Linear Logic: Its Syntax and Semantics". In J.-Y. Girard, Y. Lafont and L. Regnier (eds.), *Advances in Linear Logic.* Cambridge: Cambridge University Press. 1–42.

Girard J.-Y. (1998). "On the Meaning of Logical Rules I and II: Syntax vs. Semantics", (unpublished manuscript).

Hamblin C. L. *Fallacies.* London: Methuen. 1970.

Hintikka J. (1975). "Impossible Possible Worlds Vindicated". *Journal of Symbolic Logic, 4*:475–484. Modified and reedited in Hintikka J. and M.B., *The Logic of Epistemology and the Epistemology of Logic.* Dordrecht: Kluwer, 63–72, 1989.

— *The Principles of Mathematics Revisited*, Cambridge: Cambridge University Press. 1996.

— *Selected Papers*, Vols. I–IV. Dordrecht: Kluwer. 1996–1998.

Hintikka J. and G. Sandu. (1996). "A Revolution in Logic?", in *Nordic Journal of Philosophical Logic 1*:169–183.

Hyland, M. (1997). "Game Semantics". In A. Pitts and P. Dybjer (eds.), *Semantics and Logics of Computation*, Cambridge: Cambridge University Press. 131–182.

Johnson R. H. (1999). "The Relation Between Formal and Informal Logic", *Argumentation 13*:265–274.

Kripke S. (1965). "Semantical Analysis on Modal Logics II; non Normal Propositional Calculi". In Addison J.W., Henkin L. and Tarski A. Eds.) *The Theory of Models.* Amsterdam: North Holland Publishing Co. 206–220.

Lodder A. *Dialaw. On Legal Justification and Dialogical Models of Argumentation.* Law and Philosophy Library, Dordrecht: Kluwer. 1999.

Lorenz K. (1981). "Dialogical Logic". In W. Marciszewski (ed.), *Dictionary of Logic as Applied in the Study of Language. Concepts / Methods / Theories.* The Hague, Boston, London: Nijhoff. 117–125.

Lorenzen P. (1958). "Logik und Agon". *Arti del XII Congresso Internationale de Filosofia,* Venezia. 187–194. (Reprinted in Lorenzen and Lorenz, 1978. 1–8)

Lorenzen P. (1989). "Die Dialogische Begründung von Logikkalkülen". In Carl Friedrich Gethmann (Editor), *Theorie des Wissenschaftlichen Argumentierens.* Frankfurt: Suhrkamp.

Lorenzen P. and Lorenz K. *Dialogische Logik.* Darmstadt: WBG. 1978.

MacColl H. *Symbolic Logic and its Applications.* London. 1906.

Naess A. *Communication and Argument. Elements of Applied Semantics.* Universitaetsforlaget. London, Oslo: Allen & Unwin. 1966.

Perelman C. and Olbrechts-Tyteca L. *La nouvelle rhétorique.* Paris: PUF. 1958.

Priest G. Routley R. and Norman J. (eds.) *Paraconsistent Logic – Essays on the Inconsistent.* München: Philosophia Verlag, München. 1986.

Rahman S. *Über Dialogue, protologische Kategorien und andere Seltenheiten.* Peter Lang Verlag. 1993.

— (1999). "Ways of Understanding Hugh MacColl's Concept of Symbolic Existence". *Nordic Journal of Philosophical Logic, 3,* 1–2:35–58.

— (2001). "On Frege's Nightmare. A Combination of Intuitionistic, Free and Paraconsistent Logics". In H. Wansing (ed.), *Essays on Non-Classical Logic.* New Jersey, London, Singapore, Hong Kong: World Scientific. 61–85.

— (2002). "Un desafío para las teorías cognitivas de la competencia lógica: los fundamentos pragmáticos de la semántica de la lógica linear". In M. B. Wrigley (editor), *Dialogue, Language, Rationality. A Festschrift for Marcelo Dascal,* Special volume of *Manuscrito, XXV-2*:383–432.

— (2004). "Non-Normal Dialogics for a Wonderful World and More" in Heizmann G. (ed) *The Age of Alternative Logics: Assessing Philosophy of Logic and Mathematics Today.* Dordrecht: Springer (to appear).

Rahman S. and Carnielli W. A. (2000). "The Dialogical Approach to Paraconsistency". *Synthese 125,* 1–2:201–232.

Rahman S. and Bendegem J-**P**. van (2002). "The Dialogical Dynamics of Adaptive Paraconsistency". In A. Carnielli, M. Coniglio, I. M. Loffredo D'Ottaviano (eds.), *Paraconsistency, the Dialogical way to the Inconsistent.* New-York: Marcel Dekker. 295–sq..

Rahman S. and Rückert H. (1999). "Die pragmatischen Sinn und Geltungskriterien der Dialogischen Logik beim Beweis des Adjunktionssatzes", *Philosophia Scientiae, (3)*, 3:145–170.

Rahman S. and Rückert H. (2001a). "Dialogische Modallogik (für T, B, S4, und S5)". *Logique et analyse, 167–168*:243–282.

Rahman S. and Rückert H. (eds.) (2001b). "New Perspectives in Dialogical Logic". Special issue of *Synthèse, 127.*

Rahman S., Rückert H. and Fischmann M. (1997)."On Dialogues and Ontology. The Dialogical Approach to Free Logic". *Logique et analyse, 160*:357–374.

Ranta A. (1988)."Proposition as Games as Types". *Synthese 76*:378–394.

Rantala V. (1975). "Urn Models: a new Kind of Non-Standard Model for First-Order Logic." *Journal of Philosophical Logic, 4*:455–474.

Rao B.N. (1994). *A Semiotic Reconstruction of Ryle's Critique of Cartesianism.* De Gruyter.

Recanati F. (2001). "What is Said", *Synthese 128*:75–91.

Rückert H. (2000). "Why Dialogical Logic?" In H. Wansing (ed.), *Essays on Non-Classical Logic.* New Jersey, London, Singapore, Hong Kong: World Scientific. 165–186.

Sette, A. M. (1973). "On P1", *Mathematica Japonicae, 18(13)*:173–180.

Toulmin S. (1958). *The Uses of Argument.* Cambridge: Cambridge University Press.

Vanderveken D. (1991). *Meaning and Speech Acts.* Cambridge: Cambridge University Press.

Walton D.N. (1984) *Logical Dialogue-Games and Fallacies.* Washington D.C.: University Press of America.

— (1985). "New Directions in the Logic of Dialogue". In D.N. Walton (ed.), *The Logic of Dialogue, Synthese 63*:259–274.

Wansing H. (1994). "Sequent Calculi for Normal Modal Propositional Logics." *Journal of Logic Computation 4*:125–142.

Wansing H. *Displaying Modal Logic.* Dordrecht: D. Reidel. 1998.

Woods J. (1988). "Ideals of Rationality in Dialogic" *Argumentation, 2*:395–408.

Chapter 18

SOME GAMES LOGIC PLAYS*

Ahti-Veikko Pietarinen
University of Helsinki

Abstract This paper studies the across-the-board character of game-theoretic semantics (GTS) in coping with various logics, most notably the family of IF ('independence-friendly') logics of Hintikka. I will show how both GTS and IF logics may be pushed into new directions by seizing the notion of a semantic game by means of the theory of games. I will conclude with some ensuing issues bordering on the interplay between C.S. Peirce's pragmaticism and the science of pragmatics.

1. Introduction

1.1 What is game-theoretic semantics?

Game-theoretic semantics (GTS) is a semantic theory of rational interaction between two imaginary players, who are playing the roles of the Verifier (V) and the Falsifier (F). They undertake to show that a logical or natural-language formula is true (by the actions of the player with the role V) or false (by F's actions). This happens in a model with either partially or completely interpreted non-logical constants, or in a suitable linguistic environment given by collateral actions and the mutually acquired and agreed common ground of players. Formally, GTS agrees with Tarski semantics for traditional first-order logic on complete models, but otherwise their motivations as well as philosophical repercussions are worlds apart.[1]

*Supported by the Academy of Finland (*Dialogical and Game Semantics*, project no. 101687), this paper represents first results of future collaboration in the context of the projects *Science in Context* and *Proof* of the Maison des Sciences de l'Homme du Nord-Pas de Calais, led by Shahid Rahman.

[1]The key references are Hintikka, 1973; Hintikka, 1996; Hintikka & Sandu, 1997.

D. Vanderveken (ed.), Logic, Thought & Action, 409–431.

Semantic games provide a resourceful theory for logical and linguistic analysis. I will focus on three interrelated issues. First, The rules of semantic games are applicable 'across the logical board'. Second, the definition of truth is an outgrowth of a game-theoretic notion of a strategy. Third, depending on the language under evaluation, the semantics make generous use of tools from the general theory of games, sometimes putting game-theoretic notions into a novel perspective.

I will largely ignore considerations of natural language here, but it is worth noting that, because of such features as non-compositionality, appeal to rational actions, and proximity to graphical and diagrammatic systems, GTS fares both on logical and linguistic fronts of meaning analysis at least as well as the discourse-representation theory of Hans Kamp, or compositional dynamic theories of meaning operating more readily on the syntax/semantics than on the semantics/pragmatics interface.[2]

In fact, there is a matchless virtue in GTS: its analysis of meaning makes ample use of both the derivational record of past actions and of the multiplicity of possible actions and possible plays not realised in the actual play; all this contributes towards a full-dress context-dependent account of meaning of logical and linguistic expressions and discourse.

1.2 What is IF logic?

The other compartment that I will be discussing here is the family of IF ('independence-friendly') logics, suggested by Hintikka, 1996. The term refers to such extensions of traditional logics that accommodate the property of informational independence, which is manifested syntactically by a slash notation and brought out semantically by games of imperfect information. I will outline next the essentials of propositional, first-order and modal IF logics.[3] I will not be considering any logical properties of these IF logics here, as my concern is restricted on the relationship between GTS and any one of the IF logics that I come to be defining.

1.2.1 Propositional IF logic.

The propositional fragment of IF logic (L^{IF}) builds up by: (i) If $p \in$ PROP, the arity of p is n, and $i_1 \ldots i_n$ are indices, then $p_{i_1 \ldots i_n}$ and $\neg p_{i_1 \ldots i_n}$ are L^{IF}-formulas, (ii) if φ and ψ are L^{IF}-formulas then $\varphi \vee \psi$ and $\varphi \wedge \psi$ are L^{IF}-formulas, (iii) if φ is an L^{IF}-formula then $\forall i_n \, \varphi$ and $\exists i_n \, \varphi$ are L^{IF}-formulas, (iv) if φ is

[2] See Hintikka, 2002; Janasik & Sandu, 2002; Janasik et al., 2003; Pietarinen, 2001.

[3] See Hintikka, 1996; Hintikka, 2002; Hintikka & Sandu, 1997; Pietarinen, 2001; Sandu & Pietarinen, 2001.

an L^{IF}-formula then $(\exists i_n/U)\,\varphi$ is an L^{IF}-formula (U is a finite set of indices, $i_n \notin U$).

The notions of free and bound variables are the same as in first-order logic. In $(\exists i_n/U)\,\varphi$ the indices on the right-hand side of the slash are free. For simplicity, I will omit the clauses for dual prefixes such as $(\forall i_n/U)$. The models for the language will be of the form $M = \langle I^M, (p^M)_{p\in\mathrm{PROP}}\rangle$, where I^M is a set A with a designated individual a, and each p^M is a set of finite sequences of indices from I^M. Let us set $a = \mathsf{Left}$ and $A - \{a\} = \mathsf{Right}$. The use of quantified indices i_n enables us simultaneously to distinguish different tokens of sentential connectives and to rightfully hide choices concerning their values.

Let us also write $\forall i_1(\exists i_2/i_1)\,p_{i_1 i_2}$ as $(p_{11}\,(\vee/\wedge)\,p_{12}) \wedge (p_{21}\,(\vee/\wedge)\,p_{22})$. If we wish to represent by restricted quantifiers 'unbalanced' formulas, we would use identities between subformulas to denote coinciding indices. For example, $(p_1\,(\vee/\wedge)\,p_2) \wedge p_3$, rewritten as $\forall i_1(\exists i_1/i_2)\,p_{i_1 i_2}$, $p_{21} = p_{22}$ is balanced by applying idempotence law in their interpretation, subject to certain qualifications as soon as semantic games are implemented (see the next section).

1.2.2 IF logic with quantifiers. IF first-order logic is created thus: Let $Qx\psi, Q \in \{\forall, \exists\}$ and $\phi \circ \psi, \circ \in \{\wedge, \vee\}$ be $L_{\omega\omega}$-formulas in the scope of $Q_1x_1 \ldots Q_nx_n$, where $A = \{x_1 \ldots x_n\}$. Then the first-order language $L^{\mathrm{IF}}_{\omega\omega}$ is formed by: If $B \subseteq A$, then $(Qx/B)\,\psi$ and $\phi\ (\circ/B)\ \psi$ are wffs of $L^{\mathrm{IF}}_{\omega\omega}$.

There are options: to require $x \notin W$ (likewise $i_n \notin U$ etc.) means that the information sets of the corresponding game (to be defined below) are reflexively closed ('Eyes Open'), and to require that all $x \in W$ are in $\mathrm{Var}(\varphi)$ (the recursively-defined set of variables of φ) of any $L^{\mathrm{IF}}_{\omega\omega}$-formula φ in which W occurs means that the formulas are globally context-independent (and locally context-dependent), in other words the associated game does not make references to constants that do not occur in it ('Its All in the Game'). If we require that all $x \in W$ are in $\mathrm{BoundVar}(\varphi)$ (the recursively-defined set of bound variables of φ) of any IF formula φ in which W occurs, and there are no other free variables than those in W, then φ is an IF sentence.

2. Variety of games — variety of interpretations and implementations

There are differing perspectives as to the reality of the game theoretic component of semantics. My viewpoint is to adopt an implementation in terms of the factual theory of games.

2.1 Semantic games for IF logic

The notion of independence may be investigated either from the point of view of Skolem functions or from the point of view of the induced information structures of the correlated games in the sense of game theory.[4] The underlying insight is the same in both cases: the strategies (typically functions) have to be such that they may not only be defined on one previous history of the game, but have to work invariably for multiple such histories, in other words, independently of any particular interpretation of some previous element already encountered.

Let $A_j(h), j \in \{V, F\}$ define a set of legitimate actions $\langle a^i \rangle_{i=1}^n, n \in \omega$ (a move) for each non-terminal quasi-history (q-history) $h \in H - Z$, from the domain $|\mathfrak{A}|$ of the structure $\mathfrak{A}$, for each j. A q-history is $\langle a^1 \dots a^n \rangle \in A$. The structure of the logical formula uniquely determines the order of the elements in q-histories. Given a q-history h_n, j chooses $a^i \in A_j(h)$, and the game proceeds to $h_{n+1} := h_n \frown a^i$. A root r has no incoming actions. A play is a finite sequence $r, a_1, h_1, a_2 \dots$, from which V's as well as F's choices can be singled out. I will dispense with the use of roles and denote players directly by V and F.

Further, $P\colon (H - Z) \to \{V, F\}$ is a player function assigning to every $h \in H - Z$ a player j in $\{V, F\}$ whose turn is to move. A pseudo-IF formula is a subformula ψ of an IF formula φ in which $W \neq \emptyset$ and $x \in W, x \in \mathrm{Var}(\psi)$. Let $\mathrm{Sub}(\varphi)$ be a recursively defined set of pseudo-subformulas of an IF formula φ. The labelling function $L\colon H \to \mathrm{Sub}(\varphi)$ assigns to each $h \in H$: $L(\langle r \rangle) = \varphi$; for every terminal history $h \in Z, L(h)$ is a literal. The components of the game $G_A = \langle H, L, P, u_j \rangle$ jointly satisfy: if $L(h) = \neg\varphi$ and $P(h) = V$, then $h \frown \varphi \in H, L(h \frown \varphi) = \varphi, P(h \frown \varphi) = F$; if $L(h) = \neg\varphi$ and $P(h) = F$, then $h \frown \varphi \in H, L(h \frown \varphi) = \varphi, P(h \frown \varphi) = V$; if $L(h) = (\psi\,(\vee/W)\,\theta)$ or $L(h) = (\psi\,(\wedge/W)\,\theta)$, then $h \frown \mathsf{Left} \in H$, $h \frown \mathsf{Right} \in H$, $L(h \frown \mathsf{Left}) = \psi$, and $L(h \frown \mathsf{Right}) = \theta$; if $L(h) = (\psi\,(\vee/W)\,\theta)$, then $P(h) = V$; if $L(h) = (\psi\,(\wedge/W)\,\theta)$, then $P(h) = F$; if $L(h) = (\exists x/W)\,\varphi$ or $L(h) = (\forall x/W)\,\varphi$, then $h \frown a \in H$ for every $a \in |\mathfrak{A}|$; if $L(h) = (\exists x/W)\,\varphi$, then $P(h) = V$; if $L(h) = (\forall x/W)\,\varphi$, then $P(h) = F$. Payoffs are mappings $u_j(h) \to \{1, -1\}, h \in Z$. For every $h \in Z$, if $L(h) = St_1 \dots t_m$ and $(\mathfrak{A}, g) \models St_1 \dots t_m$, then $u_V(h) = 1$ and $u_F(h) = -1$, and if $L(h) = St_1 \dots t_m$ and $(\mathfrak{A}, g) \not\models St_1 \dots t_m$, then $u_V(h) = -1$ and $u_F(h) = 1$.

[4] For instance, $\exists f_1 f_2 \forall x_1 \dots x_n, z_1 \dots z_m\, Sx_1 \dots x_n, z_1 \dots z_m, f_1(x_1 \dots x_n), f_2(z_1 \dots z_m)$ is the Skolem normal form of $\begin{matrix} \forall x_1 \dots x_n\, \exists y \\ \forall z_1 \dots z_m\, \exists w \end{matrix}\; Sx_1 \dots x_n, z_1 \dots z_m, y, w$, for some $n, m \in \omega$.

A history is now qualified to be a prefix of the play-sequence with labels terminating at a q-history $h_n \in Z, n > 0$. By a history I customarily mean a finite pre-sequence of $(L(r), a_1, L(h_1), a_2 \ldots L(h_n)), h_n \in Z$. Histories are thus labelled with the subformulas of the formula under evaluation.

A set of plays is a game frame. A set of plays with payoffs assigned to the terminating histories gives rise to a game $G(\varphi, \mathfrak{A}, g)$ with a τ-structure $\mathfrak{A}$ for $\varphi \in L^{\mathrm{IF}}_{\omega\omega}$ and to $G(\psi, M)$ for $\psi \in L^{\mathrm{IF}}$.

A deterministic strategy has for all $h \in H - Z$ and $a^i \in A$ probability $f_j(h)(a^i) \in \{0, 1\}$. Actions are prescribed by a deterministic strategy $f_j : P^{-1}(\{j\}) \to (2^A - \emptyset), f_j(h) \in A(h)$.

Next, $\mathfrak{I}_j$ is information partition of $P^{-1}(\{j\})$ such that for all $h, h' \in S^i_j, h \frown a \in H$ if and only if $h' \frown a \in H, a \in A$. In case $W = \emptyset$, the only sets in $\mathfrak{I}_j$ are singletons.

As to $G(\varphi, \mathfrak{A}, g)$ and $G(\phi, M)$, it is required that if $h, h' \in S^j_i$ then $f_j(h) = f_j(h')$. In the terminology of extensive games, strategies are defined on the components of the information partition, viz. on information sets S^i_j. A V-partition is $\bigcup_{i=1}^n S^i_V$ and an F-partition $\bigcup_{i=1}^n S^i_F$.

Let h^j be a quasi-history produced by f_j. Then f_j is winning for j if for every $h^j \in H, u_j(h^j) = 1$. A truth (resp. falsity) is defined as an existence of a winning strategy for the player who initiated $G(\varphi, \mathfrak{A}, g)$ or $G(\phi, M)$ as V (resp. F).

There are plenty of histories not on the winning (equilibrium) path and would for that reason be assigned a zero probability. In the definition of truth it suffices to take into account only a subset of strategies that does not lead to such histories. Considerable work has been done in game theory to make solution concepts work for all positions.

2.2 Concurrency vs. sequentiality

The above imperfect-information games are sequential in the sense that there is a left-linear order of moves from the first component onwards captured by the system of game rules described above. The linear order of moves means that at each position, at most one player makes a choice. Accordingly, games in which more than one player may choose at any position are concurrent. I will call this sense of concurrency *informational.* It follows from the above definitions that the semantic games for IF logic are in this sense not concurrent.

However, an alternative sense of concurrency is that it is informationally independent moves that count as concurrent. Thus, the histories that pass through an information set (see below) prescribe a concurrent move at this information set with respect to the sets through which these histories have passed at a lesser depth. This sense may or may not induce temporal considerations. I will call this sense *actional* concurrency. The semantic games for IF logic are in this sense concurrent; they involve independent actions. The linear order of moves (sequentiality) thus means that there are singleton information sets at all histories. Such games are associated with the slash-free fragment of IF logic. An IF formula φ associated with a truly concurrent game is one in which no variable x_n in $Q\,x_n$ or an index i_n in $Q\,i_n$, $Q \in \{\forall, \exists\}$ fails to occur in some set W deeper in φ. A typical implementation of semantics for an IF formula lies between sequential and truly concurrent game.

2.3 Teams that communicate

Different IF formulas give rise to structurally different streams of information. For instance, one may hold that semantic games are ones of perfect information in the sense that the two players V and F do not lose knowledge about positions and the history of the game, and in which there is a difference in the communication of information, not between V and F but between members that constitute these two players, viewed as teams with agents $M^j = \{m_1^j \ldots m_n^j\}, j \in \{V, F\}, n \in \omega$.

Another point worth noting is that, since we deal with finite formulas, the length of communication and the amount of information transmitted is limited by the length, organisation and type of the components the formula contains. Precisely which information is picked, and thus the content of information that may need to be revealed to others, is left for the players to decide.

In team games, the available information concerning past actions is restricted for individual members. I will consider both the cases in which there is communication (coordination) between the members of the team and the cases in which there is no such communication.

If the team members are not allowed to communicate, each member chooses $a^i \in A(h)$ independently of others' decisions. Player's choice sequence in any particular play is made by $M_1^j \subseteq M^j$.

Why such games? Consider an $L_{\omega\omega}^{\mathrm{IF}}$-formula $\forall x \exists y (\exists z/x)\, Sxyz$. This is correlated with non-communicating team games, in which it is assumed that existential quantifiers are explicitly slashed for other existentially quantifiers variables occurring further out in the formula. Otherwise they are dependent, as in $\forall x \exists y (\exists z/xy)\, Sxyz$.

Such formulas are not the only ones coming together with non-communication. For instance, in $\forall x \exists y (\exists z/y)\, Sxyz$, the V-team consists of $\{m_1^V, m_2^V\}$, in which neither m_1^V nor m_2^V passes on the information they have derived from higher up.

Even a two-stage game between V and F may need $\{m_1, m_2\}$, as in $\forall x\,(S_1x\,(\vee/x)\,S_2x)$. Here V should not get the value of x, but since it goes with the disjuncts, we take m_1^V to receive this value while being blocked from communicating it to m_2^V, who gets to choose at $L(h) = (S_1x\,(\vee/x)\,S_2x)$.

However, consider also $\phi = \forall x \forall y \exists z\,(S_1x\,(\vee/x)\,S_2yz)$. The truth of ϕ is revealed in terms of extended Skolem normal form: $\exists f \forall x \forall y\,((S_1x \wedge f(y) = 0) \vee (S_2yz \wedge f(y) \neq 0))$. Yet, if we consider — move by move — what is going on in $G(\phi, \mathfrak{A}, g)$, when individuals have been distributed for x, y after the first two moves by F at r and its successors, does not V get this illicit information by observing the labels attached to histories when the fourth move of disjunction is planned? To circumvent this, I will assume that variables in the labelled subformulas carry no instantiated values, in other words, the players do not observe assignments. Ditto for y in $(\exists z/y)$ and x in $(\vee/x)$. Accordingly, semantic games may be viewed in their extensive forms, because in order to make ϕ true, V needs to know the value of x when choosing for z, and she needs to know the value of y when making a decision that would lead to either S_1x or to S_1yz. To know the values can symbolically be represented only by instantiating constants to corresponding variables, including those on the left-hand side of the slashes, which amounts to a pseudo-Skolem normal form representing only one particular play of the game with respective instantiations. For in extensive forms, all values, hidden and public, are included in histories.

Players may also try to recover the identity of the information set they are at by looking at the available choices also in cases in which there is a possibility that some of the histories within an information set are terminal. Propositionally, an example of this is $(p_1\,(\vee/\wedge)\,p_2) \wedge p_3$.

A different implementation is produced if the coordinating player gets to decide which choices are actually put forward among those proposed by his or her agents. Such strategies are two-tiered: first, member-specific strategies delineate actions, and second, F's or V's coordinating strategies pick from the team-internal, private choice sets so induced.

Yet another possibility is that a predetermined set of suboptimal agents proposes the actions, the weighed average of such (possibly randomised) actions being elected as the player's preferred, representative choice. This is related to the concept of bounded rationality popularised in interactive decision theory of agents. Further, it makes the 'small worlds' doctrine, according to which agents are able to preview only fragments of domains and states of the game, better understood.

2.4 Teams that do not communicate

If the team members are allowed to communicate, we get games that are associated with such IF formulas in which existentially (universally) quantified variables may depend on other existentially (universally) quantifier variables, as in $\forall x \exists y (\exists z/x)\, Sxyz$. The amount of intra-team communication determines the extent to which such dependence is realised. Such incestuous dependence creates channels by which a player may elicit additional information. For instance, although the existentially quantified variable z does not depend on the universally quantified variable x, the second member m_2^V of the V-team choosing a value for z gets to hear the value F chose for x via the other member, m_1^V, who chose a value for y, the choice of which being dependent on the choice of the value for x, for the sole reason that m_1^V's communication to m_2^V is not specifically blocked.

The strategies in communication games differ from non-communicative ones in that their input includes the choices of action in the histories by the other members of the team, which is not permitted in the non-communicative case.

2.5 Two ways of losing information

Let $\preccurlyeq$ be a partial order on the tree structure of extensive-form games. A game satisfies non-absentmindedness, if for any $h, h' \in H$, $h, h' \in S_j^i, i \in \{V, F\}$: if $h \preccurlyeq h'$ then $h = h'$. Let a depth $d(Q)$ of a quantifier and a connective be defined inductively in a standard way. All semantic games for IF formulas φ described here satisfy non-absentmindedness, because every Q has a unique depth $d(Q)$, and hence every subformula of φ has a unique position in the game given by its labelling. For any two subformulas of φ at $h, h' \in S_j^i$, $h \not\preccurlyeq h'$ and $h' \not\preccurlyeq h$.

Let $Z(h)$ be a set of plays that pass through any $h \in H$, if h becomes a subsequence of any $h' \in Z(h)$. Likewise, let $Z(a^i)$ be a set of plays that pass through an action $a^i \in A$ or a sequence of actions $\langle a^i \rangle_{i=1}^n$, if $a^i \in h' \in Z(a^i)$. Define a precedence relation $<^*$ between any two information sets $S_j^i, S_k^i \in \mathfrak{I}_i$ so that if $h, h' \in S_j^i \times S_k^i$ such that $h \prec h'$, then $S_j^i <^* S_k^i$. Thus $S_j^i <^* S_k^i$ says that there exists $h'' \in Z$ passing through h and h'. If non-absentmindedness holds then any $h'' \in Z$ passes through S_j^i or S_k^i at most once.

As before, let $P^{-1}(\{i\})$ be the set of histories where i moves playing a strategy f_i. Information set S_j^i is relevant for f_i, if $S_j^i \cap P^{-1}(\{i\})$ is non-empty. Now let $S_j^i \in \mathfrak{I}_i$. A game has *perfect recall*$_1$, if S_j^i is relevant for f_i implies $S_j^i \subset P^{-1}(\{i\})$ for all f_i. This says that while players move within their information sets they will have perfect recall in the sense of not forgetting *information* (or knowledge) that they possess.

There is also an alternative way of characterising perfect recall. A game has perfect recall$_2$, if $S_i^j <^* S_i^k$ implies the existence of a sequence of actions $\langle a^i \rangle_{i=1}^n \in A$ available from S_i^j such that $Z(S_i^k) \subseteq Z(\langle a^i \rangle_{i=1}^n)$ (for otherwise there will be wider information sets occurring for a player later on in the game). This says that i does not forget his or her *actions.*

Semantic games $G(\varphi, \mathfrak{A}, g)$ do not in general satisfy perfect recall$_i$ ($i = 1, 2$), because any $L_{\omega\omega}^{\text{IF}}$-formula φ that contains a subformula $\psi = Q_1 x_1 \ldots (Q_2 x_n / x_1)$, $Q_1, Q_2 = \exists$ or $Q_1, Q_2 = \forall$ and $d(Q_1) < d(Q_2)$, gives rise to a partition in which $\psi \in \text{Sub}(\varphi)$ beginning with Q_2 induces S_k^i and $\eta \in \text{Sub}(\varphi)$ beginning with Q_1 induces S_i^j, such that $S_i^j <^* S_i^k$. Thus $\langle a^i \rangle_{i=1}^n \in |\mathfrak{A}|$ that i chooses for $Q_1 x_1$ are available from S_i^j, but then clearly $Z(\langle a^i \rangle_{i=1}^n) \subset Z(S_i^k)$. On the other hand, perfect recall$_1$ depends on allowing 'non-standard' information sets that are not relevant for player's strategies f_i. But any such information set violates perfect recall$_1$, all $P^{-1}(\{i\})$ being $L(P^{-1}(\{i\}))$.

Syntactically speaking, a formula φ in $L_{\omega\omega}^{\text{IF}}$ exhibits perfect recall$_1$, if for any $(Q_1 i_1/U_1), (Q_2 i_2/U_2)$ in φ, if either $Q_1, Q_2 = \exists$ or $Q_1, Q_2 = \forall$ and $d(Q_1) < d(Q_2)$, then $i_1 \notin U_2$. It exhibits perfect recall$_2$, if for any $(Qj_1/U), (Q_1 j_2/U_1), (Q_2 j_1/U_2)$ in φ, if either $Q_1, Q_2 = \exists$ or $Q_1, Q_2 = \forall$ and $d(Q) < d(Q_1) < d(Q_2)$ and $j_1 \notin U_1$, then $j_1 \notin U_2, j_2 \notin U_1$.

The former clause is about player forgetting his or her own actions. The latter says that $Q_1, Q_2 = \exists$ or $Q_1, Q_2 = \forall$ give rise to imperfect recall even if one is not independent of the other, provided that players have acquired different information from elements higher up in a formula. One may thus study fragments of IF logic in which imperfect recall does not hold, or holds in some restricted sense. In the latter case we are dealing with aspects of bounded recall (Lehrer, 1988), as well as with imperfect monitoring of actions and information transmission.

2.6 Screening vs. signalling

Dually with imperfect recall, one may characterise information increase that is seen to happen in $\forall x (\exists y/x) \exists z \; Sxyx$. Both imperfect recall and informational increase may naturally be manifested: $\forall x (\exists y/x)(\exists z/y)$ $Sxyz$.

Such learning is similar to what happens in screening games (Rasmusen, 1989), in which the first player is uninformed of certain aspects of the game and the second player, being fully informed, can screen his or her actions, for instance via its members. In signalling games, on the other hand, connected to imperfect recall by the team perspective, the informed player moves first and may signal previous features of the game to the subsequent uninformed player. If in the former case the types of the first and the second player are the same, then screening amounts to learning, and likewise, the phenomenon of signalling means, for the two players of the same type, that information is being forgotten.

2.7 Concurrent games and Henkin quantifiers

The first sense of concurrency, namely that at each position, there may be more than one player choosing, gives rise to games that are associated with formulas with finite, partially-ordered quantifier prefixes of the form $\begin{smallmatrix}\forall x_1 \ldots x_n \exists y \\ \forall z_1 \ldots z_m \exists w\end{smallmatrix}$ (for some $n, m \in \omega$), and are interpreted via informational concurrency. This may happen at r as well as at any $h_n \in H - Z$, and thus the games do not need to form trees. Since on models with pairing functions two rows suffice to represent arbitrary parallel orderings (Krynicki, 1993), at most two moves exists at each q-history h of such concurrent games, and the order of these rows is immaterial.

2.8 Remark on scope

A logical distinction between informational concurrency and sequentiality is in terms of logical priority. Logical priority between logically active components may be defined as preferences in the evaluation of such components, which in turn is a derivative of the semantic information flow in the structure determined by the syntax of an IF formula. If semantic information flows from a component C_1 to C_2, the former is logically prior to the latter; in other words, the latter is logically dependent of the former. But in case information does not transmit between C_1 and C_2, they are not ordered by priority, and thus C_1 and C_2 are independent of one another in the sense that their strategic evaluation makes no use of the semantic attributes assigned to them — provided, of course, that there is no illicit communication. Traditional logic assumes that logical priority goes by the recursively defined subformula relation, which no longer holds in IF extensions.

Logical priority need to be distinguished from binding, which encodes the extent to which a quantifier and its quantified variable reaches in assigning the same values to other occurrences of the same variable in the formula (Hintikka, 1997). This distinction has significant repercussions in graphical (heterogeneous) logics of diagrams, for instance, mandating a move from two to three-dimensional spaces in which such graphs are described (Pietarinen, 2003c).

2.9 Non-partitional information structures

The information structure defined above is typically taken to be partitional in that its cells are composed of histories closed under equivalence relations. For the purposes of IF logic, this may be too ideal, as $\phi = (p_1 (\vee/\wedge) p_2) \wedge (p_3 \vee p_4)$ witnesses. After F has chosen a conjunct, only if the action was Left will V be able to fully process that she does not know whether continuation occurs in $L(h') = (p_1 (\vee/\wedge) p_2)$ or $L(h) = (p_3 \vee p_4)$. Rather than saying that V's strategy f_V is defined on S_i^V in which $h \sim_V h'$, this is implemented by defining three relations $Rhh, Rh'h'$ and Rhh'. Thus S_i^V contains antisymmetric relations.

Such games are still non-determined, since by setting $u_V(h_1) = u_V(h_4) = -1$, $L(h_1){=}p_1$, $L(h_4){=}p_4$, $u_F(h_2){=}u_F(h_3){=}{-1}$, $L(h_2){=}p_2$, $L(h_3){=}p_3$ for ϕ there are no winning strategies for V nor for F. In fact, the same payoff distributions for $G(\phi, g)$ and $G(\phi', g)$, $\phi' = (p_1 (\vee/\wedge) p_2) \wedge (p_3 (\vee/\wedge) p_4)$, the latter being closed under symmetric relations, give rise to non-determined games. The sole difference is that the player whose actions lead to non-partitional cells has fewer options to prevent the opponent from winning. For the opponent who in planning his or her moves in such cells there is no difference, and if it is costly to process all accessible histories, non-partitional models would in fact be preferable.

The specialty introduced by IF logic is that even the fundamental question of whether a player has available a strategy that gets a non-singleton information set as an input may depend on actions made in previous parts of the game. It is indeed unrealistic to expect players to fully process the information concerning the reasons that led to their uncertainties. Such uncertainties themselves may in this sense be partially realised. As noted, players may then also lack self-awareness of their own actions by taking information sets to be non-reflexive.

To make games even more realistic, players knowing that they will lose information at some point on amounts to self-awareness of their bounded memory resources. This is one of the motivations for introducing players as multi-agent teams in the first place.

3. Modal extensions

3.1 IF modal logic, propositional case

In modal logic, it may easily happen that the domain (the subset U of the set of possible worlds $\mathcal{W}$) from which the players are to pick values is such that $A(h) \neq A(h')$ for $h, h' \in H - Z, h, h' \in S_j^i$. This violates the traditional assumption in imperfect-information games according to which for any such h, h' within the same information set, $A(h) = A(h')$, and the reason is that otherwise a player could, by virtue of perfect foresight, detect differences in the histories, thus infer his or her actual location, and thus derive some information that was supposed to be hidden.

Prima facie, such restrictions will have to go in IF modal logic, because in arbitrary frames of a possible-worlds structure, there may be any composition of accessible worlds from any world that a play has reached. And if so, IF modal logic appears to dispense with perfect foresight, making it to look an idealisation of similar dubious standing as decision-maker's hyper-rationality. Furthermore, this conclusion would, in any case, backed by the simple observation that in perfect-information games with singleton information sets, no such coincidence of available actions is needed.

Let us define ϕ to be a formula of a propositional modal logic $\phi := p \mid \Box_j \varphi \mid \Diamond_j \varphi \mid \varphi_1 \wedge \varphi_2 \mid \varphi_1 \vee \varphi_2 \mid \neg\varphi$. An IF modal logic is got by letting $\triangle^k_j \psi \, (\triangle^k_j \in \{\Box^k_j, \Diamond^k_j\}, j, k \in \omega)$ be in $\mathrm{Sub}(\varphi)$, and letting $A = \{\triangle^1_1 \ldots \triangle^k_n\}$ be such that all $\triangle^k_i \in A, i \in \omega$ occur in φ and $\triangle^k_j$ occur after $\triangle^k_i, j \neq i$. Now, if $B \subseteq A$, then $\{(\triangle^k_j/B)\,\psi, (\phi\,(\circ/B)\,\psi)\} \in ML^{\mathrm{IF}}$. For example, $\Box^1_1 \Diamond^1_1 (\Diamond^2_2/\Box^1_1)\,\psi$ and $\Box^1_1(\varphi\,(\wedge/\Box^1_1)\,(\Diamond^1_1/\Box^1_1)\,\psi)$ are wffs of ML^{IF}.

However, since the notion of independence may mean different things, to keep its semantic definition general, one may impose further restrictions on the meaning of independence as need arises. Indeed, precisely what does the slash in $\Box^1_i(\Diamond^2_j/\Box^1_i)\,\varphi$ reflect? Its general meaning is that the choice of a world by V for $\Diamond^2_j$ has to be made in ignorance of the choices made by F for $\Box^1_i$. The phrase 'in ignorance of' can be interpreted in several ways. It may mean that (i) there is an equivalence relation $(h_1 \frown w_1) \sim_i (h_2 \frown w_2)$ linking any worlds w_1, w_2 that a player cannot distinguish, and consequently he or she loses track of some of the choices in the past that lead to those worlds, (ii) the choices for several modal operators are actionally concurrent, (iii) the play backtracks to the world from which the worlds chosen for $\Box^1_i$ departed.

The general definition needs to dispense with the accessibility relations $\{\rho_1 \ldots \rho_n\}$ (defined for each agent $i = 1 \ldots n$, $\rho_i \subseteq \mathcal{W} \times \mathcal{W}$, $w_1 \in [w_0]_{\rho_i}$ meaning that w_1 is i-accessible from w_0), which normally guide player's choices from worlds that the game has reached. Accordingly, player's position in the game traversing the possible-worlds structure and his or her available choices are no longer strictly correlated. When encountering an expression of the form $(\Diamond^i_j/W)\,\varphi$ in ψ, in which W is a sequence $\triangle^1_1 \ldots \triangle^m_n$ of operators already occurring in ψ, V (likewise F for $(\Box^i_j/W)\,\varphi$) chooses $w \in U \subseteq \mathcal{W}$. Reasonable condition for the subsethood is that at w, only the worlds chosen prior to reaching w ('regret') and the immediately available ones $w' \in [w]_{\rho_j}$ ('no long-distance foresight') are the candidates to be included in U. When the play backtracks from $h \in H$ to one of its subsequences h', it only means that the subdomain U is copied from $A(h')$ to be $A(h)$. This needs no equivalence relations between histories. In case $h, h' \in S^i_j$ and $A(h) \neq A(h')$, to satisfy perfect foresight, those worlds are added from $A(h)$ to $A(h')$ that do not occur in $A(h')$ and vice versa. Since these dummy copies are never reached by rational players' strategies, they do no harm.[5]

The game rules and winning are now such that given a modal model $M = \langle W, \langle \rho_1 \ldots \rho_n \rangle, g \rangle$ (g is a valuation of atomic formulas) and a game $G(\varphi, M, w)$, if $w \in g(p)$, $L(h) = p, h \in Z$ for p atomic, $u_V(h) = 1$. Otherwise $u_F(h) = 1$. Rules for conjunction and disjunction are usual.

As to the sense (i) of sect. 1, if $h, h' \in S^i_j$, then $f_i(h) = f(h')$. As to the sense (ii), an actionally concurrent game is associated with partially ordered modalities $\psi = \begin{matrix} \triangle^{1_1}_{1_1} \ldots \triangle^{1_h}_{1_m} \\ \vdots \qquad \vdots \\ \triangle^{l_1}_{j_1} \ldots \triangle^{l_k}_{j_m} \end{matrix} \varphi$, in which each column defines a state s_i at which members $m^V_1 \ldots m^V_n \subseteq V$ and $m^F_1 \ldots m^F_n \subseteq F$ choose for $\triangle^{l_k}_{j_m}$ before the game proceeds to s_{i+1}. To make sense of this interpretation, a finite sequence of distinct designated worlds $w'_0, w''_0 \ldots$ are assumed in M on which $G(\psi, M, \langle w'_0, w''_0 \ldots \rangle)$ may be initiated, and the strategies defined on histories of the partially-ordered structure of the game. As to the sense (iii), there is no informational restriction on strategies per se, which are mappings $f_i \colon H' \rightarrow A(h), H' \subseteq H, H'$ a set of histories reached with a non-zero probability in $G(\varphi, M, w), \varphi \in ML^{\mathrm{IF}}$.

Given one of the above interpretations of independence in $(\triangle^i_j / W)\, \varphi$, ψ is true (resp. false) in M (with respect to the above senses) if and only if there exists a winning strategy for V (resp. F) in the respective game.

It is clear that games for ML^{IF} are not in general determined, by defining M such that $w_1 \in [w_0]_{\rho_1}, w_2 \in [w_0]_{\rho_1}, w_1 \in [w_1]_{\rho_2}, w_2 \in [w_2]_{\rho_2}, w_1 \in [w_2]_{\rho_2}, w_2 \in [w_1]_{\rho_2}$, and $g(\psi, w_1) = \mathsf{True}$, $g(\psi, w_2) = \mathsf{False}$. Since V's information set does not distinguish between w_1 and w_2, she cannot choose the world such that $g(\psi, w_1) = \mathsf{True}$. V does not have a winning strategy either, because $g(\psi, w_1) = \mathsf{True}$.

[5]It is furthermore assumed that the player does not distinguish between dummy and real options. One interpretation of this is that it is, in fact, players' beliefs over available actions that constitute objects of their foresight, not the real objective possibilities.

3.2 IF modal logic, first-order case

The second step is to add quantifiers to the propositional base language. This can be done either by adding quantifiers to perfect-information or to imperfect-information language. It is advisable to take the first-order epistemic logic $ML_{\omega\omega}$, consisting of a signature τ, world-relative domains D_{w_i} of structures, a logical vocabulary, and formulas $\phi := S \mid \Box_i\varphi \mid \forall x\varphi \mid \exists x\varphi \mid \varphi \vee \psi \mid \neg\varphi \mid x \simeq y$ as the starting point and extend it by applying the slash notation to it (Pietarinen, 2003b).

Let $Q\psi, Q \in \{\forall x_j, \exists y_j, \triangle_i^k\}$ be an $ML_{\omega\omega}$-formula in the syntactic scope of the elements in $A = \{\triangle_1^1 \dots \triangle_n^k, \forall x_l, \exists y_l\}$. Then $ML_{\omega\omega}^{\mathrm{IF}}$ consists of the wffs of $ML_{\omega\omega}$ together with the rule: if $B \subseteq A$, then $(Q/B)\,\psi$ is an $ML_{\omega\omega}^{IF}$-formula, $Q \notin B$. Elements in B are linearly ordered. For example, in the $ML_{\omega\omega}^{\mathrm{IF}}$-formula $\Box_1^1\exists y(\exists x/\Box_1^1, y)(\Box_2^2/\Box_1^1, y)\,Sxy$ the information about the choices for $\Box_1^1$ and y is hidden in these positions in which $\exists x$ and $\Box_2^2$ are evaluated.

Skipping most of the details here (see Pietarinen, 2003b), let φ be an $ML_{\omega\omega}^{\mathrm{IF}}$-formula and let B be the set of modal operators and variables already occurred in the game $G(\varphi, \mathfrak{A}, w, g)$ when an expression of the form (Q/B) is encountered. The game rule states: (i) If $\varphi = (Q/B)\,\psi, Q \in \{\forall x, \exists x, \triangle_i^k\}$, and the game has reached w, then if $Q = \forall x$ (resp. $\exists x$), m_l^F (resp. m_l^V) chooses an individual from D_{w_1} of individuals, in which w_1 is the world from which the world chosen for the first modal operator in B departed. The next choice is in $G(\psi, \mathfrak{A}, w, g)$. (ii) If $Q = \Box_i^k$ (resp. $\Diamond_i^k$), then m_l^F (resp. m_l^V) chooses $w_1 \in \mathcal{W}$ in $\mathfrak{A}$ independently of the choices made for the elements in B, and the next choice is in $G(\psi, \mathfrak{A}, w_1, g)$. In case $\triangle_i^k \in B$, this rule takes in similar qualifications as ML^{IF}.

IF first-order modal logic is capable of distinguishing between 'de dicto' vs. 'de re' readings in a versatile way, the latter meaning that a player picks a without having exact information about the world he or she is located at, (e.g. $\Box_i^k(\exists x/\Box_i^k)\,\varphi$), whereas no such world-hiding takes place whenever $\exists x$ is chosen after $\Box_i^k$, in other words, with a full knowledge of worlds chosen for it. The information sets are thus needed to connect the world-bound manifestations of individuals across possible worlds, which accounts for identification of individuals. The distinction permits a much more versatile combination of different ways of knowing in multi-agent modalities.[6]

[6]Pietarinen, 2003b considers these cases in more detail, and Pietarinen, 2001 provides applications in relation to the classical problem of intentional identity introduced by Peter Geach, in the to multi-agent IF epistemic logic setting.

All in all, facets of independence in modal logic turn the possible-worlds semantics 'history-conscious' — an indispensable feature in clarifying the meaning of natural language expressions and logical concepts alike. One may further enrich the model by multiple assignments g for atomic p that are relative to $h \in Z$ by $g: \mathcal{W} \times H \to \{1, -1\}$. The winning conditions are given by $w \in g(p)(h)$, $L(h) = p, h \in Z$.[7]

4. More on games logic plays

4.1 Whither partiality?

The notion of partiality may arise both on the level of atomic formulas (partial models) or on the level of complex formulas. Even if models are complete, formulas may be partial with a truth-value of Undefined because for IF formulas associated with such non-determined games for which a winning strategy exists neither for V nor for F, the law of excluded middle defined by strong negation may fail (Sandu & Pietarinen, 2001).

To get partial models $M = \langle M^+, M^- \rangle$, in which $M^+ \cap M^- = \emptyset$ are disjoint subsets of the set of atomic formulas σ of a propositional language L, but not necessarily $M^+ \cup M^- = \sigma$, the assignment function $g(p), p \in \sigma$ is taken to be partial.[8] In cases of $L_{\omega\omega}, L^{\text{IF}}_{\omega\omega}, ML^{\text{IF}}_{\omega\omega}$, this may be put into a game-theoretic perspective by stipulating that V chooses proper names for constants from the domain of the structure. The denotation of the name according with the intended meaning of the predicate is then a consequence of the fact that V wins (or does not lose) against Nature who authorises such choices, whereof an atomic p is true in case V manages to get authorisation from Nature to every constant in it. In modal logic, $g(w)(p)$ may likewise be partial.

Accordingly, these examples provide a glimpse at the 'lexical semantics' side of GTS. As noted, the equivalence between GTS and Tarski semantics assumes complete models. Even if there were no slashes in an IF formula φ, the two notions of negation do not coincide if the models are not complete.

---

[7] See Pietarinen, 2001; Pietarinen, 2002; Pietarinen, 2003b for further discussion of informational independence in modal logic. Bradfield, 2000 endorses a concurrency interpretation of propositional modal logic in which accessibility relations are chosen from parallel compositions of models. In Bradfield & Fröschle, 2002, the propositional modal models contain explicit concurrency relations.

[8] The meaning of the superscripts is that M^+ denotes the set of true atomic sentences of $L(\sigma)$ and M^- denotes the set of false atomic sentences of $L(\sigma)$. In the case M^- is the complement of M^+, that is, $M^+ \cup M^- = \sigma$ and $M^+ \cap M^- = \emptyset$, the model M is a complete partial model, coinciding with a classical model.

Logics with partial models (either IF or 'slash-free') are particularly useful in dealing with inexact (fuzzy) concepts and semantic paradoxes, in virtue of the fact that the two negations (strong and weak) do not coincide (Hintikka, 2002).

Furthermore, semantic games that are not strictly competitive introduce a fourth truth-value of Over-defined. The value of Undefined may disappear for formulas on complete models under non-strict competitiveness (Pietarinen, 2003f).

4.2 Incomplete information

It has been implicit in GTS that the structure of the game is common knowledge. This may be dispensed with, at least to a degree. Not only imperfect information but *incomplete* information is conceivable. It refers to players' lack of information of the structure of the game, including payoffs. Logically, this means that players may not know which atomic formulas are wins or losses. In view of Harsanyi's (1967) result that such a lack may be implemented by a lack of information over the types randomly chosen by Nature from the type space T, we get a logical reflection of this in IF logic in which independence is extended to negations: components of a formula ψ of any of the previously-mentioned IF logics may be replaced by subformulas of ψ of the form $(E/W \cup \{\sim_1 \ldots \sim_n\})\,\varphi$, in which $\{\sim_1 \ldots \sim_n\}$ are indexed negations encountered in ψ and $E \in \{\sim_i, Qx, \triangle^k_j\}, i \notin \{1 \ldots n\}$. Via the semantics given in Pietarinen, 2003c, these incomplete-information games reduce to games of imperfect information.

One caveat is that Harsanyi transformation uses the assumption of common knowledge of priors distributed over players' types, which not only does not answer the question of the origin of such distributions, but also does not permit updates on such priors as the game goes on. Since negation gives rise to just a binary type space $T = \{v, f\}$ with equal probability assigned to its elements, problems related to defeasible beliefs in Bayesian reasoning do not arise in this context.

4.3 Further means of IFing

It is a virtue of GTS that its entire potential has not been untapped by the received notions of IF logics. Among such further topics I will list the following. (i) Associate games with 'non-stratified' information structures with formulas that bring out new kinds of independencies than just those between quantified variables and connectives (i.e. allowing information sets with histories of differing length, ones that Von Neumann & Morgenstern, 1944 did not consider). (ii) Exploit non-hyper-rational forms of decision making and information processing — not by solution concepts appealing to the notion of 'satisficing' but by adding noise and distortion to the processes of evaluation, much in the same way as there may be noise in distorted links between language and reality (Pietarinen & Sandu, 2003). (iii) Consider formulas themselves as graphs, hence manifesting all possible dependency relations between components affected by game rules. (iv) Take slashes to come with differing force, captivated by a probability distribution over their occurrences in a formula. This generalises to formulas as probabilistic nets studied in AI. Likewise, moves (sets of choices) may be associated with probability distributions that are then used in selecting actions. (v) Use semantic games in inducing probability measures. This makes them related to martingales. (vi) Consider iteration of slashes: any component C may be replaced not only by (C/W) but by $(C/(W_1/(W_2/\ldots W_n), W = \bigcup W^n_{i=1}$. Processing the right-hand side of the slash from inside-out, the meaning of the iteration is that a player i may forget information not at the time when he or she is planning a move but at some other, later location in which $-i$ is to move. Viewed as teams, this generalises to similar delayed forgetting within members of i. This is related to the possibility of choices to be readjusted during the plays.

5. Pragmatism and pragmatics

Given this catalogue of opportunities in devising and implementing different logics, it is almost as if different logics ensue from game-theoretic definitions of notions such as information and its exchange, strategic interaction, and other cognate game-theoretic concepts. Thus, the use of novel game-theoretic concepts in enriching traditional GTS is not just allegorical. It is heir to Charles S. Peirce's diagrammatic logic of existential graphs, along with their semiotic and endoporeutic method of interpretation (Pietarinen, 2003b; Pietarinen, 2003i). It is not confined to the study of truth-conditions, but accommodates much of what already Peirce perceived to be central in understanding the logic of human action, namely the pragmatic value of assertions. It is equally pragmatic as semantic, and by calling it semantic, we insinuate it not into the camp of lexical theories of the late 19th century, nor into the mathematico-symbolic circles of the semanticists such as Carnap, Tarski and their ilk, but into Peircean semiotic triad of (speculative) grammar, logic proper (critic) and (speculative, formal) rhetoric, also called methodeutic. By pragmatic, we insinuate it not into the misleadingly-termed 'psychological' programme pursued, among others, by John Austin, John Searle and H. Paul Grice, but into the much earlier pre-empiricist Peircean pragmaticism, of course not standing apart from the all-important semiotic triad.[9]

In GTS, the truth-values of sentences, be they propositional, first-order, modal or even higher-order generalised quantifiers, are determined on a model in which the meaning of non-logical constants is fixed by interpretation. In this sense, GTS falls within the genre of truth-conditional semantics. Another major compartment is defined by the role played by strategies. If we are interested in what these strategies are, what is contained in them, how they are acquired and so on, we elevate the meaning analysis from the truth-conditional study of abstract meaning to the realm of strategic meaning of expressions (Hintikka, 1987). Peirce's pragmaticism may be viewed as a partial attempt to have such theory of strategic meaning, even though he fell short of possessing a legible concept of the notion of strategy, which came into being by Emil Borel, László Kalmar, Dénes König, John von Neumann and others soon after Peirce's death.

As a case study towards realising the desiderata of strategic meaning theory, let us extend the current framework for semantic games to *hyper-extensive games*, which, among other things, allows us to represent dependency structures between strategies, not only between choices they prescribe. Given an extensive game G_A, I will define a hyper-extensive game $\mathfrak{G}$ to consist of the following. (i) A set of local states $\{l_1 \dots l_n\}$, each local state $l_j, j \in \{V, F\}$ describing the information j has at any $h \in H$, not restricted to specific histories unlike in traditional extensive games. The set l_j is built up from a set of actions $B \subseteq A$, a set of strategies $S \subseteq \mathcal{F}$, and a set of deictic individuals $E \subset \mathcal{E}$ given by the linguistic environment. The set E may also be taken to contain players' world knowledge, scripts, schemes or episodic memory symbolised, if need be, in a suitable knowledge representation language such as epistemic logic. (ii) Ordered tuples $\langle l_V, l_F \rangle$ of local states, one for each player, called global states. A global state is thus a tuple of local states. A global state captures the state of the game as viewed from outside (modeller's perspective). A global state says what the information any player possesses is at any point of the game. (iii) Functions $f\colon H \to G$ associating to any $h \in H$ a global state g, or 'information flows'. When h' is the root, the global state $g(h')$ is likely to contain only local states that are made up of the sets E. When $k \in Z$, the local state also contains the payoffs u_j associated to that terminal history k.

[9] A plethora of philosophical and historical amenities supplying those provided originally by Hintikka, 1973, are considered in Pietarinen, 2003c; Pietarinen, 2003g.

A local state l_j is thus a set $\{B, S, E, u_j\}$ of actions, strategies, environmental elements and, for terminal nodes, payoffs that the player with l_j is aware of (or has an access to). Since there are just two teams, each global state at any $h \in H$ consists of tuples of local states. A game is essentially just the set of information flows. The notion of a strategy is likewise generalised in the sense that it gets as input the local states whenever a player is planning his or her decisions. Thus a strategic decision may involve an assessment of those other strategies to which a player according to a local state has an access. A strategy $s_j \in \mathcal{F}$ is now a functional from a local state l_j to the set of actions in A.

Let us confine ourselves to hyper-extensive games of perfect information. Even so, we need to capture the notion of players 'remembering' the strategies in the game: Given $P(h) = P(h') = j$, j remembers a strategy $s_j \in F$ at $h \in H$, if $g(h) \sim_j g(h')$ then $h = h'$. That is, the player remembers the history h because there is nothing to distinguish it from h', in other words the equivalence relation $\sim_j$ does not do any work. This generalises the approach presented in Fagin et al, 1995, and broadens its application to linguistic issues.

This is not the only, and probably not even the most common, case of remembering strategies in anaphoric discourse. Sometimes the strategy is relegated to the local state associated with the history that emanates from the different part of the split discourse:

Every man carried a gun. Most of them used it. (18.1)

The reason for the split is just the same as in simple anaphora, namely that the choice for *every* prompts a move by F. The hyper-extensive games capture this by including the relevant strategies that arise from the functional dependency in the former clause to the specification of the player's local states at the history in which the latter clause is evaluated. For instance, in (18.1) *most* prompts a move by V from the set of men carrying a gun, and *it* is interpreted by applying the same strategy that V used in the subgame at the history in which she had chosen for the indefinite *a gun*.

Precisely how difficult it is to make do with abstract meaning alone is shown by anaphora that appears to exhibit functional dependency, but in which what is expressed by the posterior clause is not a consequence but an antecedent of the fact given in the former clause:

Yesterday, every student failed an examination. The brains just did not work. (18.2)

Furthermore, it is not inconceivable to even have functional cataphora:

> Most students did not get high grades. But everyone passed a math examination last week. (18.3)

It is possible to read (18.3) so that the functional dependency is of a reversed sort: Whatever *most students* denotes has to be chosen among those individuals who passed a math exam last week. How this is done in hyper-extensive games is such that discourse splits in two even if the universal clause exists in the anterior clause. It then gets evaluated, and the function induced in the anterior is included to the local state of V choosing for *most* in the antecedent.

Because a number of strategies from which linguistic meaning is derived are not just abstract, global options up for grabs but refer to subjective and epistemic elements, we can never be absolutely precise about the processes and the linguistic mechanisms that are responsible for the transmission of certain strategies from some parts of discourse to other, anaphoric ones. The transmission may, among other things, be constrained by things like agent's range of attention and awareness, short-term memory concerning text processing, or any other capacity in retrieving strategies linked with other parts of the game. Thus, what it means that certain strategy is 'remembered', actually subsumes a range of phenomena. Variables to be instantiated are rather like memory registers with pointers. By not assuming too much on the relation between the registers and pointers, we leave ample space for further consideration on strategic aspects of anaphora and the theory of strategic meaning. The phenomena that could be analysed from the game-theoretic perspective of strategic meaning include salience, choice functions, and the topic/focus contrast. Choice functions in particular are weak forms of strategy functions incapable of reproducing the dependence structure of variables.

To return to the relationship between pragmatics and its elder philosophical brethren, what, then, are the key differences between Peirce's sense and the programme of the same name that came to be popularised by Grice? Both undertook to study assertions, or language in action, and both admitted ample contextual dependence in the determination of what assertions mean. But whereas Grice promoted a study based on a set of invariants (maxims) preserved in rational conversation, Peirce was primarily interested in how semantic relations between language and the world, or if you prefer a more phaneroscopical elucidation, between signs and objects, are created in continuous transformations by either real or imaginary language users, and how they constantly evolve by virtue of such transformations. What he came to advocate was the context-dependent use of assertions with all kinds of indexical signs, along with their intentional character (the spectra of different notions of interpretants), spelled out in terms of actions taken by two imaginary parties, the Utterer and the Interpreter. The first level of such actions is the collaborative model-building task, in which the parties take turns in proposing and authorising elementary assertions that may then be put forward (or scribed in terms of graphical logic). The second level refers to the competitive task of interpreting complex assertions by assigning semantic attributes picked from the domain of the model to their components. Both levels presuppose not the existence of any constant, immutable domain, but collateral observation (that is, both factual and conceptual information provided by such observation) together with mutual knowledge of such collaterality.

Thus, a reasonable amount of common ground between the parties begins to germinate in Peirce's logic. This fact may be exploited in linguistic arguments appealing to the notion of 'salience', ranging from the semantic content of supposed 'specific indefinites' to inductive arguments in language evolution.

Despite dissimilarities, a good number of concepts that Grice is typically said to have introduced may be found in Peirce's pragmatic study of assertions. The first notion in Peirce's triadic spectra of intentional-effectual–communicational interpretants antedates Grice's utterer' meaning. It is also related to Grice's surprising terminology of "straightforward interpretant", as distinct from the "non-straightforward interpretant" arising in certain maxim-flouting contexts of getting in conversational implicatures (Grice, 1989).[10]

References

Bradfield J. (2000). "Independence: Logic and Concurrency", in P.G. Clote and H. Schwichtenberg, (eds) *Proceedings of the 14th International Workshop on Computer Science Logic, Lecture Notes in Computer Science* 1862, Berlin: Springer-Verlag, 247–261.

Bradfield J. and Fröschle S. (2002). "Independence-Friendly Modal Logic and True Concurrency, *Nordic Journal of Computing 9*, 102–117.

Fagin R., Halpern J.Y., Moses Y. and Vardi M.Y. (1995). *Reasoning About Knowledge*, Cambridge, Mass.: MIT Press.

Grice P. (1989). *Studies in the Way of Words.* Cambridge, MA: Harvard University Press.

[10]Further commonalties between the pragmatic stances of Peirce and Grice are investigated in Pietarinen, 2003h.

Harsanyi J. (1967). "Games with Incomplete Information Played by 'Bayesian' Players. Part I: The Basic Model", *Management Science 14*, 159–182.

Hintikka J. (1973). *Logic, Language Games and Information.* Oxford: Oxford University Press.

— (1987). "Language Understanding and Strategic Meaning".*Synthese 73*:497–529.

— (1996). *The Principles of Mathematics Revisited.* New York: Cambridge University Press.

— (1997). "No Scope for Scope?". *Linguistics and Philosophy 20*:515–544.

— (2002). "Hyperclassical Logic (aka IF logic) and its Implications for Logical Theory". *Bulletin of Symbolic Logic 8*:404–423.

— (2002). "Negation in Logic and in Natural Language". *Linguistics and Philosophy 25*:585–600.

Hintikka J. and Sandu G. (1997). "Game-Theoretical Semantics", in J. van Benthem, and A. ter Meulen (eds), *Handbook of Logic and Language.* Amsterdam: Elsevier, 361–410.

Janasik T. and Sandu G. (2003). "Dynamic Game Semantics", in Peregrin, J., (ed.), *Meaning: The Dynamic Turn.* Dordrecht: Kluwer.

Janasik T., Pietarinen A.-V. and Sandu G. (2003). "Anaphora and Extensive Games", in M. Andronis et al. (eds), *Papers from the 38th Meeting of the Chicago Linguistic Society.* Chicago: Chicago Linguistic Society.

Krynicki M. (1993). "Hierarchies of Partially Ordered Connectives and Quantifiers", *Mathematical Logic Quarterly 39*:287–294.

Lehrer E. (1988). "Repeated Games with Stationary Bounded Recall", *Journal of Economic Theory 46*:130–144.

von Neumann J. and Morgenstern O. (1944). *Theory of Games and Economic Behavior.* New York: John Wiley.

Pietarinen A.-V. (2000). "Logic and Coherence in the Light of Competitive Games", *Logique et Analyse 43*: 371–391.

— (2001a). "Intentional Identity Revisited", *Nordic Journal of Philosophical Logic 6*:144–188.

— (2001b). "Most even Budged yet: Some Cases for Game-Theoretic Semantics in Natural Language", in *Theoretical Linguistics 27*:20–54.

— (2002). "Knowledge Constructions for Artificial Intelligence", in Hacid, M.-S., Ras, Z.W., Zighed, D.A., Kodratoff, Y., (eds), *Foundations of Intelligent Systems, Lecture Notes in Artificial Intelligence 2366.* Springer, 303–311.

— (2003a). "What do Epistemic Logic and Cognitive Science Have to do with Each Other?", *Cognitive Systems Research 4*:169-190.

— (2003b). "Peirce's Game-Theoretic Ideas in Logic", *Semiotica, 144*:33–47.

— (2003c). "Peirce's Theory of Communication and its Contemporary Relevance", in Nyíri, Kristóf (ed.) *Mobile Learning: Essays on Philosophy, Psychology and Education.* Vienna: Passagen Verlag. 81–98.

— (2003d). "Diagrammatic Logic and Game-Playing", to appear in Malcolm, Grant (ed.) *Multidisciplinary Studies of Visual Representations and Interpretations.* Elsevier.

— (2003e). "IF Logic and Incomplete Information", to appear.

— (2003f). "Moving Pictures of Thought", to appear.

— (2003g). "Dialogue Foundations and Informal Logic", to appear.

— (2003h). "Pragmatics from Peirce to Grice and Beyond", to appear.

Pietarinen A.-V. and Sandu G. (2003). "IF Logic, Game-Theoretical Semantics and new Prospects for the Philosophy of Science", in Rahman, Shahid, Symons, John, Gabbay, Dov and van Bendegem, Jean-Paul, (eds), *Logic, Epistemology and the Unity of Science.* Dordrecht: Kluwer.

Rasmusen E. (1994). *Games and Information* (first edition 1989). Cambridge, Mass.: Blackwell.

Sandu G. and Pietarinen A.-V. (2001). "Partiality and Games: Propositional Logic", *Logic Journal of the IGPL 9*:107–127.

Chapter 19

BACKWARD INDUCTION WITHOUT TEARS?

Jordan Howard Sobel
University of Toronto

1. Introduction

Players *resolve* games by moves that determine outcomes. Theorists *solve* games: they demonstrate that players who satisfy stated rationality- and belief-conditions resolve games in certain outcomes. A *weak solution* of a game shows that players who satisfy certain conditions resolve it somehow or other in an outcome. A *strong solution* shows additionally how, by what deliberations, these players reach that outcome.

The principal game studied is explained in Section 2. It is a 'backward-induction-terminating game'. A strong solution of this game for players who satisfy a controversial rationality and information condition is given in Section 3. The condition, in its call for robust beliefs in future rationality regardless of track records of irrationality, has been said to be a condition of players who are 'not especially reasonable' (Rabinowicz 1997), to be a 'dubious' condition (Broome and Rabinowicz 1999), and to be 'highly controversial (to say the least)' (Rabinowicz 1999). Theorists want backward-induction solutions that are based on less demanding conditions. Section 4 delivers a *weak* solution for the backward-induction outcome of our principal game that assumes less demanding idealizing conditions. Section 5, hankering after a *strong* solution, wonders *how* agents who satisfy these less demanding conditions would reach the backward-induction outcome of the game, by what deliberations. One answer is that they would satisfy in addition the controversial condition of Section 3, and reach the backward-induction outcome in the way indicated there. Other answers that have been suggested would not make these agents especially reasonable or well-informed, and so are of limited interest. Left will be the question whether conditions sig-

D. Vanderveken (ed.), Logic, Thought & Action, 433-461.

nificantly different from my controversial condition are both "definitely consistent with the traditional idealisations in game theory" (Broome and Rabinowicz 1999, p.239), and sufficient for a strong solution for the backward-induction outcome of the game. I would not care, if there are not. In my view that demanding condition, or one very like it, is right for especially reasonable and informed players, for *i*deal players, of this game.[1]

2. One Coin or Two?

2.1 *The game.*[2] X and Y are at a table on which there are dollar coins. In round one, X can appropriate one coin, or two. Coins she appropriates are removed from the table to be delivered when the game is over. If she takes two, the game is over, she gets these, and Y gets nothing. If she takes just one, there is a second round in which Y chooses one coin or two. Depending on his choice there may be a third round in which it is X's turn to choose, and so on until a player takes two coins, or there is just one coin left and the player whose turn it is takes it. Those are the rules, and X and Y will play by the rules. They are interested not merely mainly, but only, in money for themselves. 'As X has no kindness for Y, so he has no kindness for her.'

2.2 Let a *possible round* of a game be a round the reaching of which would be consistent with the rules of the game and the number coins on the table at the outset. Let *the possible length of a game* be the number n of its possible rounds. The possible length of a game is the number of coins on the table when it begins. If n is even, and the game is played through n rounds, Y has the last turn, and X and Y split the coins. Here is an annotated game-tree for this game if n is even and at least 6.

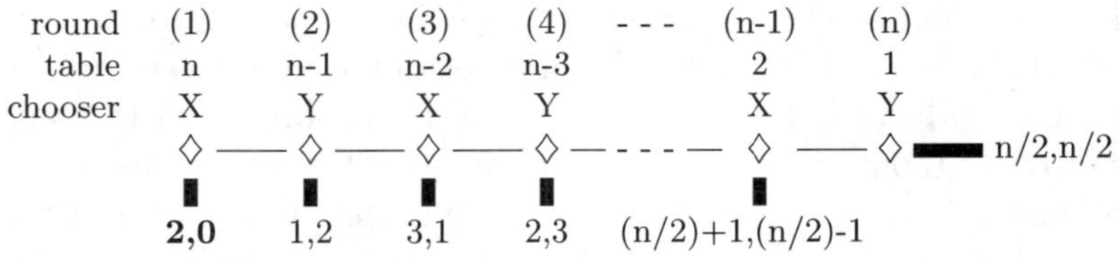

Options at a choice-node are restricted to Across and Down from this node. If n is odd, X has the last turn, and ends up with one more coin than Y.

2.3 *Backward induction.* In a *backward-induction calculation* at a choice-node of a game, unique maximizing moves from last choice-nodes are

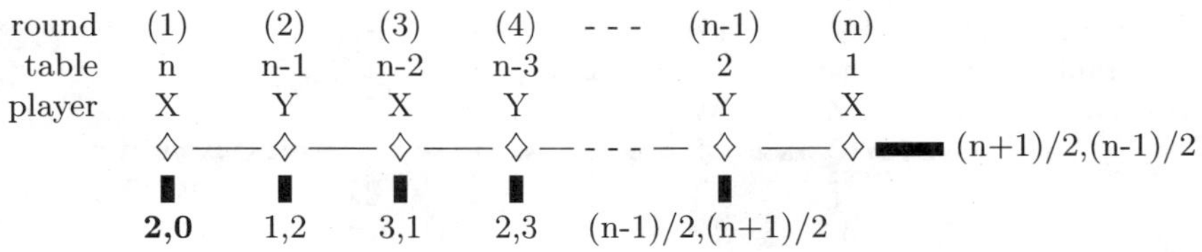

identified, branches lopped off before these nodes, and outcomes for pruned branches are set to equal outcomes for maximizing choices at what were last choice-nodes of the unpruned tree. The pruned tree is operated on in the same fashion. This process is repeated until the tree is pruned to the choice-node. A *backward-induction game* is a finite perfect information game in which backward-induction calculations identify for each choice-node a unique move, the *backward-induction move* from that node. "[T]ies... create special problems" (Rabinowicz 1995, p. 10): "to refrain from technical difficulties.... I... restrict attention to games in which each player has a strict ordering of terminal nodes" (Basu 1990, p. 40). The *backward-induction outcome* of a backward-induction game is the outcome of the game when only backward-induction moves are taken. In '*backward-induction terminating games*' (Rabinowicz 1996, p. 5), each backward-induction move is a game-terminating move.

Numbers at termini of the trees in the previous section are for player's coins as of, and values for, these termini: first numbers are for X's coins and value; second numbers are for Y's. 'BI-moves' are indicated by heavy lines. Values for the 'BI-outcome' are emphasized. In this game of selfish players, if each were to make only BI-moves, each would do less well than each would if they both made none of these moves, and played out through round n.[3] The centipedes of the previous section are BI-terminating games.

2.4 Suppose unselfish players W and Z in a game with the same rules that starts with an even number n of coins on the table. These players love each other as they love themselves. Loving W and Z do not care how the coins are divided between them. Their values for termini go by numbers of coins for the two of them as of these termini. Numbers of coins for W and Z are bracketed in this loving-couple centipede.

BI-calculation does not identify a unique move in the (n-1) round. W's options are 'tied' in this round. So this is not a 'BI-game'. But pruning the tied branches recommends itself, when they are tied not only for the chooser, but for everyone, at the branching node. With that added to the rule for 'BI-calculation', we have that W and Z would be pleased

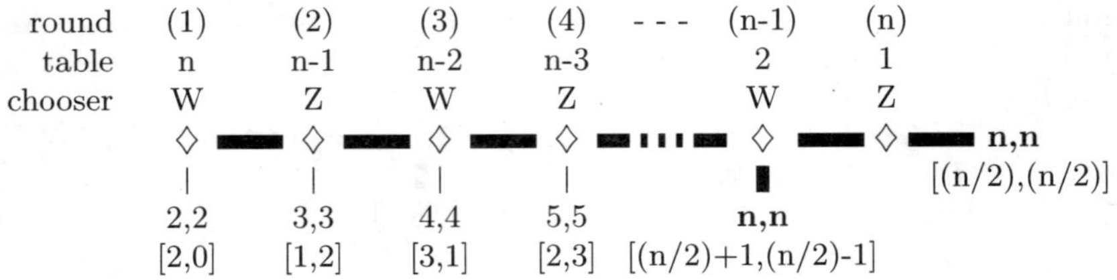

to make their BI-moves in this centipede which is *not* a 'BI-*terminating* game'. No moves would work out better for them in it.

BI-reasoning has gotten 'bad press' for trouble it can make for agents up to it, bad press that may be compared with more extensive bad press against causal maximizing (Newcomb problems), and dominance reasoning (Prisoners's dilemmas). In fact, BI-rationality is not always a bad thing for agents up to it, anymore than are causal maximizing, and dominance rationality.[4]

3. A Strong Solution for Players Who Would Be Knowledgeable and Practically Reasonable No Matter What

One Coin or Two resolves for X and Y by sound BI-reasoning, if they satisfy the following condition that has been found to make questionable their theoretical reasonableness.

3.1

> KNOWLEDGE OF RESILIENT RATIONALITY, COMPOUNDED ROBUSTLY FORWARD — **KofRR,CRF**. For a one-coin-or-two game with n coins on the table in the beginning, X knows in round (1) that [n], were round (n) reached, there would be one coin on the table, Y would know it was round (n), and Y would appropriate the coin; X knows in round (1) that [n - 1], were round (n - 1) reached, she would know it was round (n - 1), she would be maximizing-rational in it, she would know the rules and that there were two coins on the table, and she would know that [n]; X knows in round (1) that [n - 2], were round (n - 2) reached, Y would know it was round (n - 2), Y would be maximizing-rational in it, Y would know the rules and that there were three coins on the table, and Y would know that [n - 1]; *and so on* to round (1) in which X knows it is round (1), knows the rules and that there are n coins on the table, and knows that [2], and in which she is maximizing-rational.

The condition entails not only that X is *maximizing-rational* in round (1), but that X is *resiliently* maximizing-rational in round (1), where this means that X is maximizing-rational in round (1), and that, for each possible subsequent round, were it reached and it was X's turn in it, X would be maximizing-rational in it. The condition entails, indeed, that for each possible round, were it reached and it was X's turn, X would be resiliently maximizing-rational in it. The condition entails the same for Y.

"[R]esilient rationality is deep-seated and includes not only a display of rational actions, but a deeply entrenched and ineradicable disposition to such actions that would assert itself no matter what insults it had suffered." (Sobel 1994, ch. 16, p. 352.) Robert Aumann writes, "*Rationality* of a player means that he is a habitual payoff maximizer: that no matter where he finds himself [found himself] — at which vertex [in a game] — he will [would] not knowingly continue with a strategy that yields [yielded] him less than he could have gotten with a different strategy" (1995, p. 7).[5] However, while assuming something like resilient maximizing-rationality, Aumann does not explicitly assume anything like '*knowledge* of resilient maximizing-rationality *compounded robustly forward*'.[6] For another difference, my condition provides a basis for a strong solution by BI-reasoning, whereas Aumann offers only a weak solution. He argues only that "in PI games, common knowledge of rationality implies backward-induction" (Aumann 1995, p. 7), only that common knowledge of rationality implies that "the backward-induction outcome is reached" (Aumann 1995, p. 6) *somehow.* His common knowledge condition is, I think, sufficient for a strong solution only if it is meant to entail something like the subjunctive common belief-condition compounded robustly forward of (Sobel 1994, Ch. 16).[7]

3.2 *A strong solution based on KofRR,CRF.* If X and Y satisfy the condition, then the game can resolve by BI-reasoning in the choice by X in round (1) to appropriate two coins. The reasoning enabled could proceed in words somewhat like those in which I have said that "each player [in each round of an iterated prisoners's dilemma of known finite length] could reason... to the conclusion that his defection maximizes in that round" (Sobel 1994, ch. 16,p. 348).[8] The reasoning by X in One Coin or Two could be: "Were round (n) reached, Y would have no choice but to appropriate the remaining coin, and we would each get **n/2** coins: let that be proposition (i). Were round (n - 1) reached, I would know that, proposition (i), would be rational, and, maximizing, would appropriate two coins thereby terminating the game, and getting **(n/2 + 1)** coins, while Y got *(n/2 - 1)* coins: let that be proposition (ii). Were round (n - 2) reached, Y would know that, proposition (ii), would be rational, and, maximizing, would appropriate two coins, and get **n/2**, while I got (n/2 - 1): let that be proposition (iii)." And so on to: "I know now, in round (1), proposition (n -1), that were round (2) reached, Y would know proposition (n-2), and, maximizing, would appropriate two coins and leave me with only one, whereas I can appropriate two coins. Let me do that!" The game *can* resolve for ideal gameplayers in that manner in its BI-outcome. I assume that ideal gameplayers would be *resourceful* in the sense that games that can resolve for them in some outcome by sound reasoning, do resolve for them in this outcome by sound reasoning. (*Cf.*, Sobel 1994, Ch. 14, p. 304.) My conclusion is that, given the condition of *KofRR,CRF*, the game *does* resolve by sound reasoning on X's part in a choice to appropriate two coins and end the game as soon as it starts.[9]

3.3 *The controversy of KofRR,CRF. If* ideally rational and informed gameplayers would have knowledge of their resilient rationality compounded robustly through games, and would be resourceful, then One Coin or Two, could resolve by BI-reasoning in the choice by X of two coins in round (1), and would resolve by some sound reasoning or other by X in that choice. But could *ideally rational and informed* gameplayers satisfy this strong knowledge-condition? The condition implies that if X and Y are ideally rational and informed gameplayers, then even if a round were reached by a string of take-one-coin moves — even if a round were reached by a *long* string of such moves — the player whose turn it was to choose would be fully confident that if he were to choose just one coin in it, the other player would be resiliently rational in the next round and take two, since that player would realize that the other player would be resiliently rational in the round after that and take two coins, if there is a round after that, since *that* player would realize... and so on. All that, I repeat, would, according to the strong condition, obtain after perhaps a long string of rounds involving the same people, table, and rules, in which rounds just one coin had been taken. That can seem to be 'not especially reasonable'. "Would they *never learn*?!"[10] critics may complain.

Consider the state of X's opinions in round (1). Rabinowicz might say: "[she] expects to... keep an unbroken belief in [her] own [and the other's] future rationality even if [her, and his] intermediate behaviour, contrary to [her] expectations, would turn out to be irrational.... Such a stubborn self-confidence [and confidence in the other] in the face of conflicting evidence does not seem to be especially reasonable." Quoted in (Rabinowicz 1996, p. 1) from (Rabinowicz 1995). Rabinowicz adds in 1997: "Nor does it seem reasonable to *expect* the players to be so stubbornly confident in their beliefs **or to be incorruptible in their dispositions to rational behaviour**." (August 10, 1997, emphasis added.) I respond: It would *be* especially reasonable for an agent *to* *be* incorruptible in his or her disposition to rational behaviour, and to be resiliently rational, so that "past irrationality [would *not*] exert a corrupting influence on present play" (Broome and Rabinowicz 1999, p. 238).[11] *Questionable* is only the reasonableness of the players's stubborn *confidence* in their resilient rationality, and stubborn confidence in that stubborn confidence, that my condition implies in abundance.[12] The unquestionable reasonableness of resilient rationality can, however, be worked to deal with the questionable reasonableness of confidence therein. It is the thin end of a wedge, the fat end of which is knowledge of this resilient rationality, of knowledge of this knowledge, and so on, compounded robustly forward. For what is the alternative, given the *reality* of resilient rationality? *Ignorance* in some round of this rationality then, or ignorance then of knowledge of the players's compounded forward knowledge of it? But ignorance does not make for 'special reasonableness', and is furthermore not consistent with "traditional idealisations in game theory" (as Broome and Rabinowicz are concerned that assumptions should be — p. 239): traditional idealisations would have players know themselves and their fellows very well.

Suppose ideally rational and informed players *were* to be irrational for some rounds and in error about themselves. Suppose additionally that after those rounds they finally came to their senses, acted rationally, and were resiliently rational at last. How long should it take them to tumble to their realized virtue? How many rounds should it take for each to know that he was himself resiliently rational? How many rounds to recognize the resilient rationality of his fellows? The fewer the better for an *ideal*. The fewer rounds that that would take, the closer to his being an *ideally perceptive* game player. The fewest rounds that could take would be *no rounds*. And that seems the right answer for an *ideally* rational and well informed player. Ignorance of one's *own* practical qualities, while ofcourse *human* does not seem '*especially* reasonable', and is surely not consistent with what would be the reasonableness and information of an *ideal player*. Similarly for ignorance of the practical qualities of those with whom one would interact who would share one's own excellent qualities.[13]

It is not a matter of *stubborn confidence*, but of what would be *very rapid appreciations* of supposed late actualisations of full-blown resilient rationality. Similarly, *mutatis mutandis*, for knowledge of knowledge thereof compounded forward: "Ideal players would always, no matter what, know themselves and each other [perfectly]" (Sobel 1994, Ch. 16, p. 355). "[W]hatever they knew that they had done, they would still at sight know each other for the resiliently rational players that — even if only at last, and for the first time — they in fact would be" (p. 357), and similarly for this knowledge that they would — even if only at last — in fact have. There is no reason to say that they would 'turn a blind eye to past deficiencies' (Basu 1990, p. 33). One should suppose instead that though aware of their past deficiencies, they would have something like 'the present evidence of their senses' that all that was behind them. "It would be a strange model that supposed that no matter how long and how near to completely it had been dormant, rationality would always assert itself firmly and forevermore, but [contrary to the condition of robust knowledge compounded forward]... allowed that that it had asserted itself would not always be appreciated... at least not right away" (Sobel 1994, Ch. 16, p. 355).[14] *Ideal* players, hyperrational maximizers in games, would "know each other too well to teach each other what to expect" or to "learn what to expect of each other from experience or by induction" (Sobel 1994, Ch. 15, p. 336).[15] They would at any node know themselves and their fellows in their perhaps just gain practical perfections *immediately*. I am not persuaded by the opposition to strong conditions of resilient rationality and robust knowledge thereof compounded forward that are sufficient for sundry BI-resolutions. Still, 'it is of some interest' (Aumann 1998) that less controversial assumptions are sufficient for 'the backward induction outcome's resulting' (p. 98) somehow in some games. In the next section such assumptions are shown to be sufficient for a weak solution of the game. Following that I take up the possibility that something like my conditions are necessary for a strong solution.

4. A Weak Solution for Players Who Are Always Knowledgeable and Reasonable

4.1 'The game' shall henceforth be short for 'One Coin or Two played by X and Y with an even number of coins greater than 6'. It can be demonstrated that, given several *indicative* stipulations for 'ideal rationality and information in the game' that are plausibly analytic of that, if X and Y are ideally rational and informed players, X will appropriate two coins in round (1) thereby terminating the game. The stipulated conditions concern rounds *in which* both players are ideally rational and informed, *that have been reached* without *prior* irrationality or information-deficiency: They concern ideally rational and well-informed developments of the game.[16] Conditions to come include that movers in such rounds are minimally rational in a maximizing sense (Section 4.3.1), and that they believe that movers in next rounds, if reached, will be ideally rational and informed in it (Section 4.3.2). Other conditions go to knowledge of the game, of necessities, of the state of play, and of moves just before they make them (Sections 4.3.3).[17]

4.2 *Pillars of the argument.*

> *The Main Result.* When the game is played by 'ideally rational and informed players', as partly defined by conditions detailed in Section 4.3, the BI-outcome is reached.

This weak solution follows from

> The *Trivial Lemma.* When the game is played, round (1) is reached and there are more than two coins on the table, *and* when it is played by 'ideally rational and informed players,' the players are 'ideally rational and informed' in round (1).

together with

> *The Main Theorem.* When the game is played by players who are 'ideally rational and informed,' (for each number r) [if round (r) of the game is reached without irrationality or information-deficiency in prior rounds, and the players are 'ideally rational and informed' in round (r), then (if there are in (r) at least two coins on the table, the player whose turn it is in round (r) chooses two coins and terminates the game)].

'Numbers' throughout are positive integers. 'Rounds' are possible rounds as defined in Section 2.1. 'Information-deficiency' is short for 'lack of information that would be had by ideal gameplayers'.

4.3 *Ideally rational and informed players* — conditions analytic thereof

4.3.1 *Rationality.* First a definition, then the condition.

> DEFINITION OF MINIMAL RATIONALITY (**DefMinR**). A player is *minimally rational* in a round in which it is his turn to choose, if and only if the choice he makes is not such that 'he believes just before he makes it'[18] that another choice would lead to his getting more coins when the game is terminated.

Cf.: "*Rationality* of a player means that... he will not knowingly continue with a strategy that yields him less than he could have gotten with a different strategy." (Aumann 1995, p. 7.) This 'rationality' is *practical* rationality only.

> CONDITIONAL MINIMAL RATIONALITY (**ConMinR**). It is by stipulation analytically necessary of *the game when it is played by ideally rational and informed players* that: at each round that is reached in the game when it is played by ideally rational and informed players, if it is reached without irrationality or information-deficiency in prior rounds and the players are ideally rational and informed in this round, then the player whose turn it is to choose in this round is minimally rational.

[@]*IdealGamePlayers* [IdealGamePlayers ⊃

(r)(p)([Reached(r) & ReachedWithoutPrIrr∨InfDef(r) & PlayersIR&I(r)]
⊃ [(Mover(p,r) ⊃ MinRat(p,r)])].

Abbreviations

—**[@]:** it is analytically necessary (perhaps by stipulation) of — that:
IdealGamePlayers: the game is played by ideally rational and informed players
Reached(r): r is reached
ReachedWithoutPrIrr∨InfDef(r): r is reached without prior irrationality or information-deficiency; **PlayersIdealR&I**(r): players are ideally rational and informed in round (r)
Mover(p,r): p is the player whose turn it is to move in round (r)
MinRat(p,r): p is minimally rational in round (r).

4.3.2 *Information*

4.3.2.1 *The main information-condition*

Beliefs in Ideal Rationality and Information (**BinIdealR&I**). It is by stipulation analytically necessary of *the game when it is played by ideally rational and informed players* that, at each round that is reached without irrationality or information-deficiency in prior rounds in the game when it is played by ideally rational and informed players, if the players are ideally rational and informed in this round, then the player whose turn it is to choose in this round believes, just before he or she makes his or her choice, that the game is being played by ideally rational and informed players, and that if the next round is reached, it will be reached without irrationality or information-deficiency in prior rounds and the players will still be ideally rational and informed in it.

[@] *IdealGamePlayers* [IdealGamePlayers ⊃
(r)(p)([Reached(r) & ReachedWithoutPrIrr∨InfDef(r) & PlayersIdealR&I(r)] ⊃
(Mover(p,r) ⊃ *Believes*(p,r)(IdealGamePlayers & [Reached(r+1) ⊃
ReachedWithoutPrIrr ∨ InfDef(r+1) & PlayersIdealR&I(r+1)])))].

Additional abbreviation — for a person p, round r, and sentence ϕ that expresses a proposition, the scheme, **Believes**(p,r) ϕ, abbreviates the scheme, p believes just before his or her choice in r that ϕ. Now come arguments for the plausibility of BinIdealR&I, that is, for its appropriateness as stipulation for 'ideal gameplayers in the game'.

4.3.2.2 Here is a *subjunctive* condition that entails BinIdealR & I:

[@]*IdealGamePlayers* [IdealGamePlayers ⊃
(r)(p)([Reached(r) & ReachedWithoutPrIrr ∨ InfDef(r) & PlayersIdealR&I(r)] □→
(Mover(p,r) ⊃ *Believes*(p,r)(IdealGamePlayers & [Reached(r+1) □→

ReachedWithoutPrIrr ∨ InfDef(r+1) & PlayersIdealR&I(r+1)]))]).

It is plausible that when the game is played by ideally rational and informed players, were a round reached without irrationality or information-deficiency, in which round all were still ideally rational and informed, the mover in this round would believe, (i), that they all were ideally rational and informed players, and, (ii), that, were the next round reached, it would be reached without current irrationality or information-deficiency on his part, and so without any prior irrationality or information-deficiency, and that all would still be ideally rational and informed players in it. Regarding (i), the mover in this round would, by hypothesis, have no reason to think that they were not all ideally rational and informed, for it is plausible that this mover would know that the round had been reached without irrationality of information-deficiency. Regarding (ii), that were the next round reached, this would *not* be because he had 'lost it', and behaved irrationally: it is plausible that he should believe that he would have choice-relevant beliefs concerning the behaviour of the mover in the next round that would have made *reasonable* his choice in the present round not to terminate the game, and so to give that player another turn. This subjunctive analogue of BinIdealR&I is plausible. Thoroughly indicative BinIdealR&I, which is entailed by it, must be at least as plausible.[19]

4.3.2.3 Broome and Rabinowicz defend their belief-conditions by deriving them from conditions "that seem definitely consistent with the traditional idealisations in game theory" (1999, p 239). BinIdealR&I can be derived similarly from the following several-part condition.

[@]*IdealGamePlayers* [IdealGamePlayers ⊃
(1) (IdealGamePlayers & [Reached(1) & ReachedWithoutPrIrr∨InfDef(1) & PlayersIdealR&I(1)]), &
(2) (s)([Reached(s) & ReachedWithoutPrIrr∨InfDef(s)] ⊃ PlayersIdealR&I(s)), &
(3) (s)([Reached(s) & ReachedWithoutPrIrr∨InfDef(s)] ⊃ (p)*Believes*(p,s)[(1) & (2)]), &
(4) (s)([Reached(s) & ReachedWithoutPrIrr∨InfDef(s)] ⊃ each player believes in s whatever is entailed by things he believes in s)]

Now comes a detailed derivation of BinIdealR&I from this compound condition.

To derive BinIdealR&I, assume

(5) IdealGamePlayers,

and, for arbitrarily selected r and p,

(6) Reached(r) & ReachedWithoutPrIrr∨InfDef(r) & PlayersIdealR&I(r),

(7) Mover(p,r).

to derive

(8) *Believes*(p,r)[IdealGamePlayers & (Reached(r+1) ⊃ [ReachedWithoutPrIrr∨InfDef(r+1) & PlayersIdealR&I(r+1)])].

With (5), (1) through (4). Consider that done.

(9) *Believes*(p,r)[(1) & (2)], (6), **(3)**

Next to show that proposition [(1) & (2)] entails,

proposition (*)

(r)(Reached(r) ⊃ [ReachedWithoutPrIrr∨InfDef(r) & PlayersIdealR&I(r)]).

to reach, using (4) and (6), *Believes*(p,r)(*).

For an indirect derivation of (*) from (1) and (2), assume ~(*) and observe that given this negation there must be a first round in which (*) fails. We may let round (k) be this first-failure-round so that, for additional assumptions, we have, given that (k) is a *failure*-round,

(10) Reached(k),

(11) ~[ReachedWithoutPrIrr∨InfDef(k) & PlayersIdealR&I(k)],

and, since by **(1)** $k \neq 1$, and k is a *first*-failure-round, that,

(12) Reached(k-1),

and

(13) [ReachedWithoutPrIrr∨InfDef(k-1) & PlayersIdealR&I(k-1)]).

(14) ReachedWithoutPrIrr∨InfDef(k), (10), (13)

(15) PlayersIdealR&I(k), (10), (14), **(2)**

and

(16) [ReachedWithoutPrIrr∨InfDef(k) & PlayersIdealR&I(k)]. (14, (15)

The contradiction on lines (11) and (16) completes the derivation of (*), from (1) and (2). [This indirect derivation is equivalent to a mathematical induction. Line (12) corresponds to what could have been the 'inductive hypothesis' of a mathematical induction. That the 'first-failure-round' cannot be round one corresponds to what could have been the 'basis.']

(17) Proposition [(1) & (2)] entails (*). (10) - (16)

(18) *Believes*(p,r)[(*): (r)(Reached(r) ⊃ [ReachedWithoutPrIrr∨InfDef(r) & PlayersIdealR&I(r)])], (9), (17), (6), **(4)**

(19) Proposition [(1) & (2)] entails IdealGamePlayers

(20) *Believes*(p,r)(IdealGamePlayers) (9), (19), (6), **(4)**

(21) Proposition (*) entails that [Reached(r+1) ⊃ [ReachedWithoutPrIrr∨InfDef(r+1) &PlayersIdealR&I(r+1)]]

(22) *Believes*(p,r)(Reached(r+1) ⊃ [ReachedWithoutPrIrr∨InfDef(r+1) & PlayersIdealR&I(r+1)]) (18), (21), (6), **(4)**

(23) *Believes*(p,r)[IdealGamePlayers & (Reached(r+1) ⊃ [ReachedWithoutPrIrr∨InfDef(r+1) & PlayersIdealR&I(r+1)])] (20), (22), (6), **(4)**

This defence of BinIR&I can be enhanced in a manner suggested in (Broome and Rabinowicz 1999): Clause (3) can be derived. The condition without clause (3) entails (lines (10) - (16) show the way) that

> [@]*IdealGamePlayers* (IdealGamePlayers ⊃ (s)[Reached(s) ⊃ ReachedWithoutPrIrr∨InfDef(s)])

We may plead for (3) on the ground that

> [@]*IdealGamePlayers* (IdealGamePlayers ⊃ (i) players believe in round (1) propositions (1) and (2), and have no false beliefs, (ii) players retain their beliefs in a round they reach without prior irrationality or information-deficiency, provided they can do so consistently with beliefs acquired, and (iii) players acquire only true beliefs in a round they reach without irrationality or information-deficiency in prior rounds).

When ideal gameplayers reach a round without prior irrationality or game-theoretic information-deficiency, they should remain ideally rational and acquire additional true beliefs sufficient to their remaining game-theoretically well-informed.

4.3.3 *Other information-conditions*

> KNOWLEDGE OF THE GAME (**KofGm**). It is by stipulation analytically necessary of *the game when it is played by ideally rational and informed players* that, when the game is played by ideally rational and informed players, for every proposition q, if q is entailed by rules of the game and assumptions definitive of it, then at each round that is reached without irrationality or information-deficiency in prior rounds in which round the players are ideally rational and informed, the player whose turn it is to choose knows just before his or her choice that q.

Since necessary propositions are entailed by all propositions, KofGm entails,

> KNOWLEDGE OF NECESSITIES (**KofN**). It is by stipulation analytically necessary of *the game when it is played by ideally rational and informed players* that, when the game is played by ideally rational and informed players, for every proposition q, if it is necessary that q, then at each round that is reached without irrationality or information-deficiency in prior rounds in which round players are ideally rational and informed, the player whose turn it is to choose knows just before his or her choice in this round that q.

[@](*IdealGamePlayers* ⊃ (r)(p)(q)[Reached(r) & ReachedWithoutPrIrr∨InfDef(r) & PlayersIdealR&I(r) ⊃ [Mover(p,r) & □(q) ⊃ Knows(p,r)(q)]])

Abbreviation — *Knows*(p,r): p knows just before his or her choice in r that, and letting 'q' range over propositions. *Cf.*: "each player... believes the propositions that are necessarily true" (Rabinowicz 1996, p. 4).

TABLE KNOWLEDGE (**TblKn**). It is by stipulation analytically necessary of *the game when it is played by ideally rational and informed players* that when the game is played by ideally rational and informed players, for every round that is reached without irrationality or information-deficiency in prior rounds in which round players are ideally rational and informed, the player whose turn it is to choose knows just before his or her choice in this round how many coins are on the table.

BELIEFS IN LOGICAL CONSEQUENCES (**BinLgCns**). When the game is played by ideally rational and informed players, for every proposition q, and every round that is reached without irrationality or information-deficiency in prior rounds, in which round players are ideally rational and informed, if q follows from propositions believed by the player whose turn it is to choose in this round just before his or her choice in this round, then this player believes q just before this choice.

Conditions of this section can be trimmed to require beliefs only in "propositions that have to be believed... if the argument to be proposed [in Section 4.4] is to go through" (Rabinowicz 1996, p. 4).

4.3.4

PRESCIENCE. It is by stipulation analytically necessary of *the game when it is played by ideally rational and informed players* that, when the game is played by ideally rational and informed players, for every round r reached without irrationality or information-deficiency in prior rounds in which round players are ideally rational and informed, the player whose turn it is to move in r knows, of the choice he or she will make in r, just before he or she makes it, that he or she is going to make it.[20]

Is prescience possible? Can a person know before he has made a choice, what choice he is going to make? Yes (*pace* Ginet 1962). Consider: "You will choose the green." "I believe you, though I can't imagine why I will choose it. I don't like green. But I trust you. I know you would not say that unless you knew it. So, having your word for it, I know it too, though I still wonder why I will choose it." 'Prescience', or knowledge of a choice just before it is made, is possible. And an ideally rational person would be 'prescient'. It is plausible that every *somewhat* rational person, almost always has a very good idea what choice he is about to make, just before he makes it. None of us *often* think, "Now where did that choice come from?!" It may be that no *choice*, properly so-termed, *could be* a complete surprise. Prescience says that *just before* a choice an agent is sure that he is about to make this choice. It does *not* say that he then has this opinion *fully in and before his mind.* He may well not. When deliberating what to do, one is not *wondering what one will do.* And when one is about *to make* a choice, one is generally not thinking *that one is about to make it.* But one could hardly be *unaware* that one is just about to make it, when one is just about to. The answer to the question, "What are you going to choose?", could hardly be, just before you make it, "I have no idea, I haven't thought about that!!", ever.[21]

4.4 **The Main Theorem** — (Section 4.2) symbolized using the additional abbreviations, **AtLstTwo**(r): there are *at least two coins* on the table in round (r), and **ChTwo&Term**(r): the player whose turn it is in round (r) chooses two coins and terminates the game — is

(r)[IdealGamePlayers ⊃ (r)([Reached(r) & ReachedWithoutPrIrr∨InfDef(r) & PlayersIdealR&I(r)] ⊃ [AtLstTwo(r) ⊃ ChTwo&Term(r)])]

To prove the Main Theorem it is convenient to prove,

Theorem*. (r)□[IdealGamePlayers ⊃ ([Reached(r) & ReachedWithoutPrIrr∨InfDef(r) & PlayersIdealR&I(r)]⊃ [AtLstTwo(r) ⊃ ChTwo&Term(r)])]

which entails it. For Theorem* I offer a weak mathematical induction on rounds in the game, starting with (n), the last possible round, and proceeding back towards the first round (1) (*cf.* Sobel 1994, ch. 16, p. 349). The number of possible rounds in the game under discussion is, recall, even and greater than six: that is part of the definition of 'the game'. This induction is served by the strength of its inductive hypothesis, given that not the Main Theorem, but the stronger Theorem*, is being proved.[22] The BASIS of our induction is the 'matrix' of Theorem* for the last possible round (n):

□[IdealGamePlayers ⊃ ([Reached(n) & ReachedWithoutPrIrr∨InfDef(n) & PlayersIdealR&I(n)] ⊃ [AtLstTwo(n) ⊃ ChTwo&Term(n)])]

To prove this proposition it is sufficient to derive its non-modal 'core' from conditions definitive of the game and ideally rational and informed players, for these conditions correspond to definitive stipulations which make necessities, and what follows from necessities is itself necessary. The Basis is true in virtue of rules and stipulations partly definitive of 'the game'. Stipulations concerning the players's rationality and information do not matter to it.

<<Proof of the Basis: There is just one coin on the table in round (n) in the game. It is thus false that there are at least two coins on the table in round (n). So the material conditional that is the non-modal 'core' of the Basis is true: for the consequent,

[AtLstTwo(n) ⊃ ChTwo&Term(n)], of its consequent,

([Reached(n) & ReachedWithoutPrIrr∨InfDef(n) & PlayersIdealR&I(n)] ⊃ [AtLstTwo(n) ⊃ ChTwo&Term(n)]), is true; it is true, since its antecedent, AtLstTwo(n), is false.>>

The INDUCTIVE GENERALIZATION of the mathematical induction is this proposition:
For any k such that $1 < k < n$: if the INDUCTIVE HYPOTHESIS,
$\Box$[IdealGamePlayers $\supset$ ([Reached(k) & ReachedWithoutPrIrr$\vee$InfDef(k) & PlayersIdealR&I(k)] $\supset$ [AtLstTwo(k) $\supset$ ChTwo&Term(k)])],
then the INDUCTIVE CONCLUSION,
$\Box$[IdealGamePlayers $\supset$ (Reached(k-1) & ReachedWithoutPrIrr$\vee$InfDef(k-1) & PlayersIdealR&I(k-1)] $\supset$ [AtLstTwo(k-1) $\supset$ ChTwo&Term(k-1)])].

Now comes a detailed proof of this Inductive Generalization that uses everything assumed in Section 4.3.

To prove the Inductive Generalization, it is sufficient to prove its inner conditional, (the Inductive Hypothesis $\supset$ the Inductive Conclusion). The proof of this conditional proceeds in an extension for modal sentential logic of the elegant natural deduction system for sentential logic of Donald Kalish and Richard Montague.[23] Added to provisions for conditional and indirect derivations of that system, is a provision for necessity derivations. Lines for what is to be shown by derivations — by the main derivation, or by a subsidiary derivation, precede their derivations — begin with the word '*SHOW*', and are never available for use. They are never available as premises for inferences, or as bases for concluding derivations. To remind of their unavailability, they are 'braced.' When a derivation is completed, lines under its *SHOW*-line are 'bracketed,' and what has been shown, prefaced by the word '*SHOWN*,' is entered on the line under the bracketed lines that show it. This line is, but the bracketed lines are no longer, available for use. *SHOW*-lines provide occasions for assumptions that are — by this provision that bracketed lines are no longer available for use — 'given up' when the derivations for which they are made are completed. **Rules**, a special principle of inference used once in the proof, says that whatever follows from available lines by rules of the game can be entered on a line. Taken for granted at points is that knowledge entails belief. Here, for ready reference are definitions and conditions that will be cited,

DefMinR: Definition of Minimal Rationality
ConMinR: Conditional Minimal Rationality
BinIdealR&I: Beliefs in Ideal Rationality and Information
KofGm: Knowledge of the Game
KofN: Knowledge of Necessities
TblKn: Table Knowledge
BinLgCns: Beliefs in Logical Consequences
Prescience

and abbreviations that will be used,

IdealGamePlayers: the game will be played by ideally rational and informed players
Reached(r): r is reached
ReachedWithoutPrIrr∨InfDef(r): r is reached without prior irrationality or information-deficiency
PlayersIdealR&I(r): players are ideally rational and informed in round (r)
Mover(p,r): p is the player whose turn it is to move in round (r)
MinRat(p,r): p is minimally rational in round (r)
Believes(p,r)ϕ: p believes just before his or her choice in r that ϕ

Now comes a derivation (see next page), from conditions analytically necessary of IdealGamePlayers, of the inner conditional of the Inductive Generalization.

The Basis and the Inductive Generalization entail Theorem* by the principle of weak mathematical induction. Theorem* entails the Main Theorem. <<For a derivation in an extension for quantified modal logic of the natural deduction system of Kalish and Montague for quantifiers,[25] let 'M' abbreviate the 'matrix' of the Main Theorem, this theorem is that (r)M, whereas Theorem* is that (r)□M. From Theorem* can be inferred by Universal Instantiation, □M,[26] and from that by Necessity, M, which suffices for a Universal Derivation of (r)M.>>

4.5 *Regarding this weak solution of the game*

4.5.1 Aumann writes: "The reader may wonder why we adduce such a lengthy formal proof for an argument that while not immediate, seems simple enough once one has found it. The reason is that this area is *very* tricky, and unless one is extremely careful and formal, it is easy to go astray — as we have found, to our dismay, on more than one occasion. (The same remark applies to [Aumann 1995].)" (Aumann 1996 [1998, p. 104]) I can say most of that of my case, though I cannot say that the weak solution I maintain has ever seemed simple to me.

4.5.2 The argument for the Main Theorem can be adapted to use 'subjunctive versions' of the stated indicative conditions that serve as premises, for these will entail the indicative conditions to which they correspond. The argument is, for two reasons, not readily adaptable to the subjunctive version of that theorem,

(r)(IdealGamePlayers ⊃ [Reached(r) & ReachedWithoutPrIrr∨InfDef(r) & PlayersIdealR&I(r)] □→ [AtLstTwo(r) ⊃ ChTwo&Term(r)])

(1) {*To SHOW* (the Inductive Hypothesis ⊃ the Inductive Conclusion)}
(2) the Inductive Hypothesis: □[IdealGamePlayers ⊃ ([Reached(k) & ReachedWithoutPrIrr∨InfDef(k) & PlayersIdealR&I(k)] ⊃ [AtLstTwo(k) ⊃ ChTwo&Term(k)])] assumption for conditional derivation
(3) {*To SHOW* the Inductive Conclusion: □[IdealGamePlayers ⊃ ([Reached(k-1) & ReachedWithoutPrIrr∨InfDef(k-1) & PlayersIdealR&I(k-1)] ⊃ [AtLstTwo(k-1) ⊃ ChTwo&Term(k-1)])]}
To demonstrate the Inductive Conclusion it is sufficient to derive its non-modal core from necessities, since what is entailed by necessities is itself necessary. The procedure, Necessity Derivation, requires that we take care not to 'enter from without' any non-necessities.

(4) {*To SHOW* IdealGamePlayers ⊃ ([Reached(k-1) & ReachedWithoutPrIrr∨InfDef(k-1) & PlayersIdealR&I(k-1)] ⊃ AtLstTwo(k-1) ⊃ ChTwo&Term(k-1)])}
In the derivation of (4), and thus of (3), only the necessity (2) and conditions that are analytically necessary of 'IdealGamePlayers' or 'the game' are 'entered from without.'

(5) IdealGamePlayers assumption for CD
(6) {*To SHOW* [Reached(k-1) & ReachedWithoutPrIrr∨InfDef(k-1) & PlayersIdealR&I(k-1)] ⊃ [AtLstTwo(k-1) ⊃ ChTwo&Term(k-1)]}
(7) Reached(k-1) & ReachedWithoutPrIrr∨InfDef(k-1) & PlayersIdealR&I(k-1) assumption for CD
(8) {*To SHOW* AtLstTwo(k-1) ⊃ ChTwo&Term(k-1)}
(9) AtLstTwo(k-1) assumption for CD
Let 'W' abbreviate 'the player whose turn it is to move in round (k - 1)', so that, by stipulation.

(10) Mover(W,k-1)
(11) MinRat(W,k-1) ConMinR[24], (5), (7), (10)
Let **ChoosesOne**(W,k-1) abbreviate 'W chooses to appropriate one coin in round (k - 1)', and **JustOne**(k-1) abbreviate 'there will be just one coin on the table in round (k - 1)'.

(12) {*To SHOW* ∼ChoosesOne(W,k-1)}
(13) ChoosesOne(W,k-1) assumption for indirect derivation
(14) *Believes*(W,k-1)[Reached(k)] (13), TblKn, (5), (7), (10), (13), Prescience, KofGm, BinLgCns
(15) *Believes*(W,k-1)[AtLstTwo(k)] v *Believes*(W,k-1)[JustOne(k)](14), KofGm, TblKn, (5), (7), (10), BinLgCns
(16) {*To SHOW* *Believes*(W,k-1)[AtLstTwo(k)] ⊃ ∼MinRat(k-1)}
(17) *Believes*(W,k-1)[AtLstTwo(k)] assumption for CD
(18) *Believes*(W,k-1)[IdealGamePlayers ⊃ ([Reached(k) & ReachedWithoutPrIrr∨InfDef(k) & PlayersIdealR&I(k-1)] ⊃[AtLstTwo(k) ⊃ ChTwo&Term(k)])]
2: the Inductive Hypothesis, KofN [this is the only use made of this principle], (5), (7), (10)
(19) *Believes*(W,k-1)[IdealGamePlayers & (Reached(k) ⊃ [ReachedWithoutPrIrr∨InfDef(k) & PlayersIdealR&I(k)])] BinIdealR&I [this is the only use made of this principle], (5), (7), (10)
(20) *Believes*(W,k-1)[ReachedWithoutPrIrr∨InfDef(k) & PlayersIdealR&I(k)]
(14), (19), BinLgCns, (5), (7), (10)
(21) *Believes*(W,k-1)[ChTwo&Term(k)] (14), (20), (18), BinLgCns, (5), (7), (10)
(22) Of the choice he makes in round (k - 1), W believes, just before he makes it, that it will lead to his getting in the end, when the game is over, exactly one more coin.
(13), Prescience, (5), (7), (10), (21), KofGm, BinLgCns
(23) W believes, just before he makes his choice in round (k - 1), that another choice open to him would give him in the end two more coins. (9), TblKn, KofGm, BinLgCns, (5), (7), (10)
(24) ∼MinRat(W,k-1) (22), (23), DefMinR

(25) *SHOWN* *Believes*(W,k-1)[AtLstTwo(k) ⊃ ∼MinRat(k-1)] (17) - (24), CD
(26) {*To SHOW* *Believes*(W,k-1)[JustOne(k)] ⊃ ∼MinRat(k-1)]}
(27) *Believes*(W,k-1)[JustOne(k)] assumption for CD
(28) Of the choice he makes in round (k - 1), W believes, just before he makes it, that it will lead to his getting in the end, when the game is over, only one more coin.
(7), (13), (5), Prescience, (27), KofGm, BinLgCns
(29) W believes, just before he makes his choice in round (k -1), that another choice open to him would give him in the end two more coins. (7), (9), (5), TblKn, KofGm, BinLgCns
(30) ∼MinRat(W,k-1) (28), (29), DefMinR

(31) *SHOWN* *Believes*(W,k-1)[JustOne(k)] ⊃ ∼MinRat(W,k-1)] (27) - (30), CD
(32) ∼MinRat(W,k-1) (15), (25), (31), Separation of Cases
(33) MinRat(W,k-1) (11)

(34) *SHOWN* ∼ChoosesOne(W,k-1) (13) - (33), ID (32 and 33 make a contradiction)
(35) ChTwo&Term(k-1) (10), (34), Rules

(36) *SHOWN* AtLstTwo(k-1) ⊃ ChTwo&Term(k-1) (9) - (35), CD

(37) *SHOWN* [Reached(k-1) & ReachedWithoutPrIrr∨InfDef(k-1) & PlayersIdealR&I(k-1)] ⊃ [AtLstTwo(k-1) ⊃ ChTwo&Term(k-1)] (7) - (36), CD

(38) *SHOWN* IdealGamePlayers ⊃ ([Reached(k-1) & ReachedWithoutPrIrr∨InfDef(k-1) & PlayersIdealR&I(k-1)] ⊃ [AtLstTwo(k-1) ⊃ ChTwo&Term(k-1)]) (5) - (37), CD
(39) the Inductive conclusion: □[IdealGamePlayers ⊃ ([Reached(k-1) & ReachedWithoutPrIrr∨InfDef(k-1) & PlayersIdealR&I(k-1)] ⊃ [AtLstTwo(k-1) ⊃ ChTwo&Term(k-1)])] (4) - (38), ND (see comment under (3))
(40) *SHOWN* (the Inductive Hypothesis ⊃ the Inductive Conclusion) (2) - (39), CD

First, the simple proof of the Basis exploits the fact that the consequent of the conditional under the necessity operator is a material conditional. Second, for the inductive step, the subsidiary derivation for the consequent of the non-modal core of the Inductive Conclusion that runs from (6) to (36) is a conditional proof. The consequent of the non-modal core of the Inductive Conclusion in a like argument for the displayed subjunctive version of the Main Theorem would be a subjunctive conditional. Conditional proof is not a valid procedure for subjunctive conditionals. The displayed subjunctive version of the Main Theorem cannot be derived in the way that the Main Theorem itself has been from the conditions of Section 4.3. Nor can it be derived in that manner from their subjunctive enhancements.

4.5.3 The Main Result (Section 4.2) follows from the Main Theorem and the Trivial Lemma (Section 4.2). This weak solution for One Coin or Two, can be adapted to every *BI-terminating game* (*cf.*, Broome and Rabinowicz 1999, p. 240). Reflection on the rôle played in it by the terminating character of One Coin or Two persuades that it *cannot* be readily adapted to non-terminating BI-games (*cf.*, Rabinowicz 1996, pp. 8, Aumann 1998, p. 98, Broome and Rabinowicz 2001, p. 241). A question left is whether *anything* that uses, if not thoroughly indicative assumptions of rationality and information, then at any rate significantly weaker assumptions than the very demanding, heavily subjunctive ones of resilient rationality and robust knowledge (Section 3.1 above), works for even a *weak* solution for any *non-terminating* BI-game. I very much doubt it.

5. Querulous Conclusions

5.1 *But why, if X and Y are ideally rational and well informed, will X choose down in the first round? How, by what manner of reasoning, will X come to make this choice?* Somehow, for it has been demonstrated that she will. But how, consistent with their being ideally rational and well informed?

5.1.1 If X and Y are ideally rational and well informed, according to the conditions of Section 4, then X can *know* not only just before, but some time before, she chooses two coins, that that is the choice she is going to make. The argument says *how* she can know long before she makes this choice that it is the choice she is going to make. "Like this," the argument says, "since you believe that you and your opponent will be ideally rational and informed just before you make your choice, with all that that entails." Left, however, by the argument's premises that lay down thoroughly indicative conditions of ideal rationality and information, is the question what will be her *reasons* for it.[27] Left is the question how this game *resolves* for ideally rational players.

Broome and Rabinowicz say that their argument is more transparent than the argument in (Aumann 1996), because "it spells out *how* X acquires the belief that the game will end at the next turn" (1999, p. 242), if there is a next turn. Does that mean their argument makes plain her reason for choosing down so that there is not a next turn, a reason implied furthermore by the weak indicative conditions with which they work to the conclusion that she will choose down? No, for their *at-choice* conditions cannot explain even how X acquires the belief that the game will end at the next turn, if there is a next turn, when she needs it, if it is to provide, or so much as contribute, to a reason for her choice. For that she needs that belief that the game will end at the next turn, if there is a next turn, *before* her choice.

My pre-choice indicative conditions are disadvantaged in part somewhat similarly. They do entail her belief, just before her choice in round 1, that the game will end in round 2, if there is a round 2,

$$Bel[\text{Reached}(2) \supset \text{End}(2)];$$

but they entail *also* her belief then that the game will *not* end in round 2, if there is a round 2,

$$Bel[\text{Reached}(2) \supset \sim\text{End}(2)].$$

Why? Because, since they include Prescience, they entail her belief just before her choice in round 1 that she will choose down. And so — since they include Knowledge of the Game, and Beliefs in Logical Consequences — they entail her belief that round 2 is not reached, $Bel[\sim\text{Reached}(2)]$. Furthermore, to come to the way in which my conditions are similarly disadvantaged, they do not entail that she has any of these beliefs before she knows what choice she is going to make, and so at times when they could be parts of her reason for her making it: times for beliefs that can be parts of her reasons for making her choice are presumably all times before she knows she is going to make it.[28] For a last point, which could have been first, her belief in the material conditional that *the game will end in round 2, if there is a round 2*, $[\text{Reached}(2) \supset \text{End}(2)]$, only 'sounds' like a reason for her seeing to it that there is not a round 2, namely, the reason her belief in the subjunctive conditional, $[\text{Reached}(2) \,\Box\!\!\rightarrow \text{End}(2)]$, could provide. Neither my thoroughly indicative assumptions, nor those of Broome and Rabinowicz, entail that she believes that either before or at the time of her choice in round (1).

5.2 My argument establishes that if X and Y are ideally rational and informed, X will not choose one coin. It establishes that she will play as she would were she to choose pursuant to BI-reasoning. It establishes that if they are ideally rational and informed, then she *has* beliefs sufficient for a choice of two coins, for it establishes that she *makes* that choice. Being ideally rational she needs to have a good reason for making it. But the assumptions I make for my argument do not suggest what this reason could be. They do not provide her with beliefs sufficient for her game-terminating choice. We may wonder what assumptions, consistent with those already in place, would do that.

Ignoring briefly the condition that X and Y are ideal game players, there are many conditions considerably weaker than that demanding condition of mine — that condition of knowledge compounded robustly forward of resilient rationality — that would enable reasoning on X's part to her choice to take both coins and end the game. She could reason to that choice, if she believed that were she instead to choose one coin, Y would choose two coins in round (2) leaving her with only one coin. *Cf.*: "One possibility would be... that X... believes... that Y would make his backward-induction move... at the earliest possible occasion" (Rabinowicz 1996, p. 10). And why might she believe that? What assumptions could secure that belief? There are many. She might, as far as those conditions go, have no reasons for that belief. She might, as dogmatists say, 'just happen to believe that.' *But — recalling the condition briefly ignored — that would not make her especially reasonable.* Or — ignoring again that condition — she might believe that, given a chance, Y would take two coins in round (2) and terminate the game, because she believes that *he* would, against the evidence of round (1), just happen to believe in round (2) that were he to give her a second chance, she would in round (3) choose two coins and terminate the game. Or she might believe that he would choose two greedily, without thought of what would happen were he to choose across. But — recalling the condition ignore in these sentences — these reasons for that resolving belief of hers would make her ignorant, and not especially informed, if she was wrong about him, and would make him not especially reasonable if she was right about him. Again, she might reason in round (1) to the down-choice, because she believed that while Y would probably choose down in round (2), there is a chance he would not, though in that case she, in round (3), would return the favour, only for Y to choose down in round (4). But if she is resiliently rational, then she would in this opinion be mistaken about how she would behave in round (3), and so either she would be not especially reasonable, or not especially well-informed about herself

Rabinowicz offers ways such as these in which "the first player's going down and terminating that game may be rationalized" (Rabinowicz 1996, p. 10) to show that he is not committed, on pain otherwise of saying that she is not an expected utility maximizer, to my awful conditions for BI-resolutions. The rationalizations Rabinowicz sketches are sufficient to this point that he wants them to make. For this, it is not required that he produce reasons for that choice that are consistent with his players's being especially reasonable and informed. It is enough that, consistently with the weak rationality- and belief-conditions he has his players satisfy, the game might resolve for his players only thanks to their imperfections, for example, thanks to mistaken beliefs of some order or practical irrationality. That leaves the hard question — it was not Rabinowicz's question — what conditions, *consistent with X and Y being ideally rational and informed in this game*, entail that X will, maximizing, resolve the game for them in its BI-outcome?

It seems to us — to Broome, Rabinowicz, and me — that the game *must* resolve in this outcome, if X and Y are ideally rational and informed. We have relatively undemanding assumptions that we are satisfied entail this outcome for them. One wonders whether assumptions significantly different and less demanding than those of robust knowledge compounded forward of resilient rationality entail that, if X and Y are ideally rational and informed, X has a maximizing reason for moving to that outcome. *Perhaps* players who satisfy our relatively undemanding indicative conditions that are sufficient for a *weak* solution of the game, *must*, if they are to be ideally rational and informed, satisfy conditions like the demanding subjunctive ones that are sufficient for a *strong* solution of the game. That is what I think. I think, though I have no proof of this, that there are no significantly less demanding conditions sufficient for a strong solution of the game. That seems to be what some critics of backward induction think, who may say that *ideally* rational and informed players cannot be in and play One Coin or Two. And it is what some friends of this reasoning, including Rabinowicz (see below), at least sometimes suspect (fear?). We, in our weak solutions to the game, may have *proofs*, though Broome and Rabinowicz would not like to say so, sufficient to incline intellects to accept that ideally rational and well-informed players in this game would satisfy demanding subjunctive conditions not significantly different from, or more palatable than, knowledge compounded robustly forward of resilient rationality.

5.3 Perhaps, however, this *modus ponens* is properly a *modus tollens* to the effect that players in the game, though they can satisfy our weak indicative conditions of rationality and information, cannot be *ideally* rational and well-informed. Perhaps the *lesson* is that the theory of this game is properly only several theories for players of variously *bounded* rationality and information, with no theory at the centre for players of unbounded rationality and information. Perhaps *unboundedly* and *ideally* rational informed players cannot be in and play these games, and 'bounded' in 'players of variously *bounded* rationality and information' is otiose for BI-terminating games.

5.4 The decisive issues raised are, I think, meta-issues that go to the conduct of theory for games with BI-outcomes, and whether it is best served by a theory for ideal players at the centre with theories for variously bounded players around it. If it is, then, as the theory of games for ideal players is assembled, explaining what is wrong with conditions like the demanding subjunctive conditions that I have stated, will be difficult. Aumann writes: "Backward induction, the oldest idea in game theory, has maintained its centrality to this day" (Aumann 1995, p. 6). Traditional idealisations were assumed to enable such resolutions. They were *supposed* to enable such resolutions. A *general* theory of BI-resolutions *seems* to require conditions very like mine. Even a theory of *resolutions* of BI-*terminating* games seems to require such conditions. Rabinowicz has indicated an inclination to agree: "I have no proof that all this really is needed. Maybe weaker assumptions would suffice [in general for at least 'weak solutions'].... But...." (Rabinowicz 1999).[29] The condition of Knowledge of Resilient Rationality, Compounded Robustly Forward is, I think, "definitely consistent with the traditional idealisations in game theory" (Broome and Rabinowicz 2000). It is intelligible, and, by contributing to a central theory for unboundedly rational players, it is useful. So why not that condition, or a condition or conditions very like it? We are not slaves to tradition. We can, if we choose, reject parts of the tradition of the theory of games for ideal players. We can indeed reject the whole idea of such a theory, do only theories for variously 'bounded' agents, and give up on that 'oldest idea of game' that finds solutions in BI-outcomes. However, before anything so radical, we may recall that BI-resolutions can themselves be for good outcomes that players welcome, and perhaps review grounds for resistance to conditions such as mine that would enable these resolutions, even for, indeed especially for, *ideal* game players.[30]

Notes

1. This paper takes up an extensive form game. "Hyperrational Games," (Sobel 1994, ch. 14) is about normal form games played by 'hyperrational' players. It is shown that games these players are in can have only certain kinds of outcomes. 'Strong solutions', not called that, in such outcomes are explained for some games, thereby showing that these players can be in at least these games. It is shown that they cannot be in certain other games. Robert Stalnaker delivers a weak solution for the normal-form reduction or 'supergame' (Luce and Raiffa 1957, 51-3, and99) of an extensive finite iterated prisoners's dilemma (Stalnaker 1996, 157-60). (Sobel 1994, ch. 14, sec. 2.2) is relevant to the problem of upgrading this weak solution to a strong solution for hyperrational players. Stalnaker 'broadstrokes' an enrichment of his model theory in which he promises "[q]uestions about the relationship between normal and extensive forms of games... be made precise... and answered" (p. 158n20). The questions accommodated all concern, I suspect, only weak solutions.

2. Thanks to John Broome for the idea of this game.

3. "In recent years, Rosenthal's (1982) centipede game has become a touchstone of the theory of PI [perfect information] games." (Aumann 1996 [1998, p. 98]). There are two perfect information games whose trees are 'centipedes' in (Rosenthal 1981, p. 96). In each round of each game there are just two options, one of which terminates the game. Only the second of these games is a *BI*-terminating game. Neither is said to be a 'centipede'.

4. Our loving couple do not need to be up to backward induction to reach their best outcomes. W and Z could work to a good result using *forward dominance reasoning.* In each round, Across works out best no matter what moves are made in later rounds. A money pump in which saving forward reasoning is not an option, and it is *only* 'backward induction to the rescue', is discussed in (Sobel 2001).

5. "It is assumed that it is common knowledge that each player... would act rationally at each of his vertices... even when [he] knows that [it] will not be reached.... We called this condition *substantive* rationality." (Aumann 1998, p.97).

6. He recalls in 1998 that he assumes in 1995 only common knowledge *at the start of play* of that rationality (Aumann 1998, p. 98).

7. Rabinowicz worries the exact form of Aumann's conditions in relation to my resilient rationality and robust knowledge conditions (Rabinowicz 1996, p. 2n3).

8. One Coin or Two games and iterated prisoners's dilemmas of known finite lengths are finite extensive form games. One difference is that the former, but not the latter, are 'perfect information' games. In each of a sequence of prisoner's dilemmas, even if players's moves are sequenced, the player who moves second is not informed of the move that the other player has made. An important difference connected to that one is that 'tortuous labyrinths' of deliberation 'beckon' even players of resilient hyperrationality and robust-compounded-forward knowledge thereof in iterated prisoners's dilemmas, because they beckon in one-shot dilemmas notwithstanding that there are strongly dominant strategies in them (*cf.*, Sobel 1994, ch. 14, pp. 303-6 and 311-3). This problem for iterated prisoner's dilemmas is ignored in (Sobel 1994, ch. 16). When players of resilient hyperrationality and robust-compounded-forwarded knowledge thereof are in a BI-terminating game, the reasoning of the mover in round 1 to the conclusion that his terminating move would be maximizing is not similarly bothered. He can, by BI-reasoning, figure out what the mover in round 2 would do, were the game not terminated in round 1. Going back to the loving couple's game of Section 2.3, its BI-resolution is less problematic than its resolution by dominance arguments. The BI-resolution stays away from those labyrinths in which expectations needed for Bayesian maximizing deliberation cannot be settled. Dominance arguments of players in hyperrational games require that they decline to enter these labyrinths (Sobel 1994, ch. 14, pp. 306 and 311-3).

9. That only knowledge of rationality *compounded forward* is assumed gets around problems with assumptions of beliefs and knowledge of rationality 'throughout the game' (Sobel 1994, ch. 16, pp. 355-6). And that knowledge of rationality compounded forward *at every choice-node* is assumed makes inapplicable an objection Cristina Bicchieri makes to "theorists [who] have [as is customarily done] assumed... that mutual rationality... [is] common knowledge among players....and proceed[ed] to solve the game by backward-induction" (Bic-

chieri 1993, p. 195). "My objection to [their] argument is that it does not depict how players reason.... A player who wants to make an optimal choice needs to know the outcomes of alternative moves.... To assume that his action will have no effect on the opponent's future play means assuming, for example, that even if the opponent observes a behavior inconsistent with rationality being common knowledge, she will play 'as if' rationality were common knowledge. But this is hardly a rational behavior!" (*Ibid.*) My assumption entails that a player knows that even if the opponent were to observe behavior inconsistent with rationality *having been* common knowledge, he would know that it was *currently* common knowledge compounded robustly forward. He would know that she would not play merely 'as if' that were so. He would know that she would play with the knowledge that that was so. The behavior he expected of her would not be based on fiction and pretense. It would be quite rational.

10. Learn what? To expect, I have supposed, a string of *take-one* moves, *not* a string of *irrational* take-one moves. The moves supposed *could* all be irrational, but, assuming that what is rational is to maximize one's coins, they need not all be so. The moves, with one exception, could be *known* to be *maximizing.* Suppose the game is of length 8, and that it *will* run its full course in take-one moves. Consider now the initial string of two moves, and suppose that each is taken in the confidence that, if taken, the game will run its course in take-one moves. Each is maximizing, and neither is so only thanks to a mistaken confidence. Similarly for the next four moves. Also the eighth move, the last move, would be rational. Only the seventh move would not be maximizing. That said, I wonder *exactly why* their continuing to believe in their *resilient rationality* after many *take-one* moves — that is, after many *non-backward–induction* moves — 'would not be especially reasonable'. I will henceforth take the objection to be to a continued belief in 'resilient rationality at last' after an unbroken string of *irrational* take-one moves.

11. Bicchieri writes that "systematic mistakes would be at odds with rationality, since one would expect a rational player to learn from past actions and modify his behavior" (Bicchieri 1993, p. 135). Systematic mistakes, irrational choices, are at odds with a player's past rationality, but *not* with his having had the 'subjunctive-part' of resilient rationality, a natural basis for which could be a *disposition to learn* **from** *one's mistakes.*

12. Broome and Rabinowicz distinguish. They describe the assumption resilient rationality *merely* as 'dubious', as '*unrealistic*', I assume they mean. (Broome and Rabinowicz 1999, p. 238). That is, I am sure they would agree, consistent with its being "consistent with the traditional **idealisations** in game theory" (p. 239, emphasis added). When they describe the assumption of 'robust confidence' in this rationality as 'dubious', they explain how this confidence *might reasonably* be shaken by observations of irrational behaviour. (P. 239.) They seek to impugn the consistency of this assumption with traditional idealisations.

13. I suspect that theorists who are challenged by my conditions think that *ideal* players could learn of their own dispositions, as well as of those of other players, only from experience of their exercises, so that pending an occasion for its display they would be ignorant of new-found dispositions. "Ignorant?" I hear an ideal player protest. "*Moi?.*"

14. Philip Reny can *seem* to argue that 'common beliefs' in Bayesian rationality compounded forward are possible in very few games, and that they are not possible in Take It Or Leave It (Reny 1993, pp. 258 and 260-2). But he does not. He works with the idea that 'Bayesian rationality is common belief *upon* reaching a node' that makes relevant beliefs about ways in which this node could be *reached* (p. 263). His players are concerned with the quality of their past conduct in ways in which players with common knowledge *compounded forward* of their resilient rationality need not be (*cf.*, p. 271).

15. Would the resolution I propose be "stable... with respect to forward induction (i.e., with respect to deductions based on the opponent's past rational behaviour)" (Bicchieri 1993, p. 135)? Yes, trivially, since it says that One Coin or Two is terminated by X in the first round when there is not past behaviour of Y to consult.

16. The idea that one can base a solution on such conditions confined to rational and informed developments of a game, "originally proposed by John Broome, is investigated by [Richard] Hern" (Rabinowicz 1996, p. 15).

17. *Other somewhat similar arguments.* **Rabinowicz** develops two arguments, one assumes that beliefs *just-before-a-choice* are decisive for its rationality; the other assumes that beliefs decisive for the rationality of a choice are *at-its-time.* I run only one argument that makes the former assumption. Differences between my conditions, and his for the *at-choice* case, include that he requires that the player who is to move first is minimally rational in his choices, and says nothing about the other player's rationality in rounds, if reached, in which it is that player's move (Rabinowicz 1996, p. 4). He *requires*, for this point of view, that the player with the first move should have a stack of beliefs of many orders. He is to have a belief about the rationality and information of the next mover, if there is one, about the beliefs at the time of that mover's move of this next mover, if there is one, about the rationality and information of the *next* mover, if there is one, *and so on.* My conditions 'spin off' stacks of beliefs not unlike those that Rabinowicz assumes for both of his exercises, but the higher order beliefs play no role in my arguments. Rabinowicz's conditions for his *at-choice* exercise are non-subjunctive, as are my conditions for the before-choice exercise I conduct. For his *pre-choice* exercise, however, he uses *subjunctive* rationality- and belief-conditions. A difference between our main *arguments* is that mine in Section 4.4 is a 'weak' mathematical induction that takes possible rounds of a game in reverse order, starting with the last and 'counting down' toward the first. His (Rabinowicz 1996, pp. 7, 11, and 16) are 'strong' mathematical inductions on possible lengths of versions of the game he studies, starting with the shortest, and 'counting up' to longest ones.
Aumann uses in his argument, that "at the start of play there is common knowledge of *ex post* material rationality" (Aumann 1996).One gathers that he assumes that there is such common knowledge at every vertex that is reached. Material rationality "stipulates that i act rationally at those of his vertices that are actually reached," and "a player i [is] *ex post* rational at a vertex v if there is no strategy that he knows at v would yield him a higher payoff... than the strategy he is using" (1998, pp. 97-8). [The pre-choice/at-choice distinction is different from Aumann's *ex post*/*ex ante* rationality at v distinction: the former relates to what the agent "knows at v"; the latter to what he "knows already at the beginning of the game" (Aumann 1996). 'Knows at v' is indeterminate between 'knows just before his choice at v' and 'knows upon or just after his choice at v.' Aumann adopts the *ex ante* line in (Aumann 1995, p. 13), and the *ex post* line in (Aumann 1996).]
Broome and Rabinowicz say that "Aumann's remarkable proof" (Broome and Rabinowicz 1999, p. 241) is an "elegant abridgment" of theirs. Converting it to their terms and expressing it roughly, they make it a *reductio*, that I think turns on the implicit premise hat, if the game terminates in some round, that it does so is *demonstrable*, and known by the players in every round reached, including the round, if any, before the one in which it terminates. They write that "[f]rom the perspective of this [penultimate] round, the game will end at the next round"(p. 242). The idea of the assumption that I think they make, is that ideal *players* of the game are ideal *game-theorists* of *it* (*Cf.*, Sugden 1991, p. 765). They describe their own argument as a "simplified [less detailed and precise] version of one of the arguments [the at-choice one]... in Rabinowicz 1998" (1999, pp. 238n and 239n).

18. *x believes just before time t that p* if and only if there is a time earlier than t such that, for every time t', later than that time and earlier than t, x believes at t' that p.

19. '$\Box\!\!\rightarrow$' is here a connective for 'centered subjunctive conditionals' in the sense of (Lewis 1973). Such conditionals entail corresponding material conditionals.

20. In a proof adapted to the 'at-choice' point of view, instead of Prescience, 'Introspection' would be cited. A sentence explanatory of Introspection comes from that for Prescience by replacing 'the player whose turn it is to move in r knows, of the choice he or she will make in r, just before he or she makes it, that he or she is going to make it' by 'the player whose turn it is to move in r knows, of the choice he or she makes r, when he or she is making, that he or she is making it'.

21. If Prescience is a part of ideal game-theoretic rationality, then games are possible for ideally rational and informed agents only if, were they in them they could have, before they chose what to do, beliefs that were settled concerning consequences of what they were about to choose to do, then there are games that are not possible for such agents. For example, the one-person game or decision problem 'Appointment in Samarra' is not then possible for an ideal agent. Such a person could not settle, before he chose to flee to Samarra or to stay

in Baghdad, his opinion concerning his choice, for that would settle his opinion concerning where Death awaited him, and make him choose to go to the other place. In (Sobel 1994, ch. 14) 'hyperrational agents' are cast as ideally rational and informed gameplayers, and it is maintained that there are "games in which such agents could not do anything at all" (p. 315),games that are not possible for such agents. It is, however, said there that the Appointment in Samarra is *not* such a game (p. 301). That is because I failed to make hyperrational agents 'prescient'. Had I done so I would have seen that a game is possible for them if and only if it is possible for what is there termed 'superrational players' (p. 329n7) who are like hyperrational players save that they do acts only if they are not only 'pre-choice' maximizing (which is all that I required for hyperrationals), but also 'ratifiable.' "But can agents who would *maximize expected value* be prescient? Aren't you forgetting Jeffrey's problem (Jeffrey 1990, p. 85) with probabilities of acts? If a Bayesian agent knew before he made his choice what choice he was going to make, say A rather than B, then his *grounds* for that choice would be vitiated — *prob*(B) would be 0, and thus *des*(B) not be defined." That is a problem for Bayesian agents who would maximize '*evidential* expected value' spelled out in terms of *standard conditional probabilities* for worlds on acts, which probabilities are not defined for 0-probability acts. There is no such problem for my hyperrational agents. They would maximize '*causal* expected value' defined in terms of something like probabilities of act-world causal conditionals, e.g., *prob*(B □→ w). Acts of 0 probability are not special relative to these probabilities: that *prob*(B) = 0, is consistent with *prob*(B □→ w) being 0, 1, $\frac{1}{2}$, or whatever, including what it was when the agent began to deliberate, and did not know what he was going to choose and do. Causal decision theory does not have Jeffrey's reason for adopting "the strict point of view, in which the agent can only try to make... true" what he chooses, and should never, before an act, assign a probability of 1 to it (Jeffrey 1990, p. 85).

22. Theorem* is not — **Necessary Main Theorem**: □(r)[IdealGamePlayers ⊃ ([Reached(r) & ReachedWithoutPrIrr∨InfDef(r) & PlayersIdealR&I(r)] ⊃ [AtLstTwo(r) ⊃ ChTwo&Term(r)])]. Necessary Main Theorem, since not a universal generalization, is not a candidate for proof by mathematical induction. However, since 'r' are 'possible rounds' (as defined in Section 2.1) of the game, and these are the same in every possible world, Theorem* is equivalent to Necessary Main Theorem: The mathematical induction of the former thus establishes as well the latter.

23. This extension is detailed in an appendix to *Logic and Theism*, Chapter 3, which is linked the web page `http://www.scar.utoronto.ca/~sobel/`

24. In a fully spelled out derivation, the necessity sentence ConMinR would be entered on a line. Then, using a rule of Necessity, the nonmodalized core of ConMinR would be inferred. From this and (5) a universal generalization would be inferred by *modus ponens*. The first quantifier would be instantiated to W, and the second to (k-1). Then (7) and (10) would be conjoined, and another inference by *modus ponens* made to (11). Similar compressions are made when other conditions are used.

25. This extension is detailed in the appendix to Chapter 3 of *Logic and Theism* cited in note 23 above.

26. For 'r' ranges over 'possible rounds' of the game, and these, as said in note 22, are the same in every possible world

27. Robert Sugden and Cristina Bicchieri see that their purely indicative solutions to BI-terminating centipedes leave this question: (Sugden 1991, p. 772) and (Bicchieri 1993, p. 134).

28. Similarly, while my conditions entail that she believes just before her choice not only that it is the choice she will make, but that making it will be 'minimally rational', they do not entail that she believes this 'when' *this* could be a part of her reason for making it, if *it* could ever be, for her, *ideally rational* agent that she is, a part of her reason for her making it.

29. K. Basu should agree, and say that he has demonstrated the point. "Standard solution concepts, like subgame perfection, implicitly require that players turn a blind eye to another player's 'irrationality' even if this has been revealed by virtue of having reached a node that could not have been reached had this player behaved rationally.... The aim of this paper is... a formal impossibility theorem. The theorem shows that a definition of rational

behaviour which is applicable to all extensive games and which does not suffer from [that] problem of unreached nodes... does not exist. This is because such a definition would run into difficulty with... the [repeated] Prisoner's Dilemma and... games described by Rosenthal (1981) and Reny 1986)." (Basu 1990, p. 33.) According to that theorem, there does not exist a solution concept for all extensive form games that satisfies a weak backward induction B, a rationality condition U*, and another rationality condition S (p. 39). Of axiom U that "asserts... that a player who has once been observed behaving irrationally must be, then onwards, treated as completely unpredictable," Basu says: "I feel this is a better assumption than the traditional one which would ignore the revealed irrationality and continue to treat the play as completely rational." (p. 41, added) Untraditional U* replaces 'completely unpredictable' by "less predictable than a rational player" (p. 38).

I have argued that the *traditional* assumption, tuned up a bit, is better than U* for a theory for unboundedly rational players in extensive form games of perfect information. I have argued that it is *right* for this theory. Basu says that untraditional U axiom, or U*, is better, but he does not say *for what* it is better, or therefore indicate whyand how it is better. Perhaps his view is that U* is better for a theory of unboundedly rational players in extensive form games of perfect information, notwithstanding that this means that its main theorem is that no games of that kind that have come up for discussion have solutions for such players. He says that he is "inclined to go along with" the position that "treat[s]... U* as reasonable, and reject[s] the view that every game has a solution" (p.43). But why is he so inclined? If that is his view, it could be that he does not give reasons why U* is better than traditional assumptions for unboundedly rational players, because he thinks this is obvious once the 'blind-eye problem' of traditional assumptions with unreached nodes is articulated. It is not obvious to me.

30. A draft of this paper was prepared during my tenure as research fellow at SCASSS (Swedish Collegium for Advanced Study in the Social Sciences) for presentation in a workshop conducted on February 21 and 22, 1998 at SCASSS on backward induction. Participants included John Broome, Martin Dufwenberg, Wlodek Rabinowicz, Rysiek Sliwinski, and Arnis Vilks. I am grateful to SCASSS for its wonderfully stimulating atmosphere and research-facilitating management, to the other participants in this workshop for their instructive presentations and comments, to comments of Robert Sugden, and as always to comments and criticisms of Willa Fowler Freeman Sobel.

References

Aumann R. (1995). "Backward Induction and Common Knowledge of Rationality," *Games and Economic Behavior 8*:6–19.

— (1996). "Deriving Backward Induction in the Centipede Game without Assuming Rationality at Unreached Vertices," for presentation at Workshop on Game Theory, August 15, 1996. (Dated July 2, 1996, this paper agrees with "On the Centipede Game," *Games and Economic Behaviour 23*:97–105, 1998, received July 10, 1996.)

Basu K. (1990). "On the Non-Existence of a Rationality Definition for Extensive Games," *International Journal of Game Theory 19*:33–44.

Bicchieri C. (1993). *Rationality and Coordination*, Cambridge: Cambridge University Press.

Broome J. and Rabinowicz W. (1999). "Backwards Induction in the Centipede Game," *Analysis 59*:237–242.

Ginet C. (1962). "Can the Will Be Caused?" *Philosophical Review 71*:49–55.

Jeffrey R. (1990). *The Logic of Decision: Second Edition* (paperback edition). Chicago: University of Chicago Press.

Lewis D. (1970). *Counterfactuals.* Cambridge, MA: Harvard University Press.

Rabinowicz W. (1995). "To Have One's Cake and Eat It, Too: Sequential Choice and Expected-utility Violations," *Journal of Philosophy 92*:586–620.

— (1996). "Grappling with the Centipede: Defence of Backward Induction for BI-terminating Games," *Lund Philosophy Reports.* (Revised August 20, 1997.)

— (1999). "Some Remarks on Thorsten Clausing's *The Logical Modelling of Reasoning Processes in Games* (manuscript).

Reny P. (1986). *Rationality, Common Knowledge and the Theory of Games*, Ph.D. Dissertation, Princeton University.

— (1993). *Journal of Economic Theory 59*:257–274.

Rosenthal R. (1981). "Games of Perfect Information, Predatory Pricing and Chain-Store Paradox," *Journal of Economic Theory 25*:92–100.

Sobel J.H. (1970). "Utilitarianisms: Simple and General," in *Inquiry 13*:394–449.

— (1994). Chapter 11 "Maximization, Stability of Decision, and Actions in Accordance with Reason," Chapter 14, "Hyperrational Games," and Chapter 16, "Backward Induction Arguments: A Paradox Regained," in *Taking Chances: Essays on Rational Choice*, Cambridge: Cambridge University Press 1994 (Chapter 11 revised from *Philosophy of Science* 1990, and Chapter 16 revised from *Philosophy of Science* 1993 — a working revised version of Chapter 14 is linked to my home page, `http://www.scar.utoronto.ca/~sobel/`).

— (2001). "Money Pumps," *Philosophy of Science*, 242–257.

Stalnaker R. (1996). "Knowledge, Belief and Counterfactual Reasoning in Games," *Economics and Philosophy 12*, 133–63.

Sugden R. (1991). "Rational Choice: A Survey of Contributions from Economics and Philosophy," *The Economic Journal 101*: 751-85.

V

REASONING AND COGNITION IN LOGIC AND ARTIFICIAL INTELLIGENCE

Chapter 20

ON THE USEFULNESS OF PARACONSISTENT LOGIC

Newton C.A. da Costa,[1] Jean-Yves Béziau,[2] and Otávio Bueno[3]

[1] *University of São Paulo*

[2] *Institut de Logique, Université de Neuchâtel*

[3] *University of South Carolina*

Abstract In this paper, we examine some intuitive motivations to develop a paraconsistent logic. These motivations are formally developed using semantic ideas, and we employ, in particular, bivaluations and truth-tables to characterise this logic. After discussing these ideas, we examine some applications of paraconsistent logic to various domains. With these motivations and applications in hand, the usefulness of paraconsistent logic becomes hard to deny.

> *If geometrical space were a framework imposed on each of our representations considered individually, it would be impossible to represent to ourselves an image without this framework, and we should be quite unable to change our geometry. But this is not the case; geometry is only the summary of the laws by which these images succeed each other.*
>
> —Henri Poincaré [1905], p. 64.

D. Vanderveken (ed.), Logic, Thought & Action, 465–478.

1. Introduction

All of us, at some point, have heard questions about the usefulness of some branch of knowledge. What use is mathematics? Or topology? Or, for that matter, what use is logic? Of course, depending on the context in which such questions appear (for instance, a mathematician trying to understand a bit more of his or her own field, or a student upset with his or her final grades in mathematics), the particular features of the answer will change. What may not change, in a sense, is the *nature* of the answer. In most cases, it will indicate certain traits of the 'pragmatics' of the field under consideration, spelling out some of the connections between the theories formulated in such a field and their users, as well as the targets and constraints of the latter. In the course of such an investigation, some of the *applications* of such theories (either to their standard domain or to others) may be discussed and presented as reasons for their usefulness. Put in very general terms, these reasons can be understood in terms of the problem-solving resources disclosed by the theories (including their explicative power, the conceptual systematisation they supply, and the tools for the representation and analysis of the relevant phenomena).

To a certain extent, the same holds for logic. Taken in a very strict sense, *applied* logic is concerned (among other issues) with the study of structures that can be employed to understand the formal features of our reasoning processes.[1] At this level, just as with empirical theories, applied logic has its particular domain, being appropriate for representing certain kinds of phenomena, and hopeless for the examination of others (for instance, classical logic is by no means adequate for a constructive study of constructive mathematical thought, but can be seen as an idealised perspective on the representation of certain inferences usually found in classical mathematics). To the extent that the structures employed in a domain are appropriate to model the relevant features of it (thus 'saving the phenomena', as it were), we can claim that a particular applied logic has 'explanatory power'; it indicates, after all, how such 'phenomena' can be understood in terms of the structures supplied by such a logic. Moreover, similarly to empirical theories, applied logic also offers a conceptual systematisation of inferences that are allowed in a certain domain, in particular spelling out the constraints imposed by them. Consider, for instance, the differences between constructive math-

[1]For a development of this theme, with special emphasis on paraconsistency, see da Costa and Bueno [2001], and da Costa and Bueno [1996].

ematics (with the restrictions it brings to 'classical' mathematics)[2] and paraconsistent mathematics (with the extensions and new structures it brings to 'classical' mathematics).[3] Of course, the differences in the conceptual systematisation presented are due, in good part, to the different tools that each applied logic under consideration supply.

In a sense, the very first step that would subsequently lead to such differences was taken by changing basic features of classical logic. Each logic, just as each geometry in Poincaré's view (see the epigraph above), supplies a possible perspective for systematising our ways of representing certain phenomena ('images' in the case of geometry, and inferences in the case of applied logic). And if this is so, new perspectives can be offered by changing the logic (or, for that matter, the geometry). This straightforward fact already suggests a hint of the usefulness of non-classical logics.

In the present note, we wish to consider an instance of this general question, namely: what use is *paraconsistent logic*? In order to do so, we will first present such a logic from a perspective that can be easily understandable even by those who have little knowledge of the technical aspects of logic, namely, the semantic point of view. We will then consider, in connection with the preceding discussion, some straightforward *applications* of it, and this will convey at least a partial answer to our question. Finally, we shall briefly discuss some philosophical issues generated by such applications.

2. Remarks on language

We consider the usual language of propositional logic P with the connectives $\neg$, $\wedge$, $\vee$, $\rightarrow$. Before we construct a semantics for this language, these four connectives are nothing but: a unary connective (the first one) and three binary connectives (the remaining ones). There is, *a priori*, no justification to call them negation, conjunction, disjunction and implication.

In paraconsistent logic, we will denote by the symbol $\neg$, and call it negation, a connective that is not the same as classical negation. There are those who criticise such an abuse of language. Let us note, however, that it is difficult to claim that there is only one negation, let us say, classical negation (which would exactly model the negation of natural language or mathematics). In the literature, the word negation and the

[2]See, for instance, Bishop [1967], Heyting [1971], and Dummett [1977].

[3]Cf., for example, da Costa [1989], da Costa [2000], Mortensen [1995], and, for a discussion of the latter, da Costa and Bueno [1997].

corresponding symbols have long been used to denote different concepts, such as classical negation, intuitionistic negation, Johansson's minimal negation, Curry's negation etc.

Of course, a unitary operator must have some basic properties to be called a 'negation'. For instance, no one will call the necessity operator a negation. Nevertheless, until now, no common agreement on what the basic properties are that a unitary operator should obey to be called a 'negation' has been achieved. We will not claim that the paraconsistent negation presented here should absolutely be called as such. But we will try to convince the reader that it has enough interesting properties to deserve this name.

3. Remarks on 0-1 semantics

As is known, it is possible to construct a wide range of logics taking as basic notions two truth-values, the false and the true, designated for convenience by 0 and 1 (see da Costa and Béziau [1994]). Even the so-called 'many-valued' logics can be treated in this way. For instance, Suszko has given a 0-1 semantics for Lukasiewicz three-valued logic (see Suszko [1975]). This apparent paradox is solved by the distinction between truth-functional semantics and non truth-functional semantics. Suszko's 0-1 semantics is not truth-functional, neither will be the 0-1 semantics that we present now.[4]

Given the standard propositional language P, a 0-1 semantics for P is a set B of functions from P to $\{0, 1\}$, called bivaluations. A 0-1 semantics induces a logic in the following way: given a set B of bivaluations, we say that an object a of P (called a *formula*) is a *consequence* of a set T of objects of P (called a *theory*) iff for every bivaluation $\beta \in$ B, if β gives the value 1 to every element of T, it also gives the value 1 to the formula a.

In other words, a 0-1 semantics defines a binary relation on the Cartesian product of the power set of P and P, that we shall denote by $\models$. And we will write T $\models a$ to say that $\langle T, a\rangle \in \models$, i.e. that a is a consequence of T. In our view, a logic consists basically in presenting a set of formulas and a semantic consequence relation for that set.

It is easy to note that if a set of bivaluations B1 contains a set of bivaluations B2, the corresponding logics L1 and L2 are 'inversely proportional', i.e. the consequence relation generated by B2 contains the consequence relation generated by B1. This has a direct consequence

[4]For more details on this subject, see da Costa, Béziau and Bueno [1996]; for a related discussion of the concept of semantics, see da Costa, Bueno and Béziau [1995].

that we will use below. Let BC be the set of classical bivaluations. Then any set of bivaluations B that contains BC will generate a logic that is included in classical propositional logic LC and in which there are theories that are non-trivial. (A theory T is called *non-trivial* if there is at least one formula of the language that is not a consequence of T.)

4. Definition of the set of paraconsistent bivaluations[6]

We will consider the following set BP of bivaluations. A function β from P to {0, 1} is in BP iff it obeys the following conditions:

[C] $\beta[a \wedge b] = 1$ iff $\beta[a] = 1$ and $\beta[b] = 1$;
[D] $\beta[a \vee b] = 0$ iff $\beta[a] = 0$ and $\beta[b] = 0$;
[I] $\beta[a \rightarrow b] = 0$ iff $\beta[a] = 1$ and $\beta[b] = 0$;
[EM] if $\beta[a] = 0$, then $\beta[\neg a] = 1$;
[SN] if $\beta[a \wedge \neg a] = 1$, then $\beta[\neg(a \wedge \neg a)] = 0$;
[PN/N] if $\beta[a] = 0$, then $\beta[\neg\neg a] = 0$;
[PN/I] if $\beta[a \rightarrow b] = 1$ and $\beta[a] = 0$ or $\beta[\neg a] = 0$ or $\beta[b] = 0$ or $\beta[\neg b] = 0$, then $\beta[\neg(a \rightarrow b)] = 0$;

[PN/C], [PN/D] are conditions similar to [PN/I], when the formula is a conjunction or a disjunction.

For simplicity, we will call here LP the logic induced by BP. This logic has also been called C_1^+ elsewhere, and is an improvement, due to Béziau (see Béziau [1990]), of the logic C_1 of da Costa (see da Costa [1963]).

It is easy to see that BC is included in BP. Thus LP is included in LC. Note that generally BC is represented in terms of attributions of truth-values to atomic formulas. This can be done because, in classical logic, the set of bivaluations is freely generated by the set of bivaluations restricted to atomic formulas. But this is not the case with BP. (This property is connected with truth-functionality.)

The conditions for conjunction, disjunction and implication *mutatis mutandis* are the standard ones. The condition [EM] can be interpreted as a semantic version of the principle of the excluded middle.

What is the intuitive explanation of the other conditions for paraconsistent negation? The idea is as follows. We will say that a formula obeys the principle of contradiction for a given bivaluation iff it cannot have the value 1 simultaneously with its negation. That is to say, as in the classical case, it is true iff its negation is false.

[6]For a different presentation of paraconsistent logic, see da Costa, Béziau and Bueno [1995***a***]. A historical perspective on paraconsistent logic can be found in Arruda [1980], Arruda [1989], D'Ottaviano [1990], and da Costa, Béziau and Bueno [1995*b*].

Now, the condition [SN], interpreted in this way, states that for any formula a, $a \wedge \neg a$ obeys the principle of contradiction. This condition allows us to define a compound connective, $\neg^* a = \neg a \wedge \neg(a \wedge \neg a)$, which has all the properties of classical negation; in particular, a is true iff $\neg^* a$ is false. Consequently, this allows us to 'translate' classical logic into the paraconsistent logic LP. The translation consists in replacing the 'weak' negation $\neg$ by the strong negation $\neg^*$. We therefore have a situation that is similar to the case of intuitionistic logic. In one sense, LP is weaker than LC; in another sense, LP is stronger than LC.

The conditions [PN] state that a sufficient condition for a formula to obey the principle of contradiction is that one of its direct subformulas obeys this principle. This intuitive *preservation principle* will give to paraconsistent negations interesting properties, such as parts of De Morgan's laws.

5. Truth-tables

It is possible to adapt the classical method of truth-tables for the case of LP. The basic difference is that, in the case of LP, we must introduce in the table of a formula a not only its subformulas, but also some negations of its proper subformulas. To simplify, we will put together its subformulas and all the negations of its proper subformulas; this set will be called the *sphere* of a. Then a table for a is a set of functions from its sphere to $\{0, 1\}$, such that: (a) every such a function is the restriction of an element of BP to this sphere, and (b) every restriction of BP to this sphere appears. It is not difficult to construct these kinds of tables following the conditions that define BP (for details, see da Costa and Alves [1977]).

Example 1. The first example of a truth-table shows that the formula

$$((a \wedge \neg a) \wedge (a \rightarrow b)) \rightarrow (\neg a \rightarrow \neg b) \tag{20.1}$$

is not a tautology of LP.

a	$\neg a$	b	$\neg b$	$a \wedge \neg a$	$a \rightarrow b$	$(a \wedge \neg a) \wedge (a \rightarrow b)$	$\neg a \rightarrow \neg b$	(1.1)
0	1	0	1	0	1	0	1	1
0	1	1	0	0	1	0	0	1
0	1	1	1	0	1	0	1	1
1	0	0	1	0	0	0	1	1
1	0	1	0	0	1	0	1	1
1	0	1	1	0	1	0	1	1
1	1	0	1	1	0	0	1	1
1	1	1	0	1	1	1	0	0
1	1	1	1	1	1	1	1	1

Example 2. The following table shows that

$$\neg(a \wedge \neg b) \rightarrow (a \rightarrow b) \tag{20.2}$$

is a tautology of LP.

a	$\neg a$	b	$\neg b$	$a \wedge \neg b$	$\neg(a \wedge \neg b)$	$(a \rightarrow b)$	(1.2)
0	1	0	1	0	1	1	1
0	1	1	0	0	1	1	1
0	1	1	1	0	1	1	1
1	0	0	1	1	0	0	1
1	0	1	0	0	1	1	1
1	0	1	1	1	0	1	1
1	1	0	1	1	0	0	1
1	1	1	0	0	1	1	1
1	1	1	1	1	0	1	1
1	1	1	1	1	1	1	1

6. A change of paradigm

The main feature of paraconsistent logic is that, as opposed to classical logic, *inconsistency* and *triviality* cease to coincide. In LP, there are some inconsistent *theories* (theories in which a formula and its negation are both consequences) that are not trivial (not every formula is a consequence). Such theories are called *paraconsistent theories.*

For example, as the method of tables shows, the atomic formula b is not a consequence of the inconsistent theory constituted by the atomic formula a and its negation $\neg a$. Such a theory, therefore, is not trivial.

It is clear that the concept of *triviality* is more fundamental than the one of *inconsistency.* Moreover, it is more abstract in the sense that its definition does not depend upon the particular connectives (in particular, the negation).

7. Inconsistency and reasoning

In everyday life, it is quite common for one to face contradictions. Such contradictions may not be real contradictions, whatever this means, but in several cases they cannot be trivially eliminated and must be dealt with. In the mechanical treatment of information, contradictions also often appear. In both cases, classical logic, because it merges inconsistency with triviality, is a useless tool. Let us see now how paraconsistent logic can be useful when classical logic fails.

Imagine that we are to construct an expert system. In order to do so, we start collecting the opinion of several hundreds of experts in a particular subject. The information we get comes from reliable sources, and there is no way to tell 'good' information from 'bad' one. After interviewing all these experts, there is no way to avoid the incompatibilities found, for they in fact express opposite opinions.[7] Among such bits of information, let us suppose that a group of experts, called X_1, asserts that:

The price of chocolate will raise

and that a second group of experts, X_2, states that:

The price of chocolate will not raise.

We are therefore facing a contradiction.

Firstly, let us note that, using paraconsistent logic LP, as opposed to the classical case, we cannot derive from this contradiction any statement whatsoever. For example, we cannot derive from this contradiction the following claim (which, in particular, is not a classical tautology):

If someone eats lots of chocolate, he or she will grow enormously fat.

Secondly, in the presence of this contradiction, all the bits of reasoning that are not valid in classical logic are still not valid in LP. For instance, from such a contradiction and the following statement on which both X_1 and X_2 agree

If the price of chocolate raises, people will buy less chocolate

we cannot infer that

If the price of chocolate does not raise, people will not buy less chocolate

(see the first truth-table above).

Now, let us see a positive reasoning that we can perform in LP. Both experts X_1 and X_2 agree that

It is not the case that *the price of chocolate will raise* and *the price of chocolate cookies will not raise.*

As the second truth-table shows, it is implied by this that

If the price of chocolate raises, the price of chocolate cookies will raise.

8. A new perspective: paraconsistency

As these examples show, despite the inconsistency, paraconsistent logic allows us to draw interesting conclusions in a context where, were we to cling exclusively to the classical logic paradigm, we would get stuck, inevitably deriving anything! Thus this supplies part of our answer to the question about the usefulness of paraconsistency: it opens

[7] In certain cases, due to the huge amount of data to be taken into account, we may even not notice the existence of inconsistencies.

up an altogether different perspective to examine issues in which inconsistencies are fundamentally involved.

In a sense, faced with a contradiction, the classical paradigm will not offer any alternative but, in order to avoid trivialisation, that of rejecting (some of) the premises in terms of which the contradiction was reached. Unfortunately, this alternative may not always be open to us, since the relevant premises may in some way be entangled in our conceptual system, having such important connections with other statements of the system, that their rejection will lead to dramatic conceptual losses (see da Costa and French [1989], p. 441). And even if this were not the case, in contrast with the classical paradigm, with the employment of the tools supplied by paraconsistent logic, it is possible to take inconsistencies at face value, exploring thus the consequences that can be drawn from the system that includes them (as is clear from the examples above; see additional examples below).

Nonetheless, one can claim that, to a certain extent, there is also a second alternative within the classical paradigm. If the rejection of certain premises in some cases cannot be recommended, people working within the classical paradigm can perhaps reject the validity of (some of) the inference rules used in order to obtain the contradiction under consideration, in such a way that the latter cannot be drawn any longer. The trouble with such a move, for the classicist, is that it means changing the underlying logic (just as Poincaré's remark quoted above suggested with regard to geometry), and moving to another paradigm. In order to deal with this kind of inconsistency problem, this is exactly the suggestion we present (although, within paraconsistent logic, the change in the inference rules is *not* meant to avoid the derivation of certain contradictions, but to formulate a system in which such contradictions do not lead to a trivial system).

The paraconsistent paradigm also advances new perspectives here. In fact, in several cases, and in stark contrast with the classical paradigm, given an inconsistency, we do not need to elaborate more or less *ad hoc* strategies to reject it: we can simply accept the premises *and* the inferences that led to the contradiction in question (provided such inferences are among the ones to be found in a paraconsistent system, and that we have changed our logic to a paraconsistent one). In such a perspective, we claim, we can learn more, having a truly pluralist account of knowledge.

For someone who is classically minded, the last assertion might seem bizarre. How can a proponent of the paraconsistent view truly learn anything? After all, in a sense, part of our learning process depends upon our way of changing our beliefs, given contrary evidence. If this proponent, faced with such contrary evidence, simply adds it to the stock of his or her beliefs, claiming that 'No problem, it won't lead to trivialisation', how could he or she ever change his or her mind? How could he or she ever come to the 'saturation point', from which everything will follow?

Such questions seem to be still more pressing given the classical accounts of belief change that apparently underlie them. Put in very abstract and rough terms, such accounts will run like this. We can just keep adding any beliefs we wish to our belief system, provided we meet some consistency-preserving rule. If we fail to do that, and introduce inconsistencies into our system, it will simply be trivialised, becoming useless for any systematisation and cognitive purposes.

However, if consistency is not a necessary constraint, as is the case in the domain of paraconsistency, a different perspective on the nature of belief change will emerge. Instead of consistency preservation, the ultimate constraint now will shift to the avoidance of trivialisation. After all, within the paraconsistency paradigm, we can deal with inconsistencies, whereas triviality clearly represents cognitive bankruptcy. Indeed, while an inconsistent theory may have several interesting features, at least from a heuristic point of view (Bohr's atomic model and naive set theory are obvious examples), and we have learnt a lot from them (trying, although not exclusively, to devise consistent successors to them for instance), trivial theories are useless for any cognitive purposes. So, when paraconsistent logic, as against the classical one, clearly demarcates inconsistency from triviality, we can trace this demarcation to an epistemic distinction: between theories that, despite being inconsistent, can lead to (even inconsistent) fruitful successors, and those that are altogether hopeless for explanation, cognitive systematisation etc.

To a certain extent, part of the usefulness of paraconsistency derives from such an epistemic distinction. Inconsistent theories may be rich, interesting, full of fruitful consequences, whereas trivial theories are simply useless. With paraconsistency, the whole new domain of the inconsistent, left in complete darkness by the classical approach, is thus open to investigation. And this domain has in fact received detailed consideration since the inception of paraconsistent logic.

Let us conclude this note briefly mentioning three applications of paraconsistent logic that show this trend. They are respectively concerned with three distinct fields: mathematics, artificial intelligence, and philosophy.

(1) Cantor's naive set theory is characterised chiefly by two basic principles: the postulate of extensionality (if two sets have the same elements, then they are equal) and the postulate of comprehension (every

property determines a set). As is well known, this postulate, in the standard language of set theory, is the following scheme of formulas:

$$\exists y \forall x (x \in y \leftrightarrow \varphi(x))$$

If we replace the formula $\varphi(x)$, in the separation postulate, by $x \notin x$, Russell's paradox is immediately derived. In other words, this postulate is inconsistent. Therefore, if it is added to first-order logic, viewed as the logic of set theoretic language, we obtain a trivial theory.

Classical set theories are then constructed by imposing restrictions on the separation postulate, so that the paradoxes can be avoided. (Further axioms are then introduced in order that the resulting theory does not become too weak.) For instance, in Zermelo-Fraenkel set theory (ZF), comprehension is formulated as follows:

$$\exists y \forall x (x \in y \leftrightarrow (\varphi(x) \wedge x \in z)),$$

where the variables are subject to obvious conditions. Hence, in ZF, $\varphi(x)$ determines the subset of the elements of the set z that satisfy the formula $\varphi(x)$.

Using certain paraconsistent logics, it is possible to construct set theories in which the postulate of separation is subject either to restrictions weaker than those of the classical set theories or subject to no restrictions at all. Moreover, it is also possible to study, without trivialisation, the properties of 'inconsistent' objects, such as the Russell set, $R = \{x : x \notin x\}$. (Further details can be found, for instance, in da Costa [1986], da Costa and Bueno [2001], and da Costa, Béziau and Bueno [1998].)

(2) In certain domains, such as in the construction of expert systems, the presence of inconsistencies is almost unavoidable. In order to construct these systems, enormous knowledge bases are elaborated, aggregating the opinion of several specialists in a particular field (let us say, medicine). As one can immediately imagine, such bases are inconsistent, and one of the problems consists in how to drawn inferences from them. Some paraconsistent logics have been especially devised to deal with this problem (see, for example, da Costa and Subrahmanian [1989]).

(3) Surprisingly or not, inconsistent beliefs are frequently found, both in science and in everyday life. However, from such inconsistent belief sets, it is simply not the case that any statement whatsoever is derived. (So, apparently at least, we are not here concerned with 'trivial' systems.) In order to propose a formal framework to model some aspects of this phenomenon, certain paraconsistent doxastic logics have been constructed (see da Costa and French [1989]). In particular, the problem of

self-deception and related problems that involve the holding of contradictory beliefs, can then receive a distinct approach (see da Costa and French [1990], and da Costa and French [1988]). Moreover, the relations between rationality and consistency can also be re-evaluated. After all, one of the main arguments to the effect that consistency is a minimum condition for rationality rests on the assumption that inconsistency leads to triviality; precisely the assumption challenged by the paraconsistent approach. (For details, see French [1990], da Costa and French [1995], and da Costa, Bueno and French [1998].)

With the considerations advanced in this note, we hope to have indicated some aspects of the usefulness of paraconsistency. If we have not convinced you, gentle reader, of this point, we expect to have conveyed at least an idea of why paraconsistent logic is far from being useless.[8]

References

Arruda A.I. (1980). "A Survey of Paraconsistent Logic", in Arruda, Chuaqui, and da Costa (eds.), pp. 1–41.

Arruda A.I. (1989). "Aspects of the Historical Development of Paraconsistent Logic", in Priest, Routley and Norman (eds.), pp. 99–130.

Arruda A., Chuaqui R., and da Costa N.C.A. (eds.) (1980). *Mathematical Logic in Latin America.* Amsterdam: North-Holland.

Béziau J.-Y. (1990). "Logiques construites suivant les méthodes de da Costa", *Logique et Analyse 131–132*, pp. 259–272.

Bishop E. (1967). *Foundations of Constructive Analysis.* New York: McGraw-Hill.

da Costa N.C.A. (1963). "Calculs propositionnels pour les systèmes formels inconsistants", *Comptes-rendus de l'Académie des Sciences de Paris 257*, pp. 3790–3793.

— (1986). "On Paraconsistent Set Theory", *Logique et Analyse 115*, pp. 361–371.

da Costa N.C.A. (1989). "Mathematics and Paraconsistency (in Portuguese)", *Monografias da Sociedade Paranaense de Matemática 7.* Curitiba: UFPR.

— (2000). "Paraconsistent Mathematics", in D. Batens, C. Mortensen, G. Priest and J.-P. Van Bendegem (eds.), *Frontiers of Paraconsistency.* Dordrecht: Kluwer Academic Publishers.

da Costa N.C.A. and Alves E.H. (1977). "A Semantic Analysis of the Calculi Cn", *Notre Dame Journal of Formal Logic 16*, pp. 621–630.

[8]We wish to thank Steven French for his comments on an earlier version of this paper.

da Costa N.C.A. and Béziau, J.-Y. (1994). "Théorie de la valuation", *Logique et Analyse 146*, pp. 95–117.

da Costa, N.C.A. Béziau J.-Y. and Bueno O. (1995*a*). "Aspects of Paraconsistent Logic", *Bulletin of the Interest Group in Pure and Applied Logics 3*, pp. 597–614.

— (1995*b*). "Paraconsistent Logic in a Historical Perspective", *Logique et Analyse 150-151-152*, pp. 111–125.

— (1996). "Malinowski and Suszko on Many-Valuedness: On the Reduction of Many-Valuedness to Two-Valuedness", *Modern Logic 6*, pp. 272–299.

— (1998). *Elements of Paraconsistent Set Theory* (in Portuguese). Campinas: Coleção CLE.

da Costa N.C.A. and Bueno O. (1996). "Consistency, Paraconsistency and Truth (Logic, the Whole Logic and Nothing but the Logic)", *Ideas y Valores 100*, pp. 48–60.

— (1997). "Review of Chris Mortensen (1995)", *Journal of Symbolic Logic 62*, pp. 683–685.

— (2001). "Paraconsistency: Towards a Tentative Interpretation", *Theoria 16*, pp. 119–145.

da Costa N.C.A., Bueno O. and Béziau J.-Y. (1995). "What is Semantics? A Brief Note on a Huge Question", *Sorites - Electronic Quarterly of Analytical Philosophy 3*, pp. 43–47.

da Costa N.C.A., Bueno O. and French S. (1998). "Is There a Zande Logic?", *History and Philosophy of Logic 19*, pp. 41–54.

da Costa N.C.A. and French S. (1988). "Belief and Contradiction", *Crítica XX*, pp. 3–11.

— (1989). "On the Logic of Belief", *Philosophy and Phenomenological Research XLIX*, pp. 431–446.

— (1990). "Belief, Contradiction and the Logic of Self-Deception", *American Philosophical Quarterly 27*, pp. 179–197.

— (1995). "Partial Structures and the Logic of the Azande", *American Philosophical Quarterly 32*, pp. 325–339.

da Costa N.C.A. and Subrahmanian V.S. (1989). "Paraconsistent Logics as a Formalism for Reasoning About Inconsistent Knowledge Bases", *Artificial Intelligence in Medicine 1*, pp. 167–174.

D'Ottaviano I. (1990). "On the Development of Paraconsistent Logic and da Costa's Work", *Journal of Non-Classical Logic 7*, pp. 89–152.

Dummett M. (1977). *Elements of Intuitionism.* Oxford: Clarendon Press.

French S. (1990). "Rationality, Consistency and Truth", *Journal of Non-Classical Logic 7*, pp. 51–71.

Heyting A. (1971). *Intuitionism: An Introduction.* (3rd edition.) Amsterdam: North-Holland.

Mortensen C. (1995). *Inconsistent Mathematics.* Dordrecht: Kluwer Academic Publishers.

Poincaré H. (1905). *Science and Hypothesis.* New York: Dover.

Priest G., Routley R. and Norman J. (ed.) (1989). *Paraconsistent Logic: Essays on the Inconsistent.* Munich: Philosophia.

Suszko R. (1975). "Remarks on Lukasiewicz's Three-Valued Logic", *Bulletin of the Section of Logic 4*, pp. 87–90.

Chapter 21

ALGORITHMS FOR RELEVANT LOGIC

Paul Gochet, Pascal Gribomont and Didier Rossetto*
Université de Liège

Abstract The classical analytic tableau method has been extended successfully to modal logics (see e.g. Fitting 1983; Fitting 1993; Goré 1992) and also to relevant and paraconsistent logics Bloesch 1993a; Bloesch 1993b. The classical connection method has been extended to modal and intuitionistic logics Wallen 1990, and the purpose of this paper is to investigate whether a similar adaptation to relevant logic is possible. A hybrid method is developed for B^+, with a specific solution to the "multiplicity problem", as in the technique of modal semantic diagrams introduced in Hughes and Cresswell 1968. Proofs of soundness and completeness are also given.

1. Introduction

The sequent calculi and tableau methods suffer from three kinds of redundancies: duplication of redundant information, the consideration of reductions that do not advance the search toward finding a proof, and the need to distinguish derivations that differ in the order in which sequent rules are applied Wallen 1990, p. 82. The connection method has proved computationally more efficient for classical, modal and intuition-

*The authors thank Prof. Jean-Paul van Bendegem, editor of "Logique et Analyse", for permission to reprint "Algorithms for Relevant Logic" which was published in Logique et Analyse, 150-151-152 Special issue dedicated to the memory of Léo Apostel, pp.329-346, 1995.
We are grateful to the research team of the Automated Reasoning Project of the Australian National University (Canberra) for their invaluable help. We are very grateful to Dr. Bloesch for granting us the permission to quote his Ph.D. thesis. We received substantial help from Dr. Goré and Dr. Lugardon. We also thank Dr. van der Does and Dr. Herzig. This work was supported by a grant of the F.N.R.S. (project 8.4536.94) and an earlier version was discussed at the Centenary Conference of the Lvov-Warsaw School of Logic (Nov. 1995). We express our gratitude to our sponsors and our hosts, and to Dr. Gailly and Mrs Gailly Goffaux for several corrections. We owe the topic of this paper to the late Prof. Sylvan.

D. Vanderveken (ed.), Logic, Thought & Action, 479-496.

istic logics; it may therefore be useful to extend it to other non-classical logics. This is especially true for logics used in artificial intelligence, for which efficient theorem proving techniques are needed. Bloesch has conjectured that such an extension is feasible in several cases Bloesch 1993a, p. 24:

> "It is not clear how to create connection method style proof systems for many non classical logics. In particular the relevant and paraconsistent logics seem particularly difficult since the law of noncontradiction does not hold in most paraconsistent logics and many relevant logics. By relying heavily on the properties of classical logic, the connection method gains great efficiency but at the cost of poor flexibility. It does however seem fair to conjecture that techniques, such as using truth signs to represent exclusive semantic classes, developed in later chapters, could be applied to the connection method."

The main purpose of this paper is to investigate the connection method style for B^+, the most basic system of relevant logic, with the simplified semantics introduced in Priest and Sylvan 1992; Restall 1993. The connection method presented here draws from two sources: Wallen's connection method for modal logic K and Bloesch's tableau method for relevant and paraconsistent logics. Both methods will have to be modified to fit our purpose. B^+ is known to be decidable. We present a decision procedure which automatically supplies finite models when applied to a formula which happens to be satisfiable.

This paper goes on as follows: the axiomatic system B^+ is recalled, Bloesch's Tableau Method for B^+ is briefly presented, Wallen's Connection Method for modal logic is adapted to B^+, and the soundness and the completeness of the extension are proven.

2. The axiomatic method for relevant logic B^+

2.1 Axioms

(i) $A \to A$

(ii) $A \to (A \vee B)$, $B \to (A \vee B)$

(iii) $(A \wedge B) \to A$, $(A \wedge B) \to B$

(iv) $(A \wedge (B \vee C)) \to ((A \wedge B) \vee C)$

(v) $((A \to B) \wedge (A \to C)) \to (A \to (B \wedge C))$

(vi) $((A \to B) \wedge (B \to C)) \to ((A \vee B) \to C)$

2.2 Rules

(i) $\frac{A,\,A\to B}{B}$ (modus ponens), and its disjunctive form.

(ii) $\frac{A,\,B}{A\wedge B}$ (adjunction) and its disjunctive form

(iii) $\frac{A\to B\,,\,C\to D}{(B\to C)\to(A\to D)}$ (affixing rule) and its disjunctive form

Comment. If $\frac{A_1,\ldots,A_n}{B}$ is a rule then its disjunctive form is the rule $\frac{C\vee A_1,\ldots,C\vee A_n}{C\vee B}$.

2.3 Example

Let us prove that formula 21.1 is a theorem of B^+.

$$(p \to q) \to ((s \to q) \vee ((r \wedge p) \to q)) \qquad (21.1)$$

Proof.

(i) $(r \wedge p) \to p$ (axiom 3)

(ii) $q \to q$ (axiom 1)

(iii) $(p \to q) \to ((r \wedge p) \to q)$ (rule 3 (1, 2))

(iv) $(p \to q) \to (p \to q)$ (axiom 1)

(v) $((r \wedge p) \to q) \to ((s \to q) \vee ((r \wedge p) \to q))$ (axiom 2)

(vi) $((p \to q) \to ((r \wedge p) \to q)) \to ((p \to q) \to ((s \to q) \vee ((r \wedge p) \to q)))$ (rule 3 (4, 5))

(vii) $(p \to q) \to ((s \to q) \vee ((r \wedge p) \to q))$ (rule 1 (3, 6))

This theorem will be proven again by the method presented here.

3. Tableau method for relevant logic B^+

The principle of the semantic tableau method is rather straightforward. It is a systematic search for a model that falsifies the formula being checked for validity. If such a model does not exist, the formula is valid.

The formula φ to be proven is assumed first to be *false* in world w_0 (we write $\mathbf{F}_{w_0}\varphi$). Next it is reduced by applying rules which remove the connectives stepwise starting with the less deeply nested connective until each branch of the tableau becomes closed or gives rise to a model of $\neg\varphi$. If all the branches of the tree are closed (*i.e.* there is no model that falsifies φ), $\mathbf{F}_{w_0}\varphi$ cannot be forced and φ is valid. This process is not deterministic. Generally several tableaux can be produced depending on which order we chose when we apply the reduction rules.

The reduction rules are based on the truth conditions. For example, $A \wedge B$ is *true* iff A and B are *true*. The truth conditions of the relevant conditional bear some resemblance with those of the classical conditional. Indeed, when the relevant conditional $A \rightarrow B$ is *false* the antecedent A is *true* and the consequent B is *false*, and when the relevant conditional is *true*, the antecedent is *false* or the consequent is *true*. However, as the relevant conditional is intensional, something more is needed to capture its meaning. Just like modal formulas, the intensional formula $A \rightarrow B$ is evaluated at a world w and an accessibility relation relates the world w to the worlds w' and w'' at which A and B respectively are evaluated. The interpretation of relevant implication involves three worlds, instead of two for modal operators, so the relevant accessibility relation R is ternary.[1] The contrast between $\mathbf{F}_w A \rightarrow B$ and $\mathbf{T}_w A \rightarrow B$ matches the contrast between $\mathbf{F}_w \Box A$ and $\mathbf{T}_w \Box A$. Formula $A \rightarrow B$ is *false* at w iff w' and w'' exist such that $R_{ww'w''}$, A is *true* at w' and B is *false* at w'', just as $\Box A$ is *false* at w iff there exists a world w' such that $R_{ww'}$ and A is *false* at w'. Formula $A \rightarrow B$ is *true* iff for all w' and w'' such that $R_{ww'w''}$, A is *false* at w' or B is *true* at w''.

Taking advantage of this semantics, Bloesch Bloesch 1993a, p. 88 gives reduction rules for $\mathbf{F}_w A \rightarrow B$ and $\mathbf{T}_w A \rightarrow B$. Whenever a formula such as $\mathbf{F}_w A \rightarrow B$ appears on a branch, we record, for that branch, that $R_{ww'w''}$ and we add the formulas $\mathbf{T}_{w'} A$ and $\mathbf{F}_{w''} B$ to the branch, where w' and w'' are new worlds, *i.e.* they do not appear on the branch, and $w' = w''$ iff $w = w_0$ (*i.e.* the base world). Whenever a formula such as $\mathbf{T}_w A \rightarrow B$ appears on a branch, if we have $R_{ww'w''}$ on that branch, we may create two subbranches with the formulas $\mathbf{F}_{w'} A$ on one and $\mathbf{T}_{w''} B$ on the other.

Since the tableau rule for $\mathbf{F}_{\rightarrow}$ is a hybrid rule which bears similarity both to π-rule (modal rule dealing with $\mathbf{T}_{\Diamond}$) and to α-rule, it is appropriate to introduce the category of "$\pi\alpha$-rule". Analogously a tableau rule for $\mathbf{T}_{\rightarrow}$ is a hybrid rule which falls under the heading of "$\nu\beta$-rule" (modal ν-rule deals with $\mathbf{T}_{\Box}$).

The tableau rules based on the semantics of the connectives fall therefore in four types: α, β, $\pi\alpha$, and $\nu\beta$ (figure 1).

[1]A frame in modal logic is a graph; a frame in relevant logic is a hypergraph.

α	α_1	α_2
$\mathbf{T}_{w_i} x \wedge y$	$\mathbf{T}_{w_i} x$	$\mathbf{T}_{w_i} y$
$\mathbf{F}_{w_i} x \vee y$	$\mathbf{F}_{w_i} x$	$\mathbf{F}_{w_i} y$

β	β_1	β_2
$\mathbf{F}_{w_i} x \wedge y$	$\mathbf{F}_{w_i} x$	$\mathbf{F}_{w_i} y$
$\mathbf{T}_{w_i} x \vee y$	$\mathbf{T}_{w_i} x$	$\mathbf{T}_{w_i} y$

$\pi\alpha$	$\pi\alpha_1$	$\pi\alpha_2$
$\mathbf{F}_{w_i} x \to y$	$\mathbf{T}_{w_j} x$	$\mathbf{F}_{w_k} y$

w_j and w_k ($j = k$ iff $i = 0$) are new worlds and $R_{w_i w_j w_k}$ is added

$\nu\beta$	$\nu\beta_1$	$\nu\beta_2$
$\mathbf{T}_{w_i} x \to y$	$\mathbf{F}_{w_j} x$	$\mathbf{T}_{w_k} y$

w_j and w_k are such that $R_{w_i w_j w_k}$ was previously added

Figure 1. Reduction rules for B^+

(i) *Prolongation rules* (or α-rules). Each α-node gives rise to a prolongation of the current branch. Two successive nodes are added, one of which is labelled with α_1 and the other is labelled with α_2.

(ii) *Branching rules* (or β-rules). Each β-node splits the branch into two subbranches one of which bears a node labelled with β_1 and the other a node labelled with β_2.

(iii) *Relevant prolongation rule* (or $\pi\alpha$-rule). Each $\pi\alpha$-node gives rise to a prolongation of the branch. Two successive nodes are added, one of which is labelled with $\pi\alpha_1$ and the other is labelled with $\pi\alpha_2$.

(iv) *Relevant branching rule* (or $\nu\beta$-rule). Each $\nu\beta$-node splits the branch into two subbranches one of which bears a node labelled with $\nu\beta_1$ and the other a node labelled with $\nu\beta_2$. The $\nu\beta$-rule can be used μ times; μ, also called the multiplicity of the $\nu\beta$-formula, is the number of pairs of worlds related to w_i previously introduced by the application of a $\pi\alpha$-rule ($R_{w_i w_j w_k}$was previously added to the branch).

Comment. The parameter μ associated with a $\nu\beta$-formula has a definite value only at the end of the derivation. Here is the feature that will raise the *multiplicity problem* in the connection method.

4. Connection method extended to relevant logic

The principle of the connection method is the same as that of the tableau method: a formula is said to be proven by a connection proof whenever its attempted refutation fails. The connection method proceeds in three stages: a syntactic tree, an indexed formula tree and a path tree are successively built.

4.1 The syntactic tree

The syntactic tree or formation tree displays the structure of the formula. For example, figure 2 shows the syntactic tree of formula 21.1. The nodes are numbered by traversing the tree depthfirst, from left to right.

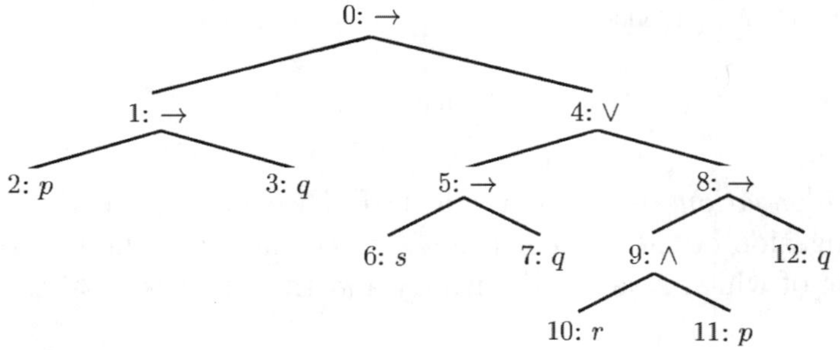

Figure 2. Syntactic Tree

4.2 The indexed formula tree

In tableau proof trees, a derivation is done by removing connectives in accordance with logical rules of elimination which we call *reduction steps.* These steps are *graphically* depicted (*e.g.* the elimination of an asserted conjunction gives rise to a prolongation of the branch, the elimination of a denied conjunction gives rise to the splitting of the branch,...). The sign of the initial formula tested for inconsistency is **F** since a refutation proof is intended. The sign of the other formulas is determined by the sign of the input formula and by the reduction rule applied to it (*e.g.* the denial of a formula whose main connective is a conjunction leads to the disjunction of the denials of each conjunct).

In the connection method, these pieces of information are encoded in the indexed formula tree. The indexed formula tree associated with a signed formula φ can be represented as an array (figure 3 shows the indexed formula tree of formula 21.1). Each line of this array is a record representing a signed subformula of φ obtained when applying recursively the reduction rules (figure 1) to φ.

The indexed formula tree has seven columns. The first column is a key identifying the line;[2] the key will be a_i, where i is the index of the

[2]The path tree will deal with these indices to represent the (sub)formulas in an efficient way.

k	$pol(k)$	$lab(k)$	$Pt(k)$	$St(k)$	$w(k)$	$h(k)$
a_0	**F**	$(p\to q)\to((s\to q)\vee((r\wedge p)\to q))$	$\pi\alpha$	–	w_0	0
a_1	**T**	$p \to q$	$\nu\beta$	$\pi\alpha_1$	w_1	1
a_2^1	**F**	p	–	$\nu\beta_1$	w_2	4
a_3^1	**T**	q	–	$\nu\beta_2$	w_3	4
a_2^2	**F**	p	–	$\nu\beta_1$	w_4	7
a_3^2	**T**	q	–	$\nu\beta_2$	w_5	7
a_4	**F**	$(s \to q) \vee ((r \wedge p) \to q)$	α	$\pi\alpha_2$	w_1	1
a_5	**F**	$s \to q$	$\pi\alpha$	α_1	w_1	2
a_6	**T**	s	–	$\pi\alpha_1$	w_2	3
a_7	**F**	q	–	$\pi\alpha_2$	w_3	3
a_8	**F**	$(r \wedge p) \to q$	$\pi\alpha$	α_2	w_1	2
a_9	**T**	$r \wedge p$	α	$\pi\alpha_1$	w_4	5
a_{10}	**T**	r	–	α_1	w_4	6
a_{11}	**T**	p	–	α_2	w_4	6
a_{12}	**F**	q	–	$\pi\alpha_2$	w_5	5

Figure 3. Indexed Formula Tree

subformula in the syntactic tree of φ. Some formulas have to be repeated (see later); superscripts are used to distinguish the multiple occurrences of these formulas, and are transmitted to their children. The second column, named $pol(k)$ (where k is the key), records the polarity of the formula (**T** or **F**). The third column, named $lab(k)$, records the label, *i.e.* the signed subformula itself. Each non atomic signed formula has a type, which can be α, β, $\pi\alpha$ or $\nu\beta$, and also two components, called *children*, as we saw in figure 1. Each child has its own type, called the primary type; its secondary type is the (primary) type of the parent formula, with the subscript 1 or 2, as indicated in figure 1. The fourth column, named $Pt(k)$, records the (primary) type of the subformula; an atomic formula has no primary type (– in the fourth column). The fifth column, named $St(k)$, records the secondary type of the subformula. The main formula φ has no secondary type (– in the fifth column). The columns "$Pt(k)$" and "$St(k)$" are introduced merely to support the understanding of the indexed formula tree. In relevant logic, as in modal logic, the polarity is ascribed to a formula relatively to a given world. The sixth column, named $w(k)$, records this piece of information; worlds are given names from the unlimited sequence $w_0, w_1, w_2, \ldots$ Indeed two identical formulas of opposite signs such as **T**A and **F**A will combine together to yield a contradiction only if they are evaluated at the same world. The last column, named $h(k)$, records the history: it contains the number of the step during which the line is created. Column 7 is also introduced to support the understanding of the indexed formula tree.

We now give the iterative procedure to construct the indexed formula tree. Throughout the execution, the procedure maintains for each line a binary variable, whose value can be "active" or "passive". Furthermore, for each line of type $\nu\beta$, a set Ω of ordered pairs of worlds is maintained.

Each step of the procedure selects an active line ℓ and generates two or more new lines $\ell_1, \ell_2, \ldots$ In every case, the label of a new line is a child of the label of the parent line, according to the type-dependent reduction rules given in figure 1. After the generation, line ℓ becomes passive, whereas lines $\ell_1, \ell_2, \ldots$ are active if their label is non atomic and passive otherwise. The set Ω associated with a new $\nu\beta$-line is empty.

Initial step 0. Create an initial line.

- The initial line has key a_0; its label is the given signed formula, to which we ascribe the initial world w_0. The history is 0. The line is made active if its (primary) type is α, β or $\pi\alpha$, and passive otherwise.

Iterative step $n > 0$. Select an active line of key k.

- If $Pt(k) = \alpha$, two lines are added to the indexed formula tree. Their labels respectively are the α_1-child and the α_2-child of $lab(k)$ determined by the α-reduction rule; their indices are extracted from the syntactic tree; their world and history are $w(k)$ and n, respectively.

- If $Pt(k) = \beta$, two lines are added to the indexed formula tree. Their labels respectively are the β_1-child and the β_2-child of $lab(k)$, determined by the β-reduction rule; their indices are extracted from the syntactic tree and their world is $w(k)$. The history of both new lines is n.

- If $Pt(k) = \pi\alpha$, two lines are added to the indexed formula tree. Their labels respectively are the $\pi\alpha_1$-child and the $\pi\alpha_2$-child of $lab(k)$, determined by the $\pi\alpha$-reduction rule; their indices are extracted from the syntactic tree. The worlds w_j and w_k associated with these two formulas must not have been used before, say the first elements of the world sequence which differ from all worlds associated with existing lines. The history of both new lines is n. Furthermore, each existing $\nu\beta$-line k' (*i.e.* $Pt(k') = \nu\beta$) such that $w(k') = w(k)$ is made active again.

- If $Pt(k) = \nu\beta$, the set E of all passive existing $\pi\alpha$-lines k' (*i.e.* $Pt(k') = \pi\alpha$) such that $w(k') = w(k)$ is determined. For each $k' \in E$ we do the following. As k' is passive, it has two children, say k'' and k''', whose associated worlds are $w(k'')$ and $w(k''')$. If $(w(k''), w(k'''))$ is not an element of the set $\Omega(k)$ associated with line k, two new lines, say ℓ and ℓ', are added to the indexed formula tree. Signed subformula $lab(\ell)$ and $lab(\ell')$ are the children of $lab(k)$, as determined by the $\nu\beta$-reduction rule; the indices ℓ and ℓ' are extracted from the syntactic tree. Remember that a superscript is used to distinguish the different pairs of children of the parent formula. The associated worlds $w(\ell)$ and $w(\ell')$ are $w(k'')$ and $w(k''')$ respectively, and the ordered pair $(w(k''), w(k'''))$ is added to $\Omega(k)$. The history of each new line is n.
  Comment. It should be emphasized that a $\nu\beta$-line k may switch several times from active to passive and conversely during the execution. Besides, each step about the $\nu\beta$-line k may introduce any number of new ordered pairs of children for k, up to the size of the set E. Last, the worlds associated with the children are not inherited from the parent line k, but from other lines, of secondary type $\pi\alpha_i$ $(i = 1, 2)$.

4.3 The path tree

The path tree (it is actually an acyclic graph, see Gribomont and Rossetto 1995 for an example) has the same role as a tableau proof tree, but its construction is computationally more efficient. The path tree comes from the following recursive definition of a path.

Basis.

- If a_0 is the root of the indexed formula tree, a_0 is a path.

Recursion.

- If S is a path containing the node α,[3] $(S \setminus \{\alpha\}) \cup \{\alpha_1\} \cup \{\alpha_2\}$ is a path.
- If S is a path containing the node β, $(S \setminus \{\beta\}) \cup \{\beta_1\}$ and $(S \setminus \{\beta\}) \cup \{\beta_2\}$ are paths.
- If S is a path containing the node $\pi\alpha$, $(S \setminus \{\pi\alpha\}) \cup \{\pi\alpha_1\} \cup \{\pi\alpha_2\}$ is a path.
- If S is a path containing the node $\nu\beta$, $S \cup \{t_1, \ldots, t_\mu\}$, where t_i $(1 \leq i \leq \mu)$ is either $\nu\beta_1^i$ or $\nu\beta_2^i$, are paths.

[3] *I.e.* a node of the indexed formula tree whose label is a signed formula of primary type α.

Comments. There are 2^μ such paths, where μ is the multiplicity associated with $\nu\beta$. The μ ordered pairs $(\nu\beta_1, \nu\beta_2)$ used here are the μ ordered pairs (among the ordered pairs $(\nu\beta_1, \nu\beta_2)$ got during the building of the indexed formula tree) which are *frame compatible* with the nodes of the current path, i.e. $\nu\beta_1$ and $\nu\beta_2$ are associated with worlds already involved in the current path.[4]

Each step in the construction of the path tree consists of the application of a rule (α, β, $\pi\alpha$, or $\nu\beta$) to a formula. These applications generate new paths. It should be emphasized here that as soon as a the successors of a path have been determined, this path can be erased. In fact it is erased when the connection method is implemented on a computer; the path tree never resides wholly in the computer memory. Only its leaves, i.e., the atomic paths, are saved and determine models of $\neg\varphi$ if they do not contain connections. The atomic paths contain only atomic formulas and (vacuously true) $\nu\beta$-formulas.

The formula tested is a theorem if and only if each leaf of the path tree contains a *connection*, that is, two signed formulas $\mathbf{T}_{w_i}A$ and $\mathbf{F}_{w_j}B$, with $w_i = w_j$ and A identical to B.

To ensure soundness, the path tree has to respect a *reduction ordering* which combines two orderings: the *subformula ordering* and the *modal ordering*. The reduction ordering (denoted $\triangleleft$) is the transitive closure of the union of the subformula ordering (denoted $\ll$) and the modal ordering (denoted $\sqsubset_{\mathcal{M}}$), so $\triangleleft = (\ll \cup \sqsubset_{\mathcal{M}})^*$ Wallen 1990.

Therefore the recursion has to be applied in the following way.

(i) *Respecting the order of the world indices*: consider nodes related to world w_i only when all nodes attached to world w_j with $j < i$ have been considered.

(ii) *Respecting the necessity order*: consider $\nu\beta$-nodes related to world w_i only when all other nodes related to world w_i have been considered.

This is automatically done if the recursion is applied in the order recorded in the seventh column $h(k)$ of the indexed formula tree.

The reason behind the frame compatibility restriction is this: allowing the introduction in a path of formulas that are not frame compatible with the other formulas of the path would amount to allowing a move from one subbranch to another in a tableau proof tree, which would clearly be unsound.

Let us observe that if we consider the semantic tableau style of proof, a $\nu\beta$-formula could introduce μ successive branching that lead to 2^μ splits of the branch. In the path tree, all the 2^μ paths generated by

[4] A *syntactic criterion* to check frame compatibility between two formulas is the following: formula φ is frame compatible with formula ψ if and only if the primary type of their common ancestor in the syntactic tree is neither β nor $\nu\beta$.

$$
\begin{array}{c}
\{\underline{a_0}\} \\
\{a_1, \underline{a_4}\} \\
\{a_1, \underline{a_5}, a_8\} \\
\{a_1, a_6, a_7, \underline{a_8}\} \\
\{\underline{a_1}, a_6, a_7, a_9, a_{12}\}
\end{array}
$$

$$
\begin{array}{cccc}
\{a_2^1, a_2^2, a_6, & \{a_3^1, a_2^2, a_6, & \{a_2^1, a_3^2, a_6, & \{a_3^1, a_3^2, a_6, \\
a_7, \underline{a_9}, a_{12}\} & a_7, \underline{a_9}, a_{12}\} & a_7, \underline{a_9}, a_{12}\} & a_7, \underline{a_9}, a_{12}\} \\
\{a_2^1, \underline{\underline{a_2^2}}, a_6, a_7, & \{a_3^1, \underline{\underline{a_2^2}}, a_6, a_7, & \{a_2^1, \underline{\underline{a_3^2}}, a_6, a_7, & \{\underline{\underline{a_3^1}}, a_3^2, a_6, \underline{\underline{a_7}}, \\
a_{10}, \underline{\underline{a_{11}}}, a_{12}\} & a_{10}, \underline{\underline{a_{11}}}, a_{12}\} & a_{10}, a_{11}, \underline{\underline{a_{12}}}\} & a_{10}, a_{11}, a_{12}\}
\end{array}
$$

Figure 4. Path Tree

its instantiation are got in one step. Indeed, in tableau trees we do the branching and instantiation *separately.* On the contrary here we combine branching with modal instantiation when applying $\nu\beta$-rules. Our policy produces a reduction of the number of nodes since 2^μ nodes are needed instead of $\sum_{i=1}^{\mu} 2^i = 2^{\mu+1} - 2$.

When building the path tree, the same operation applies for rules of any category: we replace the parent formula by its children. Here again our technique diverges from the standard one used in modal logic, where whenever a necessity rule is applied, a child is added but the parent formula is maintained; if the nth instantiation has failed to produce a contradiction, a new instantiation takes place; multiplicity is demand driven Wallen 1990.

Demand driven policy secures completeness. However if the formula tested fails to be unsatisfiable, a loop may occur. The policy advocated in this paper rules out the risk of non termination as far as B^+ is concerned. Moreover it automatically supplies finite models whenever the formula tested for inconsistency happens to be satisfiable. Our treatment of the multiplicity problem establishes a tight correspondence between the $\nu\beta$-reduction rule and the semantics of $\nu\beta$-formulas.

The path tree corresponding to formula 21.1 is represented in figure 4. To help the reader to understand the path tree, the node developed is underlined once and the indices standing for connections are underlined twice. There are four atomic paths, each of them containing a connection, so no model exists for signed formula $\mathbf{F}_{w_0}\varphi$, and φ is a B^+-valid formula.

We see that the four atomic paths all contain a connection, therefore no frame that falsifies formula 21.1 exists, *i.e.* this formula is B^+-valid.

5. Soundness and completeness

In this section we prove that our method is a sound and complete decision procedure for relevant logic B^+.

5.1 Complexity

If A is a (signed or not) formula, its *degree* $d(A)$ is the number of internal nodes of the syntactic tree associated with A, *i.e.* the total number of connectives contained in A. In other words, the degree of an atomic formula is 0; $d(\neg A) = 1 + d(A)$ and if $\circ$ is a binary operator, $d(A \circ A') = 1 + d(A) + d(A')$. The notion of degree also applies to signed formulas: $d(\mathbf{T}A) = d(\mathbf{F}A) = d(A)$.

If $\mathcal{E}$ is a set of formulas, its *complexity* $c(\mathcal{E})$ is the sum of the degrees of its elements.

Every finite set of formulas has a finite complexity. Furthermore, with usual notation, we have the following inequalities:

- if $\mathcal{E}_1 \subseteq \mathcal{E}_2$, then $c(\mathcal{E}_1) \leq c(\mathcal{E}_2)$.
- $c(\mathcal{E} \cup \{\alpha\}) > c((\mathcal{E} \setminus \{\alpha\}) \cup \{\alpha_1, \alpha_2\})$, since $d(\alpha) > d(\alpha_1) + d(\alpha_2)$.
- $c(\mathcal{E} \cup \{\beta\}) > \max(c((\mathcal{E} \setminus \{\beta\}) \cup \{\beta_1\}), c((\mathcal{E} \setminus \{\beta\}) \cup \{\beta_2\}))$, since $d(\beta) > d(\beta_1)$ and $d(\beta) > d(\beta_2)$.
- $c(\mathcal{E} \cup \{\pi\alpha\}) > c((\mathcal{E} \setminus \{\pi\alpha\}) \cup \{\pi\alpha_1, \pi\alpha_2\})$, since $d(\pi\alpha) > d(\pi\alpha_1) + d(\pi\alpha_2)$.
- $c(\mathcal{E} \cup \{\nu\beta\}) > \max(c((\mathcal{E} \setminus \{\nu\beta\}) \cup \{\nu\beta_1\}), c((\mathcal{E} \setminus \{\nu\beta\}) \cup \{\nu\beta_2\}))$, since $d(\nu\beta) > d(\nu\beta_1)$ and $d(\nu\beta) > d(\nu\beta_2)$.

5.2 Auxiliary structure

As a preliminary result, we need to know that the execution of the construction procedure for the indexed formula tree always terminates. We first observe that every step induces the addition of finitely many new lines (or nodes). This is trivial, even for $\nu\beta$-steps, since the number of lines added in a $\nu\beta$-step cannot exceed $2n$ where n is the total number of lines introduced before this step. As a consequence, termination could be prevented only if an infinite number of steps could take place.

In order to prove that every execution involves finitely many steps, we will suppose that each step of the construction procedure updates not only the indexed formula tree, but also an auxiliary structure defined below; it will then be sufficient to prove that this structure can be updated only finitely many times. The auxiliary structure is a set of hypertrees; the nodes of the hypertrees are worlds (so each hypertree is a frame). Each node w_i is labelled with a set $\ell(w_i)$ of signed formulas.

We describe now the update induced on the auxiliary structure by each step of the construction procedure, for a signed formula φ.

Initial step 0. (Creation of the initial line.)

- The structure initially contains only one frame; this frame consists of a single node w_0, whose label contains a single element, which is the initial signed formula: $\ell(w_0) = \{\varphi\}$.

Iterative step $n > 0$. (Addition of children of line k.)
In every case, frames that contain world-node $w(k)$ with signed formula "$pol(k) : lab(k)$" in $\ell(w(k))$ already exist in the auxiliary structure.

- If $Pt(k) = \alpha$, in each frame such that the signed formula $\alpha = pol(k){:}lab(k)$ is a member of $\ell(w(k))$, we update $\ell(w(k))$ by replacing its element α by the corresponding elements α_1 and α_2.

- If $Pt(k) = \beta$, each frame of the auxiliary structure that has a node named $w(k)$ with the signed formula $\beta = pol(k){:}lab(k)$ in its label $\ell(w(k))$ is updated as follows. First, the frame is replaced by two identical frames; second, in the label $\ell(w(k))$, the element β is replaced by the corresponding element β_1 in the first frame and β_2 in the second one.

- If $Pt(k) = \pi\alpha$, each frame of the auxiliary structure that has a node named $w(k)$ with the signed formula $\pi\alpha = pol(k){:}lab(k)$ in its label $\ell(w(k))$ is updated as follows. Two nodes w and w' and the hyperarrow $(w(k), w, w')$ are added to the frame, where w and w' are the new worlds associated with the children of line k. The new nodes are labelled respectively $\ell(w) = \{\pi\alpha_1\}$ and $\ell(w') = \{\pi\alpha_2\}$.

- If $Pt(k) = \nu\beta$, each frame that contains node $w(k)$, with $\nu\beta \in \ell(w(k))$ and that has a hyperarrow $(w(k), w, w')$ for some w and w' is updated as follows, if it has not been done yet. First, the frame is replaced by two identical frames; second, the children signed formulas $\nu\beta_1$ and $\nu\beta_2$ are added to $\ell(w)$ in the first frame and to $\ell(w')$ in the second frame respectively.

We first observe that duplication of frames is induced by β and $\nu\beta$-reductions. However, a branching formula can be a subformula of another branching formula, so β-reductions and $\nu\beta$-reductions can be reused several times. As a result, the number of frame duplication is bounded by $d(\varphi)!$ and the total number of frames in an auxiliary structure cannot exceed $2^{d(\varphi)!}$.

Furthermore, each frame of the auxiliary structure is finite. First, such a frame is a finitary hypertree since the number of successors of any node of the frame is bounded by the number of $\pi\alpha$-subformulas contained in the initial formula, and therefore by its degree $d(\varphi)$. Second, the length of a hyperbranch cannot exceed $d(\varphi) = c(\{\varphi\})$, since the complexity of a successor node is strictly less than the complexity of the parent node. An upper bound for the number of nodes in a frame is $d(\varphi)^{d(\varphi)+1}$, and an upper bound for the total number of nodes in the whole structure is therefore $\Sigma_\varphi =_{def} 2^{d(\varphi)!} d(\varphi)^{d(\varphi)+1}$ (this is also an upper bound for the number of hyperarrows). Tighter bounds can be found, but we do not need them here.
Comment. The notion of auxiliary structure applies not only to a signed formula, but also to a finite (conjunctive) set of signed formulas.

5.3 Termination proofs

We can now give an upper bound for the length of any execution of the construction algorithm for the indexed formula tree. Each step induces at least one of the following operations:

(i) Extension of the auxiliary structure (β-step, $\pi\alpha$-step, useful $\nu\beta$-step)
(cannot be performed more than Σ_φ times);

(ii) Addition of a formula to a node label (useful $\nu\beta$-step, $\pi\alpha$-step)
(cannot be performed more than $d(\varphi) \times \Sigma_\varphi$ times);

(iii) Reduction of an element of a node label (α-step, β-step)
(cannot be performed more than $d(\varphi) \times \Sigma_\varphi$ times).

Comment. A $\nu\beta$-step about line k will do nothing at all if no new ordered pair of worlds accessible from $w(k)$ has been created since the last activation; however, as $\nu\beta$-lines are "reactivated" by $\pi\alpha$-steps, useless steps can occur only finitely many times.
This completes the termination proof for the indexed formula tree algorithm.

In order to prove the termination of the path tree algorithm, we observe that each step adds a finite number of paths to the path tree, which is a finitary tree. Due to König's lemma, it is sufficient to establish that a new path is always strictly less complex than its parent. However, this is not true for the notion of (multi)-set complexity introduced above; indeed,

$$c((S \setminus \{\nu\beta\}) \cup \{\nu\beta^1_{i_1}, \ldots, \nu\beta^\mu_{i_\mu}\}) < c(S), \text{where } i_j \in \{1,2\} (1 \leq j \leq \mu),$$

cannot be guaranteed.

The problem is, we know the multiplicity μ to be finite, but nothing else. To solve this, we define another well-founded ordering on paths. First, to each path S we associate $m(S)$, a multiset of natural numbers, which are the degrees of the elements of the path. Second, we show that, for some relation $\sqsubset$, the domain of multisets is well-founded. Third, we show that, if S' is produced by the algorithm from the path S, then $m(S') \sqsubset m(S)$.

A multiset of natural numbers can be represented as a decreasing sequence of numbers; the lexical ordering on these sequences is defined as follows:

$$(a_1, \ldots, a_n) \sqsubset (b_1, \ldots, b_m)$$

if there is a number $r \in \mathbf{N}$ such that $r \leq n$, $r \leq m$, $a_i = b_i$ for $i = 1, \ldots, r$ and either $r = n$ and $r < m$, or $r < n$, $r < m$ and $a_{r+1} < b_{r+1}$.
As $(\mathbf{N},<)$ is a well-ordered set, so is $(\mathbf{N}^*,\sqsubset)$. Besides, the lexical ordering $\sqsubset$ induces a (partial) ordering (also noted $\sqsubset$) on the set of paths:

$$S \sqsubset S' \ =_{def} \ m(S) \sqsubset m(S')\,,$$

and this ordering is well-founded.

Last, we observe that each reduction step for a path consists in replacing an element of the path by finitely many elements of lower degree; this always leads to a $\sqsubset$-smaller path.

5.4 Hypertree-models

The auxiliary structure is in fact a set of frames, and each frame is an interpretation of the initial signed formula φ, called a *hypertree-frame*; it is a *hypertree-model* for φ if φ holds at the initial world w_0. A hypertree-frame is *consistent* if no world label contains a pair of opposite formulas. We first prove the following lemma.
Lemma I. A (hypertree-)frame $\mathcal{S}$ is consistent if and only if $\mathcal{S}, w_i \models \psi$, for all worlds $w_i \in \mathcal{S}$ and for all signed formulas $\psi \in \ell(w_i)$.
Proof. The "if" part is trivial: if $\mathcal{S}, w_i \models \psi$ and $\mathcal{S}, w_i \models \xi$, then $\{\psi, \xi\}$ cannot be a pair of opposite signed formulas. For the "only if" part, we proceed by induction on the height of the nodes in the frame. A leaf w (a node without successor) has height $h(w) = 0$, and if w is an internal node with successors $\{(w_1, w_1'), \ldots, (w_k, w_k')\}$, then $h(w) = 1 + \max\{(h(w_1) + h(w_1')), \ldots, (h(w_k) + h(w_k'))\}$.[5]
Base case. If the world w has no successor in $\mathcal{S}$, the label $\ell(w)$ contains only atoms and $\nu\beta$-formulas. The $\nu\beta$-formulas are vacuously true, and

[5]Node w has successor (w', w'') if (w, w', w'') is a hyperarrow.

the atoms are forced to be true at w, which is possible since there is no opposite pair.

Induction case. Let w be a node of height $n > 0$. As for the base case, all atomic formulas can be forced at w. Non atomic formulas are $\nu\beta$ or $\pi\alpha$. Let $\{(w_1, w'_1), \ldots, (w_k, w'_k)\}$ be the (non empty) set of successors. All w_i and w'_i have a height less than n, so the induction hypothesis applies to them. Let $\nu\beta \in \ell(w)$; due to the construction process of the frame, either $\nu\beta_1^i \in \ell(w_i)$ or $\nu\beta_2^i \in \ell(w'_i)$, for all $i = 1, \ldots, k$, so by the induction hypothesis, either $\nu\beta_1^i$ is forced at w_i or $\nu\beta_2^i$ is forced at w'_i for all $i = 1, \ldots, k$, and so $\nu\beta$ is forced at w. Similarly, let $\pi\alpha \in \ell(w)$; due to the construction process of the frame, $\pi\alpha_1 \in \ell(w_i)$ and $\pi\alpha_2 \in \ell(w'_i)$ for some $i \in \{1, \ldots, k\}$, so by the induction hypothesis, $\pi\alpha_1$ is forced at w_i and $\pi\alpha_2$ is forced at w'_i for some i, and so $\pi\alpha$ is forced at w.

Lemma II. If Φ is a satisfiable finite set of formulas, then at least one of the hypertree-frame of the Φ-auxiliary structure is a (hypertree-)model of Φ.

Proof. We proceed by induction on the complexity of Φ.

Base case. If the complexity of Φ is 0, Φ contains only atomic signed formulas, and its hypertree-frame contains the single world w_0 where each $\varphi \in \Phi$ is forced; this is possible if and only if Φ is satisfiable. This frame clearly is a model of Φ.

Induction case. Φ contains at least one non-atomic signed formula.

If Φ contains an α-formula, then $(\Phi \setminus \{\alpha\}) \cup \{\alpha_1, \alpha_2\}$ has lower complexity than Φ and is also satisfiable; therefore one of the hypertree-frames of $(\Phi \setminus \{\alpha\}) \cup \{\alpha_1, \alpha_2\}$ is a model of $(\Phi \setminus \{\alpha\}) \cup \{\alpha_1, \alpha_2\}$, and also of Φ.

If Φ contains a β-formula, then $(\Phi \setminus \{\beta\}) \cup \{\beta_1\}$ or $(\Phi \setminus \{\beta\}) \cup \{\beta_2\}$ that has lower complexity than Φ, is also satisfiable and has a hypertree-model, which is also a model of Φ.

If the set Φ contains only $\pi\alpha$-formulas and $\nu\beta$-formulas, let $\Phi = \mathcal{P} \cup \mathcal{N}$ with $\mathcal{P} = \{\pi\alpha^1, \ldots, \pi\alpha^n\}$ and $\mathcal{N} = \{\nu\beta^1, \ldots, \nu\beta^m\}$. Furthermore, let $\mathcal{N}_{12} = \{\{\nu\beta^1_{i_1}, \ldots, \nu\beta^m_{i_m}\} : i_1, \ldots, i_m \in \{1, 2\}$ and $\nu\beta \in \mathcal{N}\}$. If $X \in \mathcal{N}_{12}$, let X_1 be the set of $\nu\beta_1$-elements of X and X_2 the set of $\nu\beta_2$-elements of X. If Φ is satisfiable, then for each ordered pair $(\pi\alpha_1^i, \pi\alpha_2^i)$ such that $\pi\alpha^i \in \mathcal{P}$, there exists an element $X \in \mathcal{N}_{12}$ such that the sets $\Phi_1^i = \{\pi\alpha_1^i\} \cup X_1$ and $\Phi_2^i = \{\pi\alpha_2^i\} \cup X_2$ are also satisfiable with a lower complexity, and therefore have a hypertree-model. The corresponding hypertree-model of Φ is obtained as follows: the root is a world w_0, with $\ell(w_0) = \Phi$, and there are n outgoing hyperarrows, leading to the $2n$ hypertree-models of the $\Phi^i_{1,2}$, $i = 1, \ldots, n$. (Renaming of worlds is used to avoid name clashes.)

Comment. The case $n = 0$ is not ruled out; a set of $\nu\beta$-formulas is always B^+-satisfiable, and every one-world frame is a model.

Theorem 1. Signed formula φ has a model if and only if some hypertree-frame associated with φ is a hypertree-model of φ.
It is an immediate consequence of lemmas I and II.

5.5 Path tree, soundness and completeness

Lemma III. The path tree associated with a finite (conjunctive) set of signed formulas is finite.
Proof. See §5.3, where the termination of the construction algorithm for the path tree has been proven.

A *line* in a hypertree-frame is a sequence $(\varphi_0, \ldots, \varphi_n)$ of signed formulas such that φ_0 is the initial formula, φ_i is a child of φ_{i-1} and φ_n has no child.
Lemma IV. If S is a path, there exists a hypertree-frame such that S contains exactly one member of every line.
Proof. This is true for the root of the path tree, and if it is true for some path, it is also true for every successor-path.
Lemma V. There is a correspondence between (hypertree-)frames and atomic paths associated with a finite set Φ of formulas; each atomic path is the set of signed atomic formulas of a frame, and the set of signed atomic formulas of each frame is an atomic path.
Proof. By induction on the degree of φ.

Theorem 2. Signed formula φ has a model if and only if (at least) one of its atomic path does not contain an opposite pair.
Proof. Signed formula φ has a model if and only if it has a hypertree-model (theorem 1). The corresponding atomic path (lemma V) is consistent and does not contain an opposite pair.

Corollary. The method is sound and complete.

Conclusion. We are half-way to an extension of the connection method to the decidable relevant logic B^+. The method introduced here inherits most of its properties from the tableau method Bloesch 1993b: soundness, completeness and termination. The next step is to obtain a true connection method, and to investigate its properties by using a matrix-characterization of B^+. Extensions to more powerful systems of relevant logic should also be possible.

References

Batens D. (2001). "A Dynamic Characterization of the Pure Logic of Relevant Implication". *Journal of philosophical logic 30*:267–280.

Bloesch A. (1993). *Signed Tableaux — A Basis for Automated Theorem Proving in non Classical Logics.* PhD thesis, University of Queensland

— (1993). “A Tableau Style Proof System for Two Paraconsistent Logics”, *Notre Dame Journal of Formal Logic 34*:2, pp.295–301.

Fitting M. (1983). *Proof Methods for Modal and Intuitionistic Logics.* Dordrecht Reidel Publishing Company.

— (1993). “Basic Modal Logic”, *Handbook of Logic in Artificial Intelligence and Logic Programming*), Ed. by D. M. Gabbay, C. J. Hogger and J. A. Robinson. Oxford Clarendon Press.

Goré R. (1992). *Cut-free Sequent and Tableau Systems for Propositional Normal Logic.* Technical report, Cambridge.

Gribomont E.P. & Rossetto D. (1995). CAVEAT: Technique and Tool for Computer Aided VErification And Transformation, *Lecture Notes in Computer Science 939, Computer Aided Verification.* Springer.

Hughes G.E. & Cresswell M.J. (1968). *An Introduction to Modal Logic.* London Methuen and Co Ltd 1968, 1972.

Priest G. & Sylvan R. (1992). “Simplified Semantics for Basic Relevant Logics”, *Journal of Philosophical Logic 21*:217–232.

Restall G. (1993). “Simplified Semantics for Relevant Logics (and some of their rivals)”, *Journal of Philosophical Logic 22*:481–511.

Wallen L. (1990). *Automated Proof Search in Non-classical Logics.* Cambridge MIT Press.

Chapter 22

LOGIC, RANDOMNESS AND COGNITION

Michel de Rougemont
Université Paris-II

Abstract Many natural intensional properties in artificial and natural languages are hard to compute. We show that randomized algorithms are often necessary to have good estimators of natural properties and to verify some specific relations. We concentrate on the reliability of queries to show the advantage of randomized algorithms in uncertain cognitive worlds.

1. Introduction

Classical studies in *Complexity Theory* consider deterministic or non deterministic algorithms on perfect data and often privilege a worst-case analysis to classify between easy and hard problems. In recent years, some important developments in theoretical Computer Science have shown the fundamental role of randomness in computing in at least three different settings.

- randomized algorithms for search and decision problems.
- models for randomized verification, i.e. given a function f and two values a, b decide if $f(a) = b$.
- average case analysis on the inputs.

We believe that new ideas are emerging that could turn out to be quite relevant to Cognitive Sciences, when we try to estimate *intensions* associated with natural or artificial languages. One fundamental aspect of computations in the context of cognitive science is the ability to deal with uncertainty. We will show that randomized techniques are quite efficient in uncertain situations.

We refer to intensions as properties other than the truth-value (or extension) of a formula and concentrate in the sequel on the notion of

D. Vanderveken (ed.), Logic, Thought & Action, 497–506.

reliability. Let us fix the Universe as a large finite structure U_n of a class **K** where n is its size, with functions, relations and higher-order objects. Let $\mathscr{L}$ be the vocabulary associated with a class **K** of such structures. If we fix a language with a denotational semantics, like the standard first-order logic ($FO(\mathscr{L})$), the truth value is well-defined but of limited interest in cognitive studies. Some other properties, usually defined inductively on a structure may be more relevant. These *intensions* are in general hard to compute (in the algorithmic sense) as we will see on some examples.

For artificial languages used in Computer Science two natural intensions are: the complexity and the reliability when we deal with uncertain data. In the sequel we concentrate on the reliability question and show how to use ramdomness to estimate a classical property: *graph reliability.* We also mention that this property may be easy on *the average* for some specific distributions (natural gaussian distributions). For other intensional properties, one would conjecture similar results.

In section 2 we introduce the reliability of a query as a basic intension. In section 3, we define random computations and describe some randomized algorithms. In section 4, we mention some classical results related to the verification of properties that are hard to compute. In section 5, we discuss the role of the average case complexity.

2. The intension of queries: reliability as an example

A query on a class **K** is a function which associates with every $U_n \in \mathbf{K}$ a relation of fixed arity on U_n. If the arity is 0, we have boolean queries which are true or false. It is also called a global relation on a class in the litterature. A query is definable in a logic $\mathscr{L}$ if there exists a formula $\psi \in \mathscr{L}$ such that for all $U_n \in \mathbf{K}$, the relation defined by the query is precisely $[\psi]^{U_n}$, i.e. the relation defined by the formula. The arity of the query is the number of free variables of the formula.

For simplicity, we concentrate on the following property of queries defined by a formula ψ, the reliability $\rho(\psi)$ introduced in dR95: given a structure U_n and a random substructure U'_n (the uncertain world) $\rho(\psi)$ is the probability that the truth-value $[\psi^{U_n}]$ coincides with the truth-value $[\psi^{U'_n}]$.

Consider a finite relational database and for the sake of simplicity we assume the database to be a finite graph $G_n = (V_n, E)$ with n nodes and $E \subseteq V_n^2$ is the set of edges. Let $\delta : E \to [0, 1]$ be the uncertainty function where we interpret $\delta(e)$ as the probability that the edge e exists. The probabilistic space induced by G_n and δ is the set of all subgraphs

$G'_n = (V, E')$ of G_n with a probability

$$\mathbb{P}rob(G') = [\Pi_{e \in E'} \delta(e)].[\Pi_{e \in E - E'}(1 - \delta(e)]$$

Let Q_δ be the random variable defining the (boolean) query Q on the probabilistic space induced by G_n and δ. We denote the mathematical expectation of this random variable by $\mathbb{E}(Q_\delta)$. A distribution μ defines a different probabilistic space: it assigns for a given n, the probability of G_n.

Definition 1

The **reliability** of a boolean query Q on a graph G_n is the function:

$$\rho(Q, G_n) = 1 - \mathbb{E}_\delta(| Q_\delta - Q |)$$

The **reliability on a distribution** μ of a query Q is the function: $\rho(Q, n) = \mathbb{E}_\mu[\rho(Q, G_n)]$

This definition consider only boolean queries but generalizes to queries of arbitrary arity.

Example: Let G_5 be the graph below with 5 nodes and Q the query defined by the first-order formula:

$$\exists x, y, z(zEx \wedge xEy \wedge yEz)$$

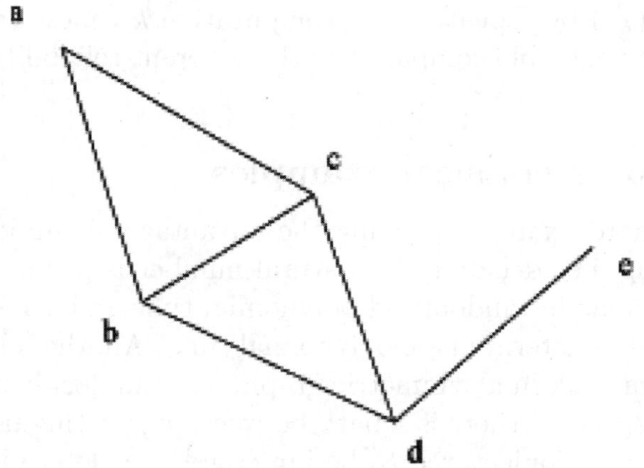

The graph G_5 with uncertain edges.

Assume $\delta(e) = \frac{1}{2}$. The value of $\rho(Q, G_5)$ is the probability that a realization, i.e. a subgraph of G_5 contains a triangle.

The reliability is hard to compute because we have to analyze all possible subgraphs G'_n, i.e. exponentially many, and check some property for each one. For many queries Q, there seems to be no better way than this exhaustive computation. The reliability of even first-order definable queries is hard to compute, not known to be computable in polynomial time.

3. Randomized Computations

There are many equivalent definitions of randomized computations. Consider a computing device, a Turing machine or a RAM (Random Access Machine) with two inputs: the real input $\mathbf{x}$ of length n and an auxiliary binary input $\mathbf{y} = y_1....y_m$, the random sequence. The probabilistic space is the set of $\mathbf{y}$ with a uniform probability $1/2^m$, i.e. each $y_i = 0$ or $y_i = 1$ is chosen with the same probability 1/2. In the case of decision problems, the machine accepts ($M(\mathbf{x}, \mathbf{y}) = 1$) or rejects (M(**x**,**y**) =0).

We say that M accepts a language $\mathscr{L}$ if

- If $\mathbf{x} \in \mathscr{L}$ then $\mathbb{P}rob_{\mathbf{y}}[M(\mathbf{x}, \mathbf{y}) = 1] \geq \frac{1}{2} + \epsilon$

- If $\mathbf{x} \notin \mathscr{L}$ then $\mathbb{P}rob_{\mathbf{y}}[M(\mathbf{x}, \mathbf{y}) = 0] \geq \frac{1}{2} + \epsilon$

The most classical complexity class (see Pap94; LdR96) is the class BPP, when M accepts or rejects deterministically in polynomial time. We can also define a probabilistic run on the input $\mathbf{x}$: it first produces $\mathbf{y}$ and then run $M(\mathbf{x}, \mathbf{y})$. Notice that the error can be made exponentially small ($\frac{1}{2^k}$) by repeating the computation k times. In particular it can be made negligible compared to the inherent reliability of hardware components.

3.1 Some classical examples

One standard example showing the advantage of randomness is primality testing, i.e. deciding if a natural number is prime or composite. This can be done in randomized polynomial time and is conjectured not to be possible in deterministic polynomial time. Another classic example is the random walk in a symmetric graph. We can decide in randomized logarithmic space[1] if there is a path between two distinguished elements s and t, but it is conjectured to be impossible for deterministic computations (in logarithmic space).

Consider a heap of needles and the associated graph where each node is the extremity of needles or the intersection of crossing needles. Edges connect nodes along the needles and there are two distinguished nodes: s and t.

[1]Logarithmic space can be understood as constant space, in the sense of a constant number of registers, each holding $\log n$ bits. To store a node in a graph with n nodes, we need to store a value i between 1 and n, requiring $\log n$ bits in the classical binary representation.

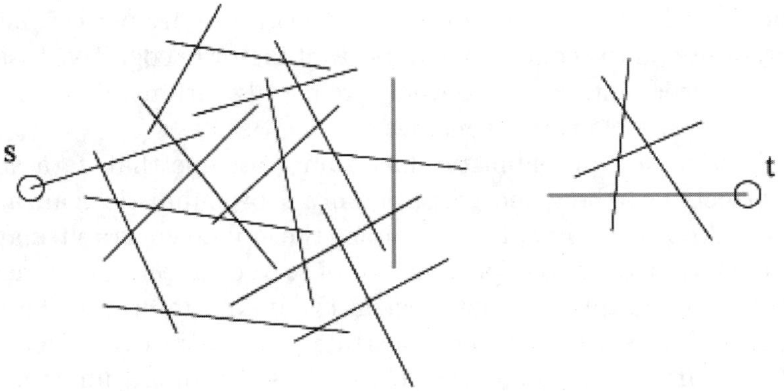

Needles: are s and t connected?

In the first question, we ask if the two distinguished points s and t are connected, i.e. if there exists a path which connects them. This is extremely easy for the human eye and for a randomized algorithm that performs a random walk from s hoping to reach t after n^2 steps. Such an algorithm generates a sequence $\mathbf{y}$ of random choices, starts in s and uses the random bits of $\mathbf{y}$ to select an adjacent node[2]. It proceeds for n^2 steps keeping only the current node. Any deterministic algorithm needs to keep track of the paths and needs polynomial space.

On the other hand, it is a much harder task if the graph is oriented and it is conjectured to be impossible to decide in randomized logarithmic space. Notice that it is also far more difficult to the human eye. We need to follow various paths edge by edge and do not have a global view of the situation, as in the previous example.

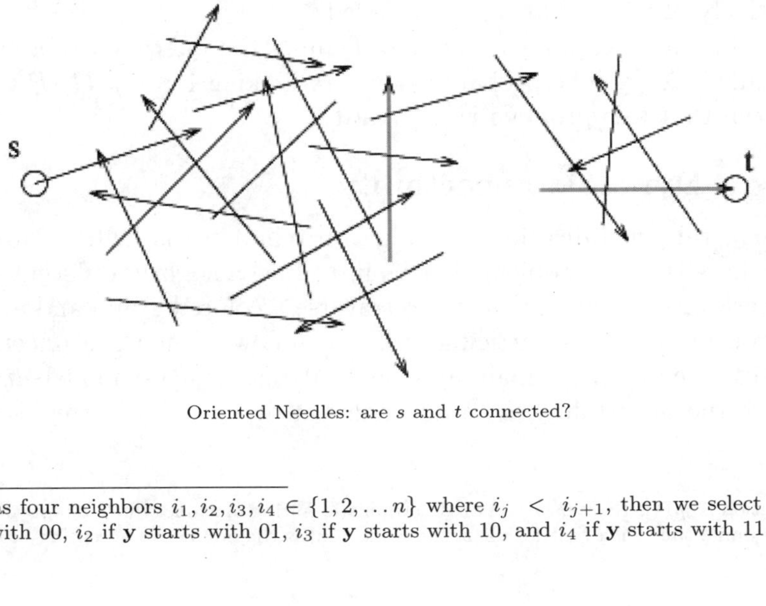

Oriented Needles: are s and t connected?

[2]If s has four neighbors $i_1, i_2, i_3, i_4 \in \{1, 2, \ldots n\}$ where $i_j < i_{j+1}$, then we select i_1 if $\mathbf{y}$ starts with 00, i_2 if $\mathbf{y}$ starts with 01, i_3 if $\mathbf{y}$ starts with 10, and i_4 if $\mathbf{y}$ starts with 11.

Notice that in the previous examples, the graphs are perfect, i.e. with no uncertainty on the edges. One important factor in cognitive tasks is to cope with uncertainty and to develop robust algorithms, i.e. procedures that are insensitive to erroneous data.

Problems with a probabilistic uncertainty assume that data are partially correct, i.e. the given graph is only a probabilistic realization of another unknown graph. For example, the unknown graph may have extra edges that do not appear in the observed graph and some edges in the observed graph may not exist in the unknown graph. An important distinction is whether the uncertainty is static (i.e. fixed as the algorithm starts) or dynamic (i.e. changes as the algorithm computes).

3.2 Static uncertainty

The probabilistic model introduced in section 2 is static in the sense that the random data is determined *before* any computation starts. For a query ψ, the computation of $\rho(\psi, G_n)$ may indeed be vary hard, in fact $\#P$ hard[3]. The standard example is the graph-reliability introduced in Val79, which is also the reliability of the query: *Are s and t connected?*. Formally the function GR is defined as follows:

GR (Graph reliability)Val79

Input: An undirected graph $G = (V, E)$ with n vertices; $s, t \in V$; and for every edge e, a rational number $\delta(e) \in [0, 1]$ representing the probability that the edge e exists (does not fail).

Output: The probability that there is a path from s to t consisting exclusively of edges that have not failed.

Consider a fixed-point formula ψ defining the query *s-t connectivity*, also called GAP. The probability we are looking for is $\rho(GAP, G_n)$. It is known that this problem is $\#P$ hard.

3.3 Dynamic uncertainty

A natural generalization of the graph reliability is DGR, Pap85 the dynamic reliability problem. Let us introduce a slightly different model of uncertainty: suppose you try to traverse a colored graph and at every step you decide on a particular edge to follow. Then the uncertainty removes some of the remaining edges. We call such a model *dynamic* because the uncertainty is an adversary at every step. In this dynamic

[3] A function is $\#P$ is there exists a non deterministic Turing machine which accepts or rejects in polynomial time such that for all x, the value $f(x)$ is the number of accepting branches.

model of uncertainty, we can show that randomized decisions can be better than deterministic ones.

This situation is *typical* in this more elaborate example BdRS96, where we try to traverse a colored graph, subject to uncertain deviations. Consider a graph supplied with additional information concerning colours of vertices, probabilities of deviations from the chosen direction, labels of edges and so on. An edge with the tail u and the head v will be denoted by uv or (u, v). Let $G = (V, E)$ be a directed graph (digraph) with the vertices V and edges E. $OUT(v) = \{e \in E : tail(e) = v\}$, $IN(v) = \{e \in E : head(e) = v\}$. $COLOURS$ is a finite set of colours, $clr : V \rightarrow COLOURS$ is the colouring function.

To model the uncertainty we introduce two functions, one is an auxiliary function of labeling which gives the local names of the edges outcoming from a given vertex, and the other, denoted by μ below, is the function describing deviations from a chosen move.

$LABELS$ is a finite set of edge labels, $lbl_v : OUT(v) \rightarrow LABELS$; lbl_v is injective without loss of generality, $\mu : E \times E \rightarrow [0, 1]$ is the function describing the *uncertainty*.

Let $e = (v, w)$ be an edge chosen to follow. Actually the motion will be along another edge $e_1 = (v, u)$ with the probability $\mu(e, e_1)$, and so the vertex w will be reached only with the probability $\mu(e, e)$. We assume

$$\sum_{e_1 \in OUT(v)} \mu(e, e_1) = 1. \qquad (22.1)$$

The input of our problem is an object of the form

$$((V,\ E,\ clr,\ lbl,\ \mu),\ s,\ t),$$

which we call *graph with uncertain deviation and source/target vertices* or UD-graph. A *strategy* is a function σ which assigns to a finite sequence of colours (the history of colours of visited points) an edge label describing uniquely the edge to follow.

$$\sigma : COLOURS^* \rightarrow LABELS\,.$$

The *semantics* of a strategy σ (or the *behaviour* due to σ) is given by the random mapping $path_\sigma : \boldsymbol{N} \rightarrow V^*$ which for every $k \in \boldsymbol{N}$ defines a random path traversed following σ in k steps. The motion starts from s. Then σ, on the basis of $clr(s)$, chooses some edge $e \in OUT(s)$ (i.e. $e = lbl_s^{-1}(\sigma(clr(s))))$, and with the probability $\mu(e, e_1)$ goes along an edge e_1 to $head(e_1)$ and so on. A path that can be a value of $path_\sigma(k)$ is called a *realization* of a strategy σ after k steps. A realization is *simple* if it contains not more than one occurrence of the target vertex. A realization is *precise* w.r.t. the target iff it is simple and has t as its last vertex.

We say that σ *leads* from s to t in k steps (with a probability $1-\theta$) if (with the probability $1-\theta$) there exists a realisation of σ of the length k with the first vertex s and the last vertex t. The general problem is to reach t from s with the maximal probability for a limited or unbounded number of steps. This motivates the following criterion of the reliability:

- $\boldsymbol{R}(\sigma,\ k) \ = \ \boldsymbol{Prob}(\sigma$ leads from s to t in not more than k steps),
- $\boldsymbol{R}_\infty(\sigma) \ = \boldsymbol{R}(\sigma) \ = \sup_k \boldsymbol{R}(\sigma, k)$.

It can be shown that computing $\boldsymbol{R}_\infty(\sigma)$ can be arbitrarly complex, and that computing $\boldsymbol{R}(\sigma,\ k)$ is $\#P$ computable. However define a **randomized strategy** as:

$$\sigma_R \ : \ COLOURS^*.\{0,1\}^* \ \rightarrow \ LABELS\,.$$

It can be shown easily that for a fixed horizon k some simple randomized strategy are better than any deterministic strategy with bounded memory dRS97. For the general problem, we can show that randomized strategies are better than deterministic ones for most finite horizon k BdRS96.

4. Randomized Verification

The verification problem for a function f can be stated as follows: given x and y, decide if $f(x) = y$. The definition of the class of functions that can be verified in randomized polynomial time was first defined by GMR85; Bab85 who introduced the class IP (Interactive proofs). It was shown that randomness and interaction could vastly increase the domain of functions verifiable in randomized polynomial time. In CDFdRS94, we gave an interactive protocol for the verification of GR, which can't be done in polynomial time but with a simple $O(n)$ interactive protocol.

Consider now the graph-traversing problem of section 3.3. Suppose two agents (programs) claim to traverse a graph with probability greater than 0.5. How do we verify their claims? How do we know that one of the program is better than the other?

It appears essential to compare strategies, or more generally cognitive tasks. An interactive proof for these problems would be extremely useful and allow us to answer these questions with a very simple randomized verification. It would lead to better strategies on specific inputs.

5. Computing on the average

The average case complexity Lev73 is another interesting approach to complexity. An algorithm A whose running time is $T_A(x)$ is computable in Average polynomial time if $\mathbb{E}_\mu[T_A(x)] \leq n^k$ for an input distribution μ. The problem $3COL$ (whether a graph is 3-colorable) is NP complete but it was shown by Gurevich that it can be solved in constant time on the average for the uniform distribution. What can we say for GR? It has been shown Sin93 that GR is not approximable but it is an open problem whether it is polynomial on the average ($Average(P)$) for the uniform distribution. Consider however the following gaussian distribution: Let μ be the gaussian distribution on the edges defined as follows:

$$\mu : e = (i, j) \longmapsto \mu[(i, j)] = \exp[-(i-j)^2]$$

A random graph for μ assumes an ordering on the vertices and the probability to join (i, j) decreases exponentially quickly with the distance $d = j - i$. For such a distribution, we showed in BdR98 that GR is computable in average polynomial time. In simple words, the algorithm works well on most inputs except on some *bad* ones which are rare.

It is important to notice how a statistical information (the distribution μ) changes the complexity of the problem and directly influences the search of randomized algorihms.

6. Conclusion

Many intensional properties associated with natural and artificial languages are difficult to compute in the classical sense. Randomized algorithms must be used for the verification or the estimation of these properties. We described the case of the reliability, a property hard to compute in general, but which can be approximated on specific inputs. We believe that many intensional properties can be approached with similar techniques, which could be useful to the cognitive sciences.

References

Babai L. (1985). Trading Group Theory for Randomness. Symposium on the Theory of Computing, 421–429.

Burago D. and de Rougemont M. (1998). "On the Average Complexity of Graph Reliability". *Fundamenta Informaticae 36*(4):307–315.

Burago D., de Rougemont M. and Slissenko A (1996). "On the Complexity of Partially Observed Markov Decision Processes". *Theoretical Computer Sciencce, (157)*:161–183.

Couveignes J.M., Diaz-Frias J.F., de Rougemont M. and Santha M (1994). "On the Interactive Complexity of Graph Reliability". FSTCS

International Symposium on Theoretical Computer Science, Madras, *LNCS 880*:1–14.

de Rougemont M. (1995). The Reliability of Queries. ACM Principles on Databases Systems, 286–191.

de Rougemont M. and Schlieder C. (1997). Spatial Navigation with Uncertain Deviations. American Asssociation for Artificial Intelligence.

Goldwasser S., Micali S. and Rackoff C. (1985). The Knowledge Complexity of Interactive Poof Systems. Symposium on the Theory of Computing, 291–304.

Lassaigne R. & de Rougemont M. (1996). *Logique et complexité*. Hermès.

Levin L. (1973). "Universal Sorting Problems". *Problems of Information Transmission 9*(3):265–266.

Papadimitriou C. (1985). "Games Against Nature". *Journal of Computer and System Sciences 31*:288–301.

Papadimitriou C. (1994). *Computational Complexity*. Addison-Wesley.

Sinclair A. (1993). *Algorithms for Random Generation and Counting.* Birhauser Verlag.

Valiant L. (1979). "The Complexity of Enumeration and Reliability Problems". *SIAM Journal of Computing 8*(3).

Chapter 23

FROM COMPUTING WITH NUMBERS TO COMPUTING WITH WORDS — FROM MANIPULATION OF MEASUREMENTS TO MANIPULATION OF PERCEPTIONS*

Lofti Zadeh
University of California

Computing, in its usual sense, is centered on manipulation of numbers and symbols. In contrast, computing with words, or CW for short, is a methodology in which the objects of computation are words and propositions drawn from a natural language, e.g., *small, large, far, heavy, not very likely, the price of gas is low and declining, Berkeley is near San Francisco, it is very unlikely that there will be a significant increase in the price of oil in the near future*, etc. Computing with words is inspired by the remarkable human capability to perform a wide variety of physical and mental tasks without any measurements and any computations. Familiar examples of such tasks are parking a car, driving in heavy traffic, playing golf, riding a bicycle, understanding speech and summarizing a story. Underlying this remarkable capability is the brain's crucial ability to manipulate perceptions – perceptions of distance, size, weight, color, speed, time, direction, force, number, truth, likelihood and other characteristics of physical and mental objects. Manipulation of perceptions plays a key role in human recognition, decision and execution processes. As a methodology, computing with words provides a foundation for a computational theory of perceptions – a theory which may have an important bearing on how humans make – and machines might make –

*This paper appeared in *IEEE Transactions on Circuits and Systems*, 105–119, 1999. We thank IEEE for granting us permission to reproduce this paper.

D. Vanderveken (ed.), Logic, Thought & Action, 507–544.

perception-based rational decisions in an environment of imprecision, uncertainty and partial truth.

A basic difference between perceptions and measurements is that, in general, measurements are crisp whereas perceptions are fuzzy. One of the fundamental aims of science has been and continues to be that of progressing from perceptions to measurements. Pursuit of this aim has led to brilliant successes. We have sent men to the moon; we can build computers that are capable of performing billions of computations per second; we have constructed telescopes that can explore the far reaches of the universe; and we can date the age of rocks that are millions of years old. But alongside the brilliant successes stand conspicuous underachievements and outright failures. We cannot build robots which can move with the agility of animals or humans; we cannot automate driving in heavy traffic; we cannot translate from one language to another at the level of a human interpreter; we cannot create programs which can summarize non-trivial stories; our ability to model the behavior of economic systems leaves much to be desired; and we cannot build machines that can compete with children in the performance of a wide variety of physical and cognitive tasks.

It may be argued that underlying the underachievements and failures is the unavailability of a methodology for reasoning and computing with perceptions rather than measurements. An outline of such a methodology – referred to as a computational theory of perceptions – is presented in this paper. The computational theory of perceptions, or CTP for short, is based on the methodology of computing with words (CW). In CTP, words play the role of labels of perceptions and, more generally, perceptions are expressed as propositions in a natural language. CW-based techniques are employed to translate propositions expressed in a natural language into what is called the Generalized Constraint Language (GCL). In this language, the meaning of a proposition is expressed as a generalized constraint, X isr R, where X is the constrained variable, R is the constraining relation and isr is a variable copula in which r is a variable whose value defines the way in which R constrains X. Among the basic types of constraints are: possibilistic, veristic, probabilistic, random set, Pawlak set, fuzzy graph and usuality. The wide variety of constraints in GCL makes GCL a much more expressive language than the language of predicate logic.

In CW, the initial and terminal data sets, IDS and TDS, are assumed to consist of propositions expressed in a natural language. These propositions are translated, respectively, into antecedent and consequent constraints. Consequent constraints are derived from antecedent constraints through the use of rules of constraint propagation. The principal constraint propagation rule is the generalized extension principle. The derived constraints are retranslated into a natural language, yielding the terminal data set (TDS). The rules of constraint propagation in CW coincide with the rules of inference in fuzzy logic. A basic problem in CW is that of explicitation of X, R and r in a generalized constraint, X isr R, which represents the meaning of a proposition, p, in a natural language.

There are two major imperatives for computing with words. First, computing with words is a necessity when the available information is too imprecise to justify the use of numbers; and second, when there is a tolerance for imprecision which can be exploited to achieve tractability, robustness, low solution cost and better rapport with reality. Exploitation of the tolerance for imprecision is an issue of central importance in CW and CTP. At this juncture, the computational theory of perceptions – which is based on CW – is in its initial stages of development. In time, it may come to play an important role in the conception, design and utilization of information/intelligent systems. The role model for CW and CTP is the human mind.

1. Introduction

In the fifties, and especially late fifties, circuit theory was at the height of importance and visibility. It played a pivotal role in the conception and design of electronic circuits and was enriched by basic contributions of Darlington, Bode, McMillan, Guillemin, Carlin, Youla, Kuh, Desoer, Sandberg and other pioneers.

However, what could be discerned at that time was that circuit theory was evolving into a more general theory – system theory – a theory in which the physical identity of the elements of a system is subordinated to a mathematical characterication of their input/output relations. This evolution was a step in the direction of greater generality and, like most generalizations, it was driven by a quest for models which make it possible to reduce the distance between an object that is modeled – the modelizand – and its model in a specified class of systems.

In a paper published in 1961 entitled "From Circuit Theory to System Theory," (Zadeh, 1961) I discussed the evolution of circuit theory into system theory and observed that the high effectiveness of system theory in dealing with mechanistic systems stood in sharp contrast to its low effectiveness in the realm of humanistic systems – systems exemplified by economic systems, biological systems, social systems, political systems and, more generally, manmachine systems of various types. In more specific terms, I wrote:

> There is a fairly wide gap between what might be regarded as "animate" system theorists and 'inanimate' system theorists at the present time, and it is not at all certain that this gap will be narrowed, much less closed, in the near future. There are some who feel that this gap reflects the fundamental inadequacy of conventional mathematics – the mathematics of precisely-defined points, functions, sets, probability measures, etc. – for coping with the analysis of biological systems, and that to deal effectively with such systems, which are generally orders of magnitude more complex than man-made systems, we need a radically different kind of mathematics, the mathematics of fuzzy or cloudy quantities which are not describable in terms of probability distributions. Indeed, the need for such mathematics is becoming increasingly apparent even in the realm of inanimate systems, for in most practical cases the *a priori* data as well as the criteria by which the performance of a man-made system are judged are far from being precisely specified or having accurately known probability distributions.

It was this observation that motivated my development of the theory of fuzzy sets, starting with the 1965 paper "Fuzzy Sets" (Zadeh, 1965), which was published in *Information and Control.*

Subsequently, in a paper published in 1973, "Outline of a New Approach to the Analysis of Complex Systems and Decision Processes," (Zadeh, 1973) I introduced the concept of a linguistic variable, that is, a variable whose values are words rather than numbers. The concept of a linguistic variable has played and is continuing to play a pivotal role in the development of fuzzy logic and its applications.

The initial reception of the concept of a linguistic variable was far from positive, largely because my advocacy of the use of words in systems and decision analysis clashed with the deep-seated tradition of respect for numbers and disrespect for words. The essence of this tradition was succinctly stated in 1883 by Lord Kelvin:

> In physical science the first essential step in the direction of learning any subject is to find principles of numerical reckoning and practicable methods for measuring some quality connected with it. I often say that when you can measure what you are speaking about and express it in numbers, you know something about it; but when you cannot measure it, when you cannot express it in numbers, your knowledge is of a meagre and unsatisfactory kind: it may be the beginning of knowledge but you have scarcely, in your thoughts, advanced to the state of science, whatever the matter may be.

The depth of scientific tradition of respect for numbers and derision for words was reflected in the intensity of hostile reaction to my ideas by some of the prominent members of the scientific elite. In commenting on my first exposition of the concept of a linguistic variable in 1972, Rudolph Kalman had this to say:

> I would like to comment briefly on Professor Zadeh's presentation. His proposals could be severely, ferociously, even brutally criticized from a technical point of view. This would be out of place here. But a blunt question remains: Is Professor Zadeh presenting important ideas or is he indulging in wishful thinking? No doubt Professor Zadeh's enthusiasm for fuzziness has been reinforced by the prevailing climate in the U.S. one of unprecedented permissiveness. 'Fuzzification' is a kind of scientific permissiveness; it tends to result in socially appealing slogans unaccompanied by the discipline of hard scientific work and patient observation.

In a similar vein, my esteemed colleague Professor William Kahan – a man with a brilliant mind – offered this assessment in 1975:

> "Fuzzy theory is wrong, wrong, and pernicious." says William Kahan, a professor of computer sciences and mathematics at Cal whose Evans Hall office is a few doors from Zadeh's. "I can not think of any problem that could not be solved better by ordinary logic." What Zadeh is saying is the same sort of things 'Technology got us into this mess and now it can't get us out.' Well, technology did not get us into this mess. Greed and weakness and ambivalence got us into this mess. What we need is more logical thinking, not less. The danger of fuzzy theory is that it will encourage the sort of imprecise thinking that has brought us so much trouble."

What Lord Kelvin, Rudolph Kalman, William Kahan and many other brilliant minds did not appreciate is the fundamental importance of the remarkable human capability to perform a wide variety of physical and mental tasks without any measurements and any computations. Familiar examples of such tasks are parking a car; driving in heavy traffic; playing golf; understanding speech and summarizing a story.

Underlying this remarkable ability is the brain's crucial ability to manipulate perceptions – perceptions of size, distance, weight, speed, time, direction, smell, color, shape, force, likelihood, truth and intent, among others. A fundamental difference between measurements and perceptions is that, in general, measurements are crisp numbers whereas perceptions are fuzzy numbers or, more generally, fuzzy granules, that is, clumps of objects in which the transition from membership to nonmembership is gradual rather than abrupt.

The fuzziness of perceptions reflects finite ability of sensory organs and the brain to resolve detail and store information. A concomitant of fuzziness of perceptions is the preponderant partiality of human concepts in the sense that the validity of most human concepts is a matter of degree. For example, we have partial knowledge, partial understanding, partial certainty, partial belief and accept partial solutions, partial truth and partial causality. Furthermore, most human concepts have a granular structure and are context-dependent.

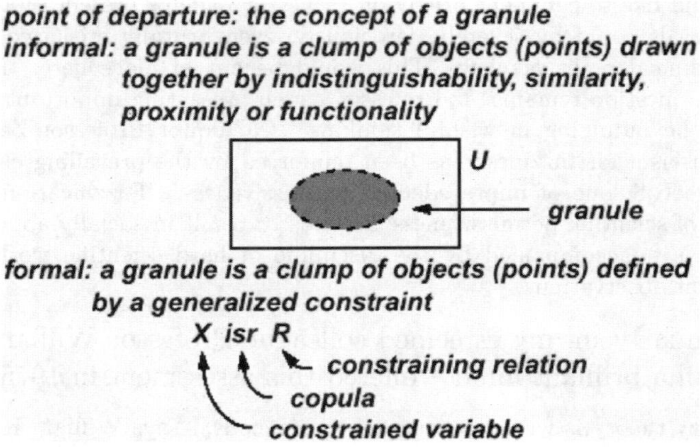

Figure 1. Informal and formal definitions of a granule.

In essence, a granule is a clump of physical or mental objects (points) drawn together by indistinguishability, similarity, proximity or functionality (Fig. 1). A granule may be crisp or fuzzy, depending on whether its boundaries are or are not sharply defined. For example, age may be granulated crisply into years and granulated fuzzily into fuzzy intervals labeled very young, young, middle-aged, old and very old (Fig. 2). A partial taxonomy of granulation is shown in Figs. 3(a) and 3(b).

body* → *head + neck + arms + chest + ... + feet
time* → *years* → *months* → *...
book* → *chapters* → *sections* → *paragraphs* → *...
age* → *young + middle-aged + old
size* → *small + medium + large
distance* → *very near + near + not.near and not.far + far + very far
force* → *weak + moderate + strong

Figure 2. Examples of crisp and fuzzy granulation.

In a very broad sense, granulation involves a partitioning of whole into parts. Modes of information granulation (IG) in which granules are crisp play important roles in a wide variety of methods, approaches and techniques. Among them are: interval analysis, quantization, chunking, rough set theory, diakoptics, divide and conquer, Dempster-Shafer theory, machine learning from examples, qualitative process theory, decision trees, semantic networks, analog-to-digital conversion, constraint programming, image segmentation, cluster analysis and many others.

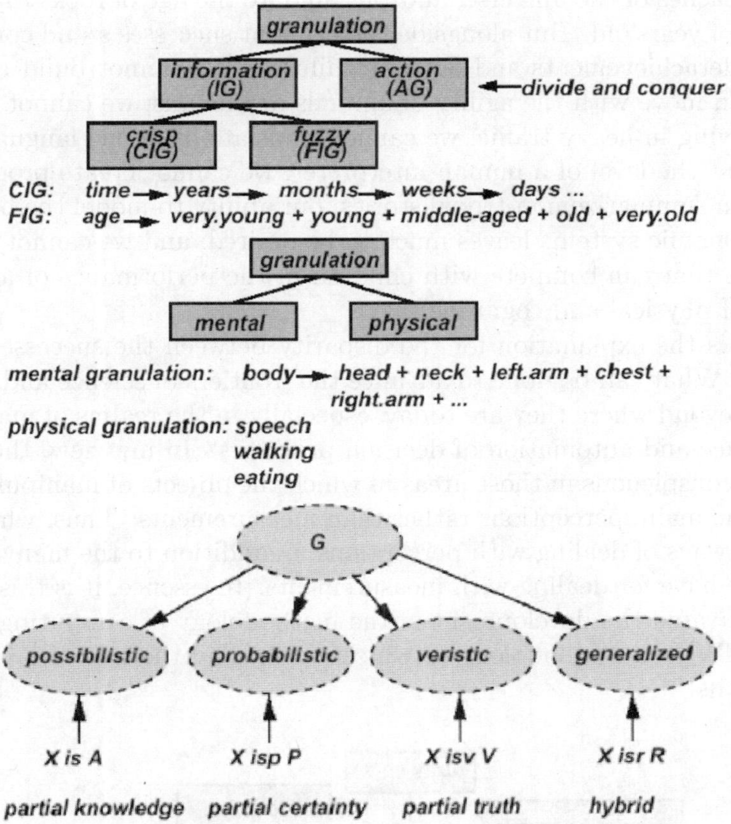

(a)

(b)

Figure 3. (a) Partial taxonomy of granulation; (b) Principal types of granules.

Important though it is, crisp IG has a major blind spot. More specifically, it fails to reflect the fact that most human perceptions are fuzzy rather than crisp. For example, when we mentally granulate the human body into fuzzy granules labeled head, neck, chest, arms, legs, etc., the length of neck is a fuzzy attribute whose value is a fuzzy number. Fuzziness of granules, their attributes and their values is characteristic of ways in which human concepts are formed, organized and manipulated. In effect, fuzzy information granulation (fuzzy IG) may be viewed as a human way of employing data compression for reasoning and, more particularly, making rational decisions in an environment of imprecision, uncertainty and partial truth.

The tradition of pursuit of crispness and precision in scientific theories can be credited with brilliant successes. We have sent men to the moon; we can build computers that are capable of performing billions of computations per second; we have constructed telescopes that can explore the far reaches of the universe; and we can date the age of rocks that are millions of years old. But alongside the brilliant successes stand conspicuous underachievements and outright failures. We cannot build robots which can move with the agility of animals or humans; we cannot automate driving in heavy traffic; we cannot translate from one language to another at the level of a human interpreter; we cannot create programs which can summarize non-trivial stories; our ability to model the behavior of economic systems leaves much to be desired; and we cannot build machines that can compete with children in the performance of a wide variety of physical and cognitive tasks.

What is the explanation for the disparity between the successes and failures? What can be done to advance the frontiers of science and technology beyond where they are today, especially in the realms of machine intelligence and automation of decision processes? In my view, the failures are conspicuous in those areas in which the objects of manipulation are, in the main, perceptions rather than measurements. Thus, what we need are ways of dealing with perceptions, in addition to the many tools which we have for dealing with measurements. In essence, it is this need that motivated the development of the methodology of computing with words (CW) – a methodology in which words play the role of labels of perceptions.

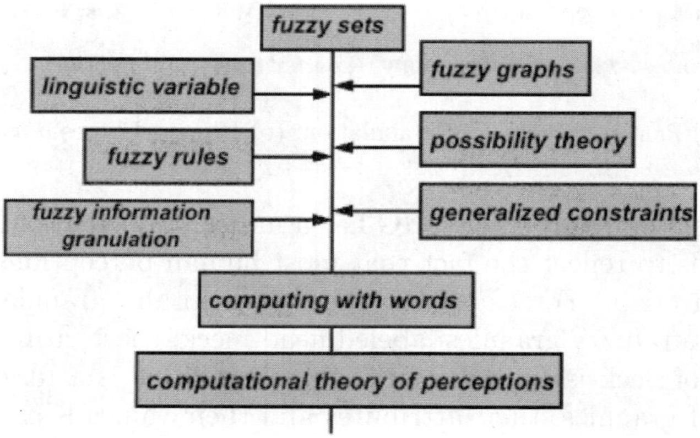

Figure 4. Conceptual structure of computational theory of perceptions.

Computing with words provides a methodology for what may be called a *computational theory of perceptions* (CTP) (Fig. 4). However, the potential impact of the methodology of computing with words is much broader. Basically, there are four principal rationales for the use of CW:

(i) *The don't know rationale.* In this case, the values of variables and/or parameters are not known with sufficient precision to justify the use of conventional methods of numerical computing. An example is decision-making with poorly defined probabilities and utilities.

(ii) *The don't need rationale.* In this case, there is a tolerance for imprecision which can be exploited to achieve tractability, robustness, low solution cost and better rapport with reality. An example is the problem of parking a car.

(iii) *The can't solve rationale.* In this case, the problem cannot be solved through the use of numerical computing. An example is the problem of automation of driving in city traffic.

(iv) *The can't define rationale.* In this case, a concept that we wish to define is too complex to admit of definition in terms of a set of numerical criteria. A case in point is concept of causality. Causality is an instance of what may be called an amorphic concept.

The basic idea underlying the relationship between CW and CTP is conceptually simple. More specifically, in CTP perceptions and queries are expressed as propositions in a natural language. Then, propositions and queries are processed by CW-based methods to yield answers to queries. Simple examples of linguistic characterization of perceptions drawn from everyday experiences are:

Robert is highly intelligent
Carol is very attractive
Hans loves wine
Overeating causes obesity
Most Swedes are tall
Berkeley is more lively than Palo Alto
It is likely to rain tomorrow
It is very unlikely that there will be a significant increase in the price of oil in the near future

Examples of correct conclusions drawn from perceptions through the use of CW-based methods are shown in Fig. 5(a). Examples of incorrect conclusions are shown in Fig. 5(b).

Perceptions have long been an object of study in psychology. However, the idea of linking perceptions to computing with words is in a different spirit. An interesting system-theoretic approach to perceptions is described in a recent work of R. Vallée (1995). A logic of perceptions has been described by H. Rasiowa (1989). These approaches are not related to the approach described in our paper.

An important point that should be noted is that classical logical systems such as propositional logic, predical logic and modal logic, as well

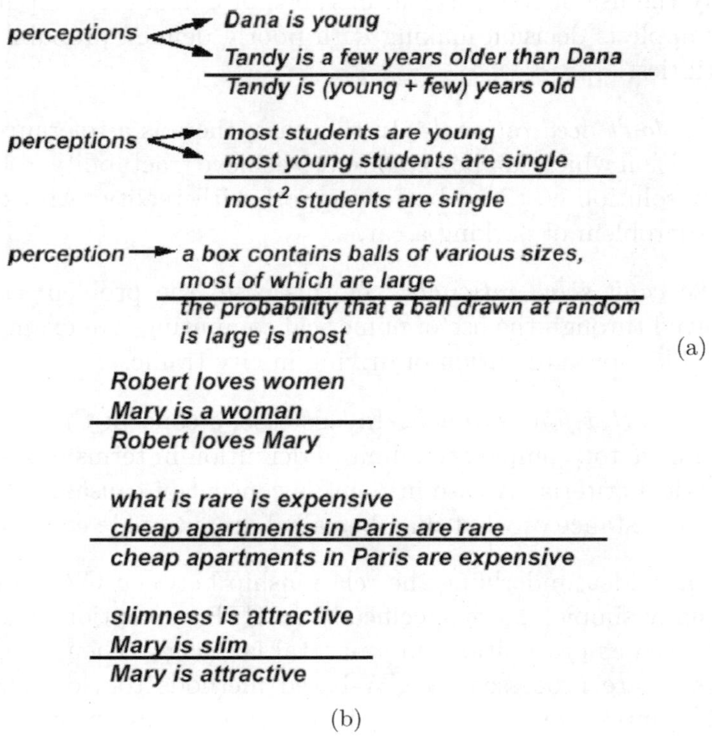

Figure 5. (a) Examples of reasoning with perceptions; (b) Examples of incorrect reasoning.

as AI-based techniques for natural language processing and knowledge representation, are concerned in a fundamental way with propositions expressed in a natural language. The main difference between such approaches and CW is that the methodology of CW – which is based on fuzzy logic – provides a much more expressive language for knowledge representation and much more versatile machinery for reasoning and computation.

In the final analysis, the role model for computing with words is the human mind and its remarkable ability to manipulate both measurements and perceptions. What should be stressed, however, is that although words are less precise than numbers, the methodology of computing with words rests on a mathematical foundation. An exposition of the basic concepts and techniques of computing with words is presented in the following sections. The linkage of CW and CTP is discussed very briefly because the computational theory of perceptions is still in its early stages of development.

2. What is CW?

In its traditional sense, computing involves for the most part manipulation of numbers and symbols. By contrast, humans employ mostly words in computing and reasoning, arriving at conclusions expressed as words from premises expressed in a natural language or having the form of mental perceptions. As used by humans, words have fuzzy denotations. The same applies to the role played by words in CW.

The concept of CW is rooted in several papers starting with my 1973 paper "Outline of a New Approach to the Analysis of Complex Systems and Decision Processes," (Zadeh, 1973) in which the concepts of a linguistic variable and granulation were introduced. The concepts of a fuzzy constraint and fuzzy constraint propagation were introduced in "Calculus of Fuzzy Restrictions," (Zadeh, 1975a), and developed more fully in "A Theory of Approximate Reasoning," (Zadeh, 1979b) and "Outline of a Computational Approach to Meaning and Knowledge Representation Based on a Concept of a Generalized Assignment Statement," (Zadeh, 1986). Application of fuzzy logic to meaning representation and its role in testscore semantics are discussed in "PRUF – A Meaning Representation Language for Natural Languages," (Zadeh, 1978b), and "Test-Score Semantics for Natural Languages and Meaning Representation via PRUF," (Zadeh, 1981). The close relationship between CW and fuzzy information granulation is discussed in "Toward a Theory of Fuzzy Information Granulation and its Centrality in Human Reasoning and Fuzzy Logic (Zadeh, 1997)."

Although the foundations of computing with words were laid some time ago, its evolution into a distinct methodology in its own right reflects many advances in our understanding of fuzzy logic and soft computing – advances which took place within the past few years. (See References and Related Papers.) A key aspect of CW is that it involves a fusion of natural languages and computation with fuzzy variables. It is this fusion that is likely to result in an evolution of CW into a basic methodology in its own right, with wide-ranging ramifications and applications.

We begin our exposition of CW with a few definitions. It should be understood that the definitions are dispositional, that is, admit of exceptions.

As was stated earlier, a concept which plays a pivotal role in CW is that of a granule. Typically, a granule is a fuzzy set of points drawn together by similarity (Fig. 1). A word may be atomic, as in *young*, or composite, as in *not very young* (Fig. 6). Unless stated to the contrary, a word will be assumed to be composite. The denotation of a word may be a higher order predicate, as in Montague grammar (Hobbs, 1978; Partee, 1976).

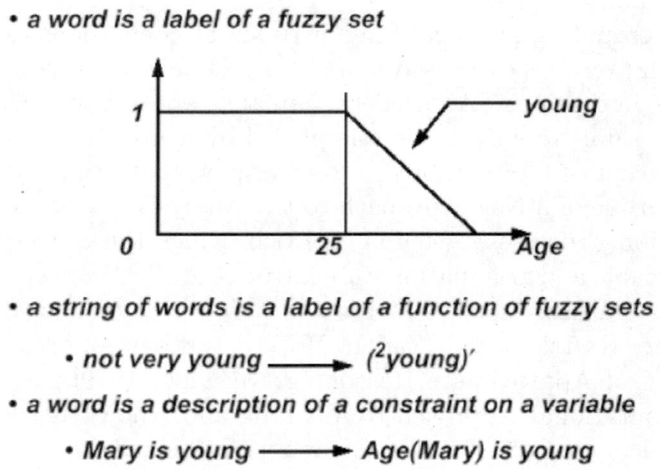

Figure 6. Words as labels of fuzzy sets.

In CW, a granule, g, which is the denotation of a word, w, is viewed as a fuzzy constraint on a variable. A pivotal role in CW is played by fuzzy constraint propagation from premises to conclusions. It should be noted that, as a basic technique, constraint propagation plays important roles in many methodologies, especially in mathematical programming, constraint programming and logic programming. (See References and Related Papers.)

As a simple illustration, consider the proposition *Mary is young*, which may be a linguistic characterization of a perception. In this case, young is the label of a granule young. (Note that for simplicity the same symbol is used both for a word and its denotation.) The fuzzy set *young* plays the role of a fuzzy constraint on the age of Mary (Fig. 6).

As a further example consider the propositions

$$p_1 = \textit{Carol lives near Mary}$$

and

$$p_2 = \textit{Mary lives near Pat.}$$

In this case, the words *lives near* in p_1 and p_2 play the role of fuzzy constraints on the distances between the residences of Carol and Mary, and Mary and Pat, respectively. If the query is: How far is Carol from Pat?, an answer yielded by fuzzy constraint propagation might be expressed as p_3, where

$$p_3 = \textit{Carol lives not far from Pat.}$$

More about fuzzy constraint propagation will be said at a later point.

A basic assumption in CW is that information is conveyed by constraining the values of variables. Furthermore, information is assumed to consist of a collection of propositions expressed in natural or synthetic language. Typically, such propositions play the role of linguistic characterization of perceptions.

A basic generic problem in CW is the following.

We are given a collection of propositions expressed in a natural language which constitute the *initial data set*, or IDS for short.

From the initial data set we wish to infer an answer to a query expressed in a natural language. The answer, also expressed in a natural language, is referred to as the *terminal data set*, or TDS for short. The problem is to derive TDS from IDS (Fig. 7).

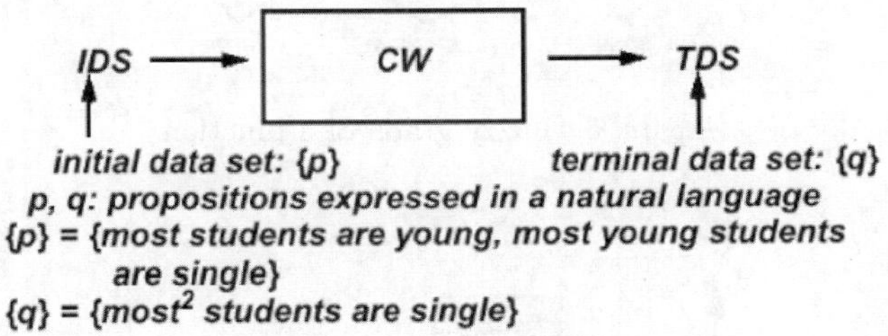

Figure 7. Computing with words as a transformation of an initial data set (IDS) into a terminal data set (TDS).

A few problems will serve to illustrate these concepts. At this juncture, the problems will be formulated but not solved.

1) Assume that a function f, $f: U \to V$, $X \in U$, $Y \in V$, is described in words by the fuzzy if-then rules

$$\begin{aligned} f: \quad & \text{if } X \text{ is } \textit{small} \text{ then } Y \text{ is } \textit{small} \\ & \text{if } X \text{is } \textit{medium} \text{ then } Y \text{ is } \textit{large} \\ & \text{if } X \text{is } \textit{large} \text{ then } Y \text{is } \textit{small} \end{aligned}$$

What this implies is that f is approximated to by a fuzzy graph f^* (Fig. 8), where

$$f^* = \textit{small} \times \textit{small} + \textit{medium} \times \textit{large} + \textit{large} \times \textit{small}$$

In f^*, $+$ and $\times$ denote respectively, the disjunction and cartesian product. An expression of the form $A \times B$, where A and B are words, will be referred to as a *Cartesian product*, *Cartesian granule*. In this sense, a fuzzy graph may be viewed as a disjunction of cartesian granules.

In essence, a fuzzy graph serves as an approximation to a function or a relation (Zadeh, 1974; 1996). Equivalently, it may be viewed as a linguistic characterization of a perception of f (Fig. 9).

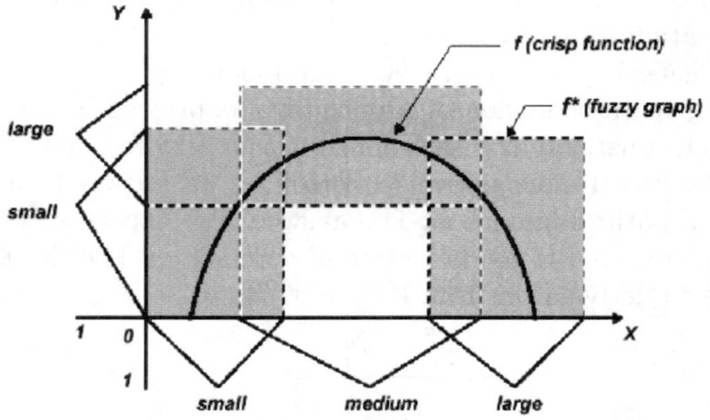

Figure 8. Fuzzy graph of a function.

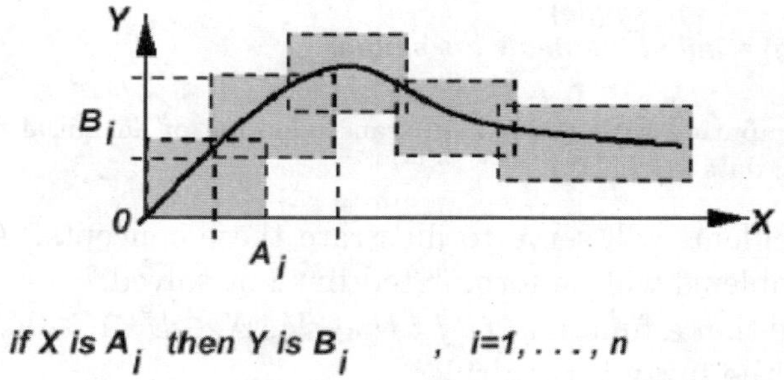

Figure 9. A fuzzy graph of a function represented by a rule-set.

In the example under consideration, the IDS consists of the fuzzy rule-set f. The query is: What is the maximum value of f (Fig. 10)? More broadly, the problem is: How can one compute an attribute of a function, f, e.g., its maximum value or its area or its roots if is described in words as a collection of fuzzy if-then rules? Determination of the maximum value will be discussed in greater detail at a later point.

2) A box contains ten balls of various sizes of which several are large and a few are small. What is the probability that a ball drawn at random is neither large nor small? In this case, the IDS is a verbal description of the contents of the box; the TDS is the desired probability.

3) A less simple example of computing with words is the following.

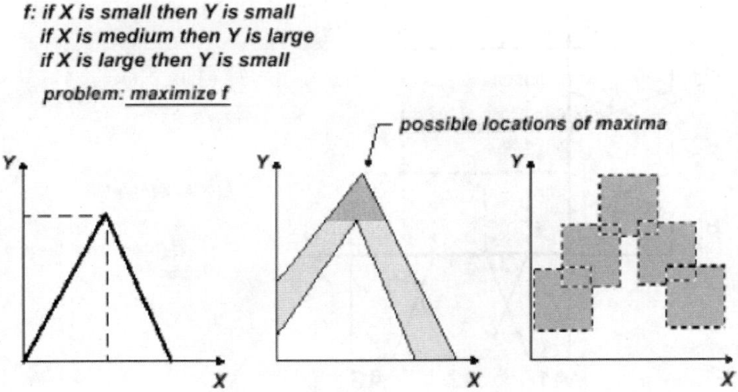

Figure 10. Fuzzy graph of a function defined by a fuzzy rule-set.

Let X and Y be independent random variables taking values in a finite set $V = \{v_1, \ldots, v_n\}$ with probabilities $p_1, \ldots, p_n$ and $q_1, \ldots, q_n$, respectively. For simplicity of notation, the same symbols will be used to denote X and Y and their generic values, with p and q denoting the probabilities of X and Y, respectively.

Assume that the probability distributions of X and Y are described in words through the fuzzy if-then rules (Fig. 11):

$$
\begin{aligned}
P: \quad & \text{if } X \text{ is } \textit{small} \text{ then } p \text{ is } \textit{small} \\
& \text{if } X \text{ is } \textit{medium} \text{ then } p \text{ is } \textit{large} \\
& \text{if } X \text{ is } \textit{large} \text{ then } p \text{ is } \textit{small}
\end{aligned}
$$

and

$$
\begin{aligned}
Q: \quad & \text{if } Y \text{ is } \textit{small} \text{ then } q \text{ is } \textit{large} \\
& \text{if } Y \text{ is } \textit{medium} \text{ then } q \text{ is } \textit{small} \\
& \text{if } Y \text{ is } \textit{large} \text{ then } q \text{ is } \textit{large}
\end{aligned}
$$

where the granules *small*, *medium* and *large* are values of linguistic variables X and Y in their respective universes of discourse. In the example under consideration, these rule-sets constitute the IDS. Note that *small* in P need not have the same meaning as *small* in Q, and likewise for *medium* and *large*.

The query is: How can we describe in words the joint probability distribution of X and Y? This probability distribution is the TDS.

For convenience, the probability distributions of X and Y may be represented as fuzzy graphs:

$$
\begin{aligned}
P: \quad & \textit{small} \times \textit{small} + \textit{medium} \times \textit{large} + \textit{large} \times \textit{small} \\
Q: \quad & \textit{small} \times \textit{large} + \textit{medium} \times \textit{small} + \textit{large} \times \textit{large}
\end{aligned}
$$

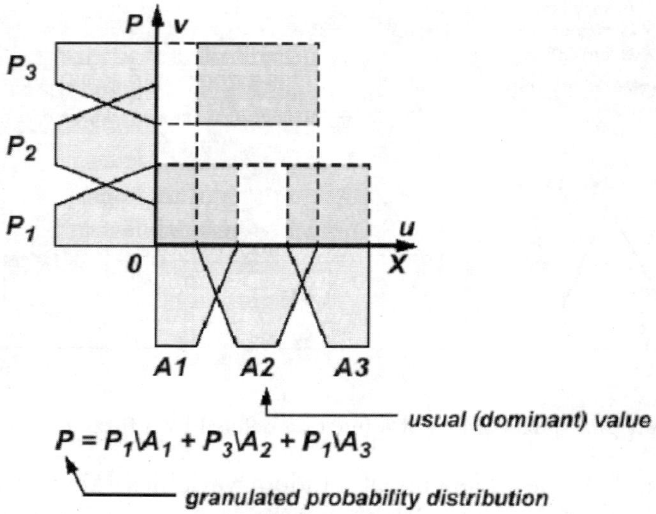

Figure 11. A fuzzy graph representation of a granulated probability distribution.

with the understanding that the underlying numerical probabilities must add up to unity.

Since X and Y are independent random variables, their joint probability distribution (P, Q) is the product of P and Q. In words, the product may be expressed as (Zadeh, 1996a):

$$\begin{aligned}(P, Q):\quad & small \times small \times (small * large) \\ & + small \times medium \times (small * small) \\ & + small \times large \times (small * large) \\ & + \cdots + large \times large \times (small * large),\end{aligned}$$

where $*$ is the arithmetic product in fuzzy arithmetic (Kaufmann and Gupta, 1985). In this example, what we have done, in effect, amounts to a derivation of a linguistic characterization of the joint probability distribution of X and Y starting with linguistic characterizations of the probability distribution of X and the probability distribution of Y.

A few comments are in order. In linguistic characterizations of variables and their dependencies, words serve as values of variables and play the role of fuzzy constraints. In this perspective, the use of words may be viewed as a form of granulation, which in turn may be regarded as a form of fuzzy quantization.

Granulation plays a key role in human cognition. For humans, it serves as a way of achieving data compression. This is one of the pivotal advantages accruing through the use of words in human, machine and man-machine communication.

The point of departure in CW is the premise that the meaning of a proposition, p, in a natural language may be represented as an implicit constraint on an implicit variable. Such a representation is referred to as a *canonical form* of p, denoted as $CF(p)$ (Fig. 12). Thus, a canonical form serves to make explicit the implicit constraint which resides in p. The concept of a canonical form is described in greater detail in the following section.

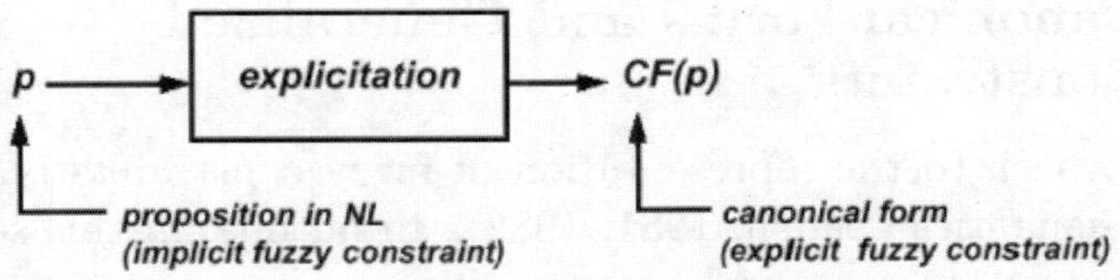

Figure 12. Canonical form of a proposition.

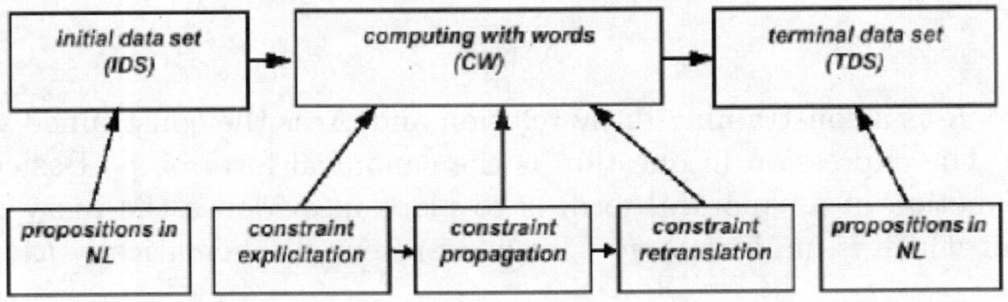

Figure 13. Conceptual structure of computing with words.

As a first step in the derivation of TDS from IDS, propositions in IDS are translated into their canonical forms, which collectively represent *antecedent* constraints. Through the use of rules for constraint propagation, antecedent constraints are transformed into *consequent* constraints. Finally, consequent constraints are translated into a natural language through the use of *linguistic approximation* (Freuder and Snow, 1990; Mamdani and Gaines, 1981), yielding the terminal data set TDS. This process is schematized in Fig. 13.

In essence, the rationale for computing with words rests on two major imperatives. First, computing with words is a necessity when the available information is too imprecise to justify the use of numbers. And second, when there is a tolerance for imprecision which can be exploited to achieve tractability, robustness, low solution cost and better rapport with reality.

In computing with words, there are two core issues that arise. First is the issue of representation of fuzzy constraints. More specifically, the question is: How can the fuzzy constraints which are implicit in propositions expressed in a natural language be made explicit. And second is the issue of fuzzy constraint propagation, that is, the question of how can fuzzy constraints in premises, i.e., antecedent constraints, be propagated to conclusions, i.e., consequent constraints.

These are the issues which are addressed in the following.

3. Representation of Fuzzy Constraints, Canonical Forms and Generalized Constraints

Our approach to the representation of fuzzy constraints is based on test-score semantics (Zadeh, 1981; 1982). In outline, in this semantics, a proposition, p, in a natural language is viewed as a network of fuzzy (elastic) constraints. Upon aggregation, the constraints which are embodied in p result in an overall fuzzy constraint which can be represented as an expression of the form

$$X \text{ is } R$$

where R is a constraining fuzzy relation and X is the constrained variable. The expression in question is the canonical form of p. Basically, the function of a canonical form is to place in evidence the fuzzy constraint which is implicit in p. This is represented schematically as

$$P \rightarrow X \text{ is } R$$

in which the arrow $\rightarrow$ denotes explicitation. The variable X may be vector-valued and/or conditioned.

In this perspective, the meaning of p is defined by two procedures. The first procedure acts on a so-called explanatory database, ED, and returns the constrained variable, X. The second procedure acts on ED and returns the constraining relation, R.

An explanatory database is a collection of relations in terms of which the meaning of p is defined. The relations are empty, that is, they consist of relation names, relations attributes and attribute domains, with no entries in the relations. When there are entries in ED, ED is said to be *instantiated* and is denoted EDI. EDI may be viewed as a description of a possible world in possible world semantics (Cresswell, 1973), while ED defines a collection of possible worlds, with each possible world in the collection corresponding to a particular instantiation of ED (Zadeh, 1982).

As a simple illustration, consider the proposition

$$p = \textit{Mary is not young.}$$

Assume that the explanatory database is chosen to be

$$\text{ED} = \text{POPULATION [Name; Age]} + \text{ YOUNG [Age; } \mu]$$

in which POPULATION is a relation with arguments Name and Age; YOUNG is a relation with arguments Age and μ; and $+$ is the disjunction. In this case, the constrained variable is the age of Mary, which in terms of ED may be expressed as

$$X = \text{Age (Mary)} = {}_{\text{Age}}\text{POPULATION [Name} = \text{Mary]}.$$

This expression specifies the procedure which acts on ED and returns X. More specifically, in this procedure, Name is instantiated to Mary and the resulting relation is projected on Age, yielding the age of Mary.

The constraining relation, R, is given by

$$R = ({}^2\text{YOUNG})'$$

which implies that the intensifier *very* is interpreted as a squaring operation, and the negation *not* as the operation of complementation (Zadeh, 1972).

Equivalently, R may be expressed as

$$R = \text{YOUNG [Age; } 1 - \mu^2].$$

As a further example, consider the proposition

$$p = \textit{Carol lives in a small city near San Francisco}$$

and assume that the explanatory database is:

$$\begin{aligned} ED = {} & \text{POPULATION [Name; Residence]} \\ & + \text{SMALL [City; } \mu] + \text{NEAR [City1; City2; } \mu] \end{aligned}$$

In this case,

$$\begin{aligned} X &= \text{Residence (Carol)} \\ &= {}_{\text{Residence}}\text{POPULATION [Name} = \text{Carol]} \end{aligned}$$

and

$$R = \textit{SMALL } [\textit{City}, \mu] \cap_{\text{City1}} \text{NEAR } [\textit{City2} = \text{San\_Francisco}]$$

In R, the first constituent is the fuzzy set of small cities; the second constituent is the fuzzy set of cities which are near San Francisco; and $\cap$ denotes the intersection of these sets.

So far we have confined our attention to constraints of the form

$$X \text{ is } R.$$

In fact, constraints can have a variety of forms. In particular, a constraint – expressed as a canonical form – may be conditional, that is, of the form

$$\text{if } X \text{ is } R \text{ then } Y \text{ is } S$$

which may also be written as

$$Y \text{ is } S \text{ if } X \text{ is } R.$$

The constraints in question will be referred to as *basic*.

For purposes of meaning representation, the richness of natural languages necessitates a wide variety of constraints in relation to which the basic constraints form an important though special class. The so-called generalized constraints (Zadeh, 1986) contain the basic constraints as a special case and are defined as follows. The need for generalized constraints becomes obvious when one attempts to represent the meaning of simple propositions such as

Robert loves women
John is very honest
checkout time is 11 am
slimness is attractive

in the language of standard logical systems.

A generalized constraint is represented as

$$X \text{ isr } R,$$

where isr, pronounced "ezar", is a variable copula which defines the way in which R constrains X. More specifically, the role of R in relation to X is defined by the value of the discrete variable r. The values of r and their interpretations are defined below:

e : equal (abbreviated to =);
d : disjunctive (possibilistic) (abbreviated to blank);
ν : veristic;
p : probabilistic;
γ : probability value;
u : usuality;
rs : random set;
rfs : random fuzzy set;
fg : fuzzy graph;
ps : rough set (Pawlak set);
... .

As an illustration, when $r = e$, the constraint is an equality constraint and is abbreviated to =. When r takes the value d, the constraint is *disjunctive* (possibilistic) and isd abbreviated to is, leading to the expression

$$X \text{ is } R$$

in which R is a fuzzy relation which constrains X by playing the role of the possibility distribution of X. More specifically, if X takes values in a universe of discourse, $U = \{u\}$, then $Poss\{X = u\} = \mu_R(u)$, where μ_R is the membership function of R, and Π_X is the possibility distribution of X, that is, the fuzzy set of its possible values (Zadeh, 1978a). In schematic form:

$$X \text{ is } R \left\{ \begin{array}{c} \Pi_X = R \\ Poss\,\{X = u\} = \mu_R(u) \end{array} \right.$$

Similarly, when r takes the value ν, the constraint is *veristic*. In the case,

$$X \text{ isv } R$$

means that if the grade of membership of u in R is μ, then $X = u$ has truth value μ. For example, a canonical form of the proposition

$$p = \textit{John is proficient in English, French and German}$$

may be expressed as

Proficiency (John) isv (1—English + 0.7—French + 0.6—German)

in which 1.0, 0.7 and 0.6 represent, respectively, the truth values of the propositions *John is proficient in English, John is proficient in French and John is proficient in German.* In a similar vein, the veristic constraint

Ethnicity (John) isv (0.5—German + 0.25—French + 0.25—Italian)

represents the meaning of the proposition *John is half German, quarter French and quarter Italian.*

When $r = p$, the constraint is *probabilistic.* In this case,

$$X \text{ isp } R$$

means that R is the probability distribution of X. For example

$$X \text{ isp } N(m, \sigma^2)$$

means that X is normally distributed with mean m and variance σ^2. Similarly,

$$X \text{ isp } \left(0.2\backslash a + 0.5\backslash b + 0.3\backslash c\right)$$

means that X is a random variable which takes the values, a, b and c with respective probabilities 0.2, 0.5 and 0.3.

The constraint

$$X \text{ isu } R$$

is an abbreviation for

$$\mathit{usually}(X \text{ is } R)$$

which in turn means that

$$\mathit{Prob}\{X \text{ is } R\} \text{ is } \mathit{usually}.$$

In this expression X is R is a fuzzy event and *usually* is its fuzzy probability, that is, the possibility distribution of its crisp probability.

The constraint

$$X \text{ isrs } P$$

is a random set constraint. This constraint is a combination of probabilistic and possibilistic constraints. More specifically, in a schematic form, it is expressed as

$$\begin{array}{l} X \text{ isp } P \\ (X, Y) \text{ is } Q \\ \hline Y \text{ isrs } R, \end{array}$$

where Q is a joint possibilitistic constraint on X and Y, and R is a random set. It is of interest to note that the Dempster-Shafer theory of evidence (Shafer, 1976) is, in essence, a theory of random set constraints.

In computing with words, the starting point is a collection of propositions which play the role of premises. In many cases, the canonical forms of these propositions are constraints of the basic, possibilistic type. In a more general setting, the constraints are of the generalized type, implying that explicitation of a proposition, p, may be represented as

$$p \rightarrow X \text{ isr } R,$$

where X isr R is the canonical form of p (Fig. 14).

As in the case of basic constraints, the canonical form of a proposition may be derived through the use of testscore semantics. In this context, the depth of p is, roughly, a measure of the effort that is needed to explicitate p, that is, to translate p into its canonical form. In this

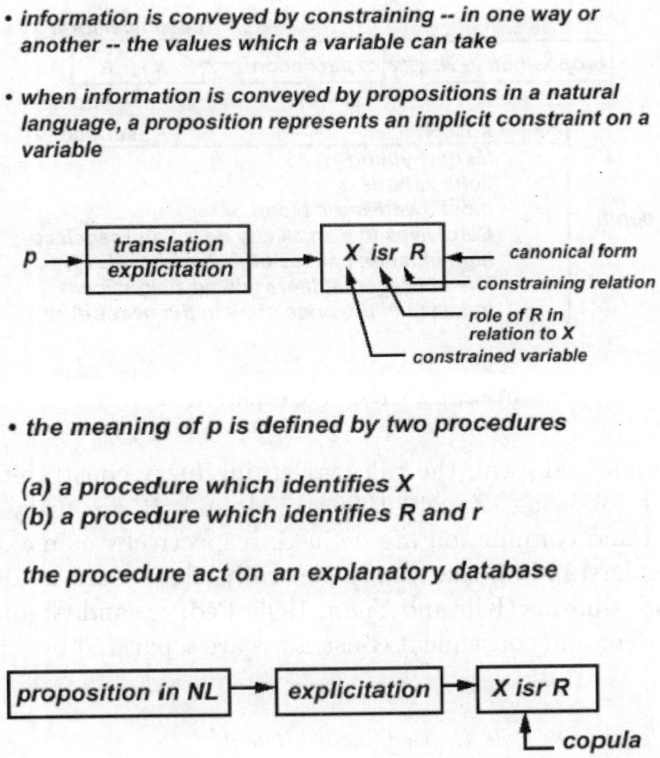

Figure 14. Representation of meaning in test-score semantics.

sense, the proposition X isr R is a surface constraint (depth=zero), with the depth of explicitation increasing in the downward direction (Fig. 15). Thus a proposition such as *Mary is young* is shallow, whereas *it is unlikely that there will be a substantial increase in the price of oil in the near future*, is not.

Once the propositions in the initial data set are expressed in their canonical forms, the groundwork is laid for fuzzy constraint propagation. This is a basic part of CW which is discussed in the following section.

4. Fuzzy Constraint Propagation and the Rules of Inference in Fuzzy Logic

The rules governing fuzzy constraint propagation are, in effect, the rules of inference in fuzzy logic. In addition to these rules, it is helpful to have rules governing fuzzy constraint modification. The latter rules will be discussed at a later point in this section.

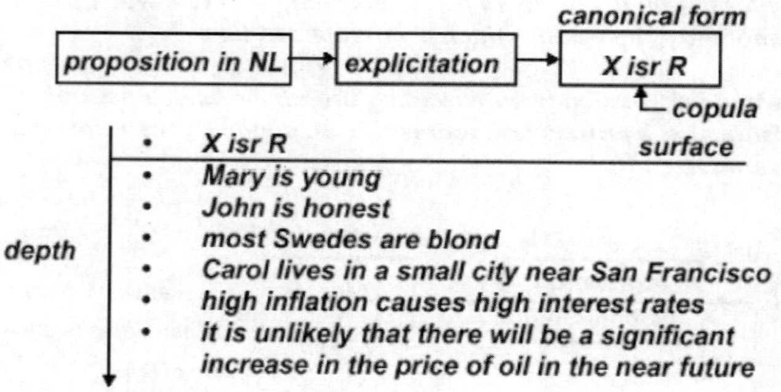

Figure 15. Depth of explicitation.

In a summarized form, the rules governing fuzzy constraint propagation are the following (Zadeh, 1996a). (A and B are fuzzy relations. Disjunction and conjunction are defined, respectively, as max and min, with the understanding that, more generally, they could be defined via t-norms and s-norms (Klir and Yuan, 1995; Pedrycz and Gomide, 1998). The antecedent and consequent constraints are separated by a horizontal line.)

Conjunctive Rule 1

$$\frac{X \text{ is } A \qquad X \text{ is } B}{X \text{ is } A \cap B}$$

Conjunctive Rule 2

$$(X \in U,\ Y \in B,\ A \subset U, B \subset V)$$

$$\frac{X \text{ is } A \qquad Y \text{ is } B}{(X,Y) \text{ is } A \times B}$$

Disjunctive Rule 1

$$\frac{X \text{ is } A \quad \text{or} \quad X \text{ is } B}{X \text{ is } A \cup B}$$

Disjunctive Rule 2

$$(A \subset U, B \subset V)$$

$$\frac{A \text{ is } A \qquad Y \text{ is } B}{(X,Y) \text{ is } A \times V \cup U \times B}$$

where $A \times V$ and $U \times B$ are cylindrical extensions of A and B, respectively.

Conjunctive Rule for isv

$$\frac{X \text{ isv } A \qquad X \text{ isv } B}{X \text{ isv } A \cup B}$$

Projective Rule

$$\frac{(X,Y) \text{ is } A}{Y \text{ is } \mathrm{proj}_V A}$$

where $\mathrm{proj}_V A = \sup_u A$.

Surjective Rule

$$\frac{X \text{ is } A}{(X,Y) \text{ is } A \times V}$$

Derived Rules

Compositional Rule

X is A
(X, Y) is B
$\overline{Y \text{ is } A \circ B}$

where $A \circ B$ denotes the composition of A and B.

Extension Principle

(mapping rule)(Zadeh, 1965; 1975)
X is A
$\overline{f(X) \text{ is } f(A)}$

where $f : U \to V$, and $f(A)$ is defined by $\mu_{f(A)}(\nu) = \sup_{u|\nu=f(u)} \mu_A(u)$.

Inverse Mapping Rule

$f(X)$ is A
$\overline{X \text{ is } f^{-1}(A)}$

where $\mu_{f^{-1}(A)}(u) = \mu_A(f(u))$.

Generalized modus ponens

X is A
if X is B then Y is C
$\overline{Y \text{ is } A \circ \big((\neg B) \oplus C\big)}$

where the bounded sum $\neg B \oplus C$ represents Lukasiewicz's definition of implication.

Generalized Extension Principle

$f(X)$ is A
$\overline{q(X) \text{ is } q\big(f^{-1}(A)\big)}$

where $\mu_q(\nu) = \sup_{u|\nu=q(u)} \mu_A(f(u))$.

The generalized extension principle plays a pivotal role in fuzzy constraint propagation. However, what is used most frequently in practical applications of fuzzy logic is the *basic interpolative rule*, which is a special case of the compositional rule of inference applied to a function which is defined by a fuzzy graph (Zadeh, 1974; 1996). More specifically, if f is defined by a fuzzy rule set

$$f : \textit{if } X \textit{is } A_i \textit{ then } X \textit{ is } B_i, \quad i = 1, \ldots, n$$

or equivalently, by a fuzzy graph

$$f \textit{ is } \sum_i A_i \times B_i$$

and its argument, X, is defined by the antecedent constraint X *is* A, then the consequent constraint on Y may be expressed as

$$Y \textit{ is } \sum_i m_i \wedge B_i,$$

where m_i is a matching coefficient, $m_i = \sup(A_i \cap A)$, which serves as a measure of the degree to which A matches A_i.

Syllogistic Rule: (Zadeh, 1984)

$$\frac{\begin{array}{l} Q_1 A\text{'s are } B\text{'s} \\ Q_2(A \text{ and } B)\text{'s are } C\text{'s} \end{array}}{(Q_1 \otimes Q_2)A\text{'s are } (B \text{ and } C)\text{'s},}$$

where Q_1 and Q_2 are fuzzy quantifiers; A, B and C are fuzzy relations; and $Q_1 \otimes Q_2$ is the product of Q_1 and Q_2 in fuzzy arithmetic.

Constraint Modification Rules: (Zadeh, 1972; 1978)

$$X \text{ is } mA \rightarrow X \text{ is } f(A),$$

where m is a modifier such as *not, very, more or less*, and $f(A)$ defines the way in which m modifies A. Specifically,

$$\begin{array}{l} \text{if } m = \textit{not} \text{ then } f(A) = A' \text{ (complement)} \\ \text{if } m = \textit{very} \text{ then } f(A) = {}^2A \text{ (left square),} \end{array}$$

where $\mu_{^2A}(u) = (\mu_A(u))^2$. This rule is a convention and should not be constructed as a realistic approximation to the way in which the modifier *very* functions in a natural language.

Probability Qualification Rule: (Zadeh, 1979b)

$$(X \text{ is } A) \text{ is } \Lambda \rightarrow P \text{ is } \Lambda,$$

where X is a random variable taking values in U with probability density $p(u)$; Λ is a linguistic probability expressed in words like *likely, not very likely*, etc.; and P is the probability of the fuzzy event X, expressed as

$$P = \int_U \mu_A(u)p(u)\,\mathrm{d}u.$$

The primary purpose of this summary is to underscore the coincidence of the principal rules governing fuzzy constraint propagation with the principal rules of inference in fuzzy logic. Of necessity, the summary is not complete and there are many specialized rules which are not included. Furthermore, most of the rules in the summary apply to constraints which are of the basic, possibilistic type. Further development of the rules governing fuzzy constraint propagation will require an extension of the rules of inference to generalized constraints.

As was alluded to in the summary, the principal rule governing constraint propagation is the generalized extension principle which in a schematic form may be represented as

$$\frac{f(X_1, \ldots, X_n) \text{ is } A}{q(X_1, \ldots, X_n) \text{ is } q(f^{-1}(A)).}$$

In this expression, $X_1, \ldots, X_n$ are database variables; the term above the line represents the constraint induced by the IDS; and the term below the line is the TDS expressed as a constraint on the query $q(X_1, \ldots, X_n)$. In the latter constraint, $f^{-1}(A)$ denotes the pre-image of the fuzzy relation A under the mapping $f : U \to V$, where A is a fuzzy subset of V and U is the domain of $f(X_1, \ldots, X_n)$.

Expressed in terms of the membership functions of A and $q(f^{-1}(A))$, the generalized extension principle reduces the derivation of the TDS to the solution of the constrained maximization problem

$$\mu_q(X_1, \ldots, X_n)(\nu) = \sup_{(u_1, \ldots, u_n)} (\mu_A(f(u_1, \ldots, u_n)))$$

in which $u_1, \ldots, u_n$ are constrained by

$$\nu = q(u_1, \ldots, u_n).$$

The generalized extension principle is simpler than it appears. An illustration of its use is provided by the following example.

The IDS is:

most Swedes are *tall*

The query is: *What is the average height of Swedes?*

The explanatory database consists of a population of N Swedes, $Name_1, \ldots, Name_N$. The database variables are $h_1, \ldots, h_N$, where h_i is the height of $Name_i$, and the grade of membership of $Name_i$ in *tall* is $\mu_{tall}(h_i)$, $i = 1, \ldots, n$.

The proportion of Swedes who are tall is given by the sigma-count (Zadeh, 1978b)

$$\sum Count\ (tall - Swedes\ /\ Swedes) = \frac{1}{N} \sum_i \mu_{tall}(h_i)$$

from which it follows that the constraint on the database variables induced by the IDS is

$$\frac{1}{N} \sum_i \mu_{tall}(h_i) \text{ is } most.$$

In terms of the database variables $h_1, \ldots, h_N$, the average height of Swedes is given by

$$h_{ave} = \frac{1}{N} \sum_i h_i.$$

Since the IDS is a fuzzy proposition, h_{ave} is a fuzzy set whose determination reduces to the constrained maximization problem

$$\mu_{h_{ave}}(\nu) = \sup_{h_1,\ldots,h_N} \left(\mu_{most}\left(\frac{1}{N}\sum_i \mu_{tall}(h_i)\right)\right)$$

subject to the constraint

$$\nu = \frac{1}{N}\sum_i h_i.$$

It is possible that approximate solutions to problems of this type might be obtainable through the use of neurocomputing or evolutionary-computing-based methods.

As a further example, we will return to a problem stated in an earlier section, namely, maximization of a function, f, which is described in words by its fuzzy graph, f^* (Fig. 10). More specifically, consider the standard problem of maximization of an objective function in decision analysis. Let us assume – as is frequently the case in real-world problems – that the objective function, f, is not well-defined and that what we know about can be expressed as a fuzzy rule-set

$$\begin{aligned} f: \quad & \textit{if } X \textit{ is } A_1 \textit{ then } Y \textit{ is } B_1 \\ & \textit{if } X \textit{ is } A_2 \textit{ then } Y \textit{ is } B_2 \\ & \cdots\cdots\cdots\cdots\cdots\cdots \\ & \textit{if } X \textit{ is } A_n \textit{ then } Y \textit{ is } B_n \end{aligned}$$

or, equivalently, as a fuzzy graph

$$f \text{ is } \sum_i A_i \times B_i.$$

The question is: What is the point or, more generally, the maximizing set (Zadeh, 1998) at which f is maximized, and what is the maximum value of f?

The problem can be solved by employing the technique of *α-cuts* (Zadeh, 1965; 1975). With refererence to Fig. 16, if A_{i_α} and B_{i_α} are *α-cuts* of A_i and B_i, respectively, then the corresponding *α-cut* of f^* is given by

$$f^*_\alpha = \sum_i A_{i_\alpha} \times B_{i_\alpha}.$$

From this expression, the maximizing fuzzy set, the maximum fuzzy set and maximum value fuzzy set can readily be derived, as shown in Figs. 16 and 17.

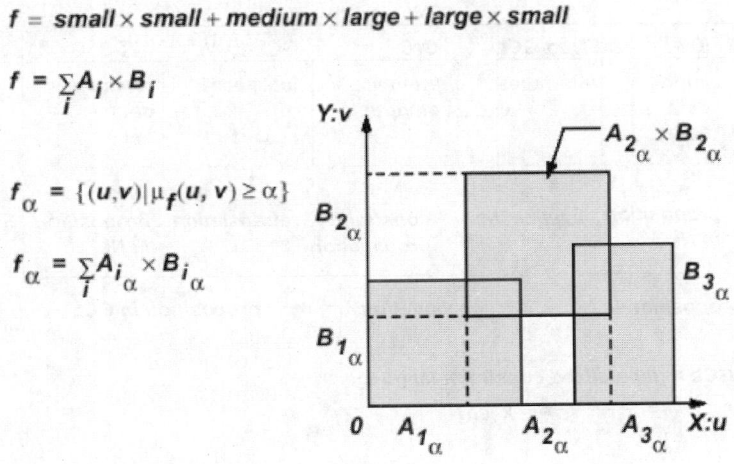

Figure 16. α-cuts of a function described by a fuzzy graph.

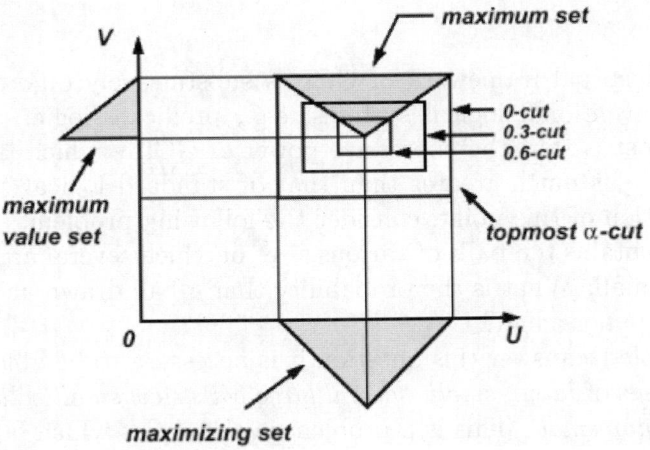

Figure 17. Computation of maximizing set, maximum set and maximum value set.

A key point which is brought out by these examples and the preceding discussion is that explicitation and constraint propagation play pivotal roles in CW. This role can be concretized by viewing explicitation and constraint propagation as translation of propositions expressed in a natural language into what might be called the *generalized constraint language* (GCL) and applying rules of constraint propagation to expressions in this language – expressions which are typically canonical forms of propositions expressed in a natural language. This process is schematized in Fig. 18.

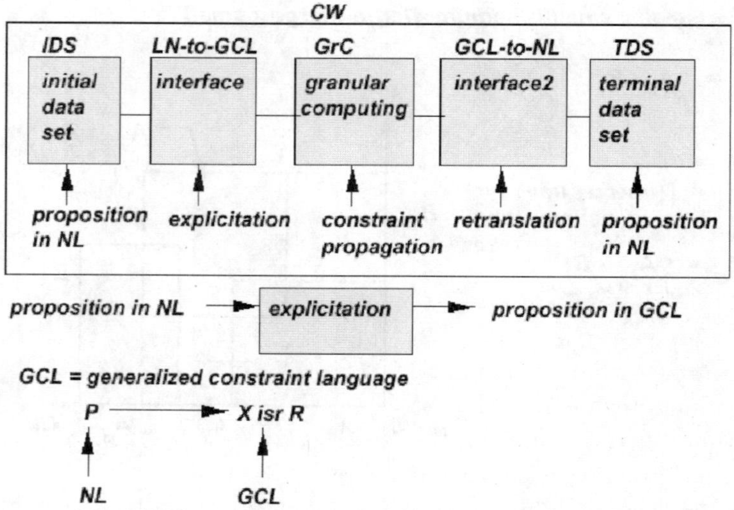

Figure 18. Conceptual structure of computing with words.

The conceptual framework of GCL is substantively differently from that of conventional logical systems, e.g., predicate logic. But what matters most is that the expressive power of GCL – which is based on fuzzy logic – is much greater than that of standard logical calculi. As an illustration of this point, consider the following problem.

A box contains ten balls of various sizes of which several are large and a few are small. What is the probability that a ball drawn at random is neither large nor small?

To be able to answer this question it is necessary to be able to define the meanings of *large, small, several large balls, few small balls* and *neither large nor small.* This is a problem in semantics which falls outside of probability theory, neurocomputing and other methodologies.

An important application area for computing with words and manipulation of perceptions is decision analysis since in most realistic settings the underlying probabilities and utilities are not known with sufficient precision to justify the use of numerical valuations. There exists an extensive literature on the use of fuzzy probabilities and fuzzy utilities in decision analysis. In what follows, we shall restrict our discussion to two very simple examples which illustrate the use of perceptions.

First, consider a box which contains black balls and white balls (Fig. 19). If we could count the number of black balls and white balls, the probability of picking a black ball at random would be equal to the proportion, r, of black balls in the box.

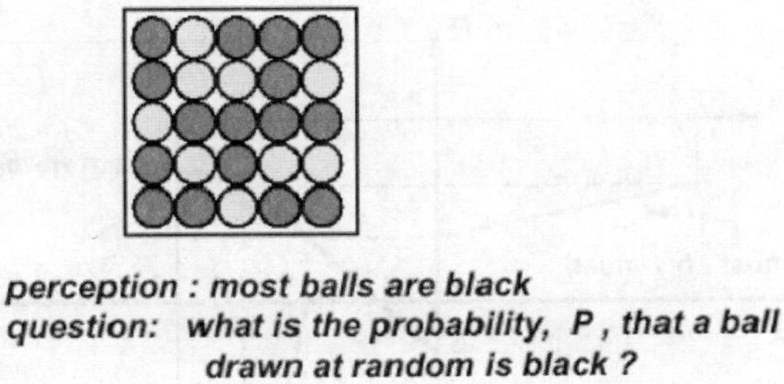

Figure 19. A box with black and white balls.

Now suppose that we cannot count the number of black balls in the box but our perception is that most of the balls are black. What, then, is the probability, p, that a ball drawn at random is black?

Assume that *most* is characterized by its possibility distribution (Fig. 20). In this case, p is a fuzzy number whose possibility distribution is *most*, that is,

$$p \text{ is } most.$$

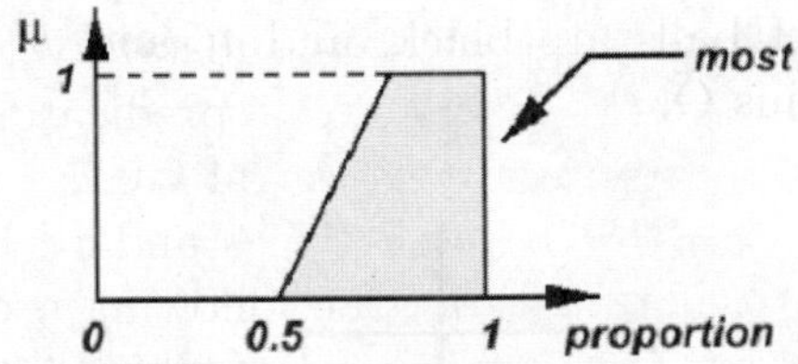

Figure 20. Membership function of *most*.

Next, assume that there is a reward of a dollars if the ball drawn at random is black and a penalty of b dollars if the ball is white. In this case, if p were known as a number, the expected value of the gain would be:

$$e = ap - b(1 - p).$$

Since we know not p but its possibility distribution, the problem is to compute the value of e when p is *most*. For this purpose, we can employ the extension principle (Zadeh, 1965; 1975), which implies that the possibility distribution, E, of e is a fuzzy number which may be expressed as

$$E = a\ most\ - b(1 - most).$$

For simplicity, assume that *most* has a trapezoidal possibility distribution (Fig. 20). In this case, the trapezoidal possibility distribution of E can be computed as shown in Fig. 21.

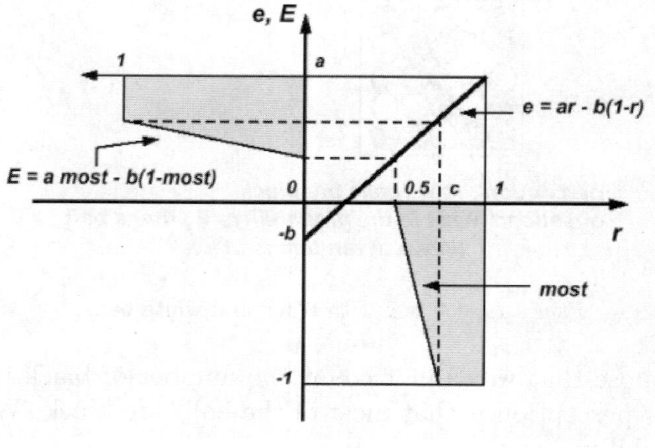

Figure 21. Computation of expectation through use of the extension principle.

It is of interest to observe that if the support of E is an interval $[\alpha, \beta]$ which straddles O (Fig. 22), then there is no non-controversial decision principle which can be employed to answer the question: Would it be advantageous to play a game in which a ball is picked at random from a box in which most balls are black, and a and b are such that the support of E contains O.

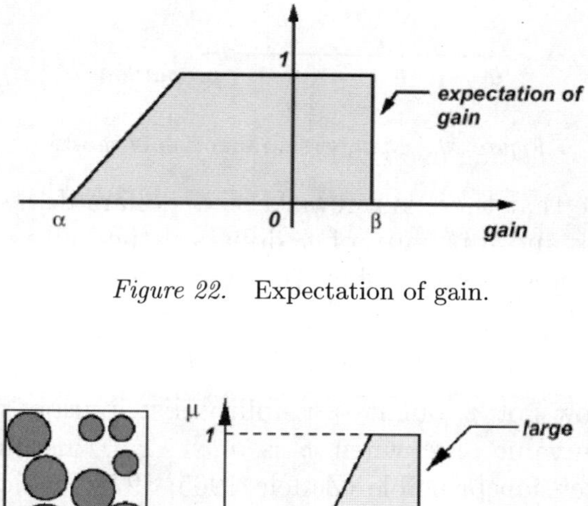

Figure 22. Expectation of gain.

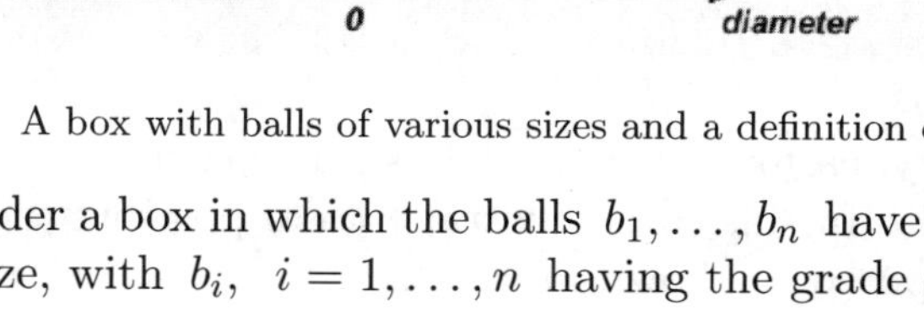

Figure 23. A box with balls of various sizes and a definition of large ball.

Next, consider a box in which the balls $b_1, \ldots, b_n$ have the same color but vary in size, with $b_i,\ i = 1, \ldots, n$ having the grade of membership

μ_i in the fuzzy set of large balls (Fig. 23). The question is: What is the probability that a ball drawn at random is large, given the perception that most balls are large?

The difference between this example and the preceding one is that the event *the ball drawn at random is large* is a fuzzy event, in contrast to the crisp event *the ball drawn at random is black.*

The probability of drawing b_i is $1/n$. Since the grade of membership of b_i in the fuzzy set of large balls is μ_i, the probability of the fuzzy event *the ball drawn at random is large* is given by (Zadeh, 1968)

$$P = \frac{1}{n}\sum \mu_i.$$

On the other hand, the proportion of large balls in the box is given by the relative sigma-count (Zadeh, 1975b; 1978)

$$\sum \text{Count } (\textit{large.balls} \ / \ \textit{balls.in.box}) = \frac{1}{n}\sum \mu_i.$$

Consequently, the canonical form of the perception *most balls are large* may be expressed as

$$\frac{1}{n}\sum \mu_i \text{ is } \textit{most}$$

which leads to the conclusion that

$$P \text{ is } \textit{most}.$$

It is of interest to observe that the possibility distribution of P is the same as in the preceding example.

If the question were: What is the probability that a ball drawn at random is *small*, the answer would be

$$P \text{ is } \frac{1}{n}\sum \nu_i$$

where ν_i, $i = 1, \ldots, n$, is the grade of membership of b_i in the fuzzy set of small balls, given that

$$\frac{1}{n}\sum \mu_i \text{ is } \textit{most}.$$

What is involved in this case is constraint propagation from the antecedent constraint on the μ_i to a consequent constraint on the ν_i. This problem reduces to the solution of a nonlinear program.

What this example points to is that in using fuzzy constraint propagation rules, application of the extension principle reduces, in general, to the solution of a nonlinear program. What we need – and do not have at present – are approximate methods of solving such programs which are capable of exploiting the tolerance for imprecision. Without such methods, the cost of solutions may be excessive in relation to the imprecision which is intrinsic in the use of words. In this connection, an intriguing possibility is to use neurocomputing and evolutionary computing techniques to arrive at approximate solutions to constrained maximization problems. The use of such techniques may provide a closer approximation to the ways in which human manipulate perceptions.

5. Concluding Remarks

In our quest for machines which have a high degree of machine intelligence (high MIQ), we are developing a better understanding of the fundamental importance of the remarkable human capacity to perform a wide variety of physical and mental tasks without any measurements and any computations. Underlying this remarkable capability is the brain's crucial ability to manipulate perceptions – perceptions of distance, size, weight, force, color, numbers, likelihood, truth and other characteristics of physical and mental objects. A basic difference between perceptions and measurements is that, in general, measurements are crisp whereas perceptions are fuzzy. In a fundamental way, this is the reason why to deal with perceptions it is necessary to employ a logical system that is fuzzy rather than crisp.

Humans employ words to describe perceptions. It is this obvious observation that is the point of departure for the theory outlined in the preceding sections.

When perceptions are described in words, manipulation of perceptions is reduced to computing with words (CW). In CW, the objects of computation are words or, more generally, propositions drawn from a natural language. A basic premise in CW is that the meaning of a proposition, p, may be expressed as a generalized constraint in which the constrained variable and the constraining relation are, in general, implicit in p.

In coming years, computing with words and perceptions is likely to emerge as an important direction in science and technology. In a reversal of long-standing attitudes, manipulation of perceptions and words which describe them is destined to gain in respectability. This is certain to happen because it is becoming increasingly clear that in dealing with real-world problems there is much to be gained by exploiting the tolerance for imprecision, uncertainty and partial truth. This is the primary motivation for the methodology of computing with words (CW) and the computational theory of perceptions (CTP) which are outlined in this paper.

Acknowledgement

The author acknowledges Prof. Michio Sugeno, who has contributed so much and in so many ways to the development of fuzzy logic and its applications.

References

Berenji H.R. (1994). "Fuzzy Reinforcement Learning and Dynamic Programming", *Fuzzy Logic in Artificial Intelligence* (A.L. Ralescu, Ed.), Proc. IJCAI'93 Workshop, Berlin: Springer-Verlag, pp. 1–9.

Black M. (1963). "Reasoning with Loose Concepts", *Dialog 2*, pp. 1–12.

Bosch P. (1978). Vagueness, Ambiguity and all the Rest, *Sprachstruktur, Individuum und Gesselschaft* (M. Van de Velde and W. Vandeweghe, Eds.). Tubingen: Niemeyer.

Bowen J., Lai R. and Bahler D. (1992a). "Fuzzy semantics and fuzzy constraint networks", *Proceedings of the 1st IEEE Conference on Fuzzy Systems*, San Francisco, pp. 1009–1016.

Bowen J., Lai R. and Bahler D. (1992b). "Lexical imprecision in fuzzy constraint networks", *Proc. Nat. Conf. Artificial Intelligence*, pp. 616–620.

Cresswell M.J. (1973). *Logic and Languages*. London: Methuen.

Dubois D., Fargier H. and Prade H. (1993). "The Calculus of Fuzzy Restrictions as a Basis for Flexible Constraint Satisfaction", *Proceedings of the 2nd IEEE International Conference on Fuzzy Systems*. San Francisco, pp. 1131–1136.

— (1994). Propagation and Satisfaction of Flexible Constraints, *Fuzzy Sets, Neural Networks, and Soft Computing* (R.R. Yager, L.A. Zadeh, Eds.) New York: Van Nostrand Reinhold, pp. 166–187.

— (1996). "Possibility Theory in Constraint Satisfaction Problems: Handling Priority, Preference and Uncertainty", *Applied Intelligence 6*:4, pp. 287–309.

Freuder E.C. and Snow P. (1990). Improved Relaxation and Search Methods for Approximate Constraint Satisfaction with a Maximin Criterion, *Proceedings of the 8th Biennial Conference of the Canadian Society for Computational Studies of Intelligence*, Ontario, pp. 227–230.

Goguen J.A. (1969). "The Logic of Inexact Concepts", *Synthese* 19, pp. 325–373.

Hobbs J.R. (1978). "Making Computation Sense of Montague's Intensional Logic", *Artificial Intelligence*, Vol. 9, pp. 287–306.

Katai O., Matsubara S., Masuichi H., Ida M., *et al.* (1992). "Synergetic Computation for Constraint Satisfaction Problems Involving Continuous and Fuzzy Variables by Using Occam, Transputer/Occam", (S. Noguchi and H. Umeo, Eds.), *Proc. 4th Transputer/Occam Int. Conf.*, Amsterdam: IOS Press, pp. 146–160.

Kaufmann A. and Gupta M.M. (1985). *Introduction to Fuzzy Arithmetic: Theory and Applications*, New York: Van Nostrand.

Klir G. and Yuan B. (1995). *Fuzzy Sets and Fuzzy Logic*, New Jersey: Prentice Hall.

Lano K. (1991). "A Constraint-Based Fuzzy Inference System". Proc. 5th Portuguese Conf. *Artificial Intelligence*, EPIA'91 (P. Barahona, L.M. Pereira, and A. Porto, Eds.), Berlin: Springer-Verlag, pp. 45–59.

Lodwick W.A. (1990). "Analysis of Structure in Fuzzy Linear Programs", *Fuzzy Sets Syst. 38*:1, pp. 15–26.

Mamdani E.H. and Gaines B.R., Eds. (1981). *Fuzzy Reasoning and its Applications.* London: Academic Press.

Mares M. (1994). *Computation Over Fuzzy Quantities*, Boca Raton: CRC Press.

Novak V. (1991). "Fuzzy Logic, Fuzzy Sets, and Natural Languages", *International Journal of General Systems 20*:1, pp. 83–97.

Novak V., Ramik M., Cerny M. and Nekola J., Eds. (1992). *Fuzzy Approach to Reasoning and Decision-Making*, Boston: Kluwer.

Oshan M.S., Saad O.M. and Hassan A.G. (1995). "On the Solution of Fuzzy Multiobjective Integer Linear Programming Problems With a Parametric Study", *Adv. Modell. Anal.* A, Vol. 24, No. 2, pp. 49–64.

Partee B. (1976). *Montague Grammar*, New York: Academic Press.

Pedrycz W. and Gomide F. (1998). *Introduction to Fuzzy Sets*, Cambridge: MIT Press, pp. 38–40.

Qi G. and Friedrich G. (1992). "Extending Constraint Satisfaction Problem Solving in Structural Design", 5th Int. Conf. *Industrial and Engineering Applications of Artificial Intelligence and Expert Systems, IEA/AIE-92* (F. Belli and F.J. Radermacher, Eds.), Berlin: Springer-Verlag, pp. 341–350.

Rasiowa H. and Marek M. (1989). "On Reaching Consensus by Groups of Intelligent Agents", *Methodologies for Intelligent Systems* (Z.W. Ras, Ed.). Amsterdam: North-Holland, pp. 234–243.

Rosenfeld A., Hummel R.A. and Zucker S.W. (1976). "Scene Labeling by Relaxation Operations", *IEEE Transactions on systems man and cybernetics* , Vol. 6, pp. 420–433.

Sakawa M., Sawada K. and Inuiguchi M. (1995). "A Fuzzy Satisficing Method for Large-Scale Linear Programming Problems with Block

Angular Structure", *European Journal of Operational Research*, Vol. 81, No. 2, pp. 399–409.

Shafer G. (1976). *A Mathematical Theory of Evidence*, Princeton: Princeton University Press.

Tong S.C. (1994). "Interval Number and Fuzzy Number Linear Programming", *Adv. Modell. Anal. A, 20*:2, pp. 51–56.

Vallée R. (1995). *Cognition et système.* Paris : l'Interdisciplinaire Système(s).

Yager R.R. (1989). "Some Extensions of Constraint Propagation of Label Sets", *International Journal of Approximate Reasoning 3*, pp. 417–435.

Zadeh L.A. (1961). "From Circuit Theory to System Theory", *Proc. of the Institute of Radio Engineers 50*, pp. 856–865.

— (1965). Fuzzy sets, *Information and Control 8*, pp. 338–353.

— (1968). "Probability Measures of Fuzzy Events", *Journal of Mathematical Analysis and Applications 23*, pp. 421–427.

— (1972). "A Fuzzy–Set-Theoretic Interpretation of Linguistic Hedges", *Journal of Cybernetics 2*, pp. 4–34.

— (1973). "Outline of a New Approach to the Analysis of Complex System and Decision Processes", *IEEE Transactions on systems man and cybernetics SMC–3*, pp. 28–44.

— (1974). *On the Analysis of Large Scale Systems, Systems Approaches and Environment Problems* (H. Gottinger, Ed.). Gottingen: Vandenhoeck and Ruprecht, pp. 23–37.

— (1975a). Calculus of Fuzzy Restrictions, *Fuzzy Sets and Their Applications to Cognitive and Decision Processes*, (L.A. Zadeh, K.S. Fu, M. Shimura, Eds.). New York: Academic Press, pp. 1–39.

— (1975b). "The Concept of a Linguistic Variable and its Application to Approximate Reasoning", Part I: *Information Sciences 8*, pp. 199–249; Part II: *Inf. Sci. 8*, pp. 301–357; Part III: *Inf. Sci. 9*, pp. 43–80.

— (1976). "A Fuzzy-Algorithmic Approach to the Definition of Complex or Imprecise Concepts", *International Journal of Man-Machine Studies 8*, pp. 249–291.

— (1978a). "Fuzzy Sets as a Basis for a Theory of Possibility", *Fuzzy Sets and Systems 1*, pp. 3–28.

— (1978b). "PRUF – A Meaning Representation Language for Natural Languages", *International Journal of Man-Machine Studies 10*, pp. 395–460.

— (1979a). Fuzzy Sets and Information Granularity, *Advances in Fuzzy Set Theory and Applications* (M. Gupta, R.Ragade and R. Yager, Eds.). Amsterdam: North-Holland, pp. 3–18.

— (1979b). "A Theory of Approximate Reasoning", *Machine Intelligence 9* (J. Hayes, D. Michie and L.I. Mikulich, Eds.). New York: Halstead Press, pp. 149–194.

— (1981). Test-Score Semantics for Natural Languages and Meaning Representation via PRUF, *Empirical Semantics* (B. Rieger, W. Germany, Eds.), Brockmeyer, pp. 281–349. Also Technical Report Memorandum 246, AI Center, SRI International, Menlo Park, CA.

— (1982). "Test-Score Semantics for Natural Languages", Proc. 9-th Int. Conf. *Computational Linguistics*, Prague, pp. 425–430.

— (1984). "Syllogistic Reasoning in Fuzzy Logic and its Application to Reasoning with Dispositions", Proc. Int. Symp. *Multiple-Valued Logic*, Winnipeg, Canada, pp. 148–153.

— (1986). "Outline of a Computational Approach to Meaning and Knowledge Representation Based on a Concept of a Generalized Assignment Statement". Proc. International Seminar on Artificial Intelligence and Man-Machine Systems (M. Thoma and A. Wyner, Eds.) Heidelberg: Springer-Verlag, pp. 198–211.

— (1994). "Fuzzy Logic, Neural Networks and Soft Computing", *Communications of the ACM 37*:3, pp. 77–84.

— (1996a). *Fuzzy Logic and the Calculi of Fuzzy Rules and Fuzzy Graphs: A Precise, Multiple Valued Logic 1*, Gordon and Breach Science Publishers, pp. 1–38.

— (1996b). "Fuzzy Logic = Computing with Words", *IEEE Transactions on Fuzzy Systems 4*, pp. 103–111.

— (1997). "Toward a Theory of Fuzzy Information Granulation and its Centrality in Human Reasoning and Fuzzy Logic", *Fuzzy Sets and Systems 90*, pp. 111–127.

— (1998). "Maximizing Sets and Fuzzy Markoff Algorithms", *IEEE Transactions on systems man and cybernetics Part C — Applications and Reviews 28*, pp. 9–15.